A Second Course in Complex Analysis

Few other books purport to be a second course in complex analysis. This book differs in that it covers more modern topics and is more geometric in focus. Most texts on complex variable theory contain the same material. However, complex analysis is a vast and diverse subject with a long history and many aspects. A second course will benefit students and introduce these new topics that they might not otherwise experience.

Lars Ahlfors alone invented many new parts of the subject; Lipman Bers made decisive contributions, and there are many others. It is easy to justify a "second course" in complex analysis. That is what this book purports to be.

Some of the topics presented here are:

- harmonic measure
- extremal length
- Riemann surfaces
- uniformization
- automorphism groups
- the Schwarz lemma and its generalizations
- analytic capacity
- the Bergman theory
- invariant metrics
- Picard's theorem
- the boundary Schwarz lemma

The goal is to expose the reader to unfamiliar parts of the subject of complex variables and perhaps to pique interest in further reading. As with the authors' other books, not only theorems and proofs are included, but also many examples and some exercises. Numerous graphics illustrate the key ideas.

Peter V. Dovbush, Dr. habil., is an Associate Professor at Moldova State University, in the Institute of Mathematics and Computer Science. He received his Ph.D. in Lomonosov Moscow State University in 1983 and Doctor of Sciences in 2003. He has published over 50 scholarly articles.

Steven G. Krantz is a Professor of Mathematics at Washington University in St. Louis. He has previously taught at UCLA, Princeton University, and Penn State University. He received his Ph.D. from Princeton University in 1974. Krantz has directed 20 Ph.D. students and eight Masters students. He has published over 130 books and over 300 scholarly articles. He is the holder of the Chauvenet Prize and the Beckenbach Book Award and the Kemper Prize. He is a Fellow of the American Mathematical Society.

Textbooks in Mathematics

Series editors:

Al Boggess, Kenneth H. Rosen

Abstract Algebra, Third Edition
An Interactive Approach
William Paulsen

Elements of Algebraic Topology, Second Edition
James R. Munkres, Steven G. Krantz, and Harold R. Parks

One Complex Variable from the Several Variable Point of View
Peter V. Dovbush and Steven G. Krantz

Math Anxiety How to Beat It
Brian Cafarella

Lectures on Differential Geometry with Maple
Mayer Humi

Numerical Analysis for Engineers
Methods and Applications
Bilal M. Ayyub and Richard H. McCuen

An Invitation to Real Analysis
Andrew D. Hwang

Fourier Series and Boundary Value Problems with Engineering Applications
Youssef N. Raffoul

A Course in Real Analysis, Second Edition
Hugo D. Junghenn

Real and Functional Analysis
Kenneth Kuttler

Mathematics for Artificial Intelligence
Jane Hawkins

Elements of Topology, Second Edition
Theory and Practice
Steven G. Krantz

Convex Analysis, Second Edition
Steven G. Krantz

A Second Course in Complex Analysis
Peter V. Dovbush and Steven G. Krantz

https://www.routledge.com/Textbooks-in-Mathematics/book-series/CANDHTEX-BOOMTH

A Second Course in Complex Analysis

Peter V. Dovbush and Steven G. Krantz

CRC Press
Taylor & Francis Group
Boca Raton London New York

CRC Press is an imprint of the
Taylor & Francis Group, an **informa** business
A CHAPMAN & HALL BOOK

Designed cover image: Pavel Dovbush

First edition published 2026
by CRC Press
2385 NW Executive Center Drive, Suite 320, Boca Raton FL 33431

and by CRC Press
4 Park Square, Milton Park, Abingdon, Oxon, OX14 4RN

CRC Press is an imprint of Taylor & Francis Group, LLC

ISBN: 978-1-041-19714-0 (hbk)
ISBN: 978-1-041-19716-4 (pbk)
ISBN: 978-1-003-71300-5 (ebk)

DOI: 10.1201/9781003713005

Typeset in Latin Modern font
by KnowledgeWorks Global Ltd.

Contents

Preface

Most texts on complex variable theory contain the same material: Cauchy theory, the Riemann mapping theorem, conformality, power series, harmonic functions, infinite products, monodromy, Runge's theorem, Hardy spaces, and so forth. But complex analysis is a vast and diverse subject with a long history and many aspects. Lars Ahlfors alone invented many new parts of the subject, Lipman Bers made decisive contributions, and there are many others. It is easy to justify a "second course" in complex analysis. That is what this book purports to be.

Of course, the authors will exercise their own taste and discretion in choosing topics here. Our own personal preferences run to geometric aspects. So some of topics that we cover are:

- harmonic measure
- extremal length
- Riemann surfaces
- uniformization
- automorphism groups
- the Schwarz lemma and its generalizations
- analytic capacity
- the Bergman theory
- invariant metrics
- Picard's theorem
- the boundary Schwarz lemma

Our goal here is to expose the reader to unfamiliar parts of the subject of complex variables, and perhaps to excite interest for further reading. The typical reader of our book will not have even heard of many of these topics.

As usual, we include not only theorems and proofs but also many examples and some exercises. We have numerous graphics to illustrate the key ideas. And plenty of exposition.

It is a pleasure to thank Pavel Dovbush for producing the technical illustrations specifically for this book and for his invaluable help with LaTeX typesetting perfection.

There are a few other books that purport to be a second course in complex analysis. Our book is different in that it contains more modern topics, and is more geometric in focus. It is also more of a textbook.

Of course, all shortcomings of the text are the responsibility of the authors. We look forward to feedback from our readership.

Peter V. Dovbush

Steven G. Krantz

Introduction

Complex analysis, traditionally known as the theory of functions of a complex variable, is the branch of mathematical analysis that investigates functions of complex numbers.

Elias M. Stein and Rami Shakarchi, in their book *Complex Analysis* [171, p. vii], write the following:

> When we begin the study of complex analysis, we enter a marvelous world full of wonderful insights. We are tempted to use the adjectives *magical*, or even *miraculous*, when describing the first theorems we learn; and in pursuing the subject, we continue to be astonished by the elegance and sweep of the results.

According to Weierstrass, a function $w = f(z)$ is called *analytic* (or *holomorphic*) in a domain $U \subset \mathbb{C}$ if it can be expanded in a power series in a neighborhood of each point $a \in U$:

$$w = f(z) = \sum_{k=0}^{\infty} c_k (z-a)^k\,; \tag{0.1}$$

in the case of several complex variables, when $U \subset \mathbb{C}^n$, $n > 1$, the series (0.1) is interpreted as a multiple power series. Holomorphic functions are precisely those that can be expressed locally as sums of power series.

Most elementary functions, including the exponential function, the trigonometric functions, and all polynomial functions, extended appropriately to complex arguments as functions $\mathbb{C} \to \mathbb{C}$, are holomorphic over the entire complex plane, making them entire functions, while rational functions P/Q, where P and Q are polynomials, are holomorphic on domains that exclude points where Q is zero. Such functions that are holomorphic everywhere except at a set of isolated points are known as meromorphic functions.

On the other hand, the functions $z \to \Re z$, $z \to |z|$, and $z \to \overline{z}$ are not holomorphic anywhere in the complex plane, as shown by their failure to satisfy the Cauchy–Riemann criterion. By the *Cauchy–Riemann criterion*, u is holomorphic (in the sense of Cauchy–Riemann) if and only if $\overline{\partial} u = 0$.

The terms "analytic" and "holomorphic" are not quite synonymous—we speak of functions holomorphic in an open set, while a function can be analytic on any sort of set.

One of the remarkable results in elementary complex analysis is that the mere existence of the derivative of u at each point of an open set implies that

u is infinitely differentiable, and indeed is locally represented by a convergent power series; that is, a function complex-differentiable in an open set must be analytic. (By way of contrast, recall that differentiability of a function on the line does not imply that the derivative is even continuous, much less differentiable, and that an infinitely differentiable function on the line need not be given by a power series.)

Chapter 1 is concerned with holomorphic and meromorphic functions and their properties. Using Stokes's theorem, we prove Cauchy–Green's formula. This plays a fundamental role in the theory of functions of a complex variable and, together with the Cauchy–Riemann equations, is now the usual way to prove Cauchy's theorem.

The Cauchy–Riemann equation $\bar{\partial}u = f$, where $\bar{\partial}$ is the Cauchy–Riemann operator and f is a $\bar{\partial}$-closed $(0, 1)$-form, is another important topic of complex analysis, and it is treated in Sections 1.4, 1.5, and 1.6. In the former, we show that the Cauchy–Riemann equation is always solvable for a domain D in $\mathbb{C}$ and when f has compact support in $\mathbb{C}$.

We are now ready to carry out basic complex analysis.

Thus, we prove that f is analytic if and only if it can be represented locally by a power series (that is, if and only if for every z in the domain of f there is a power series centered at z that converges to f at every point of the domain of f near z). We should note that in real analysis, local representability by power series is the definition of "analytic" (often written "real-analytic" to avoid confusion); for functions of a real variable, this is a much stronger condition than differentiability.

As an application, we obtain the Mittag-Leffler theorem. Mittag-Leffler's theorem concerns the existence of meromorphic functions with prescribed poles. Conversely, it can be used to express any meromorphic function as a sum of partial fractions.

So far, we have preliminary versions of Cauchy's theorem and the Cauchy integral formula, and we have used the latter to obtain the power series representation for a function holomorphic in a disk. The usefulness of the power series representation is somewhat limited because it is a local concept; it only works in disks. Nevertheless, it is a very useful tool, making many important results almost obvious.

Chapter 2 concerns extremal length. It is a relatively new conformal invariant that provides a useful way to measure the geometry of a domain in conformally invariant terms. It is considered the most important conformal invariant because it can be defined in many contexts and estimated through direct geometric arguments.

The concept of extremal length was developed by L. Ahlfors and A. Beurling in the 1940s. In [5], Ahlfors notes that Beurling invented extremal length in 1943 or 1944 but waited until 1946 to announce his results. It is an important conformal invariant that quantifies the "shape" of a domain from the complex-analytic point of view.

The concept of quasiconformal mapping—a generalization of conformal mapping—interacts naturally with extremal length.

Chapter 3 is concerned with harmonic measure. The term "harmonic measure" was introduced by R. Nevanlinna in the 1920s, but the ideas were anticipated in the work of the Riesz brothers and others. It is still studied intensively today.

Harmonic measure is a natural outgrowth of the study of the Dirichlet problem. It is a decisive tool for measuring the growth of functions and understanding how they bend. It has been used in such diverse applications as the corona problem and mapping problems. Harmonic measure has particularly interesting applications in probability theory—especially in the study of diffusions and Brownian motion.

Understanding harmonic measure requires some sophistication. And calculating the measure—working through the examples—requires some effort. But the effort is worth it, and the results are substantial. A definitive reference for the subject of harmonic measure is the book by Garnett and Marshall [57].

It includes a proof of *Hadamard's three-circles theorem* (sometimes called the "*three-lines theorem*"). It is an important sharpening of the classical maximum principle. It has proved useful in various parts of analysis, notably in proving the Riesz–Thorin interpolation theorem for linear operators. Here we shall give a thorough treatment of the three-circles theorem from the point of view of harmonic measure. Afterwards, we shall discuss the result of Riesz and Thorin.

Next, we apply the three-lines theorem to prove a result about interpolation of operators.

The theorem of F. and M. Riesz makes an important statement about the absolute continuity of harmonic measure. It is one of the celebrated results of classical function theory. There are many different renditions of the result and many different proofs. We give here a geometric version of the theorem.

Thus, functions holomorphic in the sense of Cauchy–Riemann and those holomorphic in the sense of Weierstrass are one and the same. Nevertheless, these different viewpoints led to distinct constructions of the objects—later called *Riemann surfaces*—on which multivalued functions become single-valued.

Chapter 4 concerns the global aspects of analytic functions, such as analytic continuation, global analytic functions, Riemann surfaces, and the algebraic characterization of conformal Riemann surfaces.

Chapter 4 develops the framework of *Riemann surfaces* from the point of view of functions and their analytic continuation.

Analytic continuation is treated via *Weierstrass's method*: analytic (holomorphic) elements, direct continuations, centers and discs of convergence, canonical elements, complete analytic functions, and evaluation at a point; continuation along curves addresses natural boundaries, the Weierstrass monodromy theorem, equality of canonical and complete analytic functions, derivatives, and references to Poincaré.

Function theory on Riemann surfaces follows: holomorphic and meromorphic functions, removable singularities and poles, Riemann's theorem on removable singularities, and elliptic functions, including the Weierstrass $\wp$-function and residues. Rational functions are examined alongside multiplicity, residues, meromorphicity, and the Dirichlet principle, with morphisms, isomorphisms, and automorphisms. The argument principle is formulated on Riemann surfaces. Riemann's construction is then developed: holomorphic maps and biholomorphisms, restrictions and uniqueness, local multiplicity, proper and covering maps, degrees, and the ramification and branch loci. Riemann's existence theorem is stated and applied to degrees, critical points and values, the Riemann–Hurwitz formula, branch points, and the Euler characteristic, with further applications to sheet numbers and hyperelliptic surfaces, and the algebraic relation between any two meromorphic functions (including total degree).

Algebraic functions on Riemann surfaces are examined through unbranched surfaces, critical points of polynomials, Puiseux series representations, completeness, and Picard's theorem, followed by the classical approach to the Riemann surface of an algebraic function.

The conception of Riemann surfaces, introduced in 1854 by B. Riemann, was decisive for the development of differential geometry and, in particular, of Riemannian geometry. However, they were rigorously described only in 1913 by H. Weyl as complex manifolds, in a way similar to the later definition of two-dimensional (real) differential manifolds, but with complex coordinate neighborhoods related by conformal homeomorphisms of open subsets of the complex plane rather than by homeomorphisms of open subsets of the real plane.

Chapter 5 is devoted to these issues. Riemann surfaces are defined as one-dimensional complex manifolds. Canonical examples are presented, including the Riemann sphere in homogeneous and Euclidean coordinates, projective plane algebraic curves, and closed and open surfaces together with projective invariants.

The topological underpinnings—genus and triangulations—are treated through Euclidean and equilateral triangulations, vertices, and the Euler characteristic, with results of Rado and Hurwitz, and a short proof that compact 2-manifolds admit triangulations. A brief homological framework is provided (chains, boundary operators, n-homology groups, and components), with connections to knot groups where appropriate. Chapter 5 also establishes the relationships of Riemann surfaces with the sets of zeros of polynomials in two complex variables (more precisely, nonsingular irreducible affine plane curves and algebraic curves in the complex projective plane). The chapter concludes with smooth and singular points on affine and projective curves and with Chow's theorem for projective varieties, together with the proper mapping theorem.

A compact Riemann surface is a one-dimensional compact complex manifold; that is, a manifold on which there exist no nonconstant globally defined holomorphic functions, although globally defined meromorphic functions do exist.

This naturally leads to the question of determining the dimension of the space of meromorphic functions on such a surface. The answer to this question is provided in Chapter 6 by the Riemann–Roch theorem, which establishes a fundamental relationship between the geometry of the surface and the analytical properties of functions defined on it.

In contrast to the noncompact example of a Riemann surface, namely the complex plane, it is not possible, in general, to construct a meromorphic function possessing singularities at arbitrary points of a given Riemann surface. The precise conditions under which this can be achieved are described by Abel's theorem.

Chapter 7 develops covering surfaces as a unifying framework for classical plane geometries. It begins by organizing the basic notions—number of sheets, sheet interchange, ramification points, and ramified covers—before introducing normal families and automorphisms, with Cartan's theorem providing a principal tool. A collection of basic examples, including those due to Poincaré, anchors the theory.

We then set up Riemann surfaces and covering spaces, treat coverings of Riemann surfaces, and analyze invariant metrics through quotients of $\mathbb{C}$ and of D, together with the group of covering transformations. A general discussion of covering spaces and automorphisms clarifies how these themes interact beyond the principal cases.

The narrative proceeds to compact quotients of $\mathbb{D}$ and their automorphisms, emphasizing the role of genus and the uniformization of compact Riemann surfaces theorem. Building on this, we determine structural properties of the automorphism group of a Riemann surface of genus at least 2, with attention to Hurwitz's theorem and the orbit structure (singular and nonsingular points and orbits). The chapter concludes by extending these considerations to automorphisms of multiply connected domains.

Chapter 8 assembles the potential-theoretic tools for the Uniformization theorem. The preliminary section develops the analytic framework via harmonic and subharmonic functions and the Perron family. We then introduce Green's functions and record the properties needed later. Finally, we present bipolar Green's functions as a refined construction. The presentation moves from foundational notions to the specific objects used throughout the chapter.

In Chapter 9 we discuss a classical problem in complex analysis and its relations to the rectifiability of sets in the complex plane $\mathbb{C}$. The problem is the following: which compact sets $E \subset \mathbb{C}$ are removable for bounded analytic functions in the following sense?

If U is an open set in $\mathbb{C}$ containing E and $f : U \backslash E \to \mathbb{C}$ is a bounded analytic function, then f has an analytic extension to U.

This problem has been studied for almost a century, but a geometric characterization of such removable sets is still lacking. We shall prove some partial results and discuss other results and conjectures.

Chapter 10 develops Bergman theory. It introduces the Bergman space, examines smoothness to the boundary of the Bergman kernel K_Ω, presents methods for calculating K_Ω, and develops the Poincaré–Bergman metric on the disc, with attention to its relation to the pseudohyperbolic distance. An appendix presents the biholomorphic inequivalence of the ball and the polydisc in light of these calculations.

Chapter 11 collects concise background on covering spaces, the sheaf of germs of holomorphic functions, sheaves, polynomials, and analytic subsets. The part on covering spaces reviews covering maps, evenly covered neighborhoods, sheets and fibers, morphisms of covering spaces and deck transformations, and lifts of maps; it includes the homotopy lifting property, monodromy (including along curves), conditions for the existence of lifts, and the universal covering space together with its relation to the fundamental group. The analytic background introduces germs and the sheaf of germs of holomorphic functions. The sheaf-theoretic material recalls presheaves of abelian groups with restriction homomorphisms, sheaves and stalks, the topological space associated with a presheaf, and the identity theorem for presheaves. The algebraic component summarizes the fundamental theorem of algebra, factorization results for complex polynomials, resultants and discriminants, Bézout's theorems, and basic notions on analytic subsets, including local analyticity, frontier, and reducible versus irreducible cases.

1

Preliminaries

Complex analysis has been the source of many new parts of mathematics. Complex Analysis is fundamental for the development of other areas of Mathematics, including Potential Theory and Partial Differential Equations (since the real and imaginary parts of a differentiable complex function are solutions of the Laplace partial differential equation, as for a potential associated with conservative processes in continuous media), Functional Analysis and the Calculus of Variations (since the solutions of the Laplace equation satisfy the Dirichlet Principle of minimizing the square of the norm of the respective gradient, corresponding to energy minimization for vector fields with a potential), Harmonic Analysis (since differentiable complex functions are harmonic functions, and Fourier series as well as Fourier and Laplace transforms are defined in the complex setting), Differential Geometry (where the notion of differential manifold and Riemannian Geometry found their first motivation in the Riemann surfaces initially considered in Complex Analysis to solve, by means of functions—univocal relations—the "inversion" of non-injective functions such as z^2 or $\log z$ for example), Algebraic Topology (since the notions of index or winding number of a closed path, homotopy, homology, covering, fundamental group, Euler characteristic, genus, and triangulation of a surface first appeared in the study of functions of one complex variable), Algebraic Geometry (since the level sets of complex functions are algebraic curves, and affine plane curves and projective algebraic curves z^2 are associated with Riemann surfaces), Analytic Number Theory (through the relationship between the distribution of prime numbers and the Riemann Zeta Function), and Dynamical Systems (providing the setting where this area was initiated and contributing substantially to the basic theory, including the notions of conjugacy, linearization at equilibria or periodic points, attracting basins, homoclinic orbits, hyperbolicity, the role of critical points, small denominators at neutral equilibrium points underlying KAM theory, the use of algebraic topology in dynamical systems, the notion of attractor set, and questions concerning its dimension or measure).

The origins of areas such as Algebraic Topology, Algebraic Geometry, Riemannian Geometry, Dynamical Systems, Harmonic Analysis, and Elliptic Partial Differential Equations intersect with Complex Analysis in such a way that it is most beneficial to precede the specialized study of these areas by deepening the study of Complex Analysis in those directions, since it provides a natural and relatively simple setting in which these concepts appear, thus

DOI: 10.1201/9781003713005-1

facilitating the understanding of the basic ideas and their further specialized study—this is one of the guiding principles of this book.

An example of this is Riemann's attempt to make multivalued complex functions such as the square-root function single-valued by enlarging the range. This led him to the idea of a Riemann surface. This, in turn, led him to the theory of differentiable manifolds, a mathematical subject that is the foundation of the theory of general relativity.

1.1 Holomorphic Functions

We wish to examine the notion of a "function of z" where z is a complex variable. To be sure, a complex variable can be viewed as nothing but a pair of real variables so that in one sense a function of z is nothing but a function of two real variables. This was the point of view we took in the last section in discussing continuous functions. But somehow this point of view is too general. There are some functions which are "direct" functions of $z = x + iy$ and not simply functions of the separate pieces x and y.

Consider, for example, the function $x^2 - y^2 + 2ixy$. This is a direct function of $x + iy$ since $x^2 - y^2 + 2ixy = (x + iy)^2$. On the other hand, the only slightly different-looking function $x^2 + y^2 - 2ixy$ is not expressible as a polynomial in $x + iy$. Thus we are led to distinguish a special class of functions, those given by direct, explicit, or analytic expressions in $x + iy$. When we finally do evolve a rigorous definition, these functions will be called the analytic functions.

Proposition 1.1.1. *A polynomial $P(x, y) \in \mathbb{C}[x, y]$ is analytic if and only if $P_y = iP_x$.*

Proof. The necessity of the condition can be proven in a straightforward manner. We leave the details as an exercise. To show that it is also sufficient, note that if

$$P_y = iP_x,$$

the condition must be met separately by the terms of any fixed degree. Suppose then that P has n-th degree terms of the form

$$Q(x, y) = C_0x^n + C_1x^{n-1}y + C_2x^{n-2}y^2 + \cdots + C_ny^n.$$

Since

$$\begin{gathered} Q_y = iQ_x \\ C_1x^{n-1} + 2C_2x^{n-2}y + \cdots + nC_ny^{n-1} \\ = i\left[nC_0x^{n-1} + (n-1)C_1x^{n-2}y + \cdots + C_{n-1}y^{n-1}\right] \end{gathered}$$

Comparing coefficients,

$$C_1 = inC_0 = i\binom{n}{1}C_0$$
$$C_2 = \frac{i(n-1)}{2}C_1 = i^2\frac{n(n-1)}{2}C_0 = i^2\binom{n}{2}C_0$$

and in general

$$C_k = i^k\binom{n}{k}C_0$$

so that

$$Q(x,y) = \sum_{k=0}^{n} C_k x^{n-k}y^k = C_0\sum_{k=0}^{n}\binom{n}{k}x^{n-k}(iy)^k = C_0(x+iy)^n.$$

Thus P is analytic. □

The condition $f_y = if_x$ is sometimes given in terms of the real and imaginary parts of f. That is, if $f = u + iv$, then

$$f_x = u_x + iv_x$$
$$f_y = u_y + iv_y$$

and the equation $f_y = if_x$ is equivalent to the twin equations

$$u_x = v_y; \quad u_y = -v_x.$$

These are usually called the Cauchy–Riemann equations.

Let $U \subseteq \mathbb{C}$ be an open set, and let $u : U \to \mathbb{C}$ be a C^1-function. Then we use the abbreviation

$$\overline{\partial}u = \frac{\partial u}{\partial \overline{z}} = \frac{1}{2}\left(\frac{\partial u}{\partial x} + i\frac{\partial u}{\partial y}\right),$$

where x, y are the canonical real coordinates on $\mathbb{C}$ and $z = x + iy$.

The function $\overline{\partial}u$ is called the *Cauchy–Riemann derivative* of u. By the *Cauchy–Riemann criterion*, u is holomorphic (in the sense of Cauchy–Riemann) if and only if $\overline{\partial}u = 0$.

The class of all holomorphic functions with domain U will be denoted by $\mathcal{O}(U)$.

If $\omega = udz + vd\overline{z}$ is a 1-form, then

$$d\omega = \frac{\partial u}{\partial \overline{z}}d\overline{z} \wedge dz + \frac{\partial v}{\partial z}dz \wedge d\overline{z} = \left(\frac{\partial v}{\partial z} - \frac{\partial u}{\partial \overline{z}}\right)dz \wedge d\overline{z}. \tag{1.1}$$

Of course, 1-forms can be integrated over 1-manifolds, such as the circle, and 2-forms can be integrated over 2-manifolds, such as the sphere or torus.

Finally, we recall the exterior differential operator d. In the complex setting, we can decompose $d = \partial + \overline{\partial}$, where

$$\partial\omega = \frac{\partial v}{\partial z} dz \wedge d\overline{z}$$

and

$$\overline{\partial}\omega = \frac{\partial u}{\partial \overline{z}} d\overline{z} \wedge dz\,.$$

Refer back to equation (1.1).

Theorem 1.1.2 (Stokes). *Let $U \subset \mathbb{C}$ be a bounded open domain with C^1 boundary. Let ω be a differential form of degree 1 with coefficients in $C^1(\overline{U})$. Then*

$$\int_{\partial U} U = \int_U d\omega = \int_U \partial\omega + \overline{\partial}\omega\,.$$

Standard references for Stokes's theorem are [160] and [153]. The Stokes's theorem that we have recorded here is the standard one, simply expressed in complex notation.

Cauchy–Green's formula (below) plays a fundamental role in the theory of functions of a complex variable, and together with the Cauchy–Riemann equations is now the usual way to prove Cauchy's theorem.

Now we have

Theorem 1.1.3 (Cauchy–Green formula). *Let $U \subset\subset \mathbb{C}$ be an open domain with C^1 boundary ∂D, and let f be a complex-valued continuous function on $\overline{U}$ such that $\overline{\partial} f$ is also continuous*[1] *on $\overline{U}$. Then*

$$f(z) = \frac{1}{2\pi i}\int_{\partial U} \frac{f(\zeta)d\zeta}{\zeta - z} - \frac{1}{2\pi i}\int_{\partial U} \frac{\overline{\partial} f(\zeta) \wedge d\zeta}{\zeta - z}, \quad z \in U\,. \tag{1.2}$$

Proof. Fix $z \in U$. Then, for $\zeta \in \overline{U}\backslash z$, we see that

$$d_\xi[(f(\zeta)d\zeta)/(\zeta - z)] = \overline{\partial} f(\zeta) \wedge d\zeta/(\zeta - z)$$

and, therefore, by Stokes's formula, for sufficiently small $\varepsilon > 0$

$$\frac{1}{2\pi i}\int_{|\zeta - z| = \varepsilon} \frac{f(\zeta)d\zeta}{\zeta - z} = \frac{1}{2\pi i}\int_{\zeta \in \partial U} \frac{f(\zeta)d\zeta}{\zeta - z} - \frac{1}{2\pi i}\int_{\zeta \in U, |\zeta - z| > \varepsilon} \frac{\overline{\partial} f(\zeta) \wedge d\zeta}{\zeta - z}\,.$$

It is clear that the right-hand side of this relation tends to the right-hand side of (1.2) when $\varepsilon \to 0$. Consider the left-hand side. Since $d\overline{\zeta} \wedge d\zeta/2i$ is the Lebesgue measure on $\mathbb{C}$, Stokes's formula gives

$$\int_{|\zeta - z| = \varepsilon} \frac{d\zeta}{\zeta - z} = \frac{1}{\varepsilon^2}\int_{|\zeta - z| = \varepsilon} (\overline{\zeta} - \overline{z})d\zeta = \frac{1}{\varepsilon^2}\int_{|\zeta - z| < \varepsilon} d\overline{\zeta} \wedge d\zeta = 2\pi i\,.$$

[1] This means $\partial f/\partial\overline{z}$, which is defined in the sense of distributions in U, is continuous in U and admits a continuous extension to $\overline{D}$.

Furthermore,

$$\left| \int_{|\zeta - z| = \varepsilon} \frac{f(\zeta) - f(z)}{\zeta - z} d\zeta \right| \leqq 2\pi \max_{|\zeta - z| = \varepsilon} |f(\zeta) - f(z)| \to 0 \quad \text{for } \varepsilon \to 0 .$$

Hence

$$\lim_{\varepsilon \to 0} \int_{|\zeta - z| = \varepsilon} \frac{f(\zeta) d\zeta}{\zeta - z} = 2\pi i f(z) .$$

□

It is important to note now that if U is as in the statement of the Theorem 1.1.3 and $f \in C^1(\overline{U})$ and holomorphic on the interior, then equation (1.2) simplifies to

$$f(z) = \frac{1}{2\pi i} \int_{\partial U} \frac{f(\zeta)}{\zeta - z} d\zeta . \tag{1.3}$$

This, of course, is the Cauchy integral formula that one can find in any complex analysis text. The Cauchy integral formula (1.3) shows that if f is holomorphic in a neighborhood of a closed disk, then we can recover the values of f in the interior of the disk from the values on the boundary.

Since the first integral on the right-hand side of the Cauchy–Green formula (1.2) depends holomorphically on $z \in U$, and since $\overline{\partial} f = 0$ for holomorphic functions, we obtain:

Corollary 1.1.4. *Let f be a continuous complex-valued function in an open domain $U \subseteqq \mathbb{C}$. Then f is holomorphic in U if and only if $\overline{\partial} f = 0$ in U.*

Corollary 1.1.5. *If $g \in C_0^1(\mathbb{C})$, i.e., g is continuously differentiable with compact support, then for all $z \in \mathbb{C}$ the following formula holds:*

$$g(z) = \frac{1}{2\pi i} \int_{\mathbb{C}} \frac{g_{\overline{\zeta}}(\zeta)}{\zeta - z} d\zeta \wedge d\overline{\zeta} .$$

Proof. Apply (1.2) to a disc $\mathbb{D}(0, r) \supset \operatorname{supp} g \cup \{z\}$. □

We now consider a wider class of direct functions of z, those given by infinite polynomials or "power series" in z.

We say that a function f is holomorphic at a point in the sense of Weierstrass if for every $a \in U$, there exists $r > 0$ such that in the disk $\Delta(a, r)$ the function $f(z)$ admits a power series expansion

$$f(z) = \sum_{n=0}^{\infty} c_n (z - a)^n .$$

By the corresponding theorem, the series may be differentiated term by term, and the radius of convergence does not change. From real analysis, such a series has both partial derivatives with respect to x and y:

$$\frac{\partial f}{\partial x} = \sum_{n=1}^{\infty} nc_n((x+iy)-a)^{n-1},$$

$$\frac{\partial f}{\partial y} = i\sum_{n=1}^{\infty} nc_n((x+iy)-a)^{n-1}.$$

We see that they differ only by the factor i. Recall that

$$\frac{\partial}{\partial \bar{z}} = \tfrac{1}{2}\left(\frac{\partial}{\partial x} + i\frac{\partial}{\partial y}\right).$$

Hence

$$\frac{\partial f}{\partial \bar{z}} = \tfrac{1}{2}\left(\frac{\partial f}{\partial x} + i\frac{\partial f}{\partial y}\right) = \tfrac{1}{2}\left(\frac{\partial f}{\partial x} + i^2\frac{\partial f}{\partial x}\right) = \tfrac{1}{2}\left(\frac{\partial f}{\partial x} - \frac{\partial f}{\partial x}\right) = 0,$$

which shows that f is *holomorphic in the sense of Cauchy–Riemann.*

For the converse, we use the Cauchy–Green formula (Theorem 1.1.3) to conclude that

$$\begin{aligned} f(z) &= \frac{1}{2\pi i}\oint_{|\zeta-z|=\varepsilon} \frac{f(\zeta)}{\zeta - z_0 - (z-z_0)} d\zeta \\ &= \frac{1}{2\pi i}\oint_{|\zeta-z|=\varepsilon} \frac{f(\zeta)}{\zeta - z_0} \sum_{n=0}^{\infty} \left(\frac{z-z_0}{\zeta-z_0}\right)^n d\zeta \\ &= \sum_{n=0}^{\infty} \frac{1}{2\pi i}\oint_{|\zeta-z|=\varepsilon} \frac{f(\zeta)}{(\zeta-z_0)^{n+1}} d\zeta\,(z-z_0)^n, \end{aligned}$$

where the interchange of summation and integration is justified due to uniform and absolute convergence of the series.

Thus we obtain that f is *holomorphic in the sense of Weierstrass* and, moreover,

$$f(z) = \sum_{n=0}^{\infty} \frac{f^{(n)}(z_0)}{n!}(z-z_0)^n$$

converges on $|z - z_0| < \operatorname{dist}(z_0, \partial U)$ with

$$f^{(n)}(z_0) = \frac{n!}{2\pi i}\oint_{|\zeta-z|=\varepsilon} \frac{f(\zeta)}{(\zeta-z_0)^{n+1}} d\zeta$$

for any $n \geq 0$.

Since f' is also holomorphic, we can get the values of f' at points inside the disk from the values of f' on the boundary by the same formula. Somewhat more interesting is the fact that we can get the values of f' in the interior from the values of f on the boundary:

Corollary 1.1.6 (Cauchy integral formula for derivatives). *Suppose $f \in H(V)$ and $\overline{D}(z_0, r) \subset V$. Define $\gamma : [0, 2\pi] \to V$ by $\gamma(t) = z_0 + re^{it}$ (so that γ is the boundary of the disk $D(z_0, r)$ traversed counterclockwise). Then*

$$f^{(n)}(z) = \frac{n!}{2\pi i} \int_{\gamma} \frac{f(w)dw}{(w - z)^{n+1}} \quad (n = 0, 1, 2, \ldots)$$

for all $z \in D(z_0, r)$.

Note that we have already proved Corollary 1.1.6 in the special case $z_0 = z$. We will use Cauchy's theorem to derive the general case from this special case. (Alert readers are beginning to get the idea that Cauchy's theorem is important.)

Proposition 1.1.7. *A function $f : \Omega \to \mathbb{C}$ is holomorphic if and only if for every point $z_0 \in \Omega$, there exists a neighborhood disc $U \subset \Omega$ of z_0, such that f can be written as a convergent power series $f(z) = \sum_{n=0}^{\infty} a_n (z - z_0)^n$ in U. If f is holomorphic, then $f'(z) = \sum_{n=1}^{\infty} na_n (z - z_0)^{n-1}$.*

Corollary 1.1.8. *Holomorphic functions are smooth (of class C^{∞}) that is, they have derivatives of all orders.*

Thus functions holomorphic in the sense of Riemann, holomorphic in the sense of Cauchy, and holomorphic in the sense of Weierstrass are one and the same. Nevertheless, these different viewpoints led to distinct constructions of the objects—later called *Riemann surfaces*—on which multivalued functions become single-valued.

We actually have that holomorphic functions are of class C^{ω}; that is, they are representable by power series (note that there exist functions with derivatives of all orders, which are not given by power series).

Definition 1.1.9. An *entire* function is a function holomorphic in the entire complex plane.

The definition should really say that a function is entire if it is represented by a single power series convergent in the whole plane.

We know that in the present context this is equivalent to the definition above, but the next exercise shows that the two notions are different for real-analytic functions:

Exercise for the Reader: Define $f : \mathbb{R} \to \mathbb{R}$ by $f(x) = 1/\left(1 + x^2\right)$. Show that for every $x \in \mathbb{R}$ there exists $\delta > 0$ such that f is represented by a power series convergent in $(x - \delta, x + \delta)$, although the power series at the origin does not converge on the entire line.

(Hint: This is an example of an application of complex analysis to real analysis.)

The following two results are classical and important properties of holomorphic functions that we will use frequently. In these results, we let Ω denote a connected open subset of $\mathbb{C}$.

Theorem 1.1.10 (maximum modulus theorem). *Let $f : \Omega \to \mathbb{C}$ be a holomorphic function. Assume that there exists a point $z_0 \in \Omega$ such that $|f(z)| \leq |f(z_0)|$ for all $z \in \Omega$. Then f is constant.*

An easy consequence of this theorem is that if f is holomorphic on Ω and K is a compact subset of Ω, then the maximum of $|f|$ on K is given by the value of f at a point in the boundary of K.

Theorem 1.1.11 (open mapping theorem). *Let $f : \Omega \to \mathbb{C}$ be a non-constant holomorphic function. If $U \subset \Omega$ is an open set, then $f(U)$ is an open set (in $\mathbb{C}$).*

A very important property of holomorphic functions is that they are uniquely determined by their behavior on open sets: if f and g are holomorphic on a domain U (a domain is a connected open set), and if f equals g on a nonempty open subset of U, then f equals g on all of U. To see this, we need only observe that the largest open subset of U on which $f = g$ is also closed (relative to U), since the partial derivatives which determine the power series expansion are continuous. Since U is connected, this set must be U.

Theorem 1.1.12 (the uniqueness theorem). *Let $f(z)$ and $g(z)$ be analytic in a region Ω. If the sequence of points $\{z_j\}$ in Ω where $f(z) = g(z)$ has a limit point z_0 in Ω, then $f(z) = g(z)$ for all z in Ω. In particular, if the set of zeros of $f(z)$ in Ω has a limit point in Ω, then $f(z)$ is identically zero in Ω.*

Remark 1.1.13. There are two important observations about the uniqueness theorem. First, the set Ω is assumed to be connected; otherwise, the result holds only in the connected component of Ω containing the sequence $\{z_j\}$ and the limit point z_0. Second, the limit point z_0 must belong to Ω; there exist examples of distinct holomorphic functions that take the same values at every point of a convergent sequence of points in an open set, where the limit point of such a sequence does not belong to the open set.

The next theorem tells us that the inverse of a bijective holomorphic mapping is also holomorphic. Observe that this result is not true in real analysis; for example, the function $f : \mathbb{R} \to \mathbb{R}$ given by $f(x) = x^3$ is bijective and infinitely differentiable, but the inverse function does not have a derivative at the point $x = 0$.

Theorem 1.1.14. *Let $f : \Omega \to \mathbb{C}$ be a one-to-one holomorphic function. Then $f'(z) \neq 0$ for all $z \in \Omega$, and $f^{-1} : f(\Omega) \to \mathbb{C}$ is holomorphic.*

The following local description of holomorphic functions (see [161, Theorem 10.32, p. 216]) is an easy consequence of:

Proposition 1.1.15. *If $f : \Omega \to \mathbb{C}$ is a non-constant holomorphic function, then for every point $z_0 \in \Omega$, there exists a positive integer n and a neighborhood U of z_0 in Ω, such that*

$$f(z) = (z - z_0)^n g(z),$$

where $g : U \to \mathbb{C}$ is a holomorphic function satisfying $g(z_0) \neq 0$.

Its generalization to Riemann surfaces will be a very useful tool.

Proposition 1.1.16. *Let $f : \Omega \to \mathbb{C}$ be holomorphic. Let $z_0 \in \Omega$, and let n be the order of the zero of $f(z) - f(z_0)$ at z_0. Then there exists a neighborhood U of z_0 in Ω, a positive number r, and a holomorphic function $g : U \to \mathbb{C}$ such that*

$$f(z) - f(z_0) = g(z)^n \quad \text{in } U\,.$$

The function g satisfies $|g(z)| < r$ and $g'(z) \neq 0$ for all $z \in U$.

A family $\mathcal{F}$ of holomorphic functions on a domain $\Omega \subset \mathbb{C}$ is normal in Ω if every sequence of functions $\{f_j\} \subseteq \mathcal{F}$ contains either a subsequence which converges to a limit function $f \neq \infty$ uniformly on each compact subset of Ω, or a subsequence which converges uniformly to ∞ on each compact subset.

A family $\mathcal{F}$ is said to be normal at a point $z_0 \in \Omega$ if it is normal in some neighborhood of z_0. A family of analytic functions $\mathcal{F}$ is normal in a domain Ω if and only if $\mathcal{F}$ is normal at each point of Ω.

Theorem 1.1.17 (Montel). *A family of holomorphic functions $\mathcal{A}$ on a domain Ω is normal if and only if it is uniformly bounded on compact subsets. That is, for every $K \subset \Omega$, compact, there exists a positive number M, such that $|f(z)| \leq M$, for all $z \in K$ and all $f \in \mathcal{A}$.*

The goal of [114] is to show that, in fact, most of the major elementary results in complex function theory can be derived from Montel's theorem. That will come as a surprise to many readers and, we hope, as something of a revelation.

Remark 1.1.18. A function f is holomorphic in the disk $\mathbb{D}$ if and only if it can be represented there by a convergent power series. However, convergence of the series at boundary points $|z - a| = R$ is not, in general, related to holomorphy of the function at those points.

It often happens that an explicit expression defining an analytic function is valid only for a limited range of the variable. For instance, consider the geometric series expansion

$$\frac{1}{1-z} = \sum_{n=0}^{\infty} z^n\,,$$

defines the function $(1 - z)^{-1}$ only in the disk $|z| < 1$. For other values of z, the series diverges and thus becomes meaningless, although the function itself is well defined for all values of z other than $z = 1$. The reason for this phenomenon is, of course, the singularity of the function at the point $z = 1$ which, as it were, "stops" the convergence of the series.

On the other hand, the series

$$\sum_{n=1}^{\infty} \frac{z^n}{n^2} = f(z)$$

converges at every point of the boundary circle $\{|z| = 1\}$, since it is dominated by the convergent numerical series $\sum \frac{1}{n^2}$. However, its sum f cannot be holomorphic at the point $z = 1$, because the derivative

$$f'(z) = \sum_{n=1}^{\infty} \frac{z^{n-1}}{n}$$

grows without bound as $z \to 1$ along the real axis.

As another example, we consider the function $f(z)$ defined by the integral

$$f(z) = \int_0^{\infty} e^{-zt} dt$$

where the integration is extended over the interval $(0, \infty)$. In view of

$$\left|e^{-zt}\right| = \left|e^{-(x+iy)t}\right| = e^{-xt},$$

the integral converges absolutely for $x = \mathrm{Re}(z) > 0$, and an elementary computation shows that $f(z) = z^{-1}$. The relation

$$\frac{1}{z} = \int_0^{\infty} e^{-zt} dt$$

is thus valid for $\mathrm{Re}(z) > 0$. For $\mathrm{Re}(z) \leq 0$, the integral fails to converge, although the function z^{-1} is well defined throughout this region except at the origin.

In all of these examples, the limited validity of the expressions in question presents no problems, since we possess alternative definitions that permit us to compute these functions wherever they are defined. In general, however, this is not the case. A function will be defined by means of some representation—such as a power series—which is valid only in a limited region, and no direct information is available regarding the behavior of the function outside this region. To obtain such information, it is necessary to employ methods that make it possible to "continue" the function from its original region of definition to other parts of the plane. A procedure that accomplishes this is referred to as the analytic continuation of the function.

The method of analytic continuation that will be described further (see Subsection 4.3.1) is based on the fact that, in the neighborhood of a regular point, an analytic function possesses a Taylor expansion whose circle of convergence is determined by the position of the nearest singular point of the function.

1.2 Meromorphic Functions

A point $z_0 \in \mathbb{C}$ is called an *isolated singularity* for a function f on U if f is defined and holomorphic on $U \setminus \{z_0\}$, where U is a neighborhood of z_0. Isolated singularities are of three types:

(1) *removable* if f can be extended to a holomorphic function in U;

(2) *pole* if there exists a positive integer n, such that $(z - z_0)^n f(z)$ can be extended to a holomorphic function in U;

(3) *essential* singularity if none of the above conditions are satisfied.

The following result describes the conditions under which a singularity is removable.

Theorem 1.2.1 (removable singularity theorem). *An isolated singularity z_0 of the function f is removable if and only if there exists a neighborhood V of z_0, such that f is bounded in $V \setminus \{z_0\}$. Equivalently*

$$\lim_{z \to z_0} (z - z_0) f(z) = 0$$

in $V \setminus \{z_0\}$.

Suppose z_0 is a pole of f and let n be the smallest (positive) integer such that $(z - z_0)^n f(z) = g(z)$ is holomorphic on U. The function $g(z)$ has a power series expansion in U; dividing by $(z - z_0)^n$ we get that f has the following series expression (called the *Laurent series*), valid in $U \setminus \{z_0\}$,

$$f(z) = \frac{a_{-n}}{(z - z_0)^n} + \cdots + \frac{a_{-1}}{z - z_0} + \sum_{j=0}^{\infty} a_j (z - z_0)^j$$

with $a_{-n} \neq 0$. The point z_0 is called a *pole of order* n; the coefficient a_{-1} is called the *residue* of f at z_0.

Definition 1.2.2. A function f is said to be *meromorphic* in a domain Ω if there exists a discrete subset $A \subset \Omega$ such that f is holomorphic in $\Omega \setminus A$, and has a pole at each point of A.

Definition 1.2.3. Let U be an open subset of $\mathbb{C}, g : U \to S^2$ a continuous function such that $g^{-1}(\infty)$ is a discrete subset of U. We say g is a *meromorphic function* in U if $g|_{(U \setminus g^{-1}(\infty))}$ is a holomorphic function.

In other words, if g is meromorphic in U, the set $P(g) := g^{-1}(\infty)$ is exactly the set of poles of the holomorphic function $g|_{U'}$, $U' = U \setminus P(g)$.

The collection of meromorphic functions in U will be denoted $\mathcal{M}(U)$. Let us also observe that if A is a discrete subset of an open set U in $\mathbb{C}$, $g \in \mathcal{H}(U \setminus A)$ and every point of A is either a removable singularity or a pole of g, then there is a unique continuous $\widetilde{g} : U \to S^2$ which extends g. Clearly the function $\widetilde{g}$ is meromorphic in U.

A function f which is holomorphic in the set $\mathbb{C} \setminus B(a, r)$, can be considered as a function defined in S^2 in a punctured neighborhood V of the point ∞. In that sense ∞ is an isolated singularity for this function. By introducing the change of variables $z \mapsto 1/z$ one can classify this singularity by considering the function $g(z) := f(1/z)$ defined in a punctured neighborhood of $z = 0$.

Theorem 1.2.4. *If a meromorphic function f has a pole or a removable singularity at infinity (that is, if all its singularities in $\widehat{\mathbb{C}}$ are poles) then f is a rational function.*

Proof. The number of poles of f is finite; otherwise, there would exist a limit point of poles in $\widehat{\mathbb{C}}$, since the latter is compact. Such a point would be a non-isolated singularity and hence not a pole.

Let us denote by a_ν, $\nu = 1, \ldots, n$, the finite poles of f, and define

$$g_\nu(z) = \frac{c^{(\nu)}_{-N_\nu}}{(z-a_\nu)^{N_\nu}} + \cdots + \frac{c^{(\nu)}_{-1}}{z-a_\nu}$$

as the *principal part* of f near the pole a_ν.

We also set

$$g(z) = c_1 z + \cdots + c_N z^N$$

to be the *principal part of f at infinity.* If $a = \infty$ is a removable singularity of f, we set $g = 0$.

Consider the function

$$\phi(z) = f(z) - g(z) - \sum_{\nu=1}^{n} g_\nu(z)\,.$$

This function has no singularities in $\overline{\mathbb{C}}$ and hence must be constant: $\phi(z) = c_0$. Therefore

$$f(z) = c_0 + g(z) + \sum_{\nu=1}^{n} g_\nu(z)\,, \tag{1.4}$$

which shows that f is a rational function. □

Remark 1.2.5. Expression (1.4) gives the decomposition of f into an entire part and partial fractions. Our argument thus provides a simple existence proof for such a decomposition.

We recall the concept of the Laurent principal part. Any function $f(z)$ holomorphic in an annulus

$$A = \{z \in \mathbb{C} : r < |z-a| < R\}$$

may be represented in this annulus as a sum of a converging power series

$$f(z) = \sum_{n=-\infty}^{\infty} a_n (z-a)^n\,,$$

its coefficients are determined by the formulas

$$a_n = \frac{1}{2\pi i}\int_\gamma \frac{f(w)}{(w-a)^{n+1}}\,dw, \quad n \in \mathbb{Z}\,,$$

where γ is a closed contour around a contained in A. Such representations are especially important when the inner radius is zero, that is, in punctured neighborhoods. They allow us to study functions near the singular points where they are not holomorphic.

In a neighborhood of a point $a \in \mathbb{C}$, a nonzero meromorphic function f is the sum of a Laurent series with at most finite principal part (the terms with negative index values):

$$f(z) = \sum_{n=-N}^{\infty} c_n (z-a)^n, \quad N > 0,$$

where n is an integer, and $a_{-N} \neq 0$. Again, if $N > 0$ (the sum starts with $c_{-N}\,(z-a)^{-N}$, the principal part has N terms), one has a pole of order N, and if $N \leq 0$ (the sum starts with $c_N\,(z-a)^N$, there is no principal part), one has a zero of order N.

The *residue* of a function f at an isolated singular point $a \in \mathbb{C}$ is equal to the coefficient in front of the term $(z-a)^{-1}$ in its Laurent expansion around a:

$$\mathrm{Res}_a\, f = c_{-1}\,.$$

Theorem 1.2.6. *Let the function f be holomorphic everywhere in the complex plane $\mathbb{C}$ except at a finite number of points $a_\nu, \nu = 1, \ldots, n$. Then the sum of its residues at all of its finite singular points and the residue at infinity vanishes:*

$$\sum_{\nu=1}^{n} \mathrm{Res}_{a_\nu} f + \mathrm{Res}_\infty f = 0\,. \tag{1.5}$$

Proof. We consider the circle $\gamma_R = \{z \in \mathbb{C} \mid |z| = R\}$ of such a large radius that it contains all finite singular points a_ν of f. Let γ_R be oriented counter-clockwise. The Cauchy theorem on residues implies that

$$\frac{1}{2\pi} \int_{\gamma_R} f dz = \sum_{\nu=1}^{n} \mathrm{res}_{a_\nu} f$$

while the Cauchy theorem implies that the left side does not change if R is increased further. Therefore it is equal to the negative of the residue of f at infinity. Thus the last equality is equivalent to (1.5). □

Some references for complex analysis are [4, 15, 40], and [69].

1.3 Algebraic Functions

An *algebraic function* is a function $w = w(z)$ which satisfies an algebraic equation

$$w^n + a_1(z) w^{n-1} + \cdots a_n(z) = 0$$

where the coefficients a_j are given meromorphic functions in z. A typical example is the square root $w = \sqrt{z}$, which is one of the first examples of a multi-valued function that one encounters in complex analysis.

Let $F \in \mathbb{C}[x, y]$ be an irreducible polynomial in two variables (with complex coefficients). We assume that its degree in y is ≥ 1. Recall that by Gauss's lemma, if we identify $\mathbb{C}[x, y]$ with $\mathbb{C}[x][y]$, then if F is irreducible, it is also irreducible in $\mathbb{C}(x)[y]$, the polynomial ring over the field of rational functions in x. Moreover, $\mathbb{C}[x, y]$ is a factorial ring (i.e., a unique factorization domain). An algebraic function is, intuitively, "defined" by an equation $F(x, y) = 0$ (where F is irreducible in $\mathbb{C}[x, y]$). To make this statement more precise, we begin with the following.

Theorem 1.3.1 (the implicit function theorem). *Let f be a holomorphic function of two complex variables x, y defined on*

$$\left\{(x, y) \in \mathbb{C}^2 \mid |x| < r_1, |y| < r_2\right\}, \quad r_1, r_2 > 0\,.$$

Assume that

$$f(0,0) = 0, \qquad \frac{\partial f}{\partial y}(0,0) \neq 0\,.$$

Then there exist positive numbers $\varepsilon, \delta > 0$ such that for any

$$x \in \mathbb{D}_\varepsilon = \{x \in \mathbb{C} \mid |x| < \varepsilon\}\,,$$

there is a unique solution $y(x)$ of the equation $f(x, y) = 0$ with $|y(x)| < \delta$. The function $x \mapsto y(x)$ is holomorphic on $\mathbb{D}_\varepsilon$.

Proof. Since $\frac{\partial f}{\partial y}(0,0) \neq 0$, we can choose $\delta > 0$ such that $f(0, y) \neq 0$ for $0 < |y| \leq \delta$. Now choose $\varepsilon > 0$ such that $f(x, y) \neq 0$ for $|x| \leq \varepsilon, |y| = \delta$ (this is possible since f is nonzero on the compact set $\{0\} \times \{y \mid |y| = \delta\}$). By the argument principle, if $|x| < \varepsilon$,

$$\frac{1}{2\pi i} \int_{|y|=\delta} \left\{\frac{\partial f}{\partial y}(x, y)/f(x, y)\right\} dy$$

is an integer $n(x)$ equal to the number of zeros of the function $y \mapsto f(x, y)$ in $|y| < \delta$; by our choice of δ, $n(0) = 1$. On the other hand, since $f(x, y) \neq 0$ for $|x| \leq \varepsilon, |y| = \delta$, the integrand, and thus also the integral, is a continuous function of x for $|x| < \varepsilon$. Thus $n(x) = 1$ for $|x| < \varepsilon$, which means precisely that there is a unique zero $y(x)$ of $f(x, y)$ with $|y(x)| < \delta$. That $x \mapsto y(x)$ is holomorphic follows from the formula

$$y(x) = \frac{1}{2\pi i} \int_{|y|=\delta} y \left\{\frac{\partial f}{\partial y}(x, y)/f(x, y)\right\} dy$$

(which is an immediate consequence of the residue theorem). □

For now we restrict our attention to polynomials of two variables. Let

$$F(x,y) = a_0(x)y^n + a_1(x)y^{n-1} + \cdots + a_n(x) \in \mathbb{C}[x,y] \tag{1.6}$$

be an irreducible polynomial with $n \geq 1$; the polynomials $a_0, \ldots, a_n \in \mathbb{C}[x]$ have no nonconstant common factor since F is irreducible.

If we solve the equation (1.6) for y it may happen that we obtain a multi-valued function; for instance: if $y^2 - P_d(x) = 0$ where $P_d(x)$ is a polynomial without multiple roots, we obtain the two-valued function $y = \sqrt{P_d(x)}$. (The geometric meaning of multi-valuedness of y is that (some of) the surfaces $y = \text{const}$ meet the surface $P(x,y) = 0$ in more than one point.)

The notation $y(x)$ can, therefore, only represent one *set* of complex numbers, the only notation making sense being $w \in y(x)$ and not $w = f(x)$.

Lemma 1.3.2. *Let $a \in \mathbb{C}$ be such that $a_0(a) \neq 0$ and such that there is no $b \in \mathbb{C}$ with $F(a,b) = 0 = \frac{\partial F}{\partial y}(a,b)$. Then there exists $\varepsilon > 0$ and n holomorphic functions $y_1(x), \ldots, y_n(x)$ in the disc $\{x \in \mathbb{C} \mid |x-a| < \varepsilon\}$ with the following properties:*

(i) $y_i(x) \neq y_j(x')$ if $i \neq j$, $|x-a| < \varepsilon$, $|x'-a| < \varepsilon$; moreover

$$F(x, y_i(x)) \equiv 0 \quad \textit{for} \quad |x-a| < \varepsilon, \quad i = 1, \ldots, n.$$

(ii) If $\eta \in \mathbb{C}$ and $F(x,\eta) = 0$ with $|x-a| < \varepsilon$, then $\eta = y_i(x)$ for a unique i between 1 and n.

Proof. Since $\frac{\partial F}{\partial y}(a,b) \neq 0$ for all solutions b of $F(a,b) = 0$, the polynomial $F(a,y)$ has exactly n roots $b_1, \ldots, b_n$. If $\varepsilon > 0$ is small and $y_i(x)$ is the holomorphic function on $|x-a| < \varepsilon$ with $y_i(a) = b_i$ and $F(x, y_i(x)) \equiv 0$ (which exists by the theorem above), then the y_i satisfy property (i) if ε is small enough, and property (ii), since the equation $F(x,\eta) = 0$ has at most n solutions. □

The geometric meaning of multi-valuedness of $y(x)$ is that (some of) the surfaces $x = \textit{const}$, meet the surface $f(x,y) = 0$ in more than one point.

Proposition 1.3.3. *Let $F = a_0(x)y^n + \cdots + a_n(x) \in \mathbb{C}[x,y]$ be irreducible. There are only finitely many $x \in \mathbb{C}$ such that the equations*

$$F(x,y) = 0 = \frac{\partial F}{\partial y}(x,y)$$

have a simultaneous solution $y \in \mathbb{C}$.

Proof. By the division algorithm, there are polynomials $b_i \in \mathbb{C}[x]$ $(i \geq 0)$ with

$b_0 = a_0$ and polynomials $A_j, Q_j \in \mathbb{C}[x, y]$ $(j \geq 1)$ such that

$$b_0^n F = A_1 \frac{\partial F}{\partial y} + Q_1, \quad \deg_y Q_1 < \deg_y \frac{\partial F}{\partial y} = n - 1$$

$$b_1 \frac{\partial F}{\partial y} = A_2 Q_1 + Q_2, \quad \deg_y Q_2 < \deg_y Q_1$$

$$\vdots$$

$$b_{k-1} Q_{k-2} = A_k Q_{k-1} + Q_k, \quad \deg_y Q_k < \deg_y Q_{k-1} .$$

From the equation $b_{k-2} Q_{k-3} = A_{k-1} Q_{k-2} + Q_{k-1}$, it would follow that F divides $b_{k-2} Q_{k-3}$ and hence Q_{k-3}. Repeating this argument, F would divide all the Q_j $(j \geq 1)$, hence also $\frac{\partial F}{\partial y}$ and F, contradicting the irreducibility of F. Thus $Q_k = Q_k(x) \in \mathbb{C}[x]$ is not identically zero. If now $a, b \in \mathbb{C}$ and $F(a, b) = 0 = \frac{\partial F}{\partial y}(a, b)$, we see from the above equations that $Q_1(a, b) = 0$, then $Q_2(a, b) = 0, \ldots, Q_k(a, b) = Q_k(a) = 0$. Since $Q_k \not\equiv 0$, the set

$$\left\{ x \in \mathbb{C} \,\middle|\, \exists y \in \mathbb{C} \text{ with } F(x, y) = 0 = \frac{\partial F}{\partial y}(x, y) \right\} \subset \{x \in \mathbb{C} \mid Q_k(x) = 0\}$$

is finite. □

Remark 1.3.4. We also recall that the resultant of a polynomial of one complex variable can be represented in the form of a special determinant in the coefficients of the polynomial and its derivative (see Subsection 11.4.2).

We form the *resultant* of the polynomial F, i.e., the polynomial

$$R(F, F'_y)(x) = (-1)^{n(n-1)/2} (a_0(x)(x))^{2n-2} \prod_{i \neq j} (y_i(x) - y_j(x))^2 ,$$

where $y_j(x), j = 1, \ldots, n$, are the roots of the polynomial F.

Note that, by definition, the resultant of F and $\partial F / \partial y$ is equal to 0 if and only if F and $\partial F / \partial y$ have a common root, hence if and only if $\deg R(F, \partial F / \partial y) > 0$.

In particular, the discriminant of a polynomial F is zero if and only if F has a non-trivial square factor, hence if and only if $m = \deg(F, F') > 0$. A complex polynomial of degree m has exactly m complex roots (counted with multiplicity). It follows that the set

$$\{x \in \mathbb{C} \mid R(F, F'_y)(x) = 0\}$$

is discrete and finite.

Since the roots of (1.6) are different at a point x_0, then there exist n functions $y_i(x)$, $i = 1, \ldots, n$, in the neighborhood of the point x_0 that depend on x analytically.

If x_0 is a zero of $R(F, F'_y)(x)$ and $a_0(x_0) \neq 0$ then the polynomial $F(x_0, w)$ has repeated zeros. Thus we see:

If x_0 is a not a zero of $R(F, F'_y)(x)$ then the equation $F(x_0, y) = 0$ has neither repeated finite roots, nor does it have a root at infinity.

If, however, $a_0(x_0) = 0$, then the equation $w^n F(x_0, 1/w) = 0$ has a root equal to zero. We then say that $F(x_0, w)$ has a root $w = \infty$.

Definition 1.3.5. *Critical points* $\mathcal{C} \subset \widehat{\mathbb{C}}$ are the zeros of the polynomial $R(F, F'_y)(x)$. For this reason, they are referred to as *algebraic critical points.*

Before proceeding further, we record some topological preliminaries. All topological spaces we consider will be Hausdorff.

Remark 1.3.6. A continuous map $p : X \to Y$ between locally compact spaces X, Y is proper if and only if, for any locally compact topological space Z, the product

$$p \times \mathrm{id}_Z : X \times Z \longrightarrow Y \times Z, \quad (x, x) \longmapsto (p(x), x)$$

is closed.

If X, Y have countable bases, this can be seen by using the following remark: if $\{x_1, \ldots, x_n, \ldots\}$ is a sequence of points in X without limit points, and if $\{p(x_n)\}_{n \geq 1}$ converges in Y, then the image of the closed set $\left\{\left(x_n, \frac{1}{n}\right) \mid n \geq 1\right\}$ in $X \times \mathbb{R}$ is not closed in $Y \times \mathbb{R}$.

The property in this remark can be used to define proper mappings between spaces that are not locally compact.

Remark 1.3.7. Let $p : X \to Y$ be a proper map between locally compact spaces. Let $Z \subset Y$ be a locally compact space (with the induced topology). Then $p : p^{-1}(Z) \to Z$ is again proper. In fact, a compact subset of Z is also a compact subset of Y.

Lemma 1.3.8. *Let $c_1, \ldots, c_n \in \mathbb{C}$. Let $w \in \mathbb{C}$ and suppose that $w^n + c_1 w^{n-1} + \cdots + c_n = 0$. Then*

$$|w| < 2 \max_{\nu} |c_\nu|^{1/\nu}$$

(unless $c_1 = \cdots = c_n = 0$).

Proof. Let $c = \max_\nu |c_\nu|^{1/\nu} > 0$. If $z = \frac{w}{c}$, we have $z^n + \frac{c_1}{c} z^{n-1} + \cdots + \frac{c_n}{c^n} = 0$, and since $|c_\nu| \leq c^\nu$,

$$|z|^n \leq |z|^{n-1} + \cdots + 1\,.$$

If $|z| \geq 2$, we would have $1 \leq \frac{1}{|z|} + \cdots + \frac{1}{|z|^n} \leq \frac{1}{2} + \cdots + \frac{1}{2^n} < 1$, a contradiction. Thus $|z| < 2$, i.e., $|w| < 2c$. □

Proposition 1.3.9. *Let $F \in \mathbb{C}[x, y], F(x, y) = a_0(x) y^n + \cdots + a_n(x)$, $a_0 \neq 0$. Let $V = \left\{(x, y) \in \mathbb{C}^2 \mid F(x, y) = 0\right\}$ and $S_0 = \left\{x \in \mathbb{C} \mid a_0(x) = 0\right\}$. Let $\pi : V \to \mathbb{C}$ be the projection $(x, y) \mapsto x$. Then $\pi : \pi^{-1}(\mathbb{C} \setminus S_0) \to \mathbb{C} \setminus S_0$ is proper.*

Proof. Let $K \subset \mathbb{C} - S_0$ be compact. Then there is $\delta > 0$ so that $|a_0(x)| \geq \delta$ and $|a_\nu(x)| \leq \frac{1}{\delta}$ for $x \in K$. If $(x, y) \in V$, $x \in \pi^{-1}(K)$, we have

$$y^n + \frac{a_1(x)}{a_0(x)} y^{n-1} + \cdots + \frac{a_n(x)}{a_0(x)} = 0$$

so that, by Lemma 1.3.8, $|y| \leq 2 \max_\nu \delta^{-2/\nu}$. Thus $\pi^{-1}(K)$ is bounded. Since clearly $\pi^{-1}(K) = (K \times \mathbb{C}) \cap V$ is closed in $\mathbb{C}^2$, $\pi^{-1}(K)$ is compact. $\square$

Example 1.3.10.

1. Let $\mathbb{D} = \{z \in \mathbb{C} \mid |z| < 1\}$ and $\mathbb{D}^* = \mathbb{D} \setminus \{0\}$. Then, if $n \geq 1$, the map $p_n : \mathbb{D}^* \to \mathbb{D}^*$ given by $p_n(z) = z^n$ is an n-sheeted covering. It is a standard fact in the theory of covering spaces that any connected n-sheeted covering of $\mathbb{D}^*$ is isomorphic to p_n.

2. $p : \mathbb{C} \to \mathbb{C}^*$, $p(z) = e^z$ is an infinite-sheeted covering of $\mathbb{C}^*$.

3. Let X, Y be locally compact, and let $p : X \to Y$ be a local homeomorphism (i.e., $\forall a \in X$, $\exists$ an open neighborhood U of a such that $V = p(U)$ is open in Y and $p|_U$ is a homeomorphism onto V). Then p is a finite covering if and only if it is proper.

Proof. If p is a finite covering, let $y_0 \in Y$ and let V be an open neighborhood of y_0 that is evenly covered by p. Then $p : p^{-1}(V) \to V$ is clearly proper. It follows easily that p is proper. Conversely, let p be a proper local homeomorphism, let $y_0 \in Y$, and let $p^{-1}(y_0) = \{x_1, \ldots, x_n\}$. Let U'_j be an open set with $x_j \in U'_j$ such that $p|_{U'_j}$ is a homeomorphism onto the open set $V_j = p\left(U'_j\right)$. Since p is proper and $X \setminus \bigcup_{j=1}^n U'_j$ is closed in X, the set $E = p\left(X \setminus \bigcup_{j=1}^n U'_j\right)$ is closed in Y. Clearly $y_0 \notin E$. Let $V = Y \setminus E$. Then $p^{-1}(V) \subset U'_1 \cup \cdots \cup U'_n$, and we have $V \subset V_1 \cap \cdots \cap V_n$. If we set $U_j = U'_j \cap p^{-1}(V)$, then $p^{-1}(V) = \bigcup_{j=1}^n U_j$, and $p|_{U_j}$ is a homeomorphism onto V. $\square$

Let $F \in \mathbb{C}[x, y]$ be irreducible, $F(x, y) = a_0(x) y^n + \cdots + a_n(x)$. Let $S_0 = \{x \in \mathbb{C} \mid a_0(x) = 0\}$ and

$$S_1 = \left\{ x \in \mathbb{C} \;\middle|\; \exists y \in \mathbb{C} \text{ with } F(x, y) = 0 = \frac{\partial F}{\partial y}(x, y) \right\}.$$

Then, if $V = \left\{(x, y) \in \mathbb{C}^2 \mid F(x, y) = 0\right\}$ and $\pi : V \to \mathbb{C}$ is the projection $(x, y) \mapsto x$, we have

$$\pi : \pi^{-1}\left(\mathbb{C} \setminus (S_0 \cup S_1)\right) \longrightarrow \mathbb{C} \setminus (S_0 \cup S_1)$$

is a *finite covering* (of n sheets).

This follows from Proposition 1.3.9 and the implicit function theorem. Before proceeding to show how the set V can be modified over the points of $S_0 \cup S_1 \equiv \left\{x \in \mathbb{C} \mid R(F, F'_y) = 0\right\}$ and the point at ∞ in $\mathbb{C}$ in order to define the algebraic function completely, we shall introduce the notion of a Riemann surface and some related topics.

Theorem 1.3.11. *Let X and Y be Riemann surfaces and let $p : Y \to X$ be an n sheeted branched holomorphic covering map. Let $A \subseteq X$ be a closed discrete set containing all critical values of p and set $B := p^{-1}(A)$. Let f be a holomorphic (resp. meromorphic) function on $Y \backslash B$ and let $c_1, \ldots, c_n \in \mathcal{O}(X \backslash A)$ (resp. $\in \mathcal{M}(X \backslash A)$) the elementary symmetric functions of f. Then f can be continued holomorphically (resp. meromorphically) to Y if and only if all c_j can be continued holomorphically (resp. meromorphically) to X.*

Proof. Let $a \in A$ and $p^{-1}(a) = \{b_1, \ldots, b_m\}$. Let (U, z) be a relatively compact coordinate neighborhood of a with $z(a) = 0$ and $U \cap A = \{a\}$. Then $V := p^{-1}(U)$ is a relatively compact neighborhood of $p^{-1}(a)$, since p is proper.

First suppose that f is holomorphic on $Y \backslash B$. If f can be continued holomorphically to all points b_j, then f is bounded on $V \backslash \{b_1, \ldots, b_m\}$. Hence all c_j are bounded on $U \backslash \{a\}$, and so all c_j admit a holomorphic extension to a, by Riemann's theorem on removable singularities 1.15. Conversely, if all c_j extend holomorphically to a, then all c_j are bounded on $U \backslash \{a\}$. Then also f is bounded on $V \backslash \{b_1, \ldots, b_m\}$, since

$$f(y)^n + c_1(p(y))f(y)^{n-1} + \cdots + c_n(p(y)) = 0 \quad \text{for } y \in V \backslash \{b_1, \ldots, b_m\}$$

By the Riemann's theorem on removable singularities (Theorem 5.6.6), f extends holomorphically to each b_j. Now let f be meromorphic on $Y \backslash B$. Assume that f can be continued meromorphically to all points b_j. The function $\varphi := p^* z = z \circ p$ is holomorphic on V and vanishes on all points b_j. Then $\varphi^k f$ has a holomorphic extension to all b_j provided that the integer k is chosen large enough. The elementary symmetric functions of $\varphi^k f$ are precisely $z^{kj} c_j$. By the previous paragraph, they admit a holomorphic extension to a, that is, the c_j admit a meromorphic extension to a. Conversely, suppose that all c_j extend meromorphically to a. Then all $z^{kj} c_j$ admit a holomorphic extension to a, where k is a large integer. Thus $\varphi^k f$ extends holomorphically to all b_j, whence f extends meromorphically to all b_j. □

Remark 1.3.12. For later use we remark that the proof does not use the fact that Y is connected. So in the theorem we may assume that Y is a disjoint union of finitely many Riemann surfaces.

1.4 The Mittag-Leffler Problem in an Open Planar Domain

If we restrict the domain of definition of the functions to open subsets of the plane, as in the original Mittag-Leffler problem, then we can assign polar behavior arbitrarily and always find a meromorphic function having this behavior, i.e., given any discrete set of points $\{p_i\}$, and any corresponding collection of polynomials P_i in $(z - p_i)^{-1}$ (without constant terms), there exists a globally defined meromorphic function f in the planar domain having as poles exactly the given set of points and having precisely the assigned polynomials as "principal parts" at those points, i.e., such that $f - \sum P_i$ is holomorphic in the entire plane.

The uniqueness question is much more complicated, since two functions with the same polar behavior differ by an entire function, and there is an enormous collection of entire functions defined on an open subset of the plane. In particular the space of functions with prescribed polar behavior is infinite dimensional.

In this section, we study in depth the Mittag-Leffler theorem in the complex plane, the main result which we intend to generalize to the context of Riemann surfaces.

Theorem 1.4.1. *Let $U \subset\subset \mathbb{C}$ be a bounded open domain, and let f be a bounded continuous complex-valued function in U. Then the continuous function*

$$u(z) := -\frac{1}{2\pi i}\int_U \frac{f(\zeta)d\overline{\zeta}\wedge d\zeta}{\zeta - z}, \qquad z \in U$$

is a solution of $\frac{\partial u}{\partial \overline{z}} = f$ in U.

Proof. First we consider the case that f is continuously differentiable in U. Fix $\xi \in U$. It is sufficient to prove that $\partial u/\partial\overline{z} = f$ in some neighborhood of ξ. Choose a $\mathcal{C}^\infty$-function χ on $\mathbb{C}$ such that $\chi = 1$ in a neighborhood $V_\xi \subset\subset U$ of ξ and $\chi = 0$ in some neighborhood of $\mathbb{C}\backslash U$. Then $u = u_1 + u_2$, where

$$\begin{aligned} u_1(z) &:= -\frac{1}{2\pi i}\int_U \frac{\chi(\zeta)f(\zeta)d\overline{\zeta}\wedge d\zeta}{\zeta - z}, \\ u_2(z) &:= -\frac{1}{2\pi i}\int_U \frac{(1-\chi(\zeta))f(\zeta)d\overline{\zeta}\wedge d\zeta}{\zeta - z}. \end{aligned}$$

Since $1-\chi = 0$ in V_ξ, u_2 is holomorphic in V_ξ, that is, $\overline{\partial} u_2 = 0$ in V_ξ. Therefore it remains to prove that $\partial u_1/\partial\overline{z} = f$ in V_ξ. Since $\chi = 0$ in a neighborhood of $\mathbb{C}\backslash U$, the function χf can be continued by zero to $\mathbb{C}$, and we obtain

$$u_1(z) = -\frac{1}{2\pi i}\int_{\mathbb{C}} \frac{\chi(\zeta)f(\zeta)d\overline{\zeta}\wedge d\zeta}{\zeta - z} = -\frac{1}{2\pi i}\int_{\mathbb{C}} \frac{\chi(\zeta+z)f(\zeta+z)d\overline{\zeta}\wedge d\zeta}{\zeta}.$$

Taking into account that $\partial/\partial\overline{z}[\chi(\zeta+z)f(\zeta+z)]d\overline{\zeta} = \overline{\partial}_\zeta[\chi(\zeta+z)f(\zeta+z)]$, we conclude that

$$\begin{aligned} \frac{\partial u_1(z)}{\partial \overline{z}} &= \frac{1}{2\pi i}\int_{\mathbb{C}} \frac{\overline{\partial}_\zeta[\chi(\zeta+z)f(\zeta+z)]\wedge d\zeta}{\zeta} \\ &= \frac{1}{2\pi i}\int_{\mathbb{C}} \frac{\overline{\partial}_\zeta[\chi(\zeta)f(\zeta)]\wedge d\zeta}{\zeta - z}, \qquad z \in U. \end{aligned}$$

Application of Theorem 1.1.3 with f replaced by χf and U equal to a disc containing the support of χ now gives that $\partial u_1/\partial\overline{z} = f$ in V_ξ.

Now we consider the case that f is an arbitrary continuous and bounded function on U. Choose a sequence f_k of continuously differentiable functions which converges uniformly on U to f. Then the sequence

$$u_k(z) := -\frac{1}{2\pi i}\int_U \frac{f_k(\zeta)d\overline{\zeta}\wedge d\zeta}{\zeta - z}$$

converges uniformly on U to u, and therefore we have in the sense of distributions $\partial u/\partial\overline{z} = \lim \partial u_k/\partial\overline{z} = \lim f_k = f$. □

1.5 The Compact Support Case

We begin by studying the equation $df/d\bar{z} = g \in L^1(\mathbb{C})$. An example is given for each $r > 0$ by:

$$f_r(z) = \begin{cases} 1/z & \text{if } |z| > r \\ \bar{z}/r^2 & \text{if } |z| \leq r \end{cases}$$

which satisfies $g_r = df_r/d\bar{z} = \left(1/r^2\right)\chi_{B(0,r)}(z)$. In particular $\int g_r = \pi$ is independent of r, which suggests the distributional equation:

$$\frac{d}{d\bar{z}}\frac{1}{z} = \pi\delta_0\,.$$

Using this fundamental solution (and the fact that $dxdy = (i/2)dz\wedge d\bar{z}$), we obtain:

Theorem 1.5.1. *Let $g \in C_0^k(\mathbb{C}), 1 \leq k \leq \infty$. The function*

$$u_0(z) = \frac{1}{2\pi i}\int_{\mathbb{C}} \frac{g(\zeta)}{\zeta - z} d\zeta \wedge d\overline{\zeta} \tag{1.7}$$

is $\mathcal{C}^k(\mathbb{C})$ and it is a solution of the Cauchy–Riemann equation

$$\frac{\partial u}{\partial\overline{z}} = g$$

Proof. The first proof. It is enough to check $df/d\bar{z}$ at $z = 0$. Using that fact that $df/d\bar{z} = (dg/d\bar{z}) * \frac{1}{\pi z}$, we find:

$$\begin{aligned} -2\pi i\frac{df}{dz}(0) &= \int_{\mathbb{C}} \frac{dg}{d\bar{z}}\left(\frac{-1}{z}\right) dzd\bar{z} = \int_{\mathbb{C}} d\left(\frac{g(z)dz}{z}\right) \\ &= \lim_{r\to 0}\int_{\mathbb{C}-B(0,r)} d\left(\frac{g(z)dz}{z}\right) = -\lim_{r\to 0}\int_{S^1(0,r)} g(z)\frac{dz}{z} = -2\pi i g(0) \end{aligned}$$

The penultimate minus sign comes from the fact that $S^1(0,r)$ acquires a negative orientation as $\partial(\mathbb{C} - B(0,r))$.

The second proof. Setting $w = \zeta - z$, (1.7) becomes

$$u_0(z) = \frac{1}{2\pi i}\int_{\mathbb{C}} \frac{g(w+z)}{w} dw \wedge d\overline{w}$$

Hence, we may differentiate under the integral sign, obtaining that $u_0 \in C^k(\mathbb{C})$. In particular,

$$\frac{\partial u_0}{\partial \overline{z}} = \frac{1}{2\pi i}\int_{\mathbb{C}} \frac{g_{\overline{\zeta}}}{\zeta - z} d\zeta \wedge d\overline{\zeta} = g$$

by Corollary 1.1.5. □

Corollary 1.5.2. *For any $g \in C^\infty(\mathbb{D})$, there is an $f \in C^\infty(\mathbb{D})$ with $\Delta f = g$.*

Proof. First solve $dh/dz = g$, then $df/d\overline{z} = h$. Then up to a multiplicative constant, we have $\Delta f = g$. □

Remark 1.5.3. The same results hold with $\mathbb{D}$ replaced by $\mathbb{C}$. It is easy to solve the equation $df/d\overline{z} = g$ on $\mathbb{C}$ when $g(z,\overline{z})$ is a polynomial: simply formally integrate with respect to the $\overline{z}$ variable.

In higher dimensions, $\omega = \bar{\partial} g$ should be thought of as a $(0,1)$-form, and we have the following result.

Remark 1.5.4. In general, the equation

$$\frac{\partial u}{\partial \overline{z}} = g$$

with $g \in C_0^\infty(\mathbb{C})$ does not admit compactly supported solutions. Indeed, if $v \in C_0^\infty(\mathbb{C})$ is a solution and u_0 is given by (1.7), then $\varphi = u_0 - v$ is entire and, since

$$|u_0(z)| \leq \frac{C}{d(z, \operatorname{supp} g)}$$

(X being a constant), we have $|\varphi(z)| \to 0$ when $|z| \to +\infty$. It follows that $\varphi \equiv 0$ and $u_0 = v$. Thus u_0 has compact support, consequently, in view of (1.7), g must fulfill the condition

$$\int_{\mathbb{C}} \frac{g(\zeta)}{\zeta - z} d\zeta \wedge d\overline{\zeta} = 0 \tag{1.8}$$

for $|z| \gg 0$.

Example 1.5.5. The function

$$g = \begin{cases} e^{-\frac{1}{1-|z|^2}}, & \text{for } |z| < 1 \\ 0, & \text{for } |z| \geq 1 \end{cases}$$

does not satisfy the condition (1.8).

1.6 The General Case

The proof of the existence theorem for

$$\frac{\partial u}{\partial \bar{z}} = g$$

when $g \in C^\infty$ is more involved and requires an approximation theorem.

Theorem 1.6.1 (Runge). *Let $K \subset U \subset \mathbb{C}$ be a compact set. Then $\mathcal{O}(U)$ is dense in $\mathcal{O}(K)$, with respect to the sup norm on K if and only if one of the following equivalent conditions holds*

(I) *$U \backslash K$ has no relatively compact component in D;*

(II) *K coincides with its $\mathcal{O}(U)$-envelope $\widehat{K}_U$.*

In particular, if $U = \mathbb{C}$ and $\mathbb{C}\backslash U$ is connected, $\mathbb{C}[z]$ is dense in $\mathcal{O}(K)$.

For the proof we refer to [86].

Remark 1.6.2. The Runge theorem was generalized by Mergelyan [127] (see also [186]): if condition **(I)** holds, then $\mathcal{O}(U)$ is dense, with respect to the sup norm on K, in $A(K)$, the algebra of holomorphic functions on the interior of K, continuous on K.

Theorem 1.6.3. *Let U be a domain in $\mathbb{C}$ and $g \in C^\infty(U)$. The equation*

$$\frac{\partial u}{\partial \bar{z}} = g$$

admits a solution $u \in C^\infty(U)$.

Proof. We consider at first two particular cases, namely $U = \mathbb{C}$ and $U \subsetneq \mathbb{C}$ simply connected. Let

$$U_j = \{z \in \mathbb{C} \mid |z| < j\}$$

$n \in \mathbb{N}$. For each $n \in \mathbb{N}$ let $\rho_j \in C_0^\infty(U_{j+1})$ with $\rho_j \equiv 1$ on a neighborhood of $\overline{U}_j$. Let $v_j \in C^\infty(\mathbb{C})$ be such that

$$\frac{\partial v_j}{\partial \bar{z}} = \rho_j g, \quad n \in \mathbb{N}.$$

Our goal is to modify the solutions v_j in order to have a sequence $\{v_j\}_{j \in \mathbb{N}}$ which converges to a smooth solution of the problem. Let $u_1 = v_1$. The function $v_2 - u_1$ is holomorphic on a neighborhood of $\overline{U}_1$ and so there is a polynomial $p = p(z)$ such that

$$\|(v_2 - u_1) - p\|_{\overline{U}} < \frac{1}{2}.$$

Define $u_2 = v_2 - p$. Obviously, in a neighborhood of $\overline{U}_2$ we have

$$\frac{\partial u_2}{\partial \overline{z}} = g$$

We may repeat the argument for $v_3 - u_2$ and so on. In this way we construct a sequence of functions $u_j \in C^\infty(\mathbb{C})$ with the properties

$$\frac{\partial u_j}{\partial \overline{z}} = g$$

in a neighborhood of $\overline{U}_j$ and

$$\|u_{n+1} - u_j\|_{\overline{U}_j} < \frac{1}{2^j} \tag{1.9}$$

for all $n \in \mathbb{N}$. The sequence $\{u_j\}$ converges in $C^\infty(\mathbb{C})$ to a solution u. Indeed, on each disc U_k

$$u = \lim_{j \to +\infty} u_j = u_k + \sum_{j=k}^{\infty} (u_{j+1} - u_j)$$

and the series of holomorphic functions

$$\sum_{j=k}^{\infty} (u_{j+1} - u_j)$$

converges uniformly on $\overline{U}_k$ in view of (1.9). Thus, on U_k, for every $k \in \mathbb{N}$ we have

$$\frac{\partial u}{\partial \overline{z}} = \frac{\partial u_k}{\partial \overline{z}} = g\,.$$

The case when $U \subsetneq \mathbb{C}$ is simply connected reduces to the case of the unit disc, in view of the Riemann mapping theorem. Indeed, let $f : \mathbb{D} \to U$ be a biholomorphism from the unit disc to the domain U and define

$$\tilde{g}(z') = g(f(z')) \overline{\frac{\partial f}{\partial z'}}\,.$$

If $\tilde{u}$ is a solution of

$$\frac{\partial \tilde{u}}{\partial \overline{z}} = \tilde{g}$$

on the unit disc, then the function $u = \tilde{u} \circ f^{-1}$ is a solution of

$$\frac{\partial u}{\partial z} = g$$

on U. In order to solve the problem for the unit disc, we use the same method as in the case of $\mathbb{C}$.

As for the general case, we consider an exhaustive sequence $\{K_j\}$ of compact sets such that $K_j \subset \overset{\circ}{K}_{j+1}$ and $K_j = \widehat{K}_{j\mathcal{O}(U)}$. Let $\rho_j \in C_0^\infty\left(\overset{\circ}{K}_{j+1}\right)$ such that $\rho_j \equiv 1$ on a neighborhood of K_j. The proof runs as in the first two cases using Theorem 1.6.1. □

By covering $\mathbb{C}$ with an increasing sequence of discs, we can remove from the previous result the condition of compactness of the support of g.

Theorem 1.6.4 (Dolbeault). *Let $g : \mathbb{C} \to \mathbb{C}$ be a smooth function. Then there exists a smooth function $f : \mathbb{C} \to \mathbb{C}$ with $f_{\bar{z}} = g$.*

Proof. As remarked before the statement of the theorem, the idea of the proof is to cover $\mathbb{C}$ by an increasing sequence of compact sets, on which we can solve the inhomogeneous Cauchy–Riemann equation; the limit of these "compact" solutions will be the desired function.

We start by setting $\mathbb{D}_n = \{z \in \mathbb{C} : |z| < n\}$. Since $\mathbb{C}$ is a normal space ([136, Theorem 2.3, p. 198, and Urysohn's Lemma, p. 207]) there exist smooth functions ϕ_n, with compact support in $\mathbb{D}_{n+1}$, such that $\phi_n \equiv 1$ on the disc $\{z \in \mathbb{C} \mid |z| < n + \frac{1}{2}\}$. (Observe that this disc contains the compact set $\mathbb{D}_n$.) Let $g_n = \phi_n g$. These functions have compact support, and therefore we can find smooth functions h_n, with support contained in $\mathbb{D}_{n+1}$, such that $\bar{\partial} h_n = g_n$ (in $\mathbb{D}_{n+1}$ or in $\mathbb{C}$). The functions $h_{n+1} - h_n$ are holomorphic on a neighborhood of $\mathbb{D}_n$, so they have a power series expansion that converges uniformly on $\overline{\mathbb{D}_n}$. Truncating that power series, we obtain a polynomial p_n such that

$$|(h_{n+1} - h_n - p_n)(z)| < 2^{-n}, \quad \text{for all } z \in \mathbb{D}_n$$

(this is a trivial case of Runge's theorem, [161, Theorem 113.7, p. 290]). Define a function f by

$$f = h_n + \left(\sum_{m=n}^{\infty} (h_{m+1} - h_m - p_m) \right) - (p_1 + \cdots + p_{n-1}) .$$

It is easy to see that the function f is well defined. The right-hand side of the above equality is a uniformly convergent series of holomorphic functions, so f is holomorphic and satisfies $\bar{\partial} f = \bar{\partial} h_n = g$ on X_n. □

Observe that solutions of the equation $\bar{\partial} f = g$ are not unique: if f is one solution, then so is $f + h$, where h is a holomorphic function (since $\bar{\partial} h = 0$).

Corollary 1.6.5. *The equation $\Delta f = g$ has solution in $\mathbb{C}$ for any smooth function g.*

Proof. Let h and k be smooth functions such that $h_{\bar{z}} = \bar{g}$ and $k_{\bar{z}} = \bar{h}$. Then $f = \frac{\mathrm{i}}{2} k$ satisfies $\Delta f = g$ (use the identity $\Delta = 2\mathrm{i}\partial\bar{\partial}$) □

Very often problems in the theory of several complex variables reduce to solving Cauchy–Riemann equations. This happens even in the most simple cases, as shown by the following examples.

Example 1.6.6 (Construction of a meromorphic function with given poles). Given a discrete set $F = \{z_j\}_{j \in \mathbb{N}}$ in a domain $\Omega \subset \mathbb{C}$ and a sequence $\{n_j\}_{j \in \mathbb{N}}$

of positive integers, we ask for a meromorphic function f on Ω, whose poles are the points z_j, with respective orders n_j. Consider the meromorphic functions

$$\varphi_j = \frac{1}{(z - z_j)^{n_j}},$$

$j \in \mathbb{N}$. Each φ_j is holomorphic on $\tilde{V}_j = \mathbb{C} \setminus \{z_j\}$; $\{V_j = \tilde{V}_j \cap \Omega\}_{j\in\mathbb{N}}$ is an open covering of Ω, and $\varphi_{jk} = \varphi_j - \varphi_k$ is holomorphic on $V_{jk} = V_j \cap V_k$. We may choose a locally finite refinement $\{U_j\}_{j\in\mathbb{N}}$ of $\{V_j\}_{j\in\mathbb{N}}$. Let $\{\rho_j\}_{j\in\mathbb{N}}$ be a partition of unity associated to $\{U_j\}_{j\in\mathbb{N}}$ and let

$$f_j = \sum_{k\in\mathbb{N}} \rho_k \varphi_{jk},$$

$f_j \in C^\infty(U_j)$ and $f_j - f_k = \varphi_{jk}$ is holomorphic on U_{jk}. Thus the local functions

$$g_j = \frac{\partial f_j}{\partial \bar{z}}$$

define a function $g \in C^\infty(\Omega)$ and, since the Cauchy–Riemann equation for g is solvable in Ω, there exists $u \in C^\infty(\Omega)$ such that

$$\frac{\partial u}{\partial \bar{z}} = g.$$

Then

(I) $f_j - u \in \mathcal{O}(U_j)$;

(II) On U_{jk}, $(f_j - u) - (f_k - u) = \varphi_{jk} \in \mathcal{O}(U_{jk})$, so the local functions $\varphi_j + u - f_j$ define a meromorphic global function f. On U_j we have

$$f - \varphi_j = f_j - u \in \mathcal{O}(U_j)$$

i.e., z_j is the only pole in U_j of f (of order n_j).

Example 1.6.7 (Construction of a holomorphic function with given zeros). Consider a discrete set $F = \{z_j\}_{j\in\mathbb{N}}$ in a domain $\Omega \subset \mathbb{C}$ and a sequence $\{n_j\}_{j\in\mathbb{N}}$ of positive integers. The problem is to find a holomorphic function $f : \Omega \to \mathbb{C}$, whose zeros are the points z_j, with respective orders n_j.

Consider a locally finite covering $\{U_j\}_{j\in\mathbb{N}}$ of Ω with the following properties:

(I) $U_j \cap F = \{z_j\}$;

(II) $U_{jk} = U_j \cap U_k$ is simply connected for all j and k.

In each U_j we define the holomorphic function

$$\varphi_j(z) = (z - z_j)^{n_j}.$$

Following the lines of proof in the previous example, we must find invertible holomorphic functions $h_j : U_j \to \mathbb{C}$, such that

$$\frac{\varphi_j}{\varphi_k} = \frac{h_j}{h_k} \tag{1.10}$$

on $U_{jk} \neq \emptyset$.

Since U_{jk} is simply connected and

$$\varphi_{jk} = \frac{\varphi_j}{\varphi_k} \neq 0\,,$$

there exists a logarithm $\theta'_{jk} = \log \varphi_{jk}$ of φ_{jk}. If $U_{jkl} = U_j \cap U_k \cap U_l \neq \emptyset$, then

$$\theta'_{jk} + \theta'_{kl} + \theta'_{lj} = 2\beta_{jkl}\pi i$$

with $\beta_{jkl} \in \mathbb{Z}$. From the homology theory (see [27]) we know that for every pair of indexes j and k such that $U_{jk} \neq \emptyset$, there is a $\beta_{jk} \in \mathbb{Z}$ such that

$$\beta_{jl} = -\beta_{lj} \quad \beta_{jk} = \beta_{jt} + \beta_{tj} + \beta_{jl}\,.$$

Define $\theta_{jk} = \theta'_{jk} - 2\beta_{jk}\pi i$. It follows that $\log \varphi_{jk} = \beta_{jk}$ and

$$\theta_{jk} + \theta_{kl} + \theta_{lj} = 0\,.$$

Arguing as in Example 1.6.6, there exist holomorphic functions g_j, $j \in \mathbb{N}$, such that $\theta_{jk} = g_j - g_k$ for every j, k, hence

$$\varphi_{jk} = e^{\theta_{jk}} = \frac{e^{g_j}}{e^{g_k}}\,.$$

Thus (1.10) holds with $h_j = e^{g_j}$.

From Examples 1.6.6 and 1.6.7 we derive the following

Proposition 1.6.8 (Mittag-Leffler). *Let U be a domain in $\mathbb{C}$. Let $\{z_j\}$ and $\{w_j\}$ be disjoint discrete sets in U and $\{n_j\}$ and $\{m_j\}$ sequences of positive integers. Then there exists a meromorphic function f on U such that*

(I) $\{z_j\}_{j\in\mathbb{N}}$ is the set of zeros with respective orders $\{n_j\}_{j\in\mathbb{N}}$;

(II) $\{w_j\}_{j\in\mathbb{N}}$ is the set of poles with respective orders $\{m_j\}_{j\in\mathbb{N}}$;

(III) $f(z) \in \mathbb{C}^*$, $\mathbb{C}^* = \mathbb{C} \setminus \{0\}$, for $z \neq z_j$, $z \neq w_j$, $j \in \mathbb{N}$.

Remark 1.6.9. When X is the entire Riemann sphere, the field of meromorphic functions is simply the field of rational functions in one variable over the complex field, since one can prove that any meromorphic function on the sphere is rational. (This is a special case of the so-called *GAGA principle.*)

For every Riemann surface, a meromorphic function is the same as a holomorphic function that maps to the Riemann sphere and which is not the constant function equal to ∞. The poles correspond to those complex numbers which are mapped to ∞.

On a non-compact Riemann surface, every meromorphic function can be realized as a quotient of two (globally defined) holomorphic functions. In contrast, on a compact Riemann surface, every holomorphic function is constant, while there always exist non-constant meromorphic functions.

On every non-compact Riemann surface there exists a meromorphic function with given poles $\{a_1, a_2, \ldots\}$ and given in each of them the principal part of the Laurent expansion (Mittag-Leffler theorem on the expansion of a meromorphic function).

On a compact Riemann surface (for example, on a torus) this problem is generally unsolvable—additional conditions for the compatibility of the principal parts are needed.

For the standard material from complex analysis (in particular properties of $\overline{\partial}$) used above, as well as a different arrangement of the proof of the finiteness theorem (see Chapter), see: [138].

1.7 Paths and Curves

A curve γ is a continuous map from a real interval $[a, b]$ to $\mathbb{C}$. The curve γ is said to be *closed* if $\gamma(a) = \gamma(b)$; it is said to be *open* otherwise. A *simple closed curve* is a closed curve with $\gamma(t_1) = \gamma(t_2)$ if and only if $t_1 = a$ and $t_2 = b$.

An intuitively obvious theorem about curves that turned out to be very difficult to prove is the Jordan curve theorem. This theorem is usually not necessary in complex analysis, but is useful as background.

Theorem 1.7.1 (the Jordan curve theorem). *The image of a simple closed curve (not assumed to be differentiable) separates the extended complex plane into two regions. One region is bounded (and "inside" the curve) and the other is unbounded.*

Definition 1.7.2. A *path* γ is a continuous map of an interval $[\alpha, \beta]$ of the real axis into the complex plane $\mathbb{C}$ (or $\widehat{\mathbb{C}}$). In other words, a path is a complex valued function $z = \gamma(t)$ of a real argument t, that is continuous at every point $t_0 \in [\alpha, \beta]$ in the following sense: for any $\varepsilon > 0$ there exists $\delta > 0$ so that $|\gamma(t) - \gamma(t_0)| < \varepsilon$ (or $\rho(\gamma(t), \gamma(t_0)) < \varepsilon$ if $\gamma(t_0) = \infty$) provided that $|t - t_0| < \delta$. The points $a = \gamma(\alpha)$ and $b = \gamma(\beta)$ are called the *endpoints of the path* γ. The path is closed if $\gamma(\alpha) = \gamma(\beta)$. We say that a path γ lies in *a* set M if $\gamma(t) \in M$ for all $t \in [\alpha, \beta]$.

Sometimes it is convenient to distinguish between a path and a curve. In order to introduce the latter we say that two paths

$$\gamma_1 : [\alpha_1, \beta_1] \to \widehat{\mathbb{C}} \text{ and } \gamma_2 : [\alpha_2, \beta_2] \to \widehat{\mathbb{C}}$$

are *equivalent* ($\gamma_1 \sim \gamma_2$) if there exists an increasing continuous function

$$\tau : [\alpha_1, \beta_1] \to [\alpha_2, \beta_2] \tag{1.11}$$

such that $\tau(\alpha_1) = \alpha_2, \tau(\beta_1) = \beta_2$ and so that $\gamma_1(t) = \gamma_2(\tau(t))$ for all $t \in [\alpha_1, \beta_1]$.

It is not hard to prove that the relation $\sim$ is:

- *reflexive* — $\gamma \sim \gamma$,
- *symmetric* — if $\gamma_1 \sim \gamma_2$, then $\gamma_2 \sim \gamma_1$,
- *transitive* — if $\gamma_1 \sim \gamma_2$ and $\gamma_2 \sim \gamma_3$ then $\gamma_1 \sim \gamma_3$.

Example 1.7.3. Let us consider the paths $\gamma_1(t) = t, t \in [0, 1]; \gamma_2(t) = \sin t$, $t \in [0, \pi/2]$; $\gamma_3(t) = \cos t, t \in [0, \pi/2]$ and $\gamma_4(t) = \sin t$, $t \in [0, \pi]$.

The set of values of $\gamma_j(t)$ is always the same: the interval $[0, 1]$. However, we only have $\gamma_1 \sim \gamma_2$. These two paths trace $[0, 1]$ from left to right once. The paths γ_3 and γ_4 are neither equivalent to these two, nor to each other: the interval $[0, 1]$ is traced in a different way by those paths: γ_3 traces it from right to left, while γ_4 traces $[0, 1]$ twice.

Definition 1.7.4. A *curve* is an equivalence class of paths. Sometimes, when this will cause no confusion, we will use the word "curve" to describe a set $\gamma \in \widehat{\mathbb{C}}$ that may be represented as an image of an interval $[\alpha, \beta]$ under a continuous map $z = \gamma(t)$.

Below we will introduce some restrictions on the curves and paths that we will consider. We say that $\gamma : [\alpha, \beta] \to \widehat{\mathbb{C}}$ is a *Jordan path* if the map γ is continuous and one-to-one.

A *Jordan curve* or a simple closed curve in the plane $\mathbb{R}^2$ is the image $\mathcal{C}$ of an injective continuous map of a circle into the plane, $\varphi : S^1 \to \mathbb{R}^2$.

Alternatively, a Jordan curve is the image of a continuous map $\varphi : [0, 1] \to \mathbb{R}^2$ such that $\varphi(0) = \varphi(1)$ and the restriction of φ to $[0, 1)$ is injective.

A path $\gamma : [\alpha, \beta] \to \mathbb{C}$ ($\gamma(t) = x(t) + iy(t)$) is continuously differentiable if the derivative $\gamma'(t) := x'(t) + iy'(t)$ exists for all $t \in [\alpha, \beta]$ and is continuous. A continuously differentiable path is said to be *smooth* if $\gamma'(t) \neq 0$ for all $t \in [\alpha, \beta]$. This condition is introduced in order to avoid singularities. A path is called *piecewise smooth* if $\gamma(t)$ is continuous on $[\alpha, \beta]$, and $[\alpha, \beta]$ may be divided into a finite number of closed sub-intervals so that the restriction of $\gamma(t)$ on each of them is a smooth path.

We will also use the standard notation to describe smoothness of functions and paths: the class of continuous functions is denoted C, or C^0, the class of continuously differentiable functions is denoted C^1, etc. A function that has n continuous derivatives is said to be a C^n-function.

Example 1.7.5. The paths γ_1, γ_2 and γ_3 of the previous example are Jordan, while γ_4 is not Jordan. The circle $z = e^{it}, t \in [0, 2\pi]$ is a closed smooth Jordan path; the four-petal rose $z = e^{it} \cos 2t, t \in [0, 2\pi]$ is a smooth non-Jordan path; the semi-cubic parabola $z = t^2(t+i), t \in [-1, 1]$ is a Jordan continuously differentiable piecewise smooth path. The path $z = t\left(1 + i \sin\left(\frac{1}{t}\right)\right), t \in [-1/\pi, 1/\pi]$ is a Jordan non-piecewise smooth path.

One may introduce similar notions for curves. A Jordan curve is a class of paths that are equivalent to some Jordan path (observe that since the change of variables (1.11) is one-to-one, all paths equivalent to a Jordan path are also Jordan).

The definition of a smooth curve is slightly more delicate: this notion has to be invariant with respect to a replacement of a path that represents a given curve by an equivalent one. However, a continuous monotone change of variables (1.11) may map a smooth path onto a non-smooth one unless we impose some additional conditions on the functions τ allowed in (1.11).

More precisely, a smooth curve is a class of paths that may be obtained out of a smooth path by all possible re-parameterizations (1.11) with $\tau(s)$ being a continuously differentiable function with a positive derivative. One may define a piecewise smooth curve in a similar fashion: the change of variables has to be continuous everywhere, and in addition have a continuous positive derivative except possibly at a finite number of points.

Sometimes we will use a more geometric interpretation of a curve, and say that a Jordan, or smooth, or piecewise smooth curve is a set of points $\gamma \subset \mathbb{C}$ that may be represented as the image of an interval $[\alpha, \beta]$ under a map $z = \gamma(t)$ that defines a Jordan, smooth, or piecewise smooth path.

1.8 Continuous Choice of Argument

Each non-zero complex number $z = x + iy$ can be expressed as

$$z = r(\cos\theta + i \sin\theta) = re^{i\theta},$$

where the positive number $r = |z| = \sqrt{x^2 + y^2}$ is the modulus of z. Here $r = |z|$, so it is unique. However, when $r = 0$ any value of θ leads to $z = 0$. And when $r > 0$ there are infinitely many possible values of θ, differing only by integer multiples of 2π.

Recall that the unique value of θ in the range $-\pi < \theta \leq \pi$ is called the *principal value* of the argument of z, and is denoted by

$$\arg z.$$

This defines a function

$$\arg : \mathbb{C}\backslash\{0\} :\to \mathbb{R}$$

The function arg is *not* continuous on the negative real axis. This is a result of the need to choose θ uniquely: just above the axis θ is close to π, just below it is close to $-\pi$.

Definition 1.8.1. Let $f : U \to \mathbb{C} \setminus \{0\}$ be continuous. A *continuous choice of argument* for f is a continuous map $F : U \to \mathbb{R}$ such that $f = |f|e^{iF}$.

A question that arises naturally in complex analysis is to what extent can one select an argument for each non-zero complex number so as to form a continuous function. It is well known that this cannot be achieved on the domain of non-zero complex numbers, but can be on the complex plane with a half-line excluded. The proof of this last result is what concerns us here.

For each $\alpha \in \mathbb{R}$, define the half-line H_α emanating from the origin away from $e^{i\alpha}$ by the equation

$$H_\alpha = \left\{-re^{i\alpha} \mid r \geqslant 0\right\} .$$

Let $\mathbb{C}\backslash H_\alpha$ be the result of removing the infinite ray in direction α from the plane (see Fig. 1.1). For each z in the cut plane $\mathbb{C}\backslash H_\alpha$, define the argument $\arg_\alpha(z)$ to be the unique argument of z lying in the interval $(\alpha - \pi, \alpha + \pi)$. The argument function $\arg_0$ is called the *principal argument*.

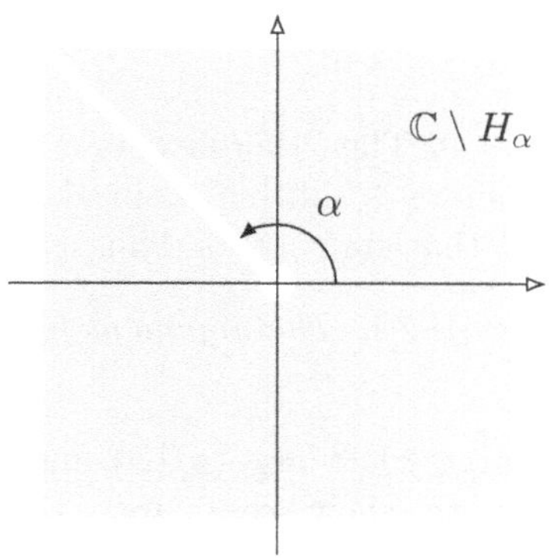

FIGURE 1.1: The cut plane $\mathbb{C} \setminus H_\alpha$.

It is geometrically clear that $\arg_\alpha$ is continuous on the cut plane $\mathbb{C}\backslash H_\alpha$, but a rigorous proof is needed. One might expect the literature to provide several simple proofs of this basic result, but such appears not to be the case. Stewart and Tall [173, p. 121] acknowledge that there is a technical difficulty in passing from the simple geometry of the situation to an analytic proof, caused by the bad behavior of the inverse trigonometric functions. They judge their own way of circumventing the difficulty, by considering a cut plane as the union of three open half-planes. References [45, 90, 146, 181] contain four more proofs of the continuity of $\arg_\alpha$ on $\mathbb{C}\backslash H_\alpha$, although none singles itself out as being definitive. The proof in [45, p. 59] is elementary, but overly intricate and uses open sets. The proofs [90, p. 46] and [181, p. 66] are sophisticated in that they employ the preservation of compactness under continuous mappings, the one in [146], of a more general result, relies upon contour integration.

The two lemmas preceding our own proof that $\arg_\alpha$ is continuous on $\mathbb{C}\backslash H_\alpha$ serve to highlight the basic facts which underlie it. The first asserts that the length of the side opposite the vertex of an isosceles triangle with equal sides of length 1 is twice the sine of half the vertex angle.

Lemma 1.8.2. *Let $\theta, \varphi \in \mathbb{R}$ with $|\theta - \varphi| \leqslant 2\pi$. Then*

$$\left|e^{i\theta} - e^{i\varphi}\right| = 2\sin\left(\frac{1}{2}\left|\theta - \varphi\right|\right).$$

Proof. We have

$$\begin{aligned}\left|e^{i\theta} - e^{i\varphi}\right| &= |(\cos\theta - \cos\varphi) + i(\sin\theta - \sin\varphi)| \\ &= \sqrt{2 - 2\cos(\theta - \varphi)} \\ &= 2\sin\left(\frac{1}{2}|\theta - \varphi|\right) \quad \left(\text{as } \frac{1}{2}|\theta - \varphi| \leqslant \pi\right).\end{aligned}$$

□

Lemma 1.8.3. *Let $x_1, x_2, \ldots, x_n, \ldots$ be a sequence in $[0, \beta]$, where $0 < \beta < \pi$, such that $\sin x_n \to 0$. Then $x_n \to 0$.*

Proof. Suppose that $x_n \not\to 0$. Then there exists $\varepsilon > 0$ such that $x_n \geqslant \varepsilon$ for infinitely many n. Hence $\sin x_n \geqslant \min\{\sin\varepsilon, \sin\beta\} > 0$ for infinitely many n, which contradicts the fact that $\sin x_n \to 0$. Thus $x_n \to 0$. □

Theorem 1.8.4 (Webster [182]). *The argument function $\arg_\alpha$ is continuous on the cut plane $\mathbb{C}\backslash H_\alpha$.*

Proof. Since, for $z \neq 0, \arg_\alpha(z) = \arg_\alpha(z/|z|)$ and the function $z \mapsto z/|z|$ is continuous, it suffices to show that $\arg_a$ is continuous on the set $A = \{z \in \mathbb{C}\backslash H_a \mid |z| = 1\}$. This we do using a sequential argument.

In A, let $z_1, z_2, \ldots, z_n, \ldots$ be a sequence converging to z_0. For $n = 0, 1, 2, \ldots$, write $\theta_n = \arg_\alpha(z_n)$. Then $z_n = e^{i\theta_n}$ and $|\theta_n - \alpha| < \pi$, so

$$|\theta_n - \theta_0| < \pi + |\theta_n - \alpha| \text{ and } \frac{1}{2}|\theta_n - \theta_0| < \alpha, \text{ where } \alpha = \frac{\pi}{2} + \frac{1}{2}|\theta_0 - \alpha| < \pi.$$

By Lemma 1.8.2,

$$|z_n - z_0| = \left|e^{i\theta_n} - e^{i\theta_0}\right| = 2\sin\left(\frac{1}{2}|\theta_n - \theta_0|\right) \to 0.$$

By Lemma 1.8.3, $\theta_n = \arg_a(z_n) \to \theta_0 = \arg_a(z_0)$. Thus $\arg_a$ is continuous on $\mathbb{C}\backslash H_\alpha$. □

Lemma 1.8.5 (the paving lemma). *Let U be an open subset of $\mathbb{C}$ and let $\gamma : [a,b] \to U$ be a continuous curve in U. There exist finitely many points $a = t_0 < t_1 < \cdots < t_n = b$ in the interval $[a,b]$ and a radius $r > 0$ such that*

$$\bigcup_{i=0}^{n} D_r(\gamma(t_i)) \subseteq U \tag{1.12}$$

and

$$\gamma([t_i, t_{i+1}]) \subseteq D_r(\gamma(t_i)), \quad i = 0, 1, \ldots, n-1 \tag{1.13}$$

Proof. By a classical theorem of analysis, γ is uniformly continuous. Given $\varepsilon = r > 0$, there exists $\delta > 0$ such that $|\gamma(t) - \gamma(s)| < r$ provided $t, s \in [a,b]$ satisfy $|t - s| \leq \delta$. We now partition the interval $[a,b]$ by finitely many points t_i as above such that $t_{i+1} - t_i \leq \delta$ for each i. Then (1.13) clearly holds. If $U = \mathbb{C}$, then (1.12) holds for any $r > 0$. The set $\gamma([a,b])$ is closed and bounded.

If $U \neq \mathbb{C}$, then there is a positive distance d between $\gamma([a,b])$ and the closed set $\mathbb{C} \backslash U$. By choosing $r < d$ both (1.12) and (1.13) hold provided $\delta > 0$ is chosen in accordance with the uniform continuity corresponding to $\varepsilon = r$. □

Theorem 1.8.6. *(i) Let $\gamma : [a,b] \to \mathbb{C}\backslash\{0\}$ be a path not passing through the origin. Then there exists a continuous choice of argument for γ.*

(ii) Any other continuous choice of argument differs from this by a constant integer multiple of 2π.

Proof. By the paving Lemma 1.8.5 we can subdivide γ into finitely many subpaths γ_k ($k = 1, \ldots, n$) such that each γ_k lies inside a disc $D_k \subseteq \mathbb{C}\backslash\{0\}$, see Fig. 1.2.

If the center of d_k is at $\rho_k e^{i\theta_k}$, then taking $\alpha_k = \theta_k + \pi$, we find that $D_k \subseteq \mathbb{C}_{\alpha_k}$, so by Theorem 1.8.4 we have $\arg_{\alpha_k}$ is continuous in D_k.

This gives a continuous choice of argument across γ_k for each k, but of course these choices need not fit together continuously. However, any choice of argument can be obtained from any other by adding a suitable integer multiple of 2π. We can therefore adjust the $\arg_{\alpha_k}$ inductively so that they do fit together continuously, as follows.

As usual, let γ_k be defined on the parametric interval $[t_{k-1}, t_k]$. There is an integer n_2 such that

$$\arg_{\alpha_1}(t_1) = \arg_{\alpha_2}(t_1) + 2n_2\pi$$

Then there is an integer n_3 such that

$$\arg_{\alpha_2}(t_2) + 2n_2\pi = \arg_{\alpha_3}(t_2) + 2n_3\pi$$

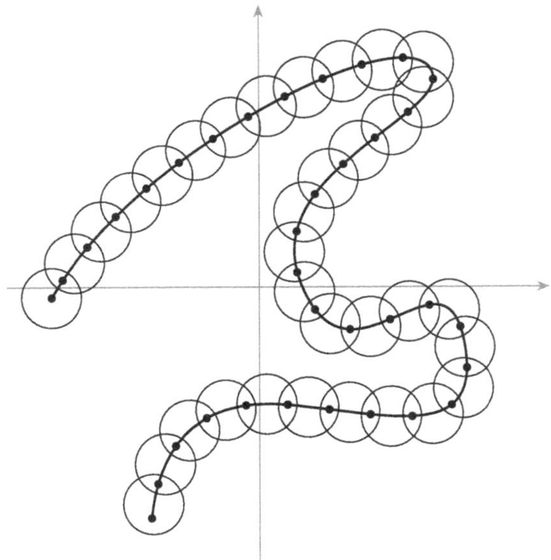

FIGURE 1.2: Paving a path.

and so on inductively, choosing n_{k+1} so that

$$\arg_{\alpha_k}(t_k) + 2n_k\pi = \arg_{\alpha_{k+1}}(t_k) + 2n_{k+1}\pi\,.$$

Then we define

$$\theta(t) = \arg_{\alpha_k}(t_k) + 2n_k 2\pi \quad (t \in [t_{k-1}, t_k])$$

with n_1 conventionally defined to be 0. Then θ is continuous. This proves that a continuous choice of argument exists. Next, suppose that θ^* is another continuous choice of argument. Then we must have

$$\theta^*(t) = \theta(t) + 2n(t)\pi$$

where $n(t)$ is an integer, possibly depending on t. But

$$n(t) = \frac{\theta^*(t) - \theta(t)}{2\pi}$$

is continuous, yet takes integer values. It is therefore constant. □

Note that a continuous choice of argument is defined on the parametric interval $[a, b]$, not on the image of γ. This means that if the path returns to the same point, with $\gamma(t_1) = \gamma(t_2)$ but $t_1 \neq t_2$, the arguments $\theta(t_1)$ and $\theta(t_2)$ may be different. Intuitively, it is clear that this will happen if the path winds round the origin a non-zero number of times between t_1 and t_2.

1.9 Homotopic Curves

A path γ in an open set Ω in $\mathbb{C}$ (or more generally a topological space, e.g., Ω a closed subset of $\mathbb{C}$) is simply a continuous map $\gamma : [0,1] \to \Omega$. We call $a = \gamma(0)$ the *starting point of the path* and $b = \gamma(1)$ the endpoint. Let us denote by $\mathcal{C}(\Omega; a, b)$ the collection of paths in Ω starting at a and ending at b. If $a = b$ we say the path is closed and a is called the *base point.* The set $\mathcal{C}(\Omega; a)$ is then the family of all closed paths (*loops* in Ω with base point a). If $z \in \Omega$ we denote by ε_z the constant path with base point $z, \varepsilon_z(t) = z (0 \leq t \leq 1)$. If $\gamma \in \mathcal{C}(\Omega; a, b)$ we denote by $\overline{\gamma}$ the inverse path to $\gamma, \overline{\gamma} \in \mathcal{C}(\Omega; b, a)$ and is defined by $\overline{\gamma}(t) = \gamma(1-t)(0 \leq t \leq 1)$. If $\gamma \in \mathcal{C}(\Omega; a)$ then $\overline{\gamma} \in \mathcal{C}(\Omega; a)$ also. Furthermore, $\overline{\varepsilon}_a = \varepsilon_a$.

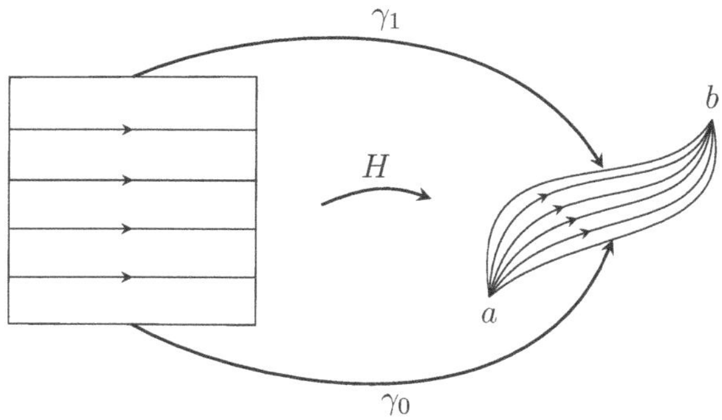

FIGURE 1.3: A homotopy with fixed endpoints. The deformation of the curve γ_0 into the curve γ_1.

If $\alpha \in \mathcal{C}(\Omega; a, b)$ and $\beta \in \mathcal{C}(\Omega; b, c)$ we denote by $\alpha\beta$ the path in $\mathcal{C}(\Omega; a, c)$ defined by:

$$\alpha\beta(t) := \begin{cases} \alpha(2t) & \text{if } 0 \leq t \leq \frac{1}{2} \\ \beta(2t-1) & \text{if } \frac{1}{2} \leq t \leq 1 \end{cases}$$

The path $\alpha\beta$ is called the *composition* of α and β. It is easy to see that $\overline{\alpha\beta} = \overline{\beta}\overline{\alpha}$. On the other hand, when $(\alpha\beta)\gamma$ is defined, then $\alpha(\beta\gamma)$ is also defined but, in general, they do not coincide.

Definition 1.9.1. Two paths $\gamma_0, \gamma_1 \in \mathcal{C}(\Omega; a, b)$ are said to be *homotopic with fixed endpoints* if there is a continuous map $H : [0,1] \times [0,1] \to \Omega$ such that $H(t,0) = \gamma_0(t)$, $H(t,1) = \gamma_1(t)$ $(0 \leq t \leq 1)$ and, $H(0,s) = a$, $H(1,s) = b (0 \leq s \leq 1)$.

One says that the homotopy H carries γ_0 into γ_1. (See Fig. 1.3.)

Proposition 1.9.2. *Homotopy is an equivalence relation in* $\mathcal{C}(\Omega; a, b)$.

Proof. Reflexivity and symmetry are clear. Let us see the transitivity. Let $\gamma_0, \gamma_1, \gamma_2 \in \mathcal{C}(\Omega; a, b), \gamma_0, \gamma_1$ homotopic by H and γ_1, γ_2 homotopic by K. One obtains a homotopy L carrying γ_0 into γ_2 by:

$$L(t,s) = \begin{cases} H(t, 2s) & 0 \leq s \leq \frac{1}{2} \\ K(t, 2s-1) & \frac{1}{2} \leq s \leq 1 \end{cases}$$

□

Denote by $[\alpha]$ the class of α under homotopy.

Notation: $\alpha \sim \beta$ (homotopy). Note that homotopic paths necessarily have the same initial points and the same endpoints.

Definition 1.9.3. A topological space X is called *path-connected* if there exists a path between any pair of points $x_0, x_1 \in X$.

We can now define loops which will play a crucial role in the following.

Definition 1.9.4. A closed curve $\alpha : [0,1] \to X$ is called a *loop* if it is homotopic to the constant curve

$$\beta(t) = \alpha(0) \quad (= \alpha(1)), t \in [0,1] \,.$$

We now formulate some simple facts. The proofs are left to the reader.

1. Parameter equivalence and homotopy equivalence are equivalence relations.

2. Parameter equivalence implies homotopy as the homotopy

$$H(t,s) = \alpha(t(1-s) + s\tau(t))$$

   shows.

3. Let

$$\alpha, \beta, \gamma : [0,1] \longrightarrow X$$

   be three curves with the property

$$\alpha(1) = \beta(0), \quad \beta(1) = \gamma(0) \,.$$

   Then

$$(\alpha \cdot \beta) \cdot \gamma \simeq \alpha \cdot (\beta \cdot \gamma), \qquad \text{and hence} \qquad (\alpha \cdot \beta) \cdot \gamma \sim \alpha \cdot (\beta \cdot \gamma)$$

4. Let α, α' and β, β' be two pairs of homotopic curves. Assume $\alpha(1) = \beta(0)$. Then the curves $\alpha \cdot \beta$ and $\alpha' \cdot \beta'$ are homotopic.

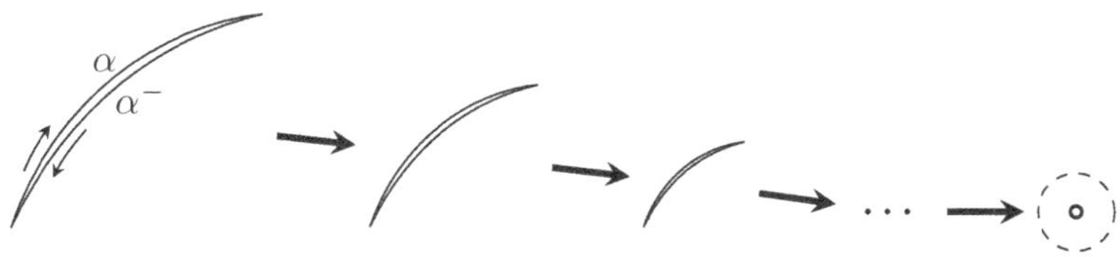

FIGURE 1.4: The curve $\alpha \cdot \alpha^-$ can be deformed to the point $\alpha(0)$.

5. The curve $\alpha \cdot \alpha^-$ is a loop, as the homotopy

$$H(t,s) = \begin{cases} \alpha(2t(1-s)), & 0 \le t \le 1/2 \\ \alpha(2(1-t)(1-s)), & 1/2 \le t \le 1 \end{cases}$$

shows.

6. Two loops α, β passing through the point z_0 are homotopic if and only if $\alpha \cdot \beta^-$ is the loop:

$$\left(\alpha \sim \beta \Longleftrightarrow \alpha \cdot \beta^- \sim \beta \cdot \beta^-\right)$$

(see Fig. 1.4).

Proposition 1.9.5. *The composition of paths and the inversion of paths are compatible with homotopies. The composition thus defined on the equivalence classes of paths is associative.*

Proof. Let F be a homotopy carrying α_0 into $\alpha_1, \alpha_0, \alpha_1 \in \mathcal{C}(\Omega; a, b)$, and G a homotopy carrying β_0 into $\beta_1, \beta_0, \beta_1 \in \mathcal{C}(\Omega; b, c)$. Then

$$H(t,s) := \begin{cases} F(2t, s), & 0 \le t \le \frac{1}{2} \\ G(2t-1, s), & \frac{1}{2} \le t \le 1, \end{cases}$$

is a homotopy carrying $\alpha_0\beta_0$ into $\alpha_1\beta_1$. If F is a homotopy carrying γ_0 into γ_1 then $(t,s) \mapsto F(1-t, s)$ is a homotopy carrying $\overline{\gamma}_0$ into $\overline{\gamma}_1$.

Finally, let $\alpha \in \mathcal{C}(\Omega; a, b), \beta \in \mathcal{C}(\Omega; b, c)$ and $\gamma \in \mathcal{C}(\Omega; c, d)$. We want to show that $\delta_0 = (\alpha\beta)\gamma$ and $\delta_1 = \alpha(\beta\gamma)$ are homotopic. Let f be the unique continuous piecewise affine map such that $f(0) = 0$, $f(1/2) = 1/4$, $f(3/4) = 1/2$, $f(1) = 1$. Then $\delta_1 = f^*(\delta_0)$, whence δ_0 and δ_1 are homotopic. □

1.10 Index of a Loop

A Jordan curve or a simple closed curve in the plane $\mathbb{R}^2$ is the image C of an injective continuous map of a circle into the plane, $\varphi : S^1 \to \mathbb{R}^2$. A Jordan arc in the plane is the image of an injective continuous map of a closed and bounded interval $[a, b]$ into the plane. It is a plane curve that is not necessarily smooth nor algebraic.

Alternatively, a Jordan curve is the image of a continuous map $\varphi : [0,1] \to \mathbb{R}^2$ such that $\varphi(0) = \varphi(1)$ and the restriction of φ to $[0,1)$ is injective. The first two conditions say that C is a continuous loop, whereas the last condition stipulates that C has no self-intersection points.

With these definitions, the Jordan curve theorem can be stated as follows:

Theorem 1.10.1 (Jordan curve theorem)**.** *Let C be a Jordan curve in the plane $\mathbb{R}^2$. Then its complement, $\mathbb{R}^2 \backslash C$, consists of exactly two connected components. One of these components is bounded (the interior) and the other is unbounded (the exterior), and the curve C is the boundary of each component.*

The Jordan curve theorem can be proved from the Brouwer fixed point theorem (in 2 dimensions) [124].

In contrast, the complement of a Jordan arc in the plane is connected:

Theorem 1.10.2. *Let $\gamma_0, \gamma_1 : [0,1] \to U$ be C^1 curves*[2] *with $\gamma_0(0) = \gamma_1(0)$ and $\gamma_0(1) = \gamma_1(1)$ (the fixed endpoint case) or $\gamma_0(0) = \gamma_0(1), \gamma_1(0) = \gamma_1(1)$ (the closed case).*

Assume that they are C^1-homotopic[3] *in the following sense: there exists a continuous map $H : [0,1] \times [0,1] \to U$ with $H(t,0) = \gamma_0(t), H(t,1) = \gamma_1(t)$ and such that $H(\cdot, s)$ is a C^1 curve for each $0 \leq s \leq 1$. Moreover, in the fixed endpoint case we assume that H freezes the endpoints, whereas in the closed case we assume that each curve from the homotopy is closed. Then*

$$\int_{\gamma_0} f(z)dz = \int_{\gamma_1} f(z)dz$$

for all $f \in \mathcal{H}(U)$. In particular, if γ is a closed curve in U, which is homotopic to a point, then

$$\oint_{\gamma} f(z)dz = 0$$

Proof. We first note the important fact that $f(z)dz$ is a closed form. Indeed,

$$d(f(z)dz) = \partial_z f(z)dz \wedge dz + \partial_{\bar{z}} f(z)d\bar{z} \wedge dz = 0$$

by the Cauchy–Riemann equation $\partial_{\bar{z}} f = 0$. Thus, Cauchy's theorem is a special case of the homotopy invariance of the integral over closed forms which

[2]This can be relaxed to piece-wise C^1, which means that we can write the curve as a finite concatenation of C^1 curves. The same comment applies to the homotopy.

[3]In light of commonly used terminology it is probably best to refer to this as homotopic through C^1 curves but for simplicity, we shall continue to abuse terminology and use C^1 homotopic.

in turn follows from Stokes's theorem. Let us briefly recall the details: since a closed form is locally exact, we first note that

$$\oint_{\eta} f(z)dz = 0$$

for all closed curves η which fall into sufficiently small disks, say. But then we can triangulate the homotopy so that

$$\int_{\gamma_0} f(z)dz - \int_{\gamma_1} f(z)dz = \sum_j \oint_{\eta_j} f(z)dz = 0$$

where the sum is over a finite collection of small loops which constitute the triangulation of the homotopy H.

One may prefer to use Green's formula (which of course follows from the Stokes theorem): provided $V \subset U$ is a sufficiently small neighborhood which is diffeomorphic to a disk, say, one can write

$$\begin{aligned}\oint_{\partial V} f(z)dz &= \oint_{\partial V} udx - vdy + i(udy + vdx) \\ &= \iint_V (-u_y - v_x)\, dxdy + i \iint_V (-v_y + u_x)\, dxdy = 0\end{aligned}$$

where the final equality sign follows from the Cauchy–Riemann equations. □

Lemma 1.10.3. *Let γ be a loop in $\mathbb{C}$. Define the map* $\text{ind}(\gamma, \cdot) : \mathbb{C}\backslash\gamma(z) \to \mathbb{C}$ *via*

$$\text{Ind}_\gamma(z) = \frac{1}{2\pi i} \int_\gamma \frac{dw}{w - z}$$

which is well defined since γ is a loop in $\mathbb{C}\backslash\{z\}$, where the integrand is holomorphic. Then the following hold.

1. $\text{Ind}_\gamma(\cdot)$ *is continuous.*

2. $\text{Ind}_\gamma(z) \in \mathbb{Z}$ *for each* $z \in \mathbb{C}\backslash\gamma(z)$.

3. $\text{Ind}(\gamma, \cdot)$ *is constant on each connected component of* $\mathbb{C}\backslash\gamma(z)$.

In particular, if $z = 0$, the degree of γ is

$$d(\gamma) = \frac{1}{2\pi i} \int_\gamma \frac{dz}{z}.$$

Proof. Fix $z \in \mathbb{C} \setminus \gamma$. The index $\text{Ind}_\gamma(z)$ is well defined since γ is a loop in $\mathbb{C} \setminus \{z\}$, and the integrand is holomorphic there.

By Theorem 1.10.2, it suffices to prove the first two items under the assumption that γ is a circle in $\mathbb{C} \setminus \{z\}$ of radius $\delta < \text{dist}(z, \gamma)$.

For $h \in \mathbb{C}$ with $|h| < \delta/2$, we have $\operatorname{dist}(z+h, \gamma) > 0$, hence $z + h \in \mathbb{C} \setminus \gamma$. Consequently,

$$|w - z - h| \geq |w - z| - |h| \geq \delta - |h| \geq \tfrac{\delta}{2},$$

for all $w \in \gamma$, and therefore

$$|\operatorname{Ind}(\gamma, z+h) - \operatorname{Ind}(\gamma, z)| = \frac{1}{2\pi}\left|\int_\gamma \left(\frac{1}{w-z-h} - \frac{1}{w-z}\right) dw\right|$$

$$= \frac{1}{2\pi}\left|\int_\gamma \frac{h}{(w-z-h)(w-z)}\, dw\right|$$

$$\leq |h|\frac{\operatorname{len}(\gamma)}{2\pi} \sup_{w\in\gamma} \frac{1}{|w-z-h||w-z|} \leq |h|\frac{\operatorname{len}(\gamma)}{\pi\delta^2}$$

$$= |h|\frac{2\pi\delta}{\delta^2} = |h|\frac{2\pi}{\delta}.$$

Thus $\operatorname{Ind}(\gamma, \cdot)$ is continuous at z.

Define $g : [0,1] \to \mathbb{C}$ by

$$g(t) = \int_0^t \frac{\gamma'(x)}{\gamma(x) - z_0}\, dx.$$

Since $z_0 \notin \gamma$, the integrand is well defined. By the fundamental theorem of calculus, g is Lipschitz on $[a,b]$ and differentiable on $[a,b] \setminus E$ for some countable set E, with

$$g'(t) = \frac{\gamma'(t)}{\gamma(t) - z_0}.$$

Define $h : [a,b] \to \mathbb{C}$ by

$$h(t) = e^{-g(t)}\, |\gamma(t) - z_0|.$$

Then h is continuous and differentiable outside a countable set, and

$$h'(t) = e^{-g(t)}\, |-g'(t)(\gamma(t) - z_0) + \gamma'(t)| = 0$$

wherever it is differentiable. Hence $h(t) = h(a) = \gamma(a) - z_0$ for all $t \in [a,1]$, which gives

$$e^{g(t)} = \frac{\gamma(t) - z_0}{\gamma(a) - z_0}, \qquad t \in [a,b].$$

In particular,

$$e^{g(1)} = \frac{\gamma(1) - z_0}{\gamma(0) - z_0} = \frac{\gamma(0) - z_0}{\gamma(1) - z_0} = 1,$$

since γ is a loop.

Therefore,

$$\int_\gamma \frac{dz}{z - z_0} = 2\pi i n, \qquad n \in \mathbb{Z}\,.$$

Finally, $d(\gamma)$ is constant on each connected component of $\mathbb{C} \setminus \gamma$, because $\mathbb{C} \setminus \gamma$ is open and its connected components coincide with its path-connected components. □

This suggests some notation:

Definition 1.10.4. For any loop γ in $\mathbb{C}$ and $z \in \mathbb{C}\backslash\gamma$ we define the *index* of γ relative to z to be

$$\mathrm{Ind}_\gamma(z) = \frac{1}{2\pi i}\int_\gamma \frac{dw}{w - z} \in \mathbb{Z}\,. \tag{1.14}$$

If γ is a loop, we say a point $z \in \mathbb{C}$ is enclosed by γ if $\mathrm{Ind}_\gamma(z) \neq 0$. We say a loop γ is counterclockwise simple if $\mathrm{Ind}_\gamma(z) \in \{0, 1\}$ for all $z \in \mathbb{C}\backslash\gamma$.

The index is sometimes called the *winding number*. This terminology is justified by the following remark.

If γ is a loop in $\mathbb{C} \setminus \{0\}$ we will define its index with respect to points $0 \in \mathbb{C}\backslash\gamma([a, b])$. It is an integer that describes how many times the loop γ turns around the point 0.

By Theorem 1.8.6, we can write $\gamma(t) = |\gamma(t)|e^{iF(t)}$, where $F(t)$ is a continuous path in $[a, b] \subset \mathbb{R}$.

Since $F(t)$ is the argument of the complex number $\gamma(t)$, equation (1.14) is usually stated as saying "the degree of the loop γ equals the variation of the argument along γ," and it is written as

$$d(\gamma) = \frac{1}{2\pi}\underset{z\in\gamma}{\Delta}\arg z = \frac{1}{2\pi i}\underset{z\in\gamma}{\Delta}\log z\,.$$

This formula is also known as the *argument principle*.

Periods of a holomorphic **1**-*form.* One can also apply Cauchy's theorem to smooth paths or loops, instead of just boundaries of regions. The fact that the form $f(z)dz$ is closed leads to:

Theorem 1.10.5. *Let $f : U \to \mathbb{C}$ be an analytic function, let b be a basepoint in U, and let $\gamma : S^1 \to U$ be a smooth loop based on b. Then*

$$p(\gamma) = \int_\gamma f(z)dz$$

depends only on the homotopy class of γ, and defines a group homomorphism

$$p : \pi_1(U, b) \to \mathbb{C}$$

called the period map.

Proof. The main point is the first statement. Let $B = S^1 \times [0, 1]$. A smooth isotopy $\gamma_t : (S^1, *) \to (U, b), t \in [0, 1]$, gives a smooth map $G : B \to U$, and $p(\gamma_0) - p(\gamma_1)$ gives the integral of $\alpha = G^*(f(z)dz)$ over ∂B. Since $f(z)$ is analytic, $\omega = f(z)dz$ is closed, and hence $\alpha = G * (\omega)$ is closed as well we have $d(\alpha) = d(G^*\omega) = G^*(d\omega) = 0$. Thus $\int_{\partial B} \alpha = \int_B d\alpha = 0$, and thus $p(\gamma_0) = p(\gamma_1)$.

On the other hand, if $1 \in U$ and $U \subset \mathbb{C}^*$ is simply connected, then there exists a unique analytic branch of the logarithm, $\log : U \to \mathbb{C}$, such that $\log(1) = 0$. □

Corollary 1.10.6. *If U is simply connected, then $F(b) = \int_a^b f(z)dz$ is well defined for all $b \in U$. Indeed, $F : U \to \mathbb{C}$ is the unique analytic function satisfying $F'(z) = f(z)$ and $F(a) = 0$.*

Example: the logarithm. For $f(z) = 1/z$ on $U = \mathbb{C}^*$, the period map gives an isomorphism, $\pi_1(U, 1) \cong 2\pi i\mathbb{Z}$. This reflects the fact that integrating dz/z gives the multi-valued function,

$$\log(z) = \log|z| + i\arg(z)$$

and $\arg(z)$ changes by 2π along a loop that winds $+1$ times around $z = 0$. On the other hand, if $1 \in U$ and $U \subset \mathbb{C}^*$ is simply connected, then there exists a unique analytic branch of the logarithm, $\log : U \to \mathbb{C}$, such that $\log(1) = 0$.

For more in this topic see [15, Chapter 1, §8].

1.11 The Fundamental Group

The distinguishing feature of a *topological space* is its continuity structure, formalized in terms of open sets or neighborhoods. A *continuous map* is a function between spaces that preserves continuity. A *homotopy* is a continuous deformation between continuous maps; two maps connected by a homotopy are said to be homotopic [80, p. 3]. The underlying idea is to disregard variations that do not affect outcomes of interest. An important example is the residue theorem in complex analysis: here, "closed curves" are continuous maps from the circle into the complex plane, and two closed curves yield the same integral if they are homotopic in the plane with the singular points removed [129].

The most intuitively evident topological invariant of a space is the number of connected components into which it decomposes. Over the past hundred years or so, we have come to realize that this primitive notion admits, in a certain sense, two higher-dimensional analogues: the homotopy and cohomology groups of the space.

Nothing could be more discrete than a finite group, and a group theorist might reasonably have felt that topology had nothing to contribute to his subject. On the other hand, he would readily concede that groups have important applications in topology. The most significant way in which discrete groups arise in topology is as fundamental groups of spaces (i.e., groups of closed paths): every topological space X, with a chosen base point, determines a discrete group G.

The homotopy groups are used in algebraic topology to classify topological spaces. The first and simplest homotopy group is the fundamental group, denoted $\pi_1(X)$, which records information about loops in a space.

Homotopy theory studies those properties of spaces and maps, which are not changed if we replace all the maps by homotopic ones.

Definition 1.11.1. The *fundamental group* of a space X with a distinguished point x_0 is its one-dimensional homotopy group

$$\pi_1(X) = \pi_b(S^1, X)\,.$$

We consider *loops* in the space X, that is, mappings

$$\varphi : [0,1] \to X$$

such that $\varphi(0) = \varphi(1) = x_0$. Two loops φ and φ' are called *homotopic* if there exists a homotopy

$$\varphi_t : [0,1] \to X$$

such that $\varphi_0 = \varphi$, $\varphi_1 = \varphi'$, and

$$\varphi_t(0) = \varphi_t(1) = x_0 \quad (0 \le t \le 1)\,.$$

The *product* $\varphi\psi$ of two loops φ and ψ is the loop χ defined by

$$\chi(t) = \begin{cases} \varphi(2t), & t \le \frac{1}{2}\,, \\ \psi(2t-1), & t \ge \frac{1}{2}\,. \end{cases}$$

In other words, the product of two loops is the loop obtained by traversing the two given loops consecutively.

It is easy to verify that this multiplication induces a well-defined multiplication on the set of *homotopy classes* of loops (one must check that if $\varphi \sim \varphi'$ and $\psi \sim \psi'$, then $\varphi\psi \sim \varphi'\psi'$). With respect to this operation, the set of homotopy classes forms a *group* (unlike the set of loops themselves, where multiplication satisfies the group axioms only up to homotopy, e.g., $\varphi(\psi\chi) \sim (\varphi\psi)\chi$, etc.,). This group is called the *fundamental group* $\pi_1(X, x_0)$.

It is clear that the inverse of the class of a loop $\varphi : [0,1] \to X$ is the class of the loop $\varphi' : [0,1] \to X$ defined by

$$\varphi'(t) = \varphi(1-t)\,.$$

A mapping $f : X \to Y$ satisfying $f(x_0) = y_0$ naturally induces a group homomorphism

$$f_* : \pi_1(X, x_0) \to \pi_1(Y, y_0)\,.$$

It is clear that if the mappings $f, g : X \to Y$ are *homotopic* (with respect to the distinguished points), then the induced homomorphisms f_* and g_* coincide.

Theorem 1.11.2 (dependence on the basepoint). *If a space X is path-connected, then*

$$\pi_1(X, x_0) \cong \pi_1(X, x_1)$$

for any points $x_0, x_1 \in X$.

Proof. Since X is path-connected, there exists a path

$$\alpha : I \to X$$

such that $\alpha(0) = x_0$ and $\alpha(1) = x_1$. We will construct a map

$$\alpha_\# : \pi_1(X, x_0) \to \pi_1(X, x_1) .$$

Let $\Phi \in \pi_1(X, x_0)$ and choose a representative loop $\varphi \in \Phi$. Define

$$\alpha_\#(\varphi) = (\alpha\varphi)\alpha^{-1} .$$

(Of course, one can concatenate not only loops but also arbitrary paths, provided the second path begins where the first one ends.) The result is a loop based at x_1.

It is clear that replacing the loop φ and the path α by homotopic ones (where path homotopy is understood in the same sense as loop homotopy, i.e., by homotopies fixing the endpoints) also replaces the loop $\alpha_\#(\varphi)$ by a homotopic one. Therefore, $\alpha_\#$ defines a map that depends only on the homotopy class of the path α; we denote this map also by the same symbol

$$\alpha_\# : \pi_1(X, x_0) \to \pi_1(X, x_1) .$$

The inverse map is constructed analogously:

$$\alpha_\#^{-1}([\psi]) = [(\alpha^{-1}\psi)\alpha] .$$

It is easy to verify that $\alpha_\#$ is a homomorphism, and hence an isomorphism.

The isomorphism $\alpha_\#$ naturally depends on the choice of α. If we replace the path α with another, non-homotopic path β, we generally obtain a different isomorphism. More precisely, if

$$\gamma = \beta\alpha^{-1} ,$$

then for any $\Phi \in \pi_1(X, x_0)$, the following equality holds:

$$\beta_\#(\Phi) = [\gamma]\, \alpha_\#(\Phi)\, [\gamma]^{-1} .$$

(Obviously, γ is a loop, so the element $[\gamma]$ is defined in the group $\pi_1(X, x_1)$.)

In particular, this implies that if the fundamental group is *abelian* (and only in that case), the isomorphism $\alpha_\#$ does not depend on the choice of α. In such a case, we can speak of *the* fundamental group of the space without specifying a basepoint.

In the general (non-abelian) case, one can refer to "the fundamental group" of a path-connected space without a basepoint only as an abstract group—that is, one can say, for instance, that it is finite or nilpotent, but not fix a specific element within it. □

The fundamental group is, in general, noncommutative. Actual computation of the fundamental group for a given space is a highly difficult and intricate task. We shall determine fundamental groups of several simple spaces as we proceed further.

Definition 1.11.3. A topological space X is said to be *arcwise connected/path-connected* if, for each pair x_1, x_2 of points in X, there exists an arc joining them, i.e., there exists some $\gamma : [0,1] \to X$ such that $\gamma(0) = x, \gamma(1) = y$. A subset $S \subset X$ is arcwise connected if the subspace S (with the induced topology) is arcwise connected.

A topological space X is said to be *locally arcwise connected* if each point of X has a basis of neighborhoods consisting of arcwise connected sets.

Definition 1.11.4. An arcwise connected topological space X is said to be *simply connected* if every closed path in X is homotopic to a constant path.

Therefore

$$(X \text{ is simply connected}) \Leftrightarrow (\pi_1(X) = \{1\}),$$

i.e., the fundamental group at any point has only one element.

The pair (X, a), where $a \in X$, is called a *pointed space*, and the distinguished point a is called the *base point*. The correspondence (a functor in the language of category theory) that assigns the fundamental group to a pointed space is an important tool in the study of various topological problems. The natural behavior of this correspondence under continuous mappings makes it even more significant.

With some deliberate inaccuracy, we shall denote this correspondence by

$$\pi_1 : \{\text{pointed topological spaces}\} \to \{\text{groups}\}.$$

We give without proof the Seifert–Van Kampen theorem [96], which is a very useful theorem for computing the fundamental groups of topological spaces. The main idea of this theorem is to decompose a topological space into simpler subspaces and describe the fundamental group of the space in terms of the invariants of the fundamental groups of its subspaces.

Theorem 1.11.5 (Seifert–Van Kampen). *Suppose X is the union of two path-connected open subspaces U and V such that $U \cap V$ is also path-connected. Choose a point $x_0 \in U \cap V$ and use it as the base point for the topological spaces X, U, V, and $U \cap V$. Suppose $i_* : \pi_1(U) \to \pi_1(X)$ and $j_* : \pi_1(V) \to \pi_1(X)$ are the homomorphisms induced by the inclusion maps. Let $\Phi : \pi_1(U) * \pi_1(V) \to \pi_1(X)$ be the map induced by i_* and j_*, where $\pi_1(U) * \pi_1(V)$ denotes the free product of the groups $\pi_1(U)$ and $\pi_1(V)$. Then Φ is a surjective map. Moreover, the kernel of this map is the smallest normal subgroup of $\pi_1(U) * \pi_1(V)$ containing all elements of the form $k_*(\gamma)l_*(\gamma)^{-1}$, where $k : U \cap V \to U$ and $l : U \cap V \to V$ are the inclusion maps.*

This result may be found in [29].

Example 1.11.6. Let B_n be the open unit disc in the plane with n points removed. We claim that the fundamental group of B_n is a free group with n generators. That is, it is isomorphic to $\underbrace{\mathbb{Z} * \mathbb{Z} * \cdots * \mathbb{Z}}_{n}$. In the case $n = 1$, B_1 is homeomorphic to $(0, 1) \times S^1$. Therefore, $\pi_1(B_1) = \mathbb{Z}$. For general n, we may decompose B_n as the union of two open sets U and V, where U is homeomorphic to B_{n-1}, V is homeomorphic to B_1, and $U \cap V$ is homeomorphic to an open disc. Therefore, the claim follows by induction on n. A closely related space to B_n is X_n, the wedge of n copies of the circle S^1. It is easy to see that there is a deformation retraction from B_n to X_n.

Definition 1.11.7. Two spaces X and Y are of the same *homotopy type* if there exist continuous maps (called homotopy equivalences) $f : X \to Y$, $g : Y \to X$ such that $gf \simeq$ identity: $X \to X$ and $fg \simeq$ identity: $Y \to Y$.

Obviously, two homeomorphic spaces are of the same homotopy type, but the converse is not true.

Example 1.11.8. Suppose Σ_g is the orientable surface of genus g. One way to construct Σ_g is to take a $4g$-gon as in Fig. 1.5, pair the edges as indicated, and then identify the edges with the same labels. To visualize why this is a genus g surface, notice that all the vertices of the polygon are identified with each other. The resulting space is a 2-sphere with $2g$ holes, where the boundaries of these holes are matched in pairs. Identifying these boundary circles in pairs is equivalent to adding g handles to the 2-sphere with $2g$ holes.

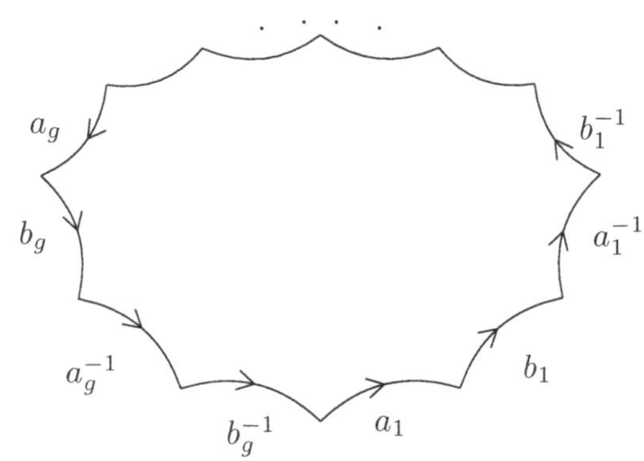

FIGURE 1.5: Σ_g is the orientable surface of genus g.

Suppose Σ_g° denotes the Riemann surface of genus g with one boundary component; that is, we remove an open disc from Σ_g to obtain Σ_g°. From the above description of Σ_g, it is clear that Σ_g° has the same homotopy type as X_{2g}, the wedge of $2g$ circles. Therefore we have

$$\pi_1(\Sigma_g^\circ) = \langle a_1, b_1, a_2, b_2, \ldots, a_g, b_g \rangle .$$

Now an application of the Seifert–Van Kampen theorem implies that

$$\pi_1(\Sigma_g) = \langle a_1, b_1, a_2, b_2, \ldots, a_g, b_g | a_1 b_1 a_1^{-1} b_1^{-1} a_2 b_2 a_2^{-1} b_2^{-1} \cdots a_g b_g a_g^{-1} b_g^{-1} = 1 \rangle .$$

Definition 1.11.9. We define a *surface group* to be the group given by the presentation

$$\Gamma_g = \left\langle a_1, b_1, a_2, b_2 \ldots a_g, b_g | a_1 b_1 a_1^{-1} b_1^{-1} a_2 b_2 a_2^{-1} b_2^{-1} \cdots a_g b_g a_g^{-1} b_g^{-1} = 1 \right\rangle .$$

Question For what integers g and h is Γ_h a subgroup of Γ_g?

As a first step, we observe the connection between the surface group Σ_g and the closed, orientable genus-g surface Γ_g.

Lemma 1.11.10. *The fundamental group* $\pi_1(\Sigma_g)$ *of the closed, orientable genus-g surface* Σ_g *is the surface group* Γ_g.

Proof. First, recall [80, Proposition 1.26] from Hatcher's Algebraic Topology regarding applications of van Kampen's theorem to cell complexes: If we attach 2-cells to a path connected space X via maps ϕ_α, making a space Y, and $N \subset \pi_1(X, x_0)$ is the normal subgroup generated by all loops $\lambda_\alpha \phi_\alpha \lambda_\alpha^{-1}$, then the inclusion $X \hookrightarrow Y$ induces a surjection $\pi_1(X, x_0) \to \pi_1(Y, x_0)$ whose kernel is N. Thus $\pi_1(Y) \approx \pi_1(X)/N$.

Consider the wedge-sum of $2g$ circles, labeled $a_1, a_2, \ldots, a_g$, and $b_1, b_2, \ldots, b_g$. This will be the 1-skeleton of a cell-decomposition of the genus-g surface. By gluing a single 2-cell along the word $a_1 b_1 a_1^{-1} b_1^{-1} \ldots a_g b_g a_g^{-1} b_g^{-1}$, we obtain the closed, orientable genus- g surface. This can be visualized, as in Fig. 1.5, by the familiar identification space of a genus- g surface as a $4g$-gon with pairs of edges, and all of the vertices, identified. □

The development of higher homotopy groups from the concept of components is deceptively simple and essentially unique.

To return to the general case, all higher homotopy groups $\pi_k(X, p)$ for $k \geq 2$ can be defined inductively by the formula

$$\pi_{k+1}(X, p) = \pi_k(\Omega_p X, \overline{p}) .$$

Moreover, if p and p' are two points of X lying in the same path component, then

$$\pi_k(X, p) \simeq \pi_k(X, p') ,$$

although the correspondence is not necessarily unique.

For Riemann surfaces such as those discussed above, all higher homotopy groups π_k with $k \geq 2$ are trivial; in part for this reason, the fundamental group π_1 suffices to classify them. In general, the groups π_k for $k \geq 2$ are Abelian.

Consequently, they were not regarded as particularly significant until the 1930s, when W. Hurewicz defined them (in the manner described above, among others) and demonstrated that, far from being trivial, they constitute the essential building blocks for describing the homotopy-theoretic properties of a space.

The relation between homotopy and homology gives the following theorem:

Theorem 1.11.11 (Hurewicz isomorphism theorem). *Let X be any path-connected topological space. Then the first nontrivial homotopy and homology occur in the same dimension and are equal, i.e., given a positive integer $n \geq 2$, if $\pi_q(X) = 0$ for $1 \leq q < n$, then $H_q(X) = 0$ for $1 \leq q < n$ and $H_n(X) = \pi_n(X)$.*

We will assume this theorem as known. Its proof may be found in, for instance, [27, pp. 225–226].

The functor H_n measures the number of "n-dimensional holes" in the space, in the sense that the n-sphere S^n has exactly one n-dimensional hole and no m-dimensional holes if $m \neq n$. A 0-dimensional hole is a pair of points in different path components, and so H_0 measures path connectedness.

1.12 Connectivity

Algebraic topology is concerned with the connectivity properties of topological spaces.

Definition 1.12.1. A space X is *connected* if when we write $X = U \cup V$ for $U, V \subseteq X, U, V$ open, $U \cap V = \emptyset$, one of U, V is empty.

Example 1.12.2. The space $\mathbb{R}$ is connected, but $\mathbb{R} \setminus \{0\}$ is not.

Corollary 1.12.3. *If $f : \mathbb{R} \to \mathbb{R}$ is continuous and there exist x, y such that $f(x), f(y)$ have opposite signs, then there exists some z such that $f(z) = 0$.*

Proof. If not, then $f^{-1}(-\infty, 0)$ and $f^{-1}(0, \infty)$ would disconnect $\mathbb{R}$. □

Remark 1.12.4. "Nice" spaces are connected if and only if they are path connected. First examples of such spaces are open subsets of $\mathbb{R}^n$ or $\mathbb{C}^n$.

Definition 1.12.5. A connected space X is *simply connected* if any two continuous maps $S^1 \to X$ can be continuously deformed into one another.

Example 1.12.6. The space $\mathbb{R}^2$ is simply connected, while $\mathbb{R}^2 \setminus \{0\}$ is not. However, the fact that $\mathbb{R}^2 \setminus \{0\}$ is not simply connected allows us to define the winding number of a curve $\gamma : S^1 \to \mathbb{R}^2 - \{0\}$. Intuitively, this is the number of times that curve wraps around the origin.

Definition 1.12.7. If $\gamma : S^1 \to \mathbb{R}^2 \setminus \{0\}$ is continuous, then we may define

$$\deg(\gamma) := \frac{1}{2\pi i} \int_\gamma \frac{dz}{z}$$

which we know from complex analysis is invariant under a homotopy γ.

Example 1.12.8. The loop $t \xrightarrow{\gamma_n} (\cos(2\pi nt), \sin(2\pi nt))$ has degree $n \in \mathbb{Z}$.

This theory allows us to prove the following impressive corollary:

Corollary 1.12.9 (fundamental theorem of algebra). *A non-constant polynomial $f \in \mathbb{C}[x]$ has a root (over $\mathbb{C}$).*

Proof. Let $f(z) = z^n + a_1 z^{n-1} + \cdots + a_n$ and further suppose that f is non-vanishing, that is $\mathrm{Im}(f) \subset \mathbb{C} \setminus \{0\}$. We can examine f along concentric circles in $\mathbb{C}$ in the following way:

$$\gamma_R(t) := f\left(Re^{2\pi it}\right) : S^1 \to \mathbb{C} \setminus \{0\}$$

and note that when $R = 0$, $\gamma_0(t)$ is constant for all t and therefore $\deg(\gamma_0) = 0$. Now, take $R > \sum |a_i| + 10$.

Let $\gamma_{R,s}(t) = z^n + s\left(a_1 z^{n-1} + \cdots + a_n\right)\big|_{Re^{2\pi it}}$ for $0 \leq s \leq 1$. This defines a homotopy of our original map f, since $\gamma_{R,1}(t) = \gamma_R(t)$.

Since we chose R to be very large, the z^n term dominates and so $\gamma_{R,1}(t)$ does in fact wrap around the origin n times. However, we may now homotope in two directions: First, we shrink s to zero, then we shrink R to zero. This gives the following:

$$0 = \deg(\gamma_0) = \deg(\gamma_R) = \deg(\gamma_{R,1}) = \deg(\gamma_{R,0}) = \deg(\gamma_n) = n$$

which is a contradiction when f is non-constant, i.e., when $n \neq 0$. □

We have an analogous notion of degree of maps of m-dimensional spheres:

Example 1.12.10. Any two maps from S^m to $\mathbb{R}^{m+1}$ can be continuously deformed into each other. But, a map

$$S^m \to \mathbb{R}^{m+1} \setminus \{0\}$$

has an integer-valued degree, invariant under deformation. The degree of the constant map is zero and the degree of the inclusion map is one.

See also [27] for more on this topic.

1.13 The Brouwer Fixed Point Theorem

Denote by $B^n = \{x \in \mathbb{R}^n \mid |x| \leq 1\}$ the unit ball in $\mathbb{R}^n$, $n > 1$.

Theorem 1.13.1 (Brouwer fixed point theorem). *Any map $f : B^n \to B^n$ has a fixed point.*

Proof. We will assume this theorem as known. Its proof may be found in, for instance, [20, p. 86]. □

Remark 1.13.2. When $n = 2$, this has a particularly elegant proof. A map without a fixed point would define a homotopy retract from $\mathbb{D}^2$ to S^1, which cannot be allowed since it would imply an isomorphism of fundamental groups.

For an elementary proof of the celebrated theorem of Brouwer see, for example, [26].

We know that

$$\pi_1\left(S^1\right) = \mathbb{Z}$$

(loops are classified by how many times they wind around). The simplest case concerns the ways that a circle (1-sphere) can be wrapped around another circle. The wrapping can be in either of two directions, and wrappings in opposite directions will cancel out after a deformation. The homotopy group $\pi_1\left(S^1\right)$ is therefore an infinite cyclic group, and is isomorphic to the group $\mathbb{Z}$ of integers under addition: a homotopy class is identified with an integer by counting the number of times a mapping in the homotopy class wraps around the circle. This integer can also be thought of as the winding number of a loop around the origin in the plane. For a proof see, for example [80, p. 29].

A space is contractible if it is homotopy equivalent to a one-point space.

Example 1.13.3. Consider $\mathbb{R}^n$ (or any contractible space). We have $\pi_i\left(\mathbb{R}^n\right) = 0$ for all $i \geq 1$, since $\mathbb{R}^n$ is homotopy equivalent to a point.

Example 1.13.4. Consider S^1. We have the usual covering map $p : \mathbb{R} \to S^1$ given by $p(t) = e^{2\pi i t}$. We already know that $\pi_1\left(S^1\right) = \mathbb{Z}$. Now if $n \geq 2$, then we have $\pi_n\left(S^1\right) = \pi_n(\mathbb{R}) = 0$.

Theorem 1.13.5. *The fundamental group of a contractible space X is trivial, $\pi_1(X, p) \cong \{\overline{p}\}$. In particular, the fundamental group of $\mathbb{R}^n$ is trivial, $\pi_1(\mathbb{R}^n, p) \cong \{\overline{p}\}$.*

Proof. A contractible space has the same fundamental group as a point $\{p\}$ and a point has a trivial fundamental group. □

Let $\mathbb{D}^2$ be the open unit disk $\left\{y + iz \in \mathbb{C} \mid y^2 + z^2 \leq 1\right\}$. Its boundary is S^1, and we let $i : S^1 \longrightarrow \mathbb{D}^2$ be the inclusion. Exactly as for $\mathbb{R}$, we see that $\pi_1\left(D^2\right) = 0$ for any choice of basepoint.

Proposition 1.13.6. *There is no continuous map $r : \mathbb{D}^2 \longrightarrow S^1$ such that* $r \circ i = \mathrm{id}$.

Proof. If there were such a map r, then the composite homomorphism

$$\pi_1\left(S^1\right) \xrightarrow{i_*} \pi_1\left(\mathbb{D}^2\right) \xrightarrow{r_*} \pi_1\left(S^1\right)$$

would be the identity. Since the identity homomorphism of $\mathbb{Z}$ does not factor through the zero group, this is impossible. $\square$

Theorem 1.13.7 (Brouwer fixed point theorem). *Any continuous map*

$$f : \mathbb{D}^2 \longrightarrow \mathbb{D}^2$$

has a fixed point.

Proof. Suppose that $f(x) \neq x$ for all x. Define $r(x) \in S^1$ to be the intersection with S^1 of the ray that starts at $f(x)$ and passes through x. Certainly $r(x) = x$ if $x \in S^1$. By writing an equation for r in terms of f, we see that r is continuous. This contradicts the proposition. $\square$

A region is a nonempty open set of $\mathbb{C}$. A domain is a non-empty connected open set of $\mathbb{C}$. When $r > 0$, the set $\{z \in \mathbb{C} \mid |z - c| < r\}$ of all complex numbers at distance strictly less than r from the center c is called the *open disk* with center c and radius r. Its closure is the *closed disk* $\{z \in \mathbb{C} \mid |z - c| \leq r\}$. Open disks are the most commonly used domains in complex analysis.

2

Extremal Length

Extremal length is a relatively new conformal invariant that provides a useful way to measure the geometry of a domain in conformally invariant terms. It is considered the most important conformal invariant because it can be defined in many contexts and estimated through direct geometric arguments.

The concept of extremal length was developed by L. Ahlfors and A. Beurling in the 1940s. In [5], Ahlfors notes that Beurling invented extremal length in 1943 or 1944 but waited until 1946 to announce his results. It is an important conformal invariant that measures the "shape" of a domain from the complex-analytic point of view.

Although calculations of extremal length are often intricate, they tend to be informative. Extremal length is a natural tool for proving, for example, that annuli with different ratios of radii are conformally inequivalent.

The concept of quasiconformal mapping—a generalization of conformal mapping—interacts naturally with extremal length.

Typically, one computes the extremal length of a family of curves within a domain. By selecting these curves judiciously, one can obtain concrete geometric information about the domain.

2.1 Some Definitions

We give here the basics of extremal length theory.

The concept of extremal length, first developed by Ahlfors and Beurling, is a useful conformal invariant. Extremal distance is a conformally invariant version of distance. As such it is a powerful tool for estimating conformal invariants like harmonic measure in terms of more geometric quantities. It is a decisive tool in the study of both conformal and quasiconformal mappings.

Let U; $U' \subseteq \mathbb{C}$ be domains and Γ be a curve family in a domain U. If $\Phi : U \to U'$ is a conformal mapping, then we can consider the image of Γ under Φ. It turns out (as we shall see in detail below) that Γ and its image have the same extremal length.

Fix a domain $U \subseteq \mathbb{C}$. Consider a positive function ρ on the domain U. We think of ρ as analogous to $|f'|$ where f is a conformal map on U. Just as the

DOI: 10.1201/9781003713005-2

image area of a set E can be computed by integrating $\int_E |f'|^2\, dxdy$, we can use ρ to define areas by $\int_E \rho^2 dxdy$.

To measure the length of the image of a path we add up the lengths of the straight segments determined by the γ-images of the end-points of subintervals. In other words, we define the length of γ to be

$$\ell(\gamma) = \int_\gamma ds\,.$$

Similarly, just as we can define $\ell(f(\gamma)) = \int_\gamma |f'(z)|\, ds$, we can define the ρ-length of a curve γ by $\int_\gamma \rho ds$. For this to make sense, we need γ to be locally rectifiable (so the arclength measure is defined) and it is convenient to assume that ρ is Borel (so that its restriction to any curve γ is also Borel and hence measurable for length measure on γ).

A function ρ, defined in a domain $U \subset \mathbb{C}$, will be called *allowable* if it satisfies the following conditions:

1. $\rho \geq 0$ and measurable.
2. $A(U,\rho) = \int\int_U \rho^2 dxdy \neq 0, \infty$.

If $\gamma : I \to U$ is any rectifiable curve in U then we let

$$L_\gamma(\rho) = \int_\gamma \rho |dz|$$

if ρ is measurable on γ, $L_\gamma(\rho) = \infty$ otherwise.[1] Here $|dz|$ is the Euclidean element of length in $\mathbb{C}$.

Put in other words, the value of $L_\rho(\gamma)$ is precisely equal to the integral of the Borel measurable function $\rho(\gamma(t))$ with respect to the Borel measure on I for which the measure of every subinterval $J \subseteq I$ is the length of the restriction of γ to J.

We introduce

$$L(\rho) = \inf_{\gamma\in\Gamma} L_\gamma(\rho)$$

and

Definition 2.1.1. The *extremal length* of Γ is

$$\lambda_U(\Gamma) = \sup_\rho \frac{L(\rho)^2}{A(U,\rho)} \tag{2.1}$$

for all allowable ρ.

An *extremal metric* for Γ is a metric ρ which attains the supremum (2.1).

---

[1] As a function of arc-length.

Because of its homogeneity, the ratio (2.1) is unchanged if the metric ρ is multiplied by a positive constant. Thus we can normalize the metric to satisfy $A(U,\rho) = 1$ so that

$$\lambda_U(\Gamma) = \sup\left\{L^2(\Gamma,\rho) : A(U,\rho) = 1\right\}$$

or we can normalize the metric by $L(\Gamma,\rho) = 1$ so that

$$\lambda_U(\Gamma)^{-1} = \inf\{A(U,\rho) : L(\Gamma,\rho) = 1\}\,.$$

The quantity $\lambda_U(\Gamma)^{-1}$ is called the *modulus or module* of Γ. Some authors emphasize moduli instead of extremal lengths.

The *extremal distance between two subsets* A and B of U is defined to be the extremal length of the collection of curves in U with one endpoint in A and the other endpoint in B.

The conformal invariance of extremal length follows immediately from the change of variables formula in the two integrals. See the next section.

2.2 The Conformal Invariance of Extremal Length

Capsule: The main reason that extremal length is important is that it is a conformal invariant. We develop that idea here.

Let $\Phi : U \to \widehat{U}$ be a conformal mapping (that is to say, a one-to-one, onto, invertible holomorphic mapping) of planar domains. Let Γ be a collection of rectifiable curves in U, and let $\widehat{\Gamma} \equiv \{\Phi \circ \gamma \mid \gamma \in \Gamma\}$ be the image curves under Φ. We claim that $\lambda_U(\Gamma) = \lambda_{\widehat{U}}(\widehat{\Gamma})$.

For the proof, suppose that $\widehat{\rho} : \widehat{U} \to [0,\infty]$ is Borel measurable. Define

$$\rho(z) = |\Phi'(x)| \cdot \widehat{\rho}(\Phi(z))$$

The change of variable $w = \Phi(z)$ gives

$$\begin{aligned} A(\rho) &= \int_U \rho(z)^2 dz d\bar{z} \\ &= \int_U \widehat{\rho}(\Phi(z))^2 \left|\Phi'(z)\right|^2 dz d\bar{z} \\ &= \int_{\widehat{U}} \widehat{\rho}(w)^2 dw d\bar{w} \\ &= A(\widehat{\rho}) \end{aligned}$$

Now let $\gamma \in \Gamma$. Set $\widehat{\gamma} = \Phi \circ \gamma$. We again use a change of variable to calculate

$$L_\rho(\gamma) = \int_\gamma \widehat{\rho}(\Phi(z))\,|\Phi'(z)|\,|dz| = \int_{\widehat{\gamma}} \rho(w)|dw| = L_{\widehat{\rho}}(\widehat{\gamma})$$

We may immediately conclude that

$$\lambda_U(\Gamma) \geq \lambda_{\widehat{U}}(\widehat{\Gamma}) .$$

But the same argument applies to the inverse mapping Φ^{-1}. That gives equality, and ends the proof.

2.3 Further Properties of Extremal Lengths

Extremal length obeys four basic rules.

Lemma 2.3.1 (the extension rule). *Let $U \subset U'$ be domains and let Γ, Γ' be path families in U. Then*

$$\lambda_{U'}(\Gamma) = \lambda_U(\Gamma) . \tag{2.2}$$

Moreover, if every $\gamma' \in \Gamma'$ contains some curve $\gamma \in \Gamma$, then

$$\lambda_{U'}(\Gamma') \geq \lambda_U(\Gamma) . \tag{2.3}$$

Proof. The curves in the family Γ' are longer and fewer than the curves in the family Γ. Therefore $\lambda_{U'}(\Gamma') \geq \lambda_{U'}(\Gamma)$ and inequality (2.3) will follow from equality (2.2).

Let ρ' be any metric on U' and set $\rho = \rho'|_U$. Then $L(\Gamma, \rho) = L(\Gamma, \rho')$ and $A(U, \rho) \leq A(U', \rho')$. Taking the supremum over ρ' yields $\lambda_{U'}(\Gamma) \leq \lambda_U(\Gamma)$. Now let ρ be any metric on U and define the metric ρ' on U' by $\rho' = \rho\chi_U$. Then $L(\Gamma, \rho) = L(\Gamma, \rho')$ and $A(U, \rho) = A(U', \rho')$. Taking the supremum over ρ, we get $\lambda_{U'}(\Gamma) \geq \lambda_U(\Gamma)$, which proves (2.3). □

Equality (2.2) says that the extremal length depends on the path family Γ and not the domain U, and for this reason we often write $\lambda(\Gamma)$ for $\lambda_U(\Gamma)$.

Lemma 2.3.2 (Grötsch principle). *If Γ_0 and Γ_1 are families of curves in disjoint domains then $\lambda(\Gamma_0 \cup \Gamma_1) = \lambda(\Gamma_0) + \lambda(\Gamma_1)$.*

Proof. Suppose that ρ_0 and ρ_1 are admissible for Γ_0 and Γ_1. Take $\rho = \rho_0$ and $\rho = \rho_1$ in their respective domains. Then it is easy to check that ρ is admissible for $\Gamma_0 \cup \Gamma_1$ and, since the domains are disjoint, $\int \rho^2 = \int \rho_1^2 + \int \rho_2^2$. Thus $\lambda(\Gamma_0 \cup \Gamma_1) \leq \lambda(\Gamma_0) + \lambda(\Gamma_1)$. By restricting an admissible metric ρ to each domain, a similar argument proves the other direction. □

The Grötsch principle and the monotonicity combine to give

Corollary 2.3.3 (parallel rule). *Suppose that Γ_0 and Γ_1 are path families in disjoint domains $\Omega_0, \Omega_1 \subset \Omega$ that connect disjoint sets E, F in $\partial\Omega$. If Γ is the path family connecting E and F in Ω, then*

$$\lambda(\Gamma) \geq \lambda(\Gamma_0) + \lambda(\Gamma_1)$$

Lemma 2.3.4 (series rule). *If Γ_0 and Γ_1 are families of curves in disjoint domains and let Γ be a path family contained in a domain $\Omega \supset \Omega_0 \cup \Omega_1$. If every curve of Γ contains a curve from both Γ_0 and Γ_1, then $\lambda(\Gamma) \geq \lambda(\Gamma_0) + \lambda(\Gamma_1)$.*

Proof. If either $\lambda(\Gamma_0)$ or $\lambda(\Gamma_1)$ is 0 or ∞, the result follows from the extension rule. Otherwise, choose metrics ρ_0 in U_0 and ρ_1 in U_1, normalized by the conditions $L(\Gamma_0, \rho_0) = A(U_0, \rho_0)$ and $L(\Gamma_1, \rho_1) = A(U_1, \rho_1)$. Then the metric $\rho = \rho_0 \chi_{U_0} + \rho_1 \chi_{U_1}$ on U satisfies $L(\Gamma, \rho) \geq L(\Gamma_0, \rho_0) + L(\Gamma_1, \rho_1)$ and $A(U, \rho) = A(U_0, \rho_0) + A(U_1, \rho_1) = L(\Gamma_0, \rho_0) + L(\Gamma_1, \rho_1)$. Thus

$$\frac{L^2(\Gamma, \rho)}{A(U, \rho)} \geq L(\Gamma_0, \rho_0) + L(\Gamma_1, \rho_1) = \frac{L^2(\Gamma_0, \rho_0)}{A(U_0, \rho_0)} + \frac{L^2(\Gamma_1, \rho_1)}{A(U_1, \rho_1)}.$$

Taking the supremum over all ρ_0 and ρ_1, we obtain $\lambda(\Gamma) \geq \lambda(\Gamma_0) + \lambda(\Gamma_1)$. □

2.4 Some Examples

Capsule: Some good examples help to illustrate the utility and significance of extremal length.

We now do some calculations to exhibit extremal length explicitly.

Example 2.4.1. Let

$$\mathcal{R} = \{(x, y) \mid 0 < x < w, 0 < y < h\}$$

be a rectangle in the complex plane. Let Γ be the set of all finite length, rectifiable curves which begin on the left edge of the rectangle and end on the right edge of the rectangle. We shall now demonstrate that

$$\lambda_{\mathcal{R}}(\Gamma) = \frac{w}{h}$$

First of all, we may take $\rho \equiv 1$ on $\mathcal{R}$. Then we see that $A(\rho) = wh$ and $L_\rho(\gamma) = w$ for any $\gamma \in \Gamma$. The definition of $\lambda_{\mathcal{R}}(\Gamma)$ as a supremum then tells us that

$$\lambda_{\mathcal{R}}(\Gamma) \geq \frac{w^2}{wh} = \frac{w}{h}.$$

For the opposite inequality, consider an arbitrary Borel measurable $\rho : \mathcal{R} \to [0, \infty]$ such that $\ell \equiv L_\rho(\Gamma) > 0$. If $y \in (0, h)$, then we let $\gamma_y(t) = iy + wt$. Then $A(\gamma_y) \in \Gamma$, so that $\ell \leq L_\rho(\gamma_y)$. We may write this last inequality as

$$\ell \leq \int_0^1 \rho(iy + wt) w dt.$$

Now integrating this inequality over $y \in (0, h)$ gives

$$h\ell \leq \int_0^h \int_0^w \rho(iy + wt)w dt dy\,.$$

We perform the change of variable $x = wt$ and then apply the Cauchy–Schwarz inequality to obtain

$$\begin{aligned} h\ell &\leq \int_0^h \int_0^w \rho(x + iy) dx dy \\ &\leq \left(\int_{\mathcal{R}} \rho^2 dx dy \int_{\mathcal{R}} dx dy \right)^{1/2} \\ &= (whA(\rho))^{1/2}\,. \end{aligned}$$

This yields that

$$\lambda_{\mathcal{R}}(\Gamma) \leq \frac{w}{h}$$

as desired.

In fact, the proof presented above shows that the extremal length of Γ is equal to the extremal length of the smaller collection of curves given by $\{\gamma_y \mid y \in (0, h)\}$.

It can be observed that the extremal length of the family of curves Γ' that connect the bottom edge of $\mathcal{R}$ to the top edge of $\mathcal{R}$ satisfies $\lambda_{\mathcal{R}}(\Gamma') = h/w$—just using the same arguments. Therefore

$$\lambda_{\mathcal{R}}(\Gamma) \cdot \lambda_{\mathcal{R}}(\Gamma') = 1\,.$$

It is reasonable to think of this equation as a duality property of extremal length. And this duality property can be useful.

For, in general, obtaining a lower bound for $\lambda_U(\Gamma)$ is easier than obtaining an upper bound—just because the lower bound involves choosing a propitious ρ and then estimating $L_\rho(\Gamma)^2/A(\rho)$ while the upper bound necessitates proving an estimate for all possible ρ. Duality, on the other hand, enables us to translate a lower bound on $\lambda_U(\Gamma')$ into an upper bound on $\lambda_U(\Gamma)$.

As Ahlfors points out in [5, p. 52], the extremal length depends on the family of curves Γ but not on the domain. To see this, suppose that $U \subseteq U'$ are planar domains. Given ρ on U, we define

$$\rho'(z) = \begin{cases} \rho(z) & \text{if } z \in U\,, \\ 0 & \text{if } z \in U' \setminus U\,. \end{cases}$$

Then ρ' is certainly Borel measurable, and

$$L_\rho(\Gamma) = L_{\rho'}(\Gamma) \quad \text{and} \quad A(\rho) = A(\rho')\,.$$

Thus

$$\lambda_{U'}(\Gamma) \geq \lambda_U(\Gamma)\,.$$

For the opposite inequality, begin with a ρ' on U' and let ρ be the restriction of ρ' to U. We see immediately that $\lambda_U(\Gamma)$ is the same for any open set U that contains all the curves γ in Γ.

As a result of the discussion in the last two paragraphs, we will henceforth write $\lambda(\Gamma)$ rather than $\lambda_U(\Gamma)$.

Example 2.4.2. Now let us examine the extremal length in an annulus. Fix $0 < r_1 < r_2 < \infty$. Define

$$\mathcal{A} = \{z \in \mathbb{C} \mid r_1 < |z| < r_2\} .$$

Further let $C_1 = \{z \in \mathbb{C} \mid |z| = r_1\}$ and $C_2 = \{z \in \mathbb{C} \mid |z| = r_2\}$. We consider the extremal length between C_1 and C_2. This is, by definition, the extremal length of the collection Γ of curves connecting C_1 to C_2.

To find a lower bound on $\lambda(\Gamma)$, we take $\rho(z) = 1/|z|$. If γ is a curve from C_1 to C_2, we see that

$$\int_\gamma |z|^{-1} ds \geq \int_\gamma |z|^{-1} d|z| = \int_\gamma d\log|z| = \log(r_2/r_1) .$$

On the other hand,

$$A(\rho) = \int_A |z|^{-2} dxdy = \int_0^{2\pi} \int_{r_1}^{r_2} r^{-2} r dr d\theta = 2\pi \log(r_2/r_2) .$$

We conclude that

$$\lambda(\Gamma) \geq \frac{\log(r_2/r2)}{2\pi} .$$

Now we shall show that this inequality is really an equality. Consider an arbitrary Borel measurable ρ such that $\ell = L_\rho(\Gamma) > 0$. For $\theta \in [0, 2\pi)$, we let $\gamma_\theta : (r_1, r_2) \to \mathcal{A}$ denote the curve $\gamma_\theta(r) = e^{i\theta} r$. Then

$$\ell \leq \int_{\gamma_\theta} \rho ds = \int_{r_1}^{r_2} \rho\left(e^{i\theta} r\right) dr .$$

We integrate in θ and apply the Cauchy–Schwarz inequality to obtain

$$2\pi\ell \leq \int_{\mathcal{A}} \rho dr d\theta \leq \left(\int_{\mathcal{A}} \rho^2 r dr d\theta\right)^{1/2} \cdot \left(\int_0^{2\pi} \int_{r_1}^{r_2} \frac{1}{r} dr d\theta\right)^{1/2} .$$

Squaring both sides gives

$$4\pi^2 \ell^2 \leq A(\rho) \cdot 2\pi \cdot \log(r_2/r_1) .$$

This inequality implies the upper bound $\lambda(\Gamma) \leq (2\pi)^{-1} \log(r_2/r_1)$. When put together with the lower bound, we obtain the final result

$$\lambda(\Gamma) = \frac{\log(r_2/r_1)}{2\pi} .$$

A nice corollary of this last example is that, if $\mathcal{A} = \{z \in \mathbb{C} \mid r_1 < |z| < r_2\}$ and $\mathcal{B} = \{z \in \mathbb{C} \mid s_1 < |z| < s_2\}$, and if $r_2/r_1 \neq s_2/s_1$, then $\mathcal{A}$ and $\mathcal{B}$ cannot be conformally equivalent.

Exercises

1. Consider the rectangular region

$$R = \{(x, y) : 0 < x < w, 0 < y < h\}.$$

How does the extremal length behave when $w \to \infty$ or $h \to \infty$?

2. Consider the annular region

$$\mathcal{A} = \{z \in \mathbb{C} : r_1 < |z| < r_2\}.$$

How does the extremal length behave when $r_2 \to \infty$ How does the extremal length behave when $r_1 \to 0$?

3. Consider the region

$$\{(r, \theta) \in \mathbf{R}^2 : 0 < t < 1, 0 < \theta < \alpha\}$$

for some $\alpha > 0$. Discuss extremal length for this region. How does your answer depend on α?

4. Can you make a statement about how extremal length compares to Lebesgue measure? Why or why not?

5. Find a definition of analytic capacity. How does extremal length compare to analytic capacity?

3

Harmonic Measure

It is unfortunate that most graduate complex analysis courses do not treat harmonic measure. Indeed, most complex analysis texts do not cover the topic.

The term "harmonic measure" was introduced by R. Nevanlinna in the 1920s. But the ideas were anticipated in work of the Riesz brothers and others. It is still studied intensively today.

Harmonic measure is a natural outgrowth of the study of the Dirichlet problem. It is a decisive tool for measuring the growth of functions, and to seeing how they bend. It has been used in such diverse applications as the corona problem and in mapping problems. Harmonic measure has particularly interesting applications in probability theory—especially in the consideration of diffusions and Brownian motion.

It requires a little sophistication to appreciate harmonic measure. And calculating the measure—working through the examples—requires some elbow grease. But the effort is worth it and the results are substantial. A definitive reference for the subject of harmonic measure is the book of Garnett and Marshall [57].

This chapter gives a basic introduction to the idea of harmonic measure.

3.1 The Idea of Harmonic Measure

Capsule: There are three conformal invariants that will be particularly important through the book: harmonic measure, extremal length, and hyperbolic distance. Of these, extremal length is the most important because it can be defined in many situations and estimated by direct geometric arguments. The other two are defined on the disk and then transferred to other domains by a conformal map.

Let $\Omega \subseteq \mathbb{C}$ be a domain with boundary consisting of finitely many Jordan curves (we call such a domain a *Jordan domain*). Let E be a finite union of arcs in $\partial\Omega$ (we allow the possibility of an entire connected boundary component—a simple, closed curve—to be one of these arcs). Then the *harmonic measure* of E at the point $z \in \Omega$ with respect to Ω is the value at z of the bounded harmonic function ω on Ω with boundary limit 1 at points of E and boundary limit 0 at points of $\partial\Omega \setminus E$ (except possibly at the endpoints of the arcs that make up E). We denote the harmonic measure by $\omega(z, \Omega, E)$.

DOI: 10.1201/9781003713005-3

Harmonic measure is a way of manipulating the data that comes from the Dirichlet problem. It is a way of measuring the geometry of a domain, and can be used to control the growth of harmonic functions.

The first question to ask about harmonic measure is that of existence and uniqueness: does harmonic measure always exist? If it does exist, is it unique? The answer to both these queries is "yes." Let us see why.

A domain U is Jordan if its boundary is a union of closed Jordan curves. Fix a *Jordan domain* Ω. It is a classical result (see [3]) that there is a conformal mapping $\Phi : \Omega \to U$ of Ω to a domain U bounded by finitely many circles. If $F \subseteq \partial U$ consists of finitely many circular arcs, then it is obvious that the harmonic measure of F exists.

It would just be the Poisson integral of the characteristic function of F, and that Poisson integral can be shown to exist by elementary conformal mapping arguments (i.e., mapping to the disc). Now a classical result of Carathéodory guarantees that Φ and its inverse extend continuously to the respective boundaries, so that the extended function is a homeomorphism of the closures. Thus we may pull the harmonic measure back from U to Ω via Φ.

As for uniqueness: if Ω is bounded, then uniqueness follows from the standard maximum principle. For the general case, we need an extended maximum principle that is due to Lindelöf:

Proposition 3.1.1. *Let Ω be a domain whose boundary is not a finite set. Let u be a real-valued harmonic function on Ω, and assume that there is a real constant $M > 0$ such that*

$$u(z) \leq M, \qquad \textit{for } z \in \Omega\,.$$

Suppose that m is a real constant and that

$$\limsup_{z \to \zeta} u(z) \leq m \tag{3.1}$$

for all except possibly finitely many points $\zeta \in \partial\Omega$. Then $u(z) \leq m$ for all $z \in \Omega$.

Remark 3.1.2. Observe that any Jordan domain certainly satisfies the hypotheses of the proposition. Furthermore, by classical results coming from potential theory (see [4]), any domain on which the Dirichlet problem is solvable will satisfy the hypotheses of the theorem. Note, in particular, that Ω certainly need not be bounded.

We conclude by noting that this last result is a variant of the famous Phragmen–Lindelöf theorem (see [161]).

Proof of Proposition 3.1.1. First assume that Ω is bounded. We will remove this extra hypothesis later. Let the diameter of Ω be d. Let the *exceptional boundary points* (at which (3.1) does not hold) be called $\zeta_1, \ldots, \zeta_k$. Let $\epsilon > 0$.

We then may apply the ordinary textbook maximum principle (see [69]) to the auxiliary function

$$h(z) = u(z) + \epsilon \sum_{j=1}^{k} \log \frac{|z - \zeta_j|}{d} .$$

Notice that h (instead of u) satisfies the hypotheses of the proposition at every boundary point. So $h \leq m$ on all of Ω. Then we let $\epsilon \to 0$ and the result follows. Now consider the case of Ω unbounded. If Ω has an exterior point (i.e., a point in the interior of the complement of the closure of Ω), then we may apply an inversion and reduce to the case in the preceding paragraph. Finally, if Ω has no exterior point, then let $R > |\zeta_j|$ for all $j = 1, \ldots, k$. Let

$$\Omega_1 = \{z \in \Omega \mid |z| < R\}$$

and

$$\Omega_2 = \{z \in \Omega \mid |z| > R\} .$$

These may not be domains (i.e., they could be disconnected), but they are certainly open sets. Let

$$S = \{z \in \Omega \mid |z| = R\} .$$

If $u \leq m$ on S, then we apply the result of the first paragraph on Ω_1 and the result of the second paragraph on Ω_2. If it is not the case that $u \leq m$ on S, then u will have a maximum N on S with $N > m$. But this will be a maximum for u on all of Ω. Since $u \leq m$ at the ends of the arcs of S, it follows that u actually achieves the maximum N on S. Therefore, by the usual maximum principle, u is identically equal to the constant N. But then the boundary condition (3.1) cannot obtain. That is a contradiction. □

It is important to realize that, if Ω has reasonably nice boundary, then $\omega(z, \Omega, E)$ is nothing other than the Poisson integral of the characteristic function of E. (We used this fact in our discussion of the existence of harmonic measure.) In any event, by the maximum principle the function ω takes real values between 0 and 1. Therefore $0 < \omega(z, \Omega, E) < 1$.

One of the reasons that harmonic measure is important is that it is a conformal invariant. Essential to this fact is Carathéodory's theorem, that a conformal map of Jordan domains will extend continuously to the closures. We now formulate the invariance idea precisely.

Proposition 3.1.3. *Let Ω_1, Ω_2 be domains with boundaries consisting of Jordan curves in $\mathbb{C}$ and $\varphi : \Omega_1 \to \Omega_2$ a conformal map. If $E_1 \subseteq \partial\Omega_1$ is an arc and $z \in \Omega_1$ then*

$$\omega(z, \Omega_1, E_1) = \omega(\varphi(z), \Omega_2, \varphi(E_1)) .$$

Here $\varphi(E_1)$ is well defined by Carathéodory's theorem.

Proof. Let h denote the harmonic function $\omega(\varphi(z), \Omega_2, \varphi(E_1))$. Then $h \circ \varphi$ equals harmonic measure for Ω_1 (by Carathéodory's theorem). □

Remark 3.1.4. It is worth recording here an important result of F. and M. Riesz (see [111, p. 72]). Let $\Omega \subseteq \mathbb{C}$ be a Jordan domain bounded by a rectifiable boundary curve. Let $\varphi : \mathbb{D} \to \Omega$ be a conformal mapping. If $E \subseteq \partial\mathbb{D}$ has Lebesgue linear measure zero, then $\varphi(E)$ also has Lebesgue linear measure zero.

3.1.1 Some Examples

An idea like harmonic measure is a bit hollow without some concrete examples. In this section, we produce several very concrete instances.

Harmonic measure helps us to understand the growth and value distribution of harmonic and holomorphic functions on Ω. It has become a powerful analytic tool. We begin to understand harmonic measure by first calculating some examples.

Example 3.1.5. Let $\mathcal{U}$ be the upper-half-plane. Let E be an interval $[-T, T]$ on the real axis, centered at 0. Let us calculate $\omega(z, \mathcal{U}, E)$ for $z = x + iy \cong (x, y) \in \Omega$. It is a standard fact (see [69]) that the Poisson kernel for $\mathcal{U}$ is

$$P(x, y) = \frac{1}{\pi} \cdot \frac{y}{x^2 + y^2} .$$

Here we use the traditional real notation for the kernel. In other words, the harmonic function Ω that we seek is given by

$$\begin{aligned} \omega(x + iy) &= \frac{1}{\pi} \int_{-\infty}^{\infty} \chi_{[-T,T]}(t) \cdot \frac{y}{(x - t)^2 + y^2} dt \\ &= \frac{1}{\pi} \int_{-T}^{T} \frac{y}{(x - t)^2 + y^2} dt . \end{aligned}$$

Now it is a simple matter to rewrite the integral as

$$\begin{aligned} \omega(x + iy, \Omega, E) &= \frac{1}{\pi} \frac{1}{y} \int_{-T}^{T} \frac{1}{\left(\frac{x-t}{y}\right)^2 + 1} dt \\ &= -\frac{1}{\pi} \operatorname{Tan}^{-1}\left(\frac{x - t}{y}\right)\Bigg|_{-T}^{T} \\ &= \frac{1}{\pi} \operatorname{Tan}^{-1}\left(\frac{x + T}{y}\right) - \frac{1}{\pi} \operatorname{Tan}^{-1}\left(\frac{x - T}{y}\right) . \end{aligned}$$

The reader may check that this function is harmonic, tends to 1 as (x, y) approaches $E \subseteq \partial\mathcal{U}$, and tends to 0 as (x, y) approaches $\partial\mathcal{U} \setminus E$. Glancing at Fig. 3.1, we see that the value of $\omega(z, \Omega, E)$ is simply α/π, where α is the angle subtended at the point (x, y) by the interval E.

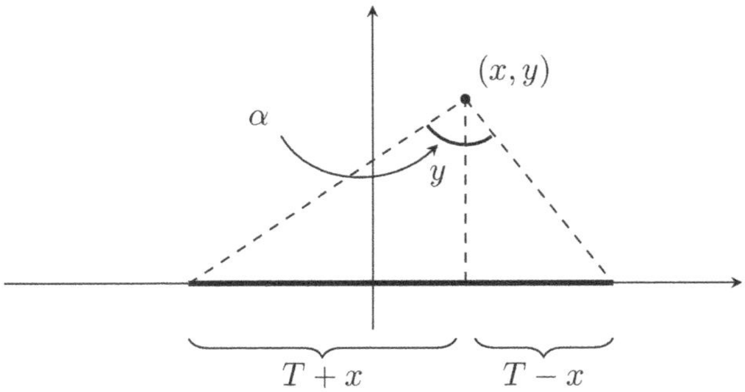

FIGURE 3.1: The value of $\omega(z, \Omega, E)$.

More generally, if E is any bounded, closed interval in the real line, then the harmonic measure $\omega(z, \mathcal{U}, E)$ will be $1/\pi$ times the angle subtended at the point (x, y) by the interval E. But we can say more. If now E is the finite disjoint union of closed, bounded intervals,

$$E = I_1 \cup \cdots \cup I_k ,$$

then the harmonic measure of E is just the sum of the harmonic measures $\omega(z, \Omega, I_j)$ for each of the individual intervals. So the harmonic measure at z is just $1/\pi$ times the sum of the angles subtended at (x, y) by each of the intervals I_j.

Example 3.1.6. Let $\mathbb{D}$ be the unit disc, and let E be an arc of the unit circle with central angle α. Then, for $z \in \mathbb{D}$,

$$\omega(z, \mathbb{D}, E) = \frac{2\theta - \alpha}{2\pi} ,$$

where θ is the angle subtended by E at the point z. See Fig. 3.2a.

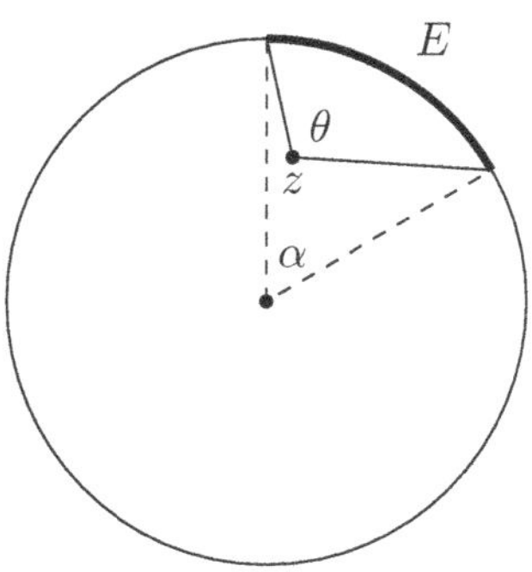

(a) The angle subtended by E at z.

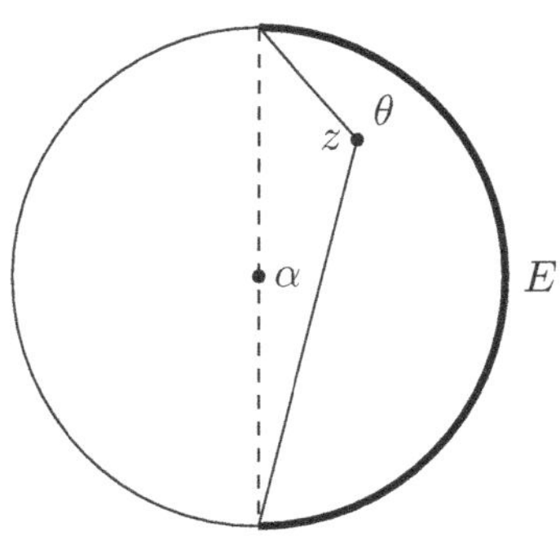

(b) The case where E is the arc from $-i$ to i.

FIGURE 3.2: Harmonic measure of an arc $E \subset \partial\mathbb{D}$ as seen from a point $z \in \mathbb{D}$.

It is convenient to first consider the special case where E is the arc from $-i$ to i; see Fig. 3.2b.

We may simply calculate the Poisson integral of $\chi(-\pi/2, \pi/2)$, where the argument of this characteristic function is the angle in radian measure on the circle. Thus, for $z = re^{i\lambda}$,

$$\begin{aligned}\omega(z, \mathbb{D}, E) &= \frac{1}{2\pi}\int_{-\pi/2}^{\pi/2} \frac{1-r^2}{1-2r\cos(\lambda - t) + r^2}\,dt \\ &= \frac{1-r^2}{2\pi}\int_{-\pi/2}^{\pi/2} \frac{dt}{(1+r^2) - 2r\cos(\lambda - t)}\,,\end{aligned}$$

At this point, it is useful to recall from calculus (and this may be calculated using the Weierstrass $w = \tan x/2$ substitution) that

$$\int \frac{dx}{a + b\cos x} = \frac{2}{\sqrt{a^2 - b^2}}\,\mathrm{Tan}^{-1}\left(\sqrt{\frac{a-b}{a+b}}\tan\frac{x}{2}\right).$$

Therefore

$$\begin{aligned}\omega(z, \mathbb{D}, E) = \frac{1}{\pi}\Bigg[&\mathrm{Tan}^{-1}\left(\frac{1+r}{1-r}\tan\frac{\lambda + \frac{\pi}{2}}{2}\right) \\ &- \mathrm{Tan}^{-1}\left(\frac{1+r}{1-r}\tan\frac{\lambda - \frac{\pi}{2}}{2}\right)\Bigg].\end{aligned}$$

Some easy but tedious manipulations now allow us to rewrite this last expression as

$$\begin{aligned}\omega(z, \mathbb{D}, E) = \frac{1}{\pi}\Bigg[&\mathrm{Tan}^{-1}\left(\left(\frac{1+r}{1-r}\right)\cdot\left(\frac{\sin\lambda + 1}{\cos\lambda}\right)\right) \\ &- \mathrm{Tan}^{-1}\left(\left(\frac{1+r}{1-r}\right)\cdot\left(\frac{\sin\lambda - 1}{\cos\lambda}\right)\right)\Bigg].\end{aligned} \tag{3.2}$$

Recall, however, that

$$\tan(\alpha - \beta) = \frac{\tan\alpha - \tan\beta}{1 + \tan\alpha\tan\beta}$$

hence

$$\mathrm{Tan}^{-1}a - \mathrm{Tan}^{-1}b = \mathrm{Tan}^{-1}\left(\frac{a-b}{1+ab}\right),$$

Applying this simple idea to (3.2) yields

$$\omega(z, \mathbb{D}, E) = \frac{1}{\pi}\,\mathrm{Tan}^{-1}\left(\frac{\left(\frac{1+r}{1-r}\right)\left(\frac{\sin\lambda+1}{\cos\lambda}\right) - \left(\frac{1+r}{1-r}\right)\left(\frac{\sin\lambda-1}{\cos\lambda}\right)}{1 + \left(\frac{1+r}{1-r}\right)\left(\frac{\sin\lambda+1}{\cos\lambda}\right)\cdot\left(\frac{1+r}{1-r}\right)\left(\frac{\sin\lambda-1}{\cos\lambda}\right)}\right).$$

Elementary simplifications finally lead to

$$\omega(z, \mathbb{D}, E) = \frac{1}{\pi}\,\mathrm{Tan}^{-1}\left(\frac{1-r^2}{2r\cos\lambda}\right).$$

Now it is helpful to further rewrite the right-hand side of this last expression as

$$\frac{2\left[\operatorname{Tan}^{-1}\left(\frac{1-r\sin\lambda}{r\cos\lambda}\right)+\operatorname{Tan}^{-1}\left(\frac{1+r\sin\lambda}{r\cos\lambda}\right)\right]-\pi}{2\pi}$$

(remembering, of course, that $\pi/2 - \operatorname{Tan}^{-1}(\gamma) = \operatorname{Tan}^{-1}(1/\gamma)$). But the preceding expression is just the formula

$$\frac{2\theta-\alpha}{2\pi},$$

for the special angle $\alpha = \pi$, which was enunciated at the start of the example. For the case of general E, we may first suppose that E is an arc centered at the point $1 \in \partial\mathbb{D}$. Second, we may reduce the general case to the one just calculated by using a Möbius transformation.

Example 3.1.7. Suppose that U is an annulus with radii $0 < r_1 < r_2 < \infty$. Then one may verify by inspection that, if E is the outer boundary circle of the annulus, then

$$\omega(z, U, E) = \frac{\log(|z|/r_1)}{\log(r_2/r_1)}.$$

In general, it is a tricky business to calculate exactly the harmonic measure for a given domain U and a given E. But one may often obtain useful estimates. The ensuing discussion will bear out this point.

3.1.2 Hadamard's Three-Circles Theorem

One of the big ideas in modern harmonic analysis is interpolation of operators. If a linear operator T maps a Banach space X to itself and another Banach space Y to itself, then is there automatically a range of "intermediate" spaces on which T is bounded? (This question goes back to M. Riesz's proof of the L^p boundedness of the Hilbert transform in 1926.)

The first genuine interpolation theorem (which treats the concept of intermediate space) is the Riesz–Thorin theorem. Its proof uses Hadamard's three-circles theorem.

Hadamard's three-circles theorem (sometimes called the "*three-lines theorem*") is an important sharpening of the classical maximum principle. It has proved useful in various parts of analysis, notably in proving the Riesz–Thorin interpolation theorem for linear operators. Here we shall give a thorough treatment of the three circles theorem from the point of view of harmonic measure. Afterwards we shall discuss the result of Riesz and Thorin.

We begin by treating some general comparison principles regarding harmonic measure. It will facilitate our discussion to first introduce a slightly more general concept of harmonic measure.

Let $U \subseteq \mathbb{C}$ be, as usual, a domain. Let A be a closed set in the extended plane $\widehat{\mathbb{C}}$. Let E denote that part of $\partial(U\backslash A)$ that lies in A. [In what follows, we will speak of pairs (U, A).] Then $\omega(z, U\backslash A, E)$ will be called the *harmonic measure* of A with respect to U, assuming that the geometry is simple enough that we can compute this number component-by-component (of $U \setminus A$). Generally speaking, in the present chapter, we assume that all boundaries that arise consist of finitely many Jordan curves. This standing hypothesis is made so that we can readily apply Lindelöf's Proposition 3.1.1. Often this standing hypothesis will go unspoken.

The first new comparison tool is called the *majorization principle.*

Prelude: We consider here a transformation property for harmonic measure.

Theorem 3.1.8. *Consider two pairs* (U, A) *and* $(\widetilde{U}, \widetilde{A})$. *Let*

$$f : U \setminus A \to \widetilde{U}$$

be holomorphic. Assume furthermore that, if $U \ni z \to A$, *then* $\widetilde{U} \ni f(z) \to \widetilde{A}$. *Then*

$$\omega(z, U, A) \leq \omega(f(z), \widetilde{U}, \widetilde{A})$$

for $z \in f^{-1}(\widetilde{U} \setminus \widetilde{A})$.

Remark 3.1.9. Of course, we consider only holomorphic f because we want a mapping that preserves harmonic functions under composition.

Proof of Theorem 3.1.8. Let us abbreviate

$$\omega = \omega(z, U, A) \quad \text{and} \quad \widetilde{\omega} = (f(z), \widetilde{U}, \widetilde{A})\,.$$

Now we apply the maximum principle to $\omega - \widetilde{\omega}$ on a connected component V of $f^{-1}(\widetilde{U} \setminus \widetilde{A})$. As z approaches the boundary of V, either z tends to a boundary point of U that is not on A, or else $f(z)$ tends to $\widetilde{A}$. In either of these circumstances, $\limsup_{z \to \partial V}(\omega - \widetilde{\omega}) \leq 0$ except when $f(z)$ tends to an endpoint of the boundary arcs of $\widetilde{U} \setminus \widetilde{A}$ that lie on $\widetilde{A}$. Since there are only finitely many such points, Lindelöf's Proposition 3.1.1 tells us that the maximum principle still remains valid. We therefore conclude that $\omega \leq \widetilde{\omega}$ on all of $f^{-1}(\widetilde{U} \setminus \widetilde{A})$. □

Corollary 3.1.10. *The function* $\omega(z, U, A)$ *increases if either* U *increases or* A *increases. That is to say, if* $U \subseteq U^*$ *and* $A \subseteq A^*$, *then*

$$\omega(z, U, A) \leq \omega\left(z, U^*, A\right) \quad \text{and} \quad \omega(z, U, A) \leq \omega\left(z, U, A^*\right)\,.$$

Proof. We prove the first statement and leave the second for the reader.

We apply the theorem with $f : U \setminus A \to U^*$ being the identity mapping. The result is now immediate. □

Now let U^* be an open disc of radius $R > 0$, and let A^* be a smaller closed disc of radius $0 < r < R$. One may see by inspection that

$$\omega\left(f(z), U^*, A^*\right) = \frac{\log(R/|f(z)|)}{\log R/r}\,.$$

We do not yet say what f is, and there is considerable latitude in practice. Nonetheless, we see immediately that the function ω is identically equal to 1 when $f(z) \in \partial A^*$ and identically equal to 0 when $f(z) \in \partial U^*$. Certainly it is harmonic.

Now we have our first result where the quantitative properties of harmonic measure play a decisive role.

Prelude: This next result is reminiscent of our arguments (below) using the three-circles theorem to establish the Riesz–Thorin theorem.

Theorem 3.1.11. *Let f be a holomorphic function on a domain U. Let A be a closed set. If $|f(z)| \leq M$ in U and $|f(z)| \leq m < M$ on A, then, for $0 \leq \theta \leq 1$,*

$$|f(z)| \leq m^\theta M^{1-\theta}$$

at points z where $\omega(z, U, A) \geq \theta$.

Remark 3.1.12. This result is of fundamental philosophical importance. It shows how the harmonic measure is a device for interpolating information about the function f.

Proof. Of course, $f : U \to \mathbb{D}(0, M)$. We take $\widetilde{A} = \mathbb{D}(0, m)$, $A = f^{-1}(\widetilde{A})$, and $\widetilde{U} = f(U)$. Then we apply the previous theorem and the remark following. So

$$\omega(z, U, A) \leq \omega(f(z), \widetilde{U}, \widetilde{A}) = \frac{\log(M/|f(z)|)}{\log M/m}\,.$$

At a point z for which $\omega(z, U, A) \geq \theta$, we have

$$\theta \leq \frac{\log(M/|f(z)|)}{\log M/m}\,.$$

This inequality is equivalent to the desired conclusion. □

If we take U and A in this last theorem to be annuli, then we can draw an important and precise conclusion. Namely, we have the following version of the three-circles theorem.

Prelude: We now have the celebrated Hadamard three-circles theorem, whose uses in analysis are many and varied.

Theorem 3.1.13 (Hadamard). *Let f be a holomorphic function on an annulus $\mathcal{A} = \{z \in \mathbb{C} : c < |z| < C\}$. For $c < \tau < C$, we let*

$$M(\tau) \equiv \max_{|z|=\tau} |f(z)| .$$

If $c < r < \rho < R < C$, then we have

$$\log M(\rho) \leq \frac{\log R - \log \rho}{\log R - \log r} \cdot \log M(r) + \frac{\log \rho - \log r}{\log R - \log r} \cdot \log M(R)$$

(This inequality says quite plainly that $\log M(s)$ is a convex function of $\log s$.)

Proof. Let

$$U = \{z \in \mathbb{C} : r < |z| < R\} ,$$

and let

$$A = \{z \in \mathbb{C} : r \leq |z| \leq r + \epsilon\}, \text{ some } \epsilon > 0 .$$

We assume that f is holomorphic on $U, |f| \leq M$ on U, and $|f| \leq m$ on A. The preceding theorem tells us that

$$|f(z)| \leq m^\theta M^{1-\theta} \tag{3.3}$$

on the set where $\omega(z, U, A) \geq \theta$, that is, on the set where

$$\frac{\log(|z|/R)}{\log((r+\epsilon)/R)} \geq \theta .$$

(Note that the roles of $r + \epsilon$ and R are reversed from their occurrence in Example 3.1.7, just because now we are creating harmonic measure for the inside circle of the annulus.) In particular, inequality (3.3) holds when $|z| = (r + \epsilon)^\theta R^{1-\theta}$. We take

$$\theta = \frac{\log R - \log \rho}{\log R - \log(r + \epsilon)}$$

and

$$|z| = (r + \epsilon)^\theta R^{1-\theta} \equiv \rho_\epsilon .$$

Then (3.3) translates to

$$M(\rho_\epsilon) \leq M(r + \epsilon)^\theta \cdot M(R)^{1-\theta} .$$

Letting $\epsilon \to 0^+$ and taking the logarithm of both sides yields the desired result. □

We note that Ahlfors [4], from which our exposition derives, likes to express the conclusion of this last result as

$$\det \begin{pmatrix} 1 & 1 & 1 \\ \log r & \log \rho & \log R \\ \log M(r) & \log M(\rho) & \log M(R) \end{pmatrix} \geq 0 .$$

We close this section by formulating the three-lines version of Hadamard's theorem. It is proved from Theorem 3.1.13 just by using conformal mapping, and we leave the details to the interested reader.

Prelude: We have the "three-lines" version of the Hadamard theorem.

Theorem 3.1.14. *Let f be a holomorphic function on the strip*

$$S = \{z \in \mathbb{C} : 0 < \operatorname{Re} z < 1\} .$$

For $0 < x < 1$ we let

$$M(x) \equiv \max_{\operatorname{Re} z = x} |f(z)| .$$

If $0 < a < \rho < b < 1$, then we have

$$\log M(\rho) \leq \frac{\log b - \log \rho}{\log b - \log a} \cdot \log M(a) + \frac{\log \rho - \log a}{\log b - \log a} \cdot \log M(b) .$$

(This inequality says quite plainly that $\log M(s)$ is a convex function of $\log s$.)

3.2 A Discussion of Interpolation of Linear Operators

Capsule: As predicted in the last **Capsule**, we now apply the three-lines theorem to prove a result about interpolation of operators. This will be a new idea, and a profound one.

Marcel Riesz discovered the idea of interpolating linear operators in the following context. Perhaps the most important linear operator in all of analysis is the Hilbert transform

$$H : f \to \int_{\mathbb{R}} \frac{f(t)}{x - t} dt .$$

This operator arises naturally in the study of the boundary behavior of conjugates of harmonic functions on the disc, in the existence and regularity theory for the Laplacian, in the summability theory of Fourier series, and, more generally, in the theory of singular integral operators. It had been an open problem for some time to show that H is a bounded operator on L^p, $1 < p < \infty$.

It turns out that the boundedness on L^2 is easy to prove and follows from Plancherel's theorem in Fourier analysis. Riesz cooked up some extremely clever tricks to derive boundedness on L^p when p is an even integer. (These ideas are explained in detail in [112].) He needed to find some way to derive therefrom the boundedness on the "intermediate" L^p spaces. Now we introduce enough language to explain precisely what the concept of "intermediate space" means.

Let X_0, X_1, Y_0, Y_1 be Banach spaces. Intuitively, we will think of a linear operator T such that

$$T : X_0 \to Y_0$$

continuously and

$$T : X_1 \to Y_1$$

continuously. We wish to posit the existence of spaces X_θ and Y_θ, $0 \leq \theta \leq 1$ such that these new spaces are natural "intermediaries" of X_0, X_1 and Y_0, Y_1 respectively. Furthermore, we want that

$$T : X_\theta \to Y_\theta$$

continuously. There is interest in knowing how the norm of T acting on X_θ depends on the norms of T acting on X_0 and X_1. One rigorous method for approaching this situation is as follows (see [17, 172], and [98] for our inspiration).

Let X_0, Y_0, X_1, Y_1 be given as in the last paragraph. Suppose that

$$T : X_0 \cap X_1 \longrightarrow Y_0 \cup Y_1$$

is a linear operator with the properties that

(i) $\|T_x\|_{Y_0} \leq C_0 \|x\|_{X_0}$ for all $x \in X_0 \cap X_1$;

(ii) $\|Tx\|_{Y_1} \leq C_0 \|x\|_{X_1}$ for all $x \in X_0 \cap X_1$.

Then we want to show that there is a collection of norms $\|\cdot\|X_\theta$ and $\|\cdot\|Y_\theta$, $0 < \theta < 1$, such that

$$\|Tx\|_{Y_\theta} \leq C_\theta \|x\|_{X_\theta} \, .$$

So the problem comes down to how to construct the "intermediate norms" $\|\cdot\|_{X_\theta}$ and $\|\cdot\|_{Y_\theta}$ from the given norms $\|\cdot\|_{X_0}, \|\cdot\|_{Y_0}, \|\cdot\|_{X_1}$, and $\|\cdot\|_{Y_1}$.

There are in fact a number of paradigms for effecting the indicated construction. Most prominent among these are the "real method" (usually attributed to Lyons and Peetre—see [121]) and the "complex method" (usually attributed to Calderón-see [31]). In this text we shall concentrate on the complex method, which was inspired by the ideas of Riesz and Hadamard that were described at the beginning of this section.

Instead of considering the general paradigm for complex interpolation, we shall concentrate on the special case that was of interest to Riesz and Thorin. We now formulate our main theorem.

Prelude: The Riesz–Thorin theorem was the very first theorem in the study of interpolation of operators. This is now a very well-developed part of analysis, and there are entire monographs devoted to the subject.

Theorem 3.2.1. *Let $1 \le p_0 < p_1 \le \infty$ and $1 \le q_0 < q_1 \le \infty$. Suppose that*

$$T : L^{p_0}(\mathbb{R}^2) \cap L^{p_1}(\mathbb{R}^2) \longrightarrow L^{q_0}(\mathbb{R}^2) \cup L^{q_1}(\mathbb{R}^2)$$

is a linear operator on Lebesgue spaces satisfying

$$\|Tf\|_{L^{q_0}} \le C_0 \|f\|_{L^{p_0}}$$

and

$$\|Tf\|_{L^{q_1}} \le C_0 \|f\|_{L^{p_1}} .$$

Define p_θ and q_θ by

$$\frac{1}{p_\theta} = (1-\theta) \cdot \frac{1}{p_0} + \theta \cdot \frac{1}{p_1}$$

and

$$\frac{1}{q_\theta} = (1-\theta) \cdot \frac{1}{q_0} + \theta \cdot \frac{1}{q_1} ,$$

with $0 \le \theta \le 1$. Then T is a bounded operator from the L^{p_θ} norm to the L^{q_θ} norm, and

$$\|Tf\|_{L^{q_\theta}} \le C_0^{1-\theta} C_1^{\theta} \|f\|_{L^{p_\theta}}$$

for all $f \in L^{p_0} \cap L^{p_1}$.

Proof. Fix a nonzero function f that is continuous and with compact support in $\mathbb{R}^2$. We will prove an a priori inequality for this f, and then extend to general f at the end. Now certainly $f \in L^{p_0} \cap L^{p_1}$. We consider the holomorphic function

$$H : \zeta \mapsto \frac{\left(\int_{\mathbb{R}^2} |Tf(x)|^{(1-\zeta)q_0 + \zeta q_1}\, dx\right)^{(1-\zeta)/q_0 + \zeta/q_1}}{\left(\int_{\mathbb{R}^2} |f(x)|^{(1-\zeta)p_0 + \zeta p_1}\, dx\right)^{(1-\zeta)/p_0 + \zeta/p_1}}$$

on the strip $S = \{\zeta \in \mathbb{C} : 0 < \operatorname{Re}\zeta < 1\}$. We define

$$M(s) = \sup_{\operatorname{Re}\zeta = s} |H(\zeta)|, 0 < s < 1 .$$

Notice that

$$\begin{aligned} M(0) &= \sup_{t\in\mathbb{R}} \left| \frac{\int_{\mathbb{R}^2} |Tf(x)|^{(1-it)q_0 + itq_1}\, dx^{(1-it)/q_0 + it/q_1}}{\int_{\mathbb{R}^2} |f(x)|^{(1-it)p_0 + itp_1}\, dx^{(1-it)/p_0 + it/p_1}} \right| \\ &\le \sup_{t\in\mathbb{R}} \frac{\int_{\mathbb{R}^2} |Tf(x)|^{q_0}\, dx^{1/q_0}}{\int_{\mathbb{R}^2} |f(x)|^{p_0}\, dx^{1/p_0}} \\ &\le C_0 . \end{aligned}$$

A similar calculation shows that

$$M(1) \le C_1 .$$

Obviously this is grist for the three-lines theorem. We may conclude that $M(s) \leq C_0^{1-s} \cdot C_1^s$, $0 < s < 1$. Now let p_ζ, q_ζ be given by

$$\frac{1}{p_\zeta} = (1-\zeta) \cdot \frac{1}{p_0} + \zeta \cdot \frac{1}{p_1}$$

and

$$\frac{1}{q_\zeta} = (1-\zeta) \cdot \frac{1}{q_0} + \zeta \cdot \frac{1}{q_1} .$$

Let $\zeta = s + it$. Then

$$\frac{\|Tf\|_{L^{q_\zeta}}}{\|f\|_{L^{p_\zeta}}} \leq M(s) \leq C_0^{1-s} C_1^s .$$

In other words,

$$\|Tf\|_{L^{q_s}} \leq C_0^{1-s} C_1^s \|f\|_{L^{p_s}} .$$

This is the desired conclusion for a function f that is continuous with compact support. The result for general $f \in L^{p_\theta}$ can be achieved with a simple approximation argument. □

3.3 The F. and M. Riesz Theorem

Capsule: One of the big ideas of twentieth century function theory is that due to the brothers F. and M. Riesz. The result has many interpretations—in terms of the absolute continuity of harmonic measure, in terms of the vanishing of negative Fourier coefficients, and many other ideas as well.

The theorem of F. and M. Riesz makes an important statement about the absolute continuity of harmonic measure. We begin with a preliminary result that captures the essence of the theorem. Our exposition here follows the lead of [57, Ch. 6].

For $0 < p < \infty$ we define the Hardy space

$$H^p(\mathbb{D}) = \left\{ f \in H(\mathbb{D}) : \sup_{0<r<1} \left(\int_0^{2\pi} \left| f\left(re^{i\theta}\right)\right|^p d\theta \right)^{1/p} \equiv \|f\|_{H^p} < \infty \right\} .$$

Hardy spaces are generalizations of the more classical space $H^\infty(\mathbb{D})$ of bounded, holomorphic functions.

The most important fact about a function in a Hardy space is that it has a boundary function. We enunciate that result here, and refer the reader to [113, Chapter 8] for the details.

Prelude: The theory of boundary limits for functions in Hardy spaces goes back to work of Fatou in 1906. Later on, Riesz and Hardy played a decisive role in fleshing out the theory and pointing in the directions that we study today.

Theorem 3.3.1. *Let* $0 < p \le \infty$ *and let* $f \in H^p(\mathbb{D})$. *Then there is a function* $f^* \in L^p(\partial\mathbb{D})$ *such that:* $\lim_{r\to 1^-} f\left(re^{i\theta}\right) = f^*\left(e^{i\theta}\right)$ *for almost every* $\theta \in [0, 2\pi)$. *Let* $f_r\left(e^{i\theta}\right) = f\left(re^{i\theta}\right)$ *for* $0 < r < 1$. *Assume that* $0 < p < \infty$. *Then* $\lim_{r\to 1^-} \|f - f_r\|_{L^p(\partial\mathbb{D})} = 0$.

In the following discussion we shall consider rectifiable curves. Basically, a rectifiable curve is a curve with finite length. The technical definition is that the curve is locally the Lipschitz image of the unit interval (see [47]). But context will make it clear that the intuitive notion of rectifiability will suffice for our purposes.

Prelude: Here it is important that we relate rectifiability for a curve with the theory of Hardy spaces. Hausdorff measure also comes into play.

Theorem 3.3.2. *Let* U *be a domain such that* $\gamma = \partial U$ *is a Jordan curve. Let*

$$\varphi : \mathbb{D} \to U$$

be a conformal map. Then the curve γ *is rectifiable if and only if* $\varphi' \in H^1$. *In case* $\varphi' \in H^1$, *then we have*

$$\|\varphi'\|_{H^1} = \text{length}(\gamma) = \mathcal{H}^1(\gamma)\,. \tag{3.4}$$

Here $\mathcal{H}^1$ *is one-dimensional Hausdorff measure.*

Remark 3.3.3. It is known (see [113, Chapter 8]) that a function in H^1 has a boundary limit function that is in $L^1(\partial\mathbb{D})$. Thus the hypothesis of the theorem says, essentially, that φ is absolutely continuous on the boundary of $\mathbb{D}$. It makes sense, then, that φ would preserve length on the boundary. Theorem 3.3.2 is essentially equivalent to the result of F. and M. Riesz that we described in Remark 3.1.4.

Proof of Theorem 3.3.2. Once again, we invoke Carathéodory's theorem; thus we know that φ and its inverse extend continuously and univalently to their respective boundaries. Let us assume that $\varphi' \in H^1$. Let

$$0 = \theta_0 < \theta_1 < \theta_2 < \cdots < \theta_k = 2\pi$$

be a partition of $[0, 2\pi]$. Then of course

$$\begin{aligned}\sum_{j=1}^{k} \left|\varphi\left(e^{i\theta_j}\right) - \varphi\left(e^{i\theta_{j-1}}\right)\right| &= \lim_{r\to 1^-} \sum_{j=1}^{k} \left|\varphi\left(re^{i\theta_j}\right) - \varphi\left(re^{i\theta_{j-1}}\right)\right| \\ &= \lim_{r\to 1^-} \sum_{j=1}^{k} \left|\int_{\theta_{j-1}}^{\theta_j} \varphi'\left(re^{i\theta}\right) ire^{i\theta} d\theta\right| \\ &\le \|\varphi'\|_{H^1}\,.\end{aligned} \tag{3.5}$$

But plainly the length of γ is the supremum, over all partitions of $[0, 2\pi)$, of the left-hand side of (3.5). We conclude that γ is rectifiable and

$$\text{length}\,(\gamma) \le \|\varphi'\|_{H^1}\,.$$

For the converse, assume that γ is rectifiable. If $0 < r < 1$ is fixed, then let $\gamma_r = \varphi(\{z \in \mathbb{C} : |z| = r\})$. Let $\epsilon > 0$. Now choose a partition $\{\theta_0, \theta_1, \ldots, \theta_k\}$ of the interval $[0, 2\pi)$ as before so that

$$\sum_{j=1}^{k} \left|\varphi\left(re^{i\theta_j}\right) - \varphi\left(re^{i\theta_{j-1}}\right)\right| \ge \text{length}\,(\gamma_r) - \epsilon\,.$$

We write

$$\eta(z) = \sum_{j=1}^{k} \left|\varphi\left(ze^{i\theta_j}\right) - \varphi\left(ze^{i\theta_{j-1}}\right)\right|\,.$$

Then η, being the sum of absolute values of holomorphic functions, is subharmonic. By Carathéodory's theorem, η is continuous on $\overline{\mathbb{D}}$. Hence

$$\sup_{\mathbb{D}} \eta(z) = \sup_{\theta} \eta\left(e^{i\theta}\right) \le \ \text{length}\,(\gamma)\,.$$

We conclude that

$$\int_0^{2\pi} \left|\varphi'\left(re^{i\theta}\right)\right| d\theta = \ \text{length}\ \ (\gamma_r) \le \eta(r) + \epsilon \le \ \text{length}\ (\gamma) + \epsilon\,.$$

It follows that $\phi' \in H^1$ and equality (3.4) is valid. We conclude by noting that, for a rectifiable curve, the ordinary notion of length and the one-dimensional Hausdorff measure are the same. □

The reason that we include Theorem 3.3.2 in the present chapter—apart from its general aesthetic interest—is that it can be interpreted in the language of harmonic measure. Let notation be as in Theorem 3.3.2. To wit, let $\gamma = \partial U$ be a rectifiable Jordan curve as above and let $F \subseteq \gamma$ be a subcurve. Let φ be a conformal mapping of U to the unit disc $\mathbb{D}$. Let $F = \varphi(E)$ and $\alpha = \varphi(0)$. Then Carathéodory's theorem, Example 3.1.6, and Hadamard's theorem tell us that

$$\omega(\alpha, U, F) = \omega(0, \mathbb{D}, E) = \frac{1}{2\pi}|E|\,,$$

where the $|\cdot|$ denotes arc length. Since F is an arc, we see that Theorem 3.3.2 and the proof of (3.4) tell us that

$$\mathcal{H}^1(F) = \mathcal{H}^1(\varphi(E)) = \lim_{r\to 1^-} \int_E \left|\varphi'\left(re^{i\theta}\right)\right| d\theta = \int_E \left|\varphi'\left(e^{i\theta}\right)\right| d\theta\,. \qquad (3.6)$$

Of course, if (3.6) holds for arcs, then, by passing to unions and intersections, we see that it holds for Borel sets F. Thus we derive the important conclusion

$$\omega(\alpha, U, F) = 0 \Rightarrow \mathcal{H}^1(F) = 0\,.$$

Conversely, since (by a standard uniqueness theorem—see [111])

$$\left|\{\theta : |\varphi'(e^{1\theta})| = 0\}\right| = 0$$

we see that

$$\mathcal{H}^1(F) = 0 \Rightarrow \omega(\alpha, U, F) = 0 .$$

What we have proved, then, is that when $\gamma = \partial U$ is rectifiable then harmonic measure for U and linear measure on γ are mutually absolutely continuous. We now summarize this result in a formally enunciated theorem.

Prelude: The *F. and M. Riesz theorem* is one of the celebrated results of classical function theory. There are many different renditions of the result, and many different proofs. We give here a geometric rendition of the theorem.

Theorem 3.3.4 (F. and M. Riesz, 1916). *Let U be a simply connected planar domain such that $\gamma = \partial U$ is a rectifiable Jordan curve. Assume that*

$$\varphi : \mathbb{D} \to U$$

is conformal. Then $\varphi' \in L^1(\partial\mathbb{D})$. For any Borel set $E \subseteq \partial\mathbb{D}$ it holds that

$$\mathcal{H}^1(\varphi(E)) = \int_E |\varphi' e^{i\theta}| d\theta ,$$

and for any Borel set $F \subset \partial\Omega$,

$$\omega(F) = 0 \Longleftrightarrow \mathcal{H}^1(F) = 0 .$$

We conclude by noting that Theorems 3.3.2 and 3.3.4 are equivalent, just because formula (3.4) is valid precisely when $\varphi' \in H^1$.

When the domain Ω has a nice boundary, harmonic measure is particularly easy to understand.

Suppose that Ω is a domain on the sphere and that $p \in \Omega$. A function $g(z;p)$ is a *Green's function* for Ω with pole (or singularity) at $p, p \neq \infty$, if

$$\begin{aligned} &g(z;p) \text{ is harmonic on } \Omega - \{p\} , \\ &g(z;p) + \log|z-p| \text{ is harmonic near } p , \\ &\lim\{g(z;p) : z \to \zeta\} = 0 \text{ for all } \zeta \in \partial\Omega . \end{aligned} \tag{3.7}$$

If $p = \infty$, then (3.7) is modified to

$$g(z;\infty) - \log|z| \text{ is harmonic near } \infty$$

Theorem 3.3.5. *Suppose that Ω is bounded by a finite number of disjoint analytic simple closed curves. For each $p \in \Omega$ we have*

$$d\omega_p = \frac{-1}{2\pi}\frac{\partial}{\partial n} g(\cdot;p) ds ,$$

where $g(\cdot;p)$ is the Green's function for Ω with pole at p, $\partial/\partial n$ is the derivative in the direction of the outward normal at $\Gamma = \partial\Omega$, and ds is arc length.

Proof. The proof is an application of Green's theorem. Let h be a smooth function on Γ and for small $\delta > 0$ let

$$\Omega_\delta = \Omega \backslash \{z : |z-p| \leqslant \delta\}.$$

Let u be the harmonic extension to Ω of h and let $v(z) = g(z;p)$. Then, by Green's theorem on Ω_δ,

$$\iint_{\Omega_\delta} [u\Delta v - v\Delta u] dxdy = \int_{\partial\Omega_\delta} \left[u\frac{\partial v}{\partial n} - v\frac{\partial u}{\partial n}\right] ds. \tag{3.8}$$

Now the left-hand side of (3.8) is zero since both u and v are harmonic on Ω_δ.

Further, $v = 0$ on Γ, so (3.8) simplifies to

$$\int_\Gamma h\frac{\partial v}{\partial n} ds = \int_{|z-p|=\delta} \left[u\frac{\partial v}{\partial n} - v\frac{\partial u}{\partial n}\right] ds.$$

Let us set $z = p + re^{it}, 0 \leqslant r \leqslant \delta$, and $0 \leqslant t \leqslant 2\pi$. Then the normal derivative is just the radial derivative and $g(z;p) = -\log r + G(z)$ where G is harmonic near p. Thus,

$$\frac{\partial g}{\partial n} = -\frac{1}{r} + \text{continuous term}.$$

Since u is also continuous near p, we have

$$\begin{aligned}\int_{|z-p|=\delta} u\frac{\partial g}{\partial n} ds &= -\int_0^{2\pi} u\left(p + \delta e^{it}\right) dt + 0(\delta) \\ &\to -2\pi u(p) \quad \text{as } \delta \to 0.\end{aligned}$$

Finally, the other term

$$-\int_{|z-p|=\delta} g\frac{\partial u}{\partial n} ds$$

behaves like $-\delta \log \delta$ as $\delta \to 0$ and so goes to zero as $\delta \to 0$. □

It is convenient to record here two more facts related to harmonic measure. Let us assume that $\Gamma = \partial\Omega$ consists of $m+1$ disjoint analytic simple closed curves. Let $p \in \Omega$ and let $h(z) = h(z;p)$ be the harmonic conjugate of $g(z;p)$, the Green's function for Ω and p. Of course, h is not single-valued since it has periods about the holes in Ω (we will explore this type of difficulty later in some detail) but locally $g + ih$ is analytic and its derivative is single-valued on Ω. We write $Q = g + ih$ and Q' for its complex derivative.

Proposition 3.3.6. *We have* $d\omega_p(\zeta) = \frac{i}{2\pi} Q'(\zeta) d\zeta$.

Proof. Let n be a unit outward normal at $\zeta \in \Gamma$, and let τ be a unit tangent at ζ. By the Cauchy–Riemann equations

$$\frac{\partial h}{\partial n}(\zeta) = \frac{\partial g}{\delta \tau}(\zeta) = 0$$

since g vanishes on Γ. Next, note that $n = -i\tau = -id\zeta/|d\zeta|$ and

$$Q'(\zeta) = \lim_{t\downarrow 0} \frac{Q(\zeta + tn) - Q(\zeta)}{tn}.$$

Further,

$$\frac{\partial g}{\partial n} = \lim_{t\downarrow 0} \frac{g(\zeta + tn) - g(\zeta)}{t}.$$

Thus

$$\begin{aligned} iQ'(\zeta)d\zeta &= i \lim_{t\downarrow 0} \frac{Q(\zeta + tn) - Q(\zeta)}{tn} d\zeta \\ &= -\lim_{t\downarrow 0} \frac{g(\zeta + tn) - g(\zeta)}{t} |d\zeta| \\ &= -\frac{\partial g}{\partial n} ds \end{aligned}$$

which establishes the result in light of Theorem 3.3.5. □

Proposition 3.3.7. *Let $\Gamma = \partial\Omega$ consist of $m+1$ disjoint analytic simple closed curves, let $p \in \Omega$, and let $Q(z) = g(z;p) + ih(z;p)$ where h is the (multiple-valued) harmonic conjugate of g. Then*

(a) Q' does not vanish on Γ,

(b) Q' has precisely m zeros in Ω, counting multiplicities.

Proof. We know that Q' has a single pole of order one at p. Further, from Proposition 3.3.6 we also know that $iQ'(z)dz$ is a non-negative measure on Γ. Thus, the total change in $1/2\pi \arg Q'(z)$ as z traverses Γ once is precisely $m-1$. But this must be exactly the number of zeros of Q' minus the number of poles, and hence (b) is proven. To see (a) note that the Cauchy–Riemann equations imply that h is increasing locally on Γ and thus Q must be one-to-one in a neighborhood of each $\zeta \in \Gamma$. □

Exercises

1. Let E be the disjoint union of two closed intervals in the real line and let $\mathcal{U}$ be the upper half-plane. Calculate the harmonic measure of E with respect to $\mathcal{U}$.

2. Let E be the disjoint union of two closed arcs in the unit circle in the plane. Let D be the unit disc. Calculate the harmonic measure of E with respect to D.

3. Interpret the three lines theorem or the three circles theorem in the language of convexity.

4. Find a definition of weak type space in the literature (the book of Stein and Weiss is a good source). Formulate the Marcinkiewicz interpolation theorem. Compare with the Riesz–Thorin theorem.

5. Formulate a version of the F. and M. Riesz theorem in the language of Hardy spaces.

6. Many important linear operators are *not* bounded on L^1 and *not* bounded on L^∞. The Hilbert transform is a good example. The Hilbert transform *is* of weak type $(1, 1)$. And it is bounded on BMO. Search the literature and give a definition of BMO. It is a fact, though a rather subtle one, that L^2 is an interpolation space between weak type $(1, 1)$ and BMO.

4

Riemann Surfaces

It is the purpose of this and the following chapter to introduce the theory of Riemann surfaces at an elementary level with the hope that the interested reader will then be able to approach any of the standard treatises (e.g., [6, 46, 95]) with significant insight.

The content of these chapters is quite classical: most of it is known since the beginning of the last century. We follow a rather geometrical and analytical approach to the subject, in order to stay "as concrete as possible". We have tried to use a language familiar to every student: the only prerequisites are some familiarity with the basic facts concerning holomorphic functions, differential forms, and manifolds.

Behind every abstraction there are very simple examples that lead to it. So we define a Riemann surface as an intuitive model for some many-valued function rather than an abstract surface. We then obtain explicit examples of Riemann surfaces. After this we will give a definition of an abstract Riemann surface.

Riemann surfaces were first studied as the natural domain of definition of (multivalued) holomorphic or meromorphic functions. They were the starting point for the development of the theory of real and complex manifolds (see Weyl [188]). Nowadays, Riemann surfaces are simply defined as one-dimensional complex manifolds (see Chapter 5).

There are many good reviews devoted to Riemann surfaces, for example, by Dieudonné [42, 59].

Riemann surfaces were, at least conceptually, first introduced by Riemann in his celebrated 1851 PhD dissertation. He considered a surface spread over $\mathbb{C}$ with several sheets lying over it, in such a way that a complex multi-valued function with branches becomes a true well-defined function, mapping each branch to a sheet. His posterity gave a rigorous definition of the (so called) *concrete* Riemann surfaces as covering spaces of $\mathbb{C}$ (for an enlightening review see [152]).

Klein was the first to move away from the covering space approach and adopt a differential geometric one: he studied complex functions defined on curved surfaces embedded in Euclidean space. In fact, this marked the first introduction of an atlas of holomorphic structures.

At that time, the term *Riemann surface* referred to a compact 2-manifold equipped with an arc-length element ds^2 and biholomorphic transition functions in its atlas.

DOI: 10.1201/9781003713005-4

Several years later—after the introduction of Cantor's second countability and Hausdorff's separation axioms (not before the 1920s)—Radó arrived at the abstract definition of Riemann surfaces using triangulations: a Hausdorff, second-countable topological surface endowed with a complex structure. The equivalence between concrete and abstract Riemann surfaces was established by a theorem of Behnke and Stein (1947–1949) [12], and later improved by Gunning and Narasimhan (1967) [77].

For a comprehensive reference in algebraic geometry ("if it is not proven there, it must be incorrect"), see Grothendieck, A. and Dieudonné, J. A. (various volumes) [123]. Note: This material is highly challenging to read without a substantial background in algebraic geometry and remains extremely demanding even for experienced readers.

4.1 The Riemann Surface of a Function

That the concept of Riemann surface is nontrivial is manifest from its connection with the theory of multivalued analytic functions. In fact, for every such function one can construct a Riemann surface on which it becomes single-valued. We restrict ourselves to algebraic functions, so the corresponding Riemann surfaces are compact.

In order to consider a complete analytic function $\mathcal{F}$ as a function in the ordinary sense, we shall associate to $\mathcal{F}$ an analytic manifold, $\mathfrak{R}$, a Riemann surfaces, on which $\mathcal{F}$ will be a single-valued function.

The key idea is that the *complete analytic function*—that is, all the values of a function obtained by analytic continuation—can be made single-valued by redefining its domain as a Riemann surface.

We should refrain from calling an object a function unless it is indeed a function. To make a complete analytic function $\mathcal{F}$ a function one must manufacture a domain (the range will be $\mathbb{C}$) and show that $\mathcal{F}$ gives a "rule." This is easy. In a sense we let $\mathcal{F}$ be its own domain; more precisely, let

$$\mathcal{R} = \{(z, f_z) : (z, f_z) \in \mathcal{F}\} .$$

Define $\mathcal{F} : \mathcal{R} \to \mathbb{C}$ by $\mathcal{F}(z, f_z) = f_z(z)$. All these values are called *values* of the analytic function at $z = a$ and a function of this type is considered as a multiple-valued function. In this way $\mathcal{F}$ becomes an "honest" function. In other words the Riemann surface of f_0 is "the largest connected Riemann surface unramified over $\mathbb{C}$ on which the germ f_0 may be extended as a holomorphic function."

Instead of first giving many examples and then abstracting a definition, we will soon state the definition of a Riemann surface directly, and then check examples against it. Before that, we need to review some preliminaries.

In the early days Riemann surfaces were only a helpful means to represent multivalued functions. Klein was the first to express the opposite opinion, cf. [102, p. 555]: "The Riemann surface is not just an illustration of the functions in question, rather it defines them." Klein also forged an intimate alliance between Riemann's ideas and invariant theory, algebra, number theory and—above all—group theory: "Verschmelzung von Riemann und Galois" (fusion of Riemann and Galois) was one of his aims.

The symmetric tilings introduced by Klein can be interpreted in modern terms as complex atlases, where each tile map functions as a local chart defining the complex structure:

Definition 4.1.1. A surface $\mathfrak{R}$ is a connected Hausdorff space which satisfies the second axiom of countability and has a basis for the open sets of sets homeomorphic to open discs in $\mathbb{C}$. A *Riemann surface* $\mathfrak{R}$ is a surface with an open cover $\{V_\alpha \mid \alpha \in A\}$, for some index set A, and homeomorphisms $\{\varphi_\alpha\}, \varphi_\alpha : V_\alpha \to D_\alpha$, where D_α is a disc in $\mathbb{C}$ and the *neighborhood relation* $\varphi_\beta \circ \varphi_\alpha^{-1}$ is biholomorphic wherever it is defined. A pair $(V_\alpha, \varphi_\alpha)$ is called a *chart* and the set $\{(V_\alpha, \varphi_\alpha) \mid \alpha \in A\}$ is called an *atlas*. The elements φ_α will be called a *local parameter*. All Riemann surfaces are orientable.

In the above definition, each V_α or $(V_\alpha, \varphi_\alpha)$ is called a *holomorphic coordinate chart* and φ_α is called a holomorphic coordinate map or function. The set $\{(V_\alpha, \varphi_\alpha)\}$ is called a *holomorphic coordinate covering* of M.

One way to define a (compact) Riemann surface is as a connected subset $X \subset \mathbb{C}^n$ such that for every point $x \in X$ there is a biholomorphism (a holomorphic map with holomorphic inverse) between X and an open subset of $\mathbb{C}$. However, that definition is not so satisfying, because it requires that we specify an embedding $X \to \mathbb{C}^n$ for some n. It would be better to have an intrinsic definition, not depending on an embedding.

Remark 4.1.2. A Riemann surface is the datum of the topological space X together with the atlas $(U_i, f_i)_{i \in I}$. In particular, it is possible to have two Riemann surfaces with the same underlying topological spaces but different atlases.

Recall that a topological space X is Hausdorff if for any two points $x, y \in X$ there are open sets U, V with $x \in U, y \in V$ so that $U \cap V = \varnothing$. A topological space is second countable if there is a countable collection of open sets U_i so that every open set is a union of such U_i. A topological space is connected if it cannot be written as a disjoint union of two nonempty open sets. A map is continuous if the preimage of an open set is open. A homeomorphism is a continuous map between two topological spaces that has a continuous inverse.

We also recall that a *topological space* T is called a second countable topological space if there is a countable collection of open subsets $\mathcal{T} = \{V_k\}_{k=1}^{\infty}$ such that for any open subset $V \subset T$ and $p \in V$, there is an element $V_{k_0} \in \mathcal{T}$ such that $p \in V_{k_0} \subset V$. The collection $\mathcal{T}$ is also called a *basis* for the topological space T.

We begin with an abstract Riemann surface. There are several equivalent definitions of a Riemann surface. Needless to say, the historical development was the exact opposite: examples first and then—after more than sixty years of "naive" but no less successful usage of the concept of a Riemann surface—came the definitions.

An alternative definition of a Riemann surface can be given using the notion of holomorphic atlas. Weyl's definition of *Riemann surfaces*, in modern language and without unnecessary hypotheses, is the following: A Riemann surface is a one-dimensional complex manifold. That is:

Definition 4.1.3. A Riemann surface is a Hausdorff, second countable, connected topological space X with an open cover by sets $U_i \subset X$ and maps $f_i : U_i \to \mathbb{C}$ so that (1) f_i is a homeomorphism onto an open subset of $\mathbb{C}$ and (2) The transition functions $f_j \circ f_i^{-1}|_{f_i(U_i \cap U_j)} : f_i(U_i \cap U_j) \to f_j(U_i \cap U_j)$ are complex analytic (i.e., holomorphic). A collection of $(U_i, f_i)_{i \in I}$ as above is called an *atlas* for X. A particular (U_i, f_i) is called a *chart*.

It is a theorem of Radó that every Riemann surface is second countable (for a proof, see [88, Section 1.3]).

4.2 Constructions of Riemann Surfaces

The beauty of the theory of Riemann surfaces is due to the fact that Riemann surfaces can be described in many completely different ways. Interrelations among these descriptions comprise an essential part of the theory.

There are a lot of constructions of Riemann surfaces.

1. *Conformal Structures*

   Let X be a C^∞ real manifold of dimension 2. Let g be any C^∞ Riemannian metric on X. A classical theorem due to Gauss when X and g are real analytic and to Korn and Lichtenstein in the C^∞ case (cf. [36] for a simple proof) says that, locally on X, the metric g is conformally flat, i.e., that X may be covered by charts U, such that there are C^∞ local coordinates (x, y) on U in terms of which g takes the form

   $$g = \lambda\left(dx^2 + dy^2\right) ,$$

   where λ is a C^∞ function from U to $\mathbb{R}_+^*$. Furthermore, suppose that X is oriented. Permuting x and y if necessary, we can suppose that the charts $(U; x, y)$ are compatible with the given orientation. Then the complex-valued charts $(U, x + iy)$ are such that the coordinate changes between them are oriented and conformal diffeomorphisms, i.e., biholomorphic transformations. Hence these charts define a structure of a complex manifold on X (of complex dimension 1).

Moreover, for any function $\mu \in \mathcal{C}^\infty(X, \mathbb{R}_+^*)$, the Riemannian metric μg clearly defines the same holomorphic structure on X as g.

Thus we see that the data of an orientation and of a conformal class of C^∞ metrics on X determines a structure of Riemann surface on X. Furthermore, one easily checks that one gets in that way a one-to-one correspondence:

$$\left\{ \begin{array}{c} \text{holomorphic structures} \\ \text{on } X \text{ compatible with} \\ \text{its } C^\infty \text{ structure} \end{array} \right\} \longleftrightarrow \left\{ \begin{array}{c} \text{conformal class of metrics} \\ \text{on } X, \text{ orientation on } X \end{array} \right\}.$$

A nice feature of this construction is that it immediately provides a large supply of Riemann surfaces: any oriented surface embedded in a Riemannian manifold (e.g., in the "physical" Euclidean three space $\mathbb{R}^3$) gives a Riemann surface!

The above correspondence shows that the study of Riemann surfaces may be seen as the study of conformally invariant properties of two-dimensional Riemannian manifolds. That is the reason why Riemann surfaces occur in recent topics of theoretical physics such as string theory or conformal field theory.

2. *Algebraic curves*

   Historically, the theory of Riemann surfaces arose from the study of algebraic functions and of their integrals. The Riemann surfaces occurring in this study are Riemann surfaces attached to algebraic curves, which are defined by the vanishing of a family of complex polynomials.

3. *Quotients of Riemann surfaces*

   A third way to get Riemann surfaces is to construct them as quotients under group actions of some other Riemann surfaces. For instance, an elliptic curve $E_\tau = \mathbb{C}/(\mathbb{Z} + \tau\mathbb{Z})$ is the quotient of the complex plane by the action by translation of the lattice $\mathbb{Z} + \tau\mathbb{Z}$.

4. Any connected, oriented Riemannian 2-manifold (X, g) has a unique compatible complex structure, such that g is a conformal metric. (This result requires isothermal coordinates or harmonic functions.)

5. *Algebraic curves.* A curve $X \subset \mathbb{C}^2$ defined by $f(x, y) = 0$ is smooth if $df \neq 0$ along X. Example: $f(x, y) = y^2 - p(x)$ is smooth iff p has no multiple roots. Non-example: $y^2 = x^3, y^2 = x^2(x+1)$. Double zeros give nodes; triple zeros give cusps. Provided X is connected, there is a unique way to make X into a Riemann surface for which x and y are

analytic functions. Indeed, one can use projection to the coordinate axes to define charts for X, and the transition functions are then analytic by the implicit function theorem.

Even when $f(x, y)$ is not smooth, there is a natural way to find a finite union of Riemann surfaces $Y = \bigcup X_i$ and an analytic map $f : Y \to X$ that is an isomorphism away from the singularities of X. If $f(x, y)$ is irreducible, then Y itself is a Riemann surface, called the *normalization* of X.

6. *Polygons* in $\mathbb{C}, \widehat{\mathbb{C}}, \mathbb{H}$, glued together by isometries, naturally form Riemannian surfaces. Polyhedra in $\mathbb{R}^3$, such as the octahedron and the icosahedron give further examples. Triangulated surfaces give even more. (Relation to billiards.)

7. If X is a Riemann surface and $\pi : Y \to X$ is a covering map, with Y connected, then there is a unique complex structure on Y such that π is holomorphic and Y is a Riemann surface in its own right.

8. *Finite branched coverings.* Let X be a compact Riemann surface and $E \subset X$ a finite set, and let $X^* = X \setminus E$ and let $\rho : \pi_1(X^*) \to G$ be a map to a finite group. Then the corresponding G-covering space $Y^* \to X^*$ completes to a give a holomorphic map $\pi : Y \to X$, branched over E.

9. *Finite quotients.* If $G \subset \text{Aut}(Y)$ is a finite group of automorphisms, then $X = Y/G$ is a topological surface, and X carries a unique complex structure such that $\pi : Y \to X = Y/G$ is holomorphic. The Weierstrass $\wp$-function gives the map $E \mapsto E/G \cong \widehat{\mathbb{C}}$ where $G = \mathbb{Z}/2$ acts by $z \mapsto -z$ in the group law on an elliptic curve E.

10. *Analytic continuation*

    Finally we should mention, at least because of its historical importance, a last way to construct Riemann surfaces: analytic continuation. The study of analytic functions in the plane leads unavoidably to the consideration of multi-valued functions. This problem was resolved, both by Weierstrass and by Riemann. We shall make some of these elementary ideas precise, and see first what Weierstrass and Riemann were thinking.

4.3 Analytic Continuation of Complex Functions

There are various ways one might show that a holomorphic function can be extended to a larger domain. The Schwarz Reflection Principle (Theorem 4.3.3) is an important example: If an open set is symmetric about the real axis then every function holomorphic in the "upper half" of the set with imaginary part tending to 0 on the real axis extends to the entire set.

To say that a holomorphic function is defined on an open set U is to specify its domain. In some cases, the function is the restriction to U of a holomorphic function defined on a larger open set. A typical example is the case of a removable isolated singularity.

Another example arises when a function is defined by a power series expansion. For instance, consider

$$f(z) = \sum_{n \geq 0} z^n \quad \text{for } z \in \mathbb{D}\,.$$

The series can be summed to obtain

$$f(z) = (1 - z)^{-1}\,,$$

which is holomorphic on $\mathbb{C} \setminus \{1\}$ and thus extends f to this larger open set.

Example 4.3.1. Consider the principal branch of the logarithmic function

$$\log(z) = \log|z| + i \arg z, \quad -\pi < \arg z \leq \pi\,.$$

Observe that such a function $\log(z)$ is the unique analytic extension to $\mathbb{C}\backslash[\infty, 0]$ of the real $\log(x)$. The function $\log(z)$ is also the unique analytic extension of the convergent power series

$$g(z) = \sum_{n=1}^{\infty} (-1)^{n-1} \frac{(z-1)^n}{n} \qquad (|z| < 1)$$

simply because $\log(z) = g(z)$ for $(|z| < 1)$.

Permanence of functional relationships. Sometimes one can show that a function satisfies a "functional equation" and then use that to do the extending.

Deriving a functional equation is often the key step in analytic continuation. A famous example is the Riemann functional equation relating the gamma function and the Riemann zeta function. A simple example of this is given by

Example 4.3.2. The *gamma function* Γ is defined to be

$$\Gamma(z) = \int_0^{\infty} t^{z-1} e^{-t} dt, \quad \mathrm{Re}(z) > 0\,.$$

The function Γ is a differentiable (and therefore analytic) function for $\mathrm{Re}(z) > 0$, where the integral is defined.

For n a positive integer the function coincides with the factorial $\Gamma(n) = (n-1)!$—this can be verified by integration by parts of the integral $\int_0^\infty t^{n-1}e^{-t}dt$.

Further by integration by parts one obtains the functional relation

$$\Gamma(z) = \int_0^\infty t^{z-1}e^{-t}dt = \frac{t^z}{z}e^{-t}\Big|_0^\infty + \frac{1}{z}\int_0^\infty t^{z-1}e^{-t}dt = \frac{\Gamma(z+1)}{z}.$$

We notice that $\Gamma(z+1)$ is a well-defined analytic function for $\mathrm{Re}(z) > -1$ and therefore the ratio $\frac{\Gamma(z+1)}{z}$ has a simple pole at $z = 0$. We conclude that the function

$$\Gamma(z) = \frac{\Gamma(z+1)}{z}$$

has an analytic extension in $\mathrm{Re}(z) > -1$ and $z \neq 0$. Repeating the integration by parts multiple times we arrive at

$$\Gamma(z) = \frac{\Gamma(z+1)}{z} = \frac{\Gamma(z+2)}{z(z+1)} = \cdots = \frac{\Gamma(z+n)}{z(z+1)\dots(z+n-1)}, \quad \forall n \in \mathbb{N}.$$

In the right-hand side of the above relation the function $\Gamma(z+n)$ is analytic for $\mathrm{Re}(z) > -n$. It follows that the function $\Gamma(z)$ has an analytic extension for $\mathrm{Re}(z) > -n$ and $z \neq 0, -1, -2, \dots, n-1$ where it has simple poles. We can conclude that the function $\Gamma(z)$ has an analytic extension to the whole $\mathbb{C}\backslash\{0, -1, -2, -3\dots\}$ and it has simple poles at the negative integers and at 0.

Leonhard Euler first introduced and studied the gamma function $\Gamma(s)$ over the reals in the first half of the eighteenth century. Bernhard Riemann's 1859 article "On the number of primes less than a given magnitude" extended the Euler definition to a complex variable, proved its meromorphic continuation and functional equation, and established a relation between its zeros and the distribution of prime numbers.

If a finite number of analytic functions in a region Ω satisfy a certain functional equation in a part of Ω that has a limit point, then that functional equation holds everywhere in Ω.

For example, the Pythagorean identity

$$\sin^2 x + \cos^2 x = 1$$

holds for all real numbers x. Thus, it holds for all complex numbers.

Similar considerations can be applied to all real identities known from elementary analysis. Once it is verified that all the functions involved are analytic, these identities can be immediately extended to complex values.

The unique continuation theorem provides a method for making the domain of an analytic function as large as possible, e.g., starting from a given real differentiable function, or starting from a convergent power series.

An important application is that real analytic functions $f : \mathbb{R} \to \mathbb{R}$ have at most one analytic extension. In any textbook on complex analysis you can find that the complex exponential function (exp, $\mathbb{C}$) defined on $\mathbb{C}$ is the unique analytic extension of the real exponential function $(\exp, \mathbb{R})$, defined on $\mathbb{R}$. The same statement holds for the trigonometric functions $\sin(x), \cos(x)$.

A quick and direct method of analytic continuation which, whenever it is applicable, can be carried out almost without computation.

Theorem 4.3.3 (Schwarz reflection principle). *Let Ω be a domain in the upper half-plane that contains a line segment L on the real axis. Let $f(z)$ be a function analytic on Ω and continuous on L. Then the function extended by defining $f(z) = f(\overline{z})$ for z in the "reflection" $\widetilde{\Omega} = \{z \mid \overline{z} \in \Omega\}$ is an analytic continuation of $f(z)$ from Ω to the larger domain $\Omega \cup \widetilde{\Omega}$.*

This result is remarkable in that we requires only that f be continuous and real on $[a, b]$. It followed automatically that f is analytic on $[a, b]$ when continued across the real axis. To help see that g (and thus h) is analytic on $\widetilde{\Omega}$, consider the map in three steps:

$$z \mapsto \overline{z}; \quad \overline{z} \mapsto f(\overline{z}); \quad f(\overline{z}) \mapsto \overline{f(\overline{z})}\,.$$

The middle map is conformal; the first and last are anticonformal in the sense that they reverse angles. Since angles are reversed twice, the net result is to preserve angles. The whole map is thus conformal.

We note that not all analytic functions have proper analytic continuations. Regretfully, the intuitive notion of extending a function across a boundary point of Ω quickly leads to difficulties, as multivalued functions may appear. For instance, the function

$$z \mapsto \log z \quad \text{in } \mathbb{C} \setminus (-\infty, 0]$$

admits one-sided extensions at the nonzero boundary points, but these extensions do not coincide with the original function. The continuation of a holomorphic function is not always a function. See Example 4.3.12 below.

It may happen that holomorphic function $f\mathcal{O}(U)$ cannot be extended to any strictly larger connected open set: in this case a U is said to be the *natural boundary* for f.

For example, when $0 < |a| < 1$, the Fredholm series

$$\sum_{m=1}^{\infty} a^n z^{n^2}$$

converges absolutely on the closed unit disk $\{z : |z| \le 1\}$ and defines an analytic function $f(z)$ on the open disk $\{z : |z| < 1\}$. However, it can be shown that $f(z)$ has no analytic continuation outside the unit disk. Roughly speaking, the Fredholm functions are not extendable because of "gaps" in the powers of z. The sharpest "gap" theorem is the following.

Theorem 4.3.4 (Fabry gap theorem). *Let*

$$f(z) = \sum_{m=0}^{\infty} a_m z^{b_m}$$

where b_m is an increasing sequence of nonnegative integers such that

$$\lim_{m\to\infty} b_m/m = \infty\,.$$

Suppose that the power series has radius of convergence ρ. Then $f(z)$ has no analytic continuation outside the disk $\{z : |z| < \rho\}$.

Remark 4.3.5. The base $\mathbb{C}$ of analytic continuation may be replaced by any Riemann surface X. Note, however, that the "maximal" analytic continuation can enlarge under an inclusion $X \hookrightarrow X'$. An instructive example is given below.

Counterintuitive extensions. Let $f : \Delta \to \mathbb{C}$ be analytic with zero set $Z(f) \subset \Delta$ accumulating at every point of S^1. Then f admits no analytic continuation in the above sense.

Nevertheless, (Δ, f) may extend as an abstract holomorphic function on a Riemann surface. That is, there may exist a proper inclusion $j : \Delta \hookrightarrow X$ and a holomorphic map $F : X \to \mathbb{C}$ with $F \circ j = f$. Necessarily, $j(Z(f))$ must be discrete in X.

For explicit constructions, one may use Riemann mappings $j : \Delta \to U \subset \mathbb{C}$ such that the radial limit of $j(z)$ is infinite on a dense subset of S^1. (An explicit instance of such a domain U is given by the Fatou set of $J(\lambda e^z)$ for sufficiently small λ.) Choose a discrete set $B \subset U$ such that $j^{-1}(B)$ accumulates at every boundary point of Δ. If $F : \mathbb{C} \to \mathbb{C}$ is an entire function with zero set B, then $f(z) = F \circ j(z)$ has zeros accumulating everywhere on S^1, but extends from Δ to $\mathbb{C}$ via j.

These examples highlight the necessity of a rigorous definition of analytic continuation in order to exclude the pathologies associated with multivalued functions. The appropriate framework is that of Riemann surfaces, within which one obtains a maximal surface—defined precisely in this setting—that serves as the natural domain of holomorphy for the function. It is in this context that the theory of Riemann surfaces originally arose.

4.3.1 Weierstrass's Method

Riemann's first publication was his doctoral dissertation of 1851 (see [155, pp. 1–41]), devoted to the foundations of the theory of functions of a complex variable. In this work, he introduced the notion of a Riemann surface, conceived as a branched covering of the complex plane, and sketched a proof that any two disc-shaped regions of the plane are not only topologically but also analytically equivalent, i.e., there exists a one-to-one analytic correspondence between them. The majority of the results contained in the dissertation are expressed within this more general framework.

Riemann's conception of analytic function differs from that of Cauchy or Weierstrass: instead of defining them by the existence at every point of a local expansion in power series, he prefers to consider them as complex functions $U+iV$ where the real functions U, V are related by the Cauchy–Riemann equations (the usual Cauchy conditions at points other than ramification points); at the latter, it is necessary first to express the function using the complex "uniformizing parameter" t that appears in the Puiseux expansions. Moreover, Riemann is the first mathematician who exploits the fact that U and V are, then, harmonic functions.

A powerful, algebraic, alternative to Riemann's geometric complex function theory was developed by Weierstrass. Weierstrass acknowledged that analytic functions, in his sense, coincide with the functions $u + iv$ that Cauchy and Riemann had characterized as solutions of the Cauchy–Riemann equations (see also Proposition 1.1.7). Nevertheless, he rejected this approach, arguing that the definition of a class of functions may rest upon an arbitrary property whose generality cannot be established *a priori*. Furthermore, the Cauchy–Riemann equations presuppose functions $u(x, y)$ and $v(x, y)$ possessing partial derivatives, which, "in the present state of science," constitute a class that cannot be precisely defined. For this reason, he regarded the definition as "ill-suited to serve as the foundation for the theory of analytic functions."

The principle of analytic continuation was formulated by Weierstrass in 1842, [184, pp. 83–84] (published only in 1894); Riemann, [154, p. 89], likewise describes this method.

We emphasize the idea of "analytic continuation," which is a fundamental motivation for Riemann surface theory. Analytic continuations arise from analytic functions by attaching to them new points. The method of analytic continuation proposed by Weierstrass is based on the systematic application of Taylor series.

We now make a significant leap in abstraction: rather than considering individual functions, we consider sets of functions and families of such sets, which we refer to as function classes.

To begin with, we define function elements or germs (for definition of germs see Section 11.2) and their analytic continuations. The concept of analytic continuation was familiar to Riemann, who used it extensively, but it became of foundational importance in the work of Weierstrass.

According to Weierstrass, a function $w = f(z)$ is called *analytic* (or *holomorphic*) in a domain $U \subset \mathbb{C}$ if it can be expanded in a power series in a neighborhood of each point $a \in U$:

$$w = f(z) = \sum_{k=0}^{\infty} c_k (z - a)^k \, ; \tag{4.1}$$

in the case of several complex variables, when $U \subset \mathbb{C}^n$, $n > 1$, the series (4.1) is interpreted as a multiple power series. Holomorphic functions are precisely those functions that can be expressed locally as sums of power series.

Let U_1 be a subdomain of a domain U_2. Let f_1 [respectively f_2] be a holomorphic function defined in U_1, [respectively U_2]. The function f_2 is said to be an *analytic continuation* (or analytic extension) of f_1 into the (larger) domain U_2 if f_1 and f_2 coincide in the (smaller) domain U_1.

To see why this does not always happen, take the function $\sqrt{z}$. It is well-defined and holomorphic in $\{z \in \mathbb{C} \mid \operatorname{Re} z > 0\}$, but has no largest domain of definition in $\mathbb{C}$. Its natural domain of definition is a two-sheeted covering space of $\mathbb{C} \setminus \{0\}$.

A very important property of holomorphic functions is that they are uniquely determined by their behavior on open sets: if f and g are holomorphic on a domain U (that is, a connected open set), and if f equals g on a nonempty open subset of U, then f and g coincide on all of U. To see this, note that the largest open subset of U on which $f = g$ is also closed (relative to U), since the partial derivatives that determine the power series expansion are continuous. Because U is connected, this set must therefore be all of U.

The property of being analytic is local: it means that a function admits a power series expansion at every point, absolutely convergent on some disc of positive radius centered at that point. We know that a function is analytic if and only if it is holomorphic. The two terms will be used interchangeably from now on.

Definition 4.3.6. Suppose we are given an analytic function f on an open connected set U. Let $U \subset \mathbb{C}$ be an open and connected domain in $\mathbb{C}$. We call $\mathcal{F} = (U, f)$ an *analytic element* or simply an *element*. Let V be open and connected, and suppose that Δ is some connected component of $U \cap V$.

The two analytic elements $\mathcal{F} = (U, f)$ and $\mathcal{G} = (V, g)$ are *direct analytic continuations* of each other through Δ if

$$f(z) \equiv g(z) \text{ for all } z \in \Delta$$

(in this case, the values of f and g in other components of the intersection $U \cap V$ do not have to coincide; see Fig. 4.1).

So far, we have mentioned only a direct analytic extension from one domain to another, but we can extend this idea from a pair of domains to any finite number of domains. The idea is that you do not insist that they all intersect—only that there is an ordering such that each member intersects with the next.

If the function $f(z)$ is holomorphic in a domain $U \subset \mathbb{C}$, then it can be expanded in a Taylor series in the neighborhood of each point a of U, and the radius of convergence of this series is not less than the shortest distance r_a from the point a to the boundary ∂U. If the radius of convergence r_a is greater than this shortest distance, then the specified series defines a function that is holomorphic in the part of the disc of convergence external to U and coincides with $f(z)$ in the disc $U_a = \{z \in \mathbb{C} \mid |z - a| < r_a\}$. This series, therefore, gives

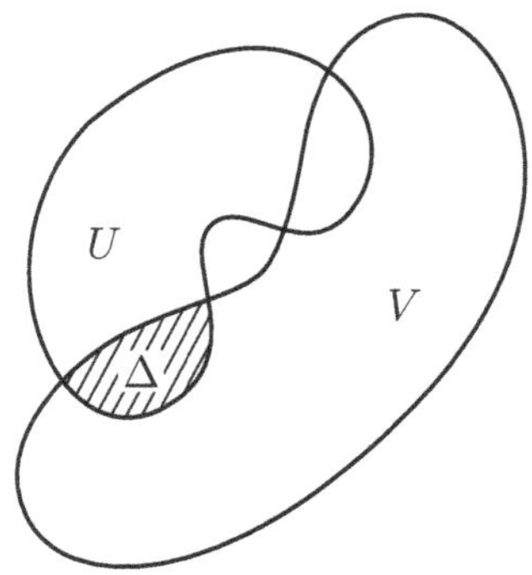

FIGURE 4.1: Analytic continuation along overlapping discs.

an extension of the function to the domain $U_a \cup U$. We also assume that if two functions are analytic in open overlapping disks and agree on the intersection, then each provides the unique continuation of the other to the union.

A pair (D, f) consisting of a domain $D \subset \widehat{\mathbb{C}}$ and a single-valued analytic, or holomorphic, function f defined on D is called an element of an analytic function or an *analytic element*, or simply just an *element*.

It is always possible when specifying an analytic function to use a Weierstrass or canonical element $(U(a, R), f_a)$, which consists for $a \neq \infty$ of a power series

$$f_a = f_a(z) = \sum_{k=0}^{\infty} c_k (z - a)^k$$

and a disc of convergence, $U(a, R) = \{z \in \widehat{\mathbb{C}} : |z - a| < R\}$, with center a and radius $R > 0$. In this case we also use the term a *regular unramified function element over a finite point* (we shall see the need for all the adjectives when we confront singular, ramified function elements). We call a the *center* of f_a and we write $Z(f_a) = a$. Note that f_a is the power series, not the function determined by the value to which it converges.

In the case $a = \infty$, a *Weierstrass element* $(U(\infty, R), f_\infty)$ consists of a series

$$f_\infty = f_\infty(z) = \sum_{k=0}^{\infty} c_k z^{-k}$$

and a domain of convergence of this series $U(\infty, R) = \{z \in \widehat{\mathbb{C}} : |z| > R\}$, $R \geq 0$.

The classical theory of analytic continuation starts with the notion of a canonical element:

Definition 4.3.7. A *canonical element* centered at $a \in \widehat{\mathbb{C}}$ is $\mathcal{F}_a = (U_a, \theta_a)$, where θ_a is the sum of a convergent power series centered at a and U_a is the disc of convergence of this series. We will call U_a the *disc of convergence* of $\mathcal{F}_a$ and denote by R_a the radius of this disc if $a \neq \infty$, or the radius of its complement if $a = \infty$.

Note that θ is the power series, not the function determined by the value to which it converges. However, we associate with θ the real number R_a, which is the radius of convergence of θ. We call a the *center of* θ and we write $Z(\theta) = a$.

If $a \in \mathbb{C}$, then the disc of convergence of the element $\mathcal{F}_a$ has the form $U_a = \{z : |z - a| < R_a\}$, where $R_a \leqslant \infty$, and the function

$$\theta_a(z) = \sum_{n=0}^{\infty} c_n (z - a)^n .$$

If $a = \infty$, then $U_a = \{z : |z| > R_a\}$, where $R_a \geqslant 0$, and

$$\theta_a(z) = \sum_{n=0}^{\infty} \frac{c_{-n}}{z^n} .$$

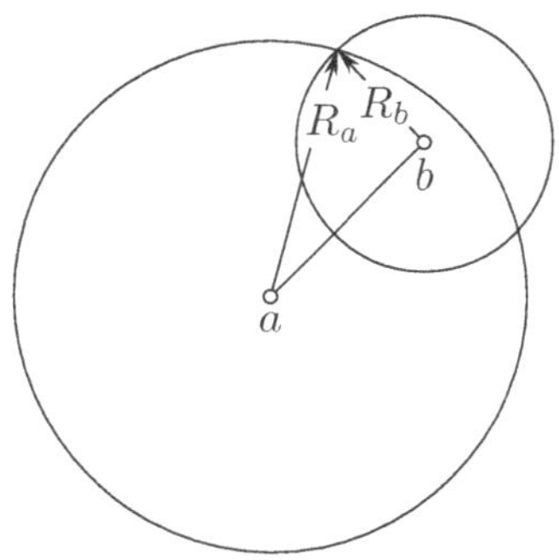

FIGURE 4.2: The two discs of convergence U_a and U_b.

If the canonical element $\mathcal{F}_b = (U_b, \theta_b)$ is a direct analytic continuation of the element $\mathcal{F}_a = (U_a, \theta_a)$ and, in addition, $b \in U_a$ (Fig. 4.2), then such a continuation is reduced simply to a rearrangement of the sum of the power series θ_a into a series in powers of $z - b$:

$$\theta_b(z) = \sum_{n=0}^{\infty} \frac{\theta_a^{(n)}(b)}{n!} (z - b)^n . \tag{4.2}$$

We also note that if two canonical elements $\mathcal{F}_a$ and $\mathcal{F}_b$ are direct analytic continuations of each other, then their discs of convergence U_a and U_b cannot compactly belong to each other (if, for example, $U_b \Subset U_a$, then θ_b is holomorphic in a disc with center b larger than U_b and, consequently, U_b cannot be a disc of convergence). Therefore, the discs U_a and U_b must have at least one common point, and from the triangle shown in Fig. 4.2 (only now it is not necessary that $b \in U_a$), we conclude that

$$|R_a - R_b| \leqslant |b - a| . \tag{4.3}$$

After Weierstrass this process of enlarging the domain in which we have the specification of an analytic function is called *analytic continuation*, and the process may be carried out repeatedly to maximize the domain in which the function is defined.

We can now formulate the Weierstrass definition of an analytic function: *The (complete) analytic function is the set $\mathcal{F}$ of all function elements obtainable from a given function element by analytic continuation.*

An analytic function is not a function in the usual sense. If $z = a$ belongs to the disc of convergence of a certain power series $f(z)$, this latter assigns a certain value $f(a)$ to $z = a$. But there may be other power series about $z = a$ equivalent to the given one which assign other values to $z = a$. All these values are called *values of the analytic function at* $z = a$ and a function of this type is considered as a multiple-valued function.

4.3.2 Continuation Along Curve

We now consider the notion of analytic continuation of a given holomorphic function f along a path. Analytic continuations arise from analytic functions by attaching to them new points.

To be more precise: Without loss of generality, we assume that for all paths considered here, the parameter varies on the interval $I = [0, 1]$.

Definition 4.3.8. An element $\mathcal{F}_0 = (U_0, \theta_0)$ is said to be *continued along a path* $\gamma : I \to \mathbb{C}$, with the origin at the center $a = \gamma(0)$ of this element, if there exists a family of elements

$$\mathcal{F}_t = (U_t, \theta_t)\,, \quad t \in I\,, \tag{4.4}$$

with centers $a_t = \gamma(t)$ and non-zero radii of convergence, satisfying the following condition: If $u\,(t_0) \subset I$ denotes a connected neighborhood of a point $t_0 \in I$ such that $\gamma(t) \in U_{t_0}$ for all $t \in u\,(t_0)$ (this neighborhood exists due to the continuity of the path), then for any $t \in u\,(t_0)$ the element $\mathcal{F}_t$ is a direct analytic continuation of $\mathcal{F}_{t_0}$ (Fig. 4.3).

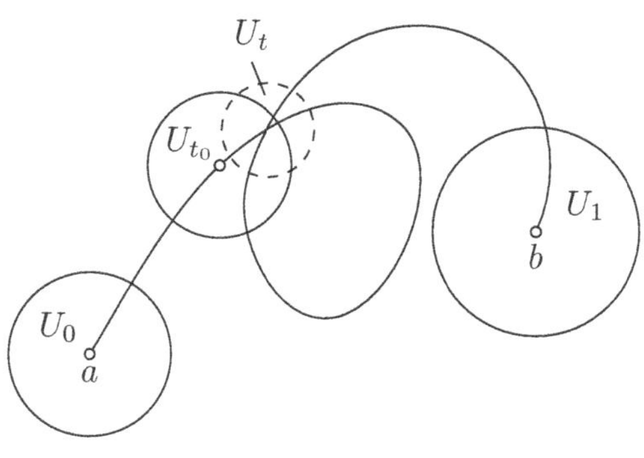

FIGURE 4.3: Analytic continuation. Many steps.

If the element $\mathcal{F}_0$ is continued along γ, then we say that the element $\mathcal{F}_1$ of the family (4.4) (with center at the end $b = \gamma(1)$ of the path) is obtained from $\mathcal{F}_0$ by analytic continuation along the path γ.

The reader may wonder why in Definitions 4.3.8 and 4.3.7 we require the regions U_t to be discs. There is no loss of generality in using discs, since any region in Σ is a union of open discs, and indeed, if we construct the continuations by means of power series, then we will be forced to use discs.

The definitions of direct analytic continuation and analytic continuation for canonical elements are somewhat simplified, because their regions (discs) intersect in connected sets and therefore there is no need to stipulate through which components of the intersections Δ the continuation is performed.

The analytic continuation along a given curve γ is unique if it is possible at all. First of all, we prove the uniqueness of the extension along the path. To do this, we agree to consider two canonical elements $\mathcal{F} = (U, f)$ and $\mathcal{G} = (V, g)$ equal ($\mathcal{F} = \mathcal{G}$) if $U = V$ and in this disk $f \equiv g$.

Theorem 4.3.9. *If the canonical element $\mathcal{F}_0$ is extended along the path γ, then as a result of its analytic extension along this path we obtain a completely definite element that does not depend on the choice of the family implementing the extension.*

Proof. Let the elements $\mathcal{F}_1$ and $\mathcal{G}_1$ be obtained by extending $\mathcal{F}_0$ along γ: the first one by means of the family of elements $\mathcal{F}_t$, and the second by means of the family $\mathcal{G}_t$ ($\mathcal{F}_0 = \mathcal{G}_0, t \in I$). Consider the set

$$\mathcal{E} = \{t \in I : \mathcal{F}_t = \mathcal{G}_t\} .$$

This set is not empty, since it contains the point $t = 0$.

This set is also open (in the topology of I).

Indeed, let $t_0 \in \mathcal{E}$, i.e., $\mathcal{F}_{t_0} = \mathcal{G}_{t_0}$; by the continuity of the path γ there exists a neighborhood $u\,(t_0) \subset I$ such that for all $t \in u\,(t_0)$ the point $\gamma(t)$ belongs to the common disc of convergence of the elements $\mathcal{F}_{t_0}$ and $\mathcal{G}_{t_0}$.

By Definition 4.3.8, for all $t \in u\,(t_0)$, both $\mathcal{F}_t$ and $\mathcal{G}_t$ are obtained by direct analytic continuation of equal elements $\mathcal{F}_{t_0} = \mathcal{G}_{t_0}$ and, therefore, coincide (see formula (4.2)).Therefore $u\,(t_0) \subset \mathcal{E}$.

But $\mathcal{E}$ is also closed. Indeed, let $t_0 \in I$ be a limit point of $\mathcal{E}$ and $u\,(t_0)$ be a neighborhood such that for all $t \in u\,(t_0)$ the points $\gamma(t)$ belong to the smaller of the discs of convergence of the elements $\mathcal{F}_{t_0}$ and $\mathcal{G}_{t_0}$ (we denote it by W). In $u\,(t_0)$ there exists a point $t_1 \in \mathcal{E}$ such that $\mathcal{F}_{t_1} = \mathcal{G}_{t_1}$. Since $\gamma\,(t_1) \in W$, then $\mathcal{F}_{t_0}$ and $\mathcal{G}_{t_0}$ are direct analytic continuations of equal elements of $\mathcal{F}_{t_1}$ and $\mathcal{G}_{t_1}$, but then by the uniqueness theorem $f_{t_0} \equiv g_{t_0}$ everywhere in W and, consequently, $t_0 = \mathcal{G}t_0$, i.e., $t_0 \in \mathcal{E}$. So, the non-empty subset $\mathcal{E} \subset I$ is both open and closed. It follows that $\mathcal{E} \equiv I$ and, in particular, that $\mathcal{F}_1 = \mathcal{G}_1$. □

Now we want to prove that continuation along a path can always be carried out in a finite number of steps, i.e., as an analytic continuation in the sense of the previous discussion.

Lemma 4.3.10. *The radius $R(t)$ of an element $\mathcal{F}$ from the family (4.4) that carries out analytic continuation along the path γ is a continuous function of t on the segment I if the discs of convergence of all $\mathcal{F}_t$ are bounded.*

Proof. For any $t_0 \in I$ we find a neighborhood $u(t_0)$ such that for $t \in u(t_0)$ the element $\mathcal{F}_t$ is a direct continuation of $\mathcal{F}_{t_0}$ (see Definition 4.3.8).

From inequality (4.3) we then obtain that

$$|R(t) - R(t_0)| \leqslant |\gamma(t) - \gamma(t_0)| \; .$$

This inequality proves the continuity of R at the point t_0. □

If the continuation of the power series f_0 along a path γ is possible, it can always be accomplished by analytic continuation along a finite chain of discs.

The radius of convergence $r_t = r(\gamma(t))$ of the series f_t is either identically infinite or is a continuous function of t and possesses a positive minimum ρ. Hence in all cases we may take a positive number ρ and a partition of the interval $0 \leqq t \leqq 1 : 0 = t_0 < t_1 < \cdots < t_n = 1$, such that $|\gamma(t_{k+1}) - \gamma(t_k)| < \rho$ for $k = 0, \ldots, n-1$. Then the discs $D_k : |z - \gamma(t_k)| < r_{t_k}$ are overlapping and the corresponding functions $f_{t_0}, \ldots, f_{t_n}$ form an analytic continuation.

Finally, we prove.

Theorem 4.3.11. *If an element $\mathcal{G}$ is obtained from $\mathcal{F}$ by analytic continuation along γ, then $\mathcal{G}$ is a direct analytic continuation of $\mathcal{F}$.*

Proof. Let $\mathcal{F}_t$, $t \in I$, be a family of elements that carry out continuation along γ. By the lemma, the radius $R(t)$ of $\mathcal{F}_t$ is continuous on I, hence there exists $\varepsilon > 0$ such that $R(t) \geq \varepsilon$ for all $t \in I$. Since the function γ is uniformly continuous on I, we can choose a finite number of points $t_\nu : t_0 = 0 < t_1 < \cdots < t_n = 1$, so that the points $z_\nu = \gamma(t_\nu)$ for $\nu = 1, \ldots, n$ satisfy the inequality $|z_\nu - z_{\nu-1}| < \varepsilon$. By Definition 4.3.8, it follows that the element $\mathcal{F}^{(\nu)} = \mathcal{F}_{t_\nu}$ is a direct analytic continuation of the element $\mathcal{F}^{(\nu-1)} = \mathcal{F}_{t_{\nu-1}}$ $(\nu = 1, \ldots, n)$. But $\mathcal{F}^{(0)} = \mathcal{F}^{\nu-1}$, and $\mathcal{F}^{(n)} = \mathcal{G}$ therefore $\mathcal{G}$ is an analytic continuation of the element $\mathcal{F}$ □

The problem of analytic continuation is as follows: given a function f analytic in a domain D, can we extend f so that it remains analytic in a domain larger than D? The major obstacle is the possibility of returning to a point where f has already been defined, but now with a new (and different) value. A typical example of this is continuing $\sqrt{z}$ around the origin.

Example 4.3.12. Consider in the disk $U = \{|z-1| < 1\}$ the function

$$f(z) = z^{1/2} \quad \left(z = re^{i\varphi}, \; -\tfrac{\pi}{2} < \varphi < \tfrac{\pi}{2}\right). \tag{4.5}$$

At each point $z \in U$ we have

$$f'(z) = \tfrac{1}{2f(z)} \, ,$$

therefore f is holomorphic in U, and $f'(z) \to \infty$ as $z \to 0$. From this it follows that the pair $\mathcal{F} = (U, f)$ forms a canonical element.

To obtain the analytic continuation of $\mathcal{F}$ along the path

$$\gamma_0 : z = e^{it}, \quad 0 \le t \le \pi$$

(the upper semicircle $\{\, |z| = 1 \,\}$), it is sufficient to consider the disk

$$U' = \{\, |z - i| < 1 \,\}$$

and to construct the analytic continuation of $\mathcal{F}$ via the chain of disks

$$U,\ U',\ V, \quad \text{where } V = \{\, |z + 1| < 1 \,\}.$$

Such a continuation can be carried out directly by extending the range of φ in formula (4.5):

1. for U': $\varphi \in [0, \pi]$,
2. for V: $\varphi \in [\frac{\pi}{2}, \frac{3\pi}{2}]$.

In the disk V we obtain the holomorphic function

$$g_0(z) = r^{1/2} e^{i\varphi/2}, \quad \tfrac{\pi}{2} < \varphi < \tfrac{3\pi}{2},$$

and the pair $\mathcal{G}_0 = (V, g_0)$ is the element that represents the required continuation of $\mathcal{F}$ along the path γ_0.

In particular, on the negative diameter of the disk V, we have $\varphi = \pi$. Hence,

$$g_0(z) = \sqrt{-x}\, e^{i\pi/2} = i\sqrt{-x}, \quad -x = |z| > 0. \tag{4.6}$$

Similarly, by means of the chain of disks

$$U,\ U'',\ V, \quad \text{where } U'' = \{\, |z + i| < 1 \,\},$$

one obtains the analytic continuation of $\mathcal{F}$ along the path

$$\gamma_1 : z = e^{it}, \quad 0 \ge t \ge -\pi.$$

This continuation consists of successively replacing in formula (4.5) the range of φ by the intervals $[-\pi, 0]$, and $[-\frac{3\pi}{2}, -\frac{\pi}{2}]$.

As a result, we obtain in the disk V the holomorphic function

$$g_1(z) = r^{1/2} e^{i\varphi/2}, \quad -\tfrac{3\pi}{2} < \varphi < -\tfrac{\pi}{2},$$

and the pair $\mathcal{G}_1 = (V, g_1)$ is the element that represents the continuation of $\mathcal{F}$ along the path γ_1.

The element $\mathcal{G}_1$ is not equal to the element $\mathcal{G}_0$: although their disks of convergence coincide, the functions differ. For example, on the negative diameter of V we now have $\varphi = -\pi$, and therefore instead of (4.6) we obtain

$$g_1(z) = \sqrt{-x}\, e^{-i\pi/2} = -i\sqrt{-x}, \quad -x = |z| > 0.$$

Thus

$$g_1(z) = -g_0(z), \quad \text{for all } z \in V.$$

Hence the study of analytic functions in the plane unavoidably leads to the consideration of multi-valued functions. It is therefore natural to try to construct an object on which a multi-valued function becomes single-valued in the usual sense.

Some of the material in this chapter may seem overly "abstract" due to the cumbersome definitions we will have to work through. The reader should therefore always try to capture the simple geometric ideas underlying these notions. To begin with, we define function elements or germs and their analytic continuations. There is a natural equivalence relation on these function elements leading to the notion of a complete analytic function.

Another topic of central importance pertains to the question whether analytic continuation along two curves γ_0 and γ_1 necessarily leads to the same terminal branch. Roughly speaking, this does happen if γ_0 can be continuously deformed into γ_1 in such a way that the continuation is possible on all intermediate curves. When formulated with precision, the result leads to the homotopy form of the monodromy theorem. It was Cauchy who first formulated the monodromy theorem, and the term monodromy is due to him.

In conclusion, we prove a theorem on the invariance of analytic continuation along a path with respect to homotopic deformations of this path. The following result shows that homotopic paths give rise to the same holomorphic continuation.

A complex analytic function (V, f) is said to be *arbitrarily continuable* in a region U if each element of the function has a continuation along every path emanating from the center of the element and lying in U. It is clear that every point of this region is the center of the same number (finite or infinite) of elements of the function.

It is sometimes impossible to extend a function analytically beyond the boundary of a region. We then call the boundary a *natural boundary.*

A fundamental connection between analytic continuation and plane topology is signaled by a result known as the "Monodromy Theorem."

Theorem 4.3.13 (Weierstrass monodromy theorem). *Let paths γ_0 and γ_1 be homotopic as paths with common endpoints, and let the element (V, f) be continued along any path $\gamma_s (s \in I)$ realizing a homotopy of $\gamma_0 \sim \gamma_1$. Then the results of continuation of (V, f) along paths γ_0 and γ_1 coincide.*

Proof. Let $\Gamma : I \times I \to \mathbb{C}$ be a function realizing a homotopy of $\gamma_0 \sim \gamma_1$, so that $\gamma_s(t) = \Gamma(s, t)$.

Denote by $\mathcal{G}^s$ $(s \in I)$ the result of the analytic continuation of the element (V, f) along the path γ_s and consider the set

$$\mathcal{E} = \left\{ s \in I : \mathcal{G}^s = \mathcal{G}^0 \right\} .$$

It is nonempty, since it contains the point $s = 0$.

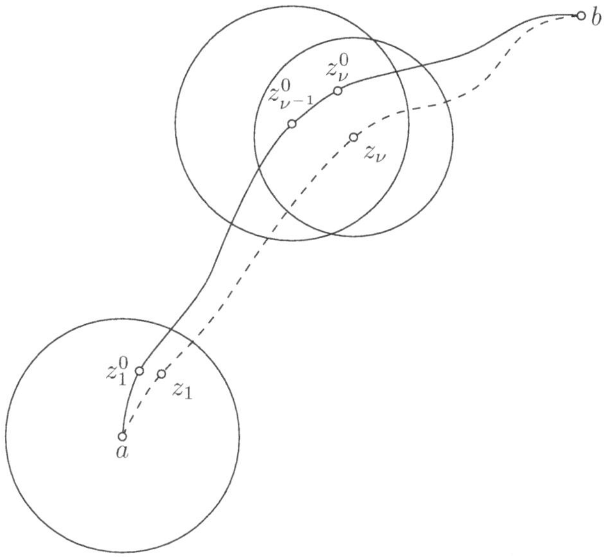

FIGURE 4.4: The homotopic paths γ_0 and γ_1 with common endpoints a and b.

Let $s_0 \in \mathcal{E}$; by the lemma, there exists $\varepsilon > 0$ such that the radii $R(t)$ of the elements $\mathcal{G}_t^{s_0}$ that carry out the continuation along γ_{s_0} are not less than ε for all $t \in I$. Since γ is uniformly continuous, there exists a neighborhood $u(s_0) \subset I$ such that for all $s \in u(s_0)$ and all $t \in I$,

$$|\gamma(s,t) - \gamma(s_0,t)| < \frac{\varepsilon}{2}. \tag{4.7}$$

By Theorem 4.3.11, we choose $t_\nu \in I$ so that the points $z_\nu^0 = \gamma(s_0, t_\nu)$ for all $v = 1, \ldots, n$ satisfy the inequality

$$\left|z_\nu^n - z_{v-1}^n\right| < \frac{\varepsilon}{2} \tag{4.8}$$

and the elements $\mathcal{G}_l^n, 1$, and $\mathcal{G}_0$, (we denote by $\mathcal{G}_\nu^0 = \mathcal{G}_{t_\nu}^{s_0}$) are direct analytic continuations of each other.

Let us also denote $z_\nu = \gamma(s, t_\nu)$, $\mathcal{G}_\nu = \mathcal{G}_{t_\nu}^s$ and note that for $s \in u(s_0)$, by virtue of (4.7) and (4.8), $\left|z_\nu - z_{\nu-1}^0\right| < \varepsilon$ for all $\nu = 1, \ldots, n$ (Fig. 4.4). The elements $\mathcal{G}_1^n$ and $\mathcal{G}_1$ are immediate analytic continuations of the element $\mathcal{F}$ and, therefore, each other (we use the fact that we have $\left|z_1^0 - a\right| < \frac{\varepsilon}{2}$ and $\left|z_1^0 - z_1\right| < \frac{\varepsilon}{2}$).

In exactly the same way, we will prove that the elements $\mathcal{G}_\nu^0$ and $\mathcal{G}_\nu$ are analytic continuations of each other ($\nu = 2, \ldots, n$). But the elements $\mathcal{G}_n^0$ and $\mathcal{G}_n$ also have a common center ($z_n^0 = z_n = b$) and therefore coincide. Since $s_0 \in \mathcal{E}$, then $\mathcal{G}_n^0 = \mathcal{G}_{l_n}^{s_0} = \mathcal{G}_0$, and therefore $\mathcal{G}_n = \mathcal{G} = \mathcal{G}^0$ i.e., $s \in \mathcal{E}$. We have proved that $u(s_0) \subset \mathcal{E}$ together with s_0, i.e., that $\mathcal{E}$ is an open set.

But at the same time $\mathcal{E}$ is also a closed set. Indeed, let s_0 be a limit point of $\mathcal{E}$; we construct a neighborhood $u(s_0)$ the same as above. In it there is a point $s \in \mathcal{E}$, so that the extension of along γ_s leads to the element $\mathcal{G}^0$.

By the same reasoning as above, the extensions along the paths γ_s and γ_{s_0} lead to equal elements, therefore $\mathcal{G}^{s_0} = \mathcal{G}^0$ and $s_0 \in \mathcal{E}$.

Thus, $\mathcal{E} = I$ and, in particular, $\mathcal{G}^1 = \mathcal{G}^0$. □

The following corollary is the most important consequence of the monodromy theorem.

Corollary 4.3.14. *Let (V, f) be a function element which admits unrestricted continuation in the simply connected region U. Then there is an analytic function $F : U \to \mathbb{C}$ such that $F(z) = f(z)$ for all z in V.*

Proof. Fix a in V and let z be any point in U. If γ is a path in U from a to z and $\{(f_t, D_t) : 0 \leq t \leq 1\}$ is an analytic continuation of (f, V) along γ then let $F(z, \gamma) = f_1(z)$. Since U is simply connected $F(z, \gamma) = F(z, \sigma)$ for any two paths γ and σ in U from a to z. Thus, $F(z) = F(z, \gamma)$ gives a well-defined function $F : G \to \mathbb{C}$. To show that F is analytic let $z \in G$ and let γ and $\{(f_t, D_t)\}$ be as above. A simple argument gives that $F(w) = f_1(w)$ for w in a neighborhood of z (Verify!); so F must be analytic. □

We end this section with an important special case of the monodromy theorem.

Theorem 4.3.15. *Let U be a simply connected open set and V be an open subset of U. Suppose that the function element (V, f) can be continued along every path from $a \in V$ to every point in U. Then the analytic continuation of (V, f) along a path from a to a point z is independent of the path, and defines an element $\mathcal{G}$.*

Proof. This is a special case of the monodromy theorem, because two paths from a to w are homotopic. □

This theorem implies that for purposes of analytic continuation we can always replace a path γ by a polygon approximating γ and which has its vertices on γ. We do not need to consider all elements on the broken line, for if the vertices are sufficiently dense it will be sufficient to use only the elements whose centers are vertices. Otherwise, we may have to interpolate intermediary points.

Now something interesting can happen in such situations. We can have a chain of analytic continuations such that it happens that the final domain is the same as the first domain, but the function we get will be completely different. What is in fact happening in this case is that these functions are two so-called *branches of an analytic function.* (So this process of analytic continuations helps you find all possible branches of an analytic function.) Note also that this process is only nontrivial and useful if there are singularities. If the function were already entire, there would be nothing to extend.

Generally speaking the phrase "analytic continuation" refers to the process of extending a holomorphic function defined in an open set to some larger domain.

By evident reasons the domains of the original and extended function are typically taken to be connected, but the domain of the extended function is not always a subset of the plane (and to be precise, the "extension" is not always literally an extension; the original domain is not always literally a subset of the domain of the extended function. Why the "extension" nonetheless deserves to be called an "extension" will become clear below; it is analogous to the fact that (with the standard definitions) the real numbers are not literally a subset of the complex numbers, but everybody always talks as though they were.)

In fact the largest extension often cannot be realized as a holomorphic function in an open subset of the plane.

If we continue f along two different paths from p to q and compare the "results," i.e., the last function $f_N \in \mathcal{O}(\Delta_N)$ (in the neighborhood of q) in each case, these need not agree. This problem only occurs, however, for non-simply-connected regions. In the case of simply connected domains, such problems do not arise—this is what the monodromy theorem tells us.

Proposition 4.3.16 (Riemann monodromy principle). *Given a region $\Omega \subseteq \mathbb{C}$ which is simply connected, i.e., $\pi_1(\Omega) = \{0\}$. Let $\Delta \subset \Omega$ be a small disk, and assume that $f \in \mathcal{O}(\Delta)$ can be analytically continued along any path $\gamma \subset \Omega$ starting at $p \in \Delta$. Then there exists $\widetilde{f} \in \mathcal{O}(\Omega)$ extending f.*

A first immediate consequence of these results concerns global uniform branches of an algebraic function. Their existence is governed by the following result, where we keep the above notation:

Proposition 4.3.17 (heredity principle). *For $F(x,y) \in \mathcal{O}(\mathbb{C}^2)$ and $f \in \mathcal{O}(\Delta)$ satisfying*

$$F(x, f(x)) = 0, \tag{4.9}$$

the analytic continuation of f along any path γ will also satisfy (4.9).

Proof. Since F and each f_i in the analytic continuation are holomorphic, so is each $F(x, f_i(x))$ (on Δ_i). But $F(x, f(x)) \equiv 0$ on $\Delta = \Delta_0$ by assumption, and since $f = f_0 \equiv f_1$ on $\Delta_0 \cap \Delta_1$, we have $F(x, f_1(x)) \equiv 0$ on $\Delta_0 \cap \Delta_1$ and therefore (by basic complex analysis) on all of Δ_1. Simply iterate this argument for $i = 1, \ldots, N$. □

To define an analytic function it is even sufficient that the convergent series (4.1) be given in a neighborhood of a single point z_0, for its values at another point z_1 and the corresponding series can be determined by the process of analytic continuation along various paths in the complex plane $\mathbb{C}$ (or in $\mathbb{C}^n$, $n > 1$) joining z_0 and z_1.

The monodromy theorem of course implies that on a simply connected domain $U \subset \mathbb{C}$ any function element (V, f), where V is a domain in U, that can be analytically continued everywhere on U defines a global analytic function on U.

Obviously, this notion does not depend on the choice of the initial element (V, f) and so you can choose the canonical element (U, f_a) as the initial function element.

Indeed, let (U_b, g_b) be any element belonging to the analytic function $\mathcal{F}$ defined by the initial element (U_a, f_a). Let (U_b, f_b) obtained from (U_a, f_a) by continuation along the path γ. But the analytic function $\mathcal{G}$ defined by the initial element (U_b, f_b) is also obtained from the inverse curve γ^- by continuation of (U_a, f_a) along the path γ. And also any other element (U_c, f_c), which is obtained from (U_a, f_a) by continuation along the path λ and from λ by continuation along the path $\gamma^- \cdot \lambda$ (for the definition of product of path see Section 1.9).

If the canonical function elements (U_a, f_a) and (U_b, f_b) are direct analytic continuations of each other, so are (U_a, f_a') and (U_b, f_b'), where the prime denotes differentiation. Hence the derivatives of the elements of an analytic function $\mathcal{F}$ are equivalent and constitute again an analytic function. This latter function is called the *derivative* $\mathcal{F}'$.

Definition 4.3.18. Two complete analytic functions defined by *canonical elements are considered equal* if they have at least one common element. (By the uniqueness theorem of continuation along a path, then all their corresponding elements are equal to each other.)

Two *complete analytic functions* defined by arbitrary elements are considered equal if any of their elements are equivalent to each other (then all their corresponding elements are equivalent).

Theorem 4.3.19. *The union of the discs of convergence of elements belonging to a complete analytic function forms a domain.*

Proof. Let U be such a union; it is open as a union of open sets (if $c \in U$, then c also belongs to the disc of convergence U_c of some element, and U_c is included in U). But U is also connected, because for any points $a, b \in U$ one can find elements belonging to the function that have these points as centers, and since these elements are obtained from each other by continuation along some path γ connecting a and b, then all points γ belong to U. Thus, U is an open connected set, i.e., domain. □

If the condition of the monodromy theorem is not satisfied, then the complete analytic function can assign several values to points $z \in U$ (they will be the values of different branches of the analytic function at the point z).

At a very early time the following question was already being posed: How many values can be assigned to a fixed point? A remarkable theorem due to Poincaré and Volterra states:

Theorem 4.3.20 (Poincaré–Voltera). *A complete analytic function can have at most a countable set of distinct elements centered at a fixed point.*

Proof. The extension from a to b along a curve can be replaced by an extension along a chain of discs. Without loss of generality, an arbitrary chain of discs can be replaced by a chain of discs with centers at rational points (i.e., both coordinates are rational) and of rational radius. The set of all such discs is countable. □

4.3.3 Examples: The Gamma and Zeta Functions

The method of analytic continuation described in the preceding section can be applied to all analytic functions, but it does not necessarily provide the easiest way of obtaining the continuation of a given function beyond its original domain of definition. By utilizing the special properties of the given function it is often possible to achieve the same purpose at a considerable saving of computational labor. We shall consider here a number of examples of this type. Our first example concerns the Gamma function, which occurs in many branches of pure and applied analysis.

Our first example concerns the Gamma function, which occurs in many branches of pure and applied analysis. For values of z such that $\mathrm{Re}(z) > 0$, this function is defined by

$$\Gamma(z) = \int_0^\infty t^{z-1} e^{-t} dt \, ,$$

where the integration is carried out along the positive axis in the t-plane.

Our next example concerns the Zeta function

$$\zeta(z) = \sum_{n=1}^{\infty} \frac{1}{n^s} \, , \tag{4.10}$$

which plays a prominent part in the analytic theory of numbers. The series (4.10) converges uniformly in any closed subdomain of the half-plane $\mathrm{Re}(z) > 1$ and thus represents there an analytic function.

A meromorphic function may have infinitely many zeros and poles. This is the case for the *gamma function*, $\Gamma(z)$, which is meromorphic in the whole complex plane, and has a simple pole at every non-positive integer.

The *Riemann zeta function* $\zeta(s)$ is also meromorphic in the whole complex plane, with a single pole of order 1 at $z = 1$. Its zeros in the left halfplane are all the negative even integers, and the Riemann hypothesis is the conjecture that all other zeros are along $\mathrm{Re}(z) = 1/2$. See [97, 131, 177], and [140] and the literature therein.

The property of being analytic is local: it means that a function has a power series expansion at every point (absolutely convergent on a disc of positive radius centered at the point).

Theorem 4.3.21. *Let γ be a path in an open set U and let g be a continuous function on γ (i.e., on the image $\gamma([a, b])$ if γ is defined on $[a, b]$). If z is not* *on γ, define*

$$f(z) = \int_\gamma \frac{g(\zeta)}{\zeta - z} d\zeta\,.$$

Then f is analytic on the complement of γ in U, and its derivatives are given by

$$f^{(n)}(z) = n! \int_\gamma \frac{g(\zeta)}{(\zeta - z)^{n+1}} d\zeta\,.$$

Proof. Let $z_0 \in U$ and z_0 not on γ. Since the image of γ is compact, there is a minimum distance between z_0 and points on γ. Select $0 < R < \text{dist}\,(z_0, \gamma)$, and take R also small enough that the closed disc $\bar{D}\,(z_0, R)$ is contained in U. Now we can apply the Cauchy formula. We may repeat the arguments of the proof. We select $0 < s < R$, and we simply replace f by g inside the integral sign.

From the formula for f, it is natural to think that one can differentiate with respect to z under the integral sign. Then one gets the integral formula also for the derivatives. □

Lemma 4.3.22 (The differentiation lemma). *Let I be an interval of real numbers, possibly infinite. Let U be an open set of complex numbers. Let $f = f(t, z)$ be a continuous function on $I \times U$. Assume:*

(i) For each compact subset K of U the integral

$$\int_I f(t, z) dt$$

is uniformly convergent for $z \in K$.

(ii) For each t the function $z \mapsto f(t, z)$ is analytic. Let

$$F(z) = \int_I f(t, z) dt\,.$$

Then F is analytic on $U, D_2 f(t, z)$ satisfies the same hypotheses as f, and

$$F'(z) = \int_I D_2 f(t, z) dt\,.$$

Proof. Let $\{I_n\}$ be a sequence of finite closed intervals, increasing to I. Let D be a disc in the z-plane whose closure is contained in U. Let γ be the circle bounding D. Then for each z in D we have

$$f(t, z) = \frac{1}{2\pi i} \int_v \frac{f(t, \zeta)}{\zeta - z} d\zeta$$

so

$$F(z) = \frac{1}{2\pi i} \int_I \int_V \frac{f(t, \zeta)}{\zeta - z} d\zeta dt\,.$$

If γ has radius R, center z_0, consider only z such that $|z - z_0| \leqq R/2$. Then

$$\left|\frac{1}{\zeta - z}\right| \leqq 2/R.$$

For each n we can define

$$F_n(z) = \frac{1}{2\pi i}\int_{I_n}\int_{\gamma}\frac{f(t,\zeta)}{\zeta - z}d\zeta dt.$$

In view of the restriction on z above, we may interchange the integrals and get

$$F_n(z) = \frac{1}{2\pi i}\int_{\gamma}\frac{1}{\zeta - z}\left[\int_{I_n} f(t,\zeta)dt\right]d\zeta.$$

Then F_n is analytic by The Differentiation Lemma. By hypothesis, the integrals over I_n converge uniformly to the integral over I. Hence F is analytic, being the uniform limit of the functions F_n for $|z - z_0| \leqq R/2$. On the other hand, $F_n'(z)$ is obtained by differentiating under the integral sign in the usual way, and converges uniformly to $F'(z)$. However

$$F_n'(z) = \frac{1}{2\pi i}\int_{I_n} D_2 f(t,z)dt.$$

This proves the theorem. □

As an application of the monodromy theorem we have that a holomorphic function is analytic. The Euler integral of the second kind is the gamma function

$$\Gamma(z) = \int_0^{\infty} e^{-t}t^{z-1}dt.$$

Then $\Gamma(z)$ fits the hypotheses of the differentiation lemma, provided

$$0 < a \leqq \operatorname{Re} z \leqq b \quad \text{if} \quad 0 < a < b \text{are real numbers.}$$

Therefore $\Gamma(z)$ is an analytic function in the right half plane. (Give the details of the proof that the above integral satisfies the hypotheses of the differentiation lemma, and write down explicitly what the derivative is.)

Another technique can also be applied.

Gamma function has a number of remarkable properties and we recommend the book [32], which is entirely dedicated to this function. In almost all every book on special functions the gamma function receives a lot of attention see, for example [176]. The classic Whittaker and Watson [191] is still an interesting source of information.

4.4 Riemann Surfaces of Functions

Although Weierstrass's theory of analytic functions, presented in the previous chapter, allows one to work with multiple-valued analytic functions successfully, it has one significant drawback: analytic functions are not functions in the usual sense of the word.

The concept of the Riemann surface of a holomorphic function is closely related to the concept of analytic continuation of a holomorphic function along a path. Since the study of analytic functions in the plane unavoidably leads to the consideration of multiple-valued functions, it is therefore natural to try to construct an object on which a multi-valued function becomes single-valued in the usual sense.

As an illustrative example, consider a complex polynomial equation of degree d in one variable. While the individual solutions may not be expressible in explicit form, it is nevertheless known that there are exactly d solutions, counted with multiplicities. The aim is to extend this perspective to polynomial equation in several variables and to investigate the types of geometric structures that can emerge in such contexts.

Riemann was clear on every occasion that he regarded an (irreducible) algebraic curve with equation $f(w, z) = 0$ of degree n in w and m in z, as an equation in complex variables that should be thought of providing an n-fold covering of the complex z-plane (or, better, z-sphere) (see Section 1.3); there are, in general, n values of w for each value of z. Points at which there are fewer than n values of z are called *branch points*. He then explained how such a covering surface could be regarded (in many different ways) as an algebraic curve in $\mathbb{C} \times \mathbb{C}$. From his standpoint a curve simply had complex points on it.

Riemann remarked that rational functions of z and w (which, by definition, are the quotient of one polynomial in z and w by another), are single-valued functions on S that are branched like z, and he proved the converse later in the paper [154] (in Werke, 88–44).

Using a construction that may be found in any elementary complex variables text, Riemann resolved the problem by constructing sheets above the complex plane. As Riemann himself repeatedly stated, he was searching for a general method capable of embracing larger classes of functions and treating them in a unified manner. It is precisely this that led him to his approach.

The method discovered by Weierstrass for studying multiple-valued functions is quite different from Riemann's. It is probably the first step in the development of the modern notions of abstract manifold and sheaf.

We shall make some of these elementary ideas precise, and see first what Weierstrass and Riemann were led to.

4.4.1 Riemann Surface of an Analytic Function

Let X be a 2-dimensional manifold (i.e., X is a Hausdorff space and any point in X has a neighborhood homeomorphic to an open set in $\mathbb{R}^2$). Consider pairs (U, φ) where U is open in X and $\varphi : U \to \varphi(U) \subset \mathbb{C}$ is a homeomorphism onto an open set in $\mathbb{C}$.

Two such pairs $(U_1, \varphi_1), (U_2, \varphi_2)$ are said to be (holomorphically) compatible if the map $\varphi_2 \circ \varphi_1^{-1} : \varphi_1(U_1 \cap U_2) \to \varphi_2(U_1, \cap U_2)$ is holomorphic; its inverse is also holomorphic by a standard result in complex analysis. A complex structure on X is a family $\mathcal{S}$ of pairs $\{(U, \varphi)\}$, which are pairwise compatible and such that $\bigcup U = X$; there is then a unique maximal family of pairs with these two properties and containing $\mathcal{S}$; we shall usually assume that the complex structure is maximal. The elements (U, φ) of this (maximal) complex structure are called charts or *coordinate neighborhoods.* In a coordinate neighborhood, we usually identify U with $\varphi(U)$ and write z for φ as one does with the usual complex variable in $\mathbb{C}$.

A Riemann surface is a connected 2-dimensional manifold X with a complex structure $\mathcal{S}$. We shall also assume that X has a countable base of open sets, although a theorem of Radó asserts that this is automatic. We will not prove this result here; see, for example [50, pp. 185–189].

If $\Omega \subset X$ is open (X is a Riemann surface) and $f : \Omega \to \mathbb{C}$ is continuous, we say that f is holomorphic if for any chart (U, φ) of X, the function $f \circ \varphi^{-1} : \varphi(\Omega \cap U) \to \mathbb{C}$ is holomorphic.

If X, Y are Riemann surfaces, $f : X \to Y$ a continuous map, f is called *holomorphic* if, for any chart (V, ψ) of Y, the function $\psi \circ f : f^{-1}(V) \to \psi(V) \subset \mathbb{C}$ is holomorphic.

Non-constant holomorphic maps between Riemann surfaces are open. Also, a bijective holomorphic map $f : X \to Y$ has a holomorphic inverse $f^{-1} : Y \to X$. Such bijective holomorphic maps are called analytic *isomorphisms* (or biholomorphic maps).

Let Y be a Riemann surface, X a connected 2-dimensional manifold, and $p : X \to Y$ a local homeomorphism. There is a unique complex structure on X for which the map p is holomorphic, obtained as follows: let U be an open set in X such that p_U is a homeomorphism onto an open set V in Y with $V \subset V_j$ for some j, where $\left\{(V_j, \psi_j)_{j \in J}\right\}$ is the given complex structure on Y. Let $\varphi_U : U \to \mathbb{C}$ be the map $\varphi_U = \psi_j \circ p$. It is easily checked that two such pairs $(U, \varphi_U), (U', \varphi_{U'})$ are holomorphically compatible, so that one obtains a complex structure on X for which p is holomorphic.

The uniqueness follows from the following remark: let $U \subset X$ be open and p_U a homeomorphism onto $V \subset Y$. Then, if p is holomorphic, the map $(p_U)^{-1} : V \to U$ is again holomorphic.

Now consider a Riemann surface X and a holomorphic map $p : X \to \mathbb{C}$, which is also a local homeomorphism. We regard $\mathbb{C}$ as the complement of $\infty \in \mathbb{P}^1$, and p as a local homeomorphism $X \to \mathbb{P}^1$.

We shall define boundary points of X. Let $\{x_\nu\}_{\nu\geq 1}$ be a sequence of points in X with the following properties:

1. $\{x_\nu\}$ is discrete (i.e., has no limit points in X);
2. $\{p(x_\nu)\}$ converges to a point $a \in \mathbb{P}^1$;
3. Let $D_\varepsilon = \{z \in \mathbb{C} \mid |z-a| < \varepsilon\}$ if $a \in \mathbb{C}$, and let $D_\varepsilon = \{z \in \mathbb{C} \mid |z| > \frac{1}{\varepsilon}\} \cup \{\infty\}$ if $a = \infty$. Then, for all sufficiently small $\varepsilon > 0$, all but finitely many of the $\{x_\nu\}$ lie in the same connected component of $p^{-1}(D_\varepsilon)$.

Two such sequences $\{x_\nu\}, \{y_\nu\}$ are called *equivalent* if the sequence

$$z_\nu = \begin{cases} x_{(\nu+1)/2}, & \text{for } \nu \text{ odd}, \\ y_{\nu/2}, & \text{for } \nu \text{ even} \end{cases}$$

again has the three properties above [i.e., $\lim p(x_\nu) = \lim p(y_\nu) = a$, and the connected components of $p^{-1}(D_\varepsilon)$ containing all but finitely many of the x_ν, y_ν respectively are the same].

A boundary point of X (relative to the map p) is then an equivalence class of sequences $\{x_\nu\}_{\nu\geq 1}$ with the three properties given above. Set $\tilde{X} = X \cup \{\text{boundary points of } X\}$.

Let P be a boundary point of X, defined by a sequence $\{x_\nu\}_{\nu\geq 1}$. We define neighborhoods of P in $\widetilde{X}$ as follows. Let $\varepsilon > 0$ be small and $D_\varepsilon = \{z \mid |z-a| < \varepsilon\}$ $(a \in \mathbb{C})$ or $D_\varepsilon = \{z \mid |z| > \frac{1}{\varepsilon}\} \cup \{\infty\}$ $(a = \infty)$, where $a = \lim p(x_\nu)$. Let Ω_ε be the connected component of $p^{-1}(D_\varepsilon)$ containing all but finitely many of the x_ν, and let $\widetilde{\Omega}_\varepsilon$ be the union of Ω_ε with those boundary points Q having the following property: if $\{y_\nu\}_{\nu\geq 1}$ defines Q, then $\{\nu \mid y_\nu \notin \Omega_\varepsilon\}$ is finite (this is independent of the sequence $\{y_\nu\}$ defining Q). The sets $\widetilde{\Omega}_\varepsilon$ $(\varepsilon > 0$ small) form a fundamental system of neighborhoods of $P \in \widetilde{X} \setminus X$.

This topology is Hausdorff: if P, Q are boundary points defined by $\{x_\nu\}, \{y_\nu\}$ respectively, and $P \neq Q$, then, by the definition of the equivalence relation, there is $\varepsilon > 0$ such that the components $\Omega_{\varepsilon,1}, \Omega_{\varepsilon,2}$ of $p^{-1}(D_\varepsilon)$ containing all but finitely many of the x_ν, y_ν respectively are distinct, and $\widetilde{\Omega}_{\varepsilon,1} \cap \widetilde{\Omega}_{\varepsilon,2} = \emptyset$. Moreover, p clearly extends to a continuous map $\widetilde{p} : \widetilde{X} \to \mathbb{P}^1$, with $\widetilde{p}(P) = a = \lim p(x_\nu)$.

A boundary point P of X is said to be algebraic if the following holds: let D_ε be a small disc around $a = \widetilde{p}(P)$ and let Ω be the connected component of $p^{-1}(D_\varepsilon)$ containing all but finitely many points of a sequence defining P; then $p(\Omega) \subset D_\varepsilon - \{a\}$ and the map $p : \Omega \to D_\varepsilon - \{a\}$ is a finite covering.

If we set $\Delta_R = \{z \in \mathbb{C} \mid |z| < R\}$ and $\Delta_R^* = \Delta_R - \{0\}$, then there is $n \geq 1$ such that the map $p : \Omega \to D_\varepsilon \setminus \{a\}$ is isomorphic to the map $p_n : \Delta^*_{\varepsilon^{1/n}} \to D_\varepsilon \setminus \{a\}$ given by $p_n(z) = a + z^n$ if $a \in \mathbb{C}$, and $p_n(z) = z^{-n}$ if $a = \infty$.

In this case, $\widetilde{\Omega} = \Omega \cup \{P\}$ is a neighborhood of P in $\widetilde{X}$ containing no other boundary points of X. Since $p_\Omega \to D_\varepsilon \setminus \{a\}$ is isomorphic to the map p_n defined above, there is a homeomorphism $\varphi : \widetilde{\Omega} \to \Delta_{\varepsilon^{1/n}}$ with $\varphi(P) = 0$ and $p \circ \varphi^{-1} = p_n$ on $\Delta^*_{\varepsilon^{1/n}}$. Clearly, φ_Ω is holomorphic.

Set $\hat{X} = X \cup \{\text{algebraic boundary points of } X\}$. We can extend the complex structure on X to one on $\hat{X}$ by taking as a chart containing an algebraic boundary point $P \in \hat{X} \setminus X$ the pair $(\widetilde{\Omega}, \varphi)$ constructed above. Let $\hat{p} = \widetilde{p}|_{\hat{X}}$. The pair $(\hat{X}, \hat{p})$ will be called the *(algebraic) completion* of (X, p). The map $\hat{p} : \hat{X} \to \mathbb{P}^1$ is holomorphic, but need not be a local homeomorphism. With the notation above, if P is an algebraic boundary point and Ω is an n-sheeted covering of $D_\varepsilon \setminus \{a\}$ with $n > 1$, then $\hat{p}$ is not a local homeomorphism at P.

This construction can be used to obtain the Riemann surface of a holomorphic function as conceived by Riemann.

We can now construct the "Riemann surface of an analytic function." The complex plane $\mathbb{C}$ is a Riemann surface, so everything stated in Section 11.2 is also valid for holomorphic functions of one complex variable. Sheaf theory provides an elegant solution to a classical problem in complex analysis, which is also one of the main origins of the theory of Riemann surfaces.

The Riemann surface X can be regarded as a completion of the universal analytic continuation of f_a, for any germ $f_a \in \mathcal{O}_a(Y)$ satisfying $(f_a, a) = 0$ at a point where the discriminant of $P(T)$ is not zero.

Let $X = \mathbb{C}$ and consider $\mathcal{O}_\mathbb{C}$. Let M be a connected component of $\mathcal{O}_\mathbb{C}$, and $p : M \to \mathbb{C} \subset \mathbb{P}^1$ the restriction to M of the map $f_a \mapsto a$ constructed earlier; the map $p : M \to \mathbb{C}$ is a local homeomorphism, so there is a unique structure of a Riemann surface on M for which p is holomorphic. Then p is a local analytic isomorphism, i.e., any $a \in M$ has an open neighborhood U such that $p|_U$ is an analytic isomorphism of U onto the open set $p(U)$.

We define a holomorphic function h on M by $h(f_a) = f_a(a)$ (if (U, f) is a pair defining f_a with U connected, then $N(U, f) \subset M$ and $h(f_a) = f(x)$ for all $x \in U$, so h is holomorphic). Intuitively, this "universal" function h describes all the germs that can be obtained by analytic continuation of a fixed germ $f_a \in M$.

Let now $\widehat{M} = M \cup \{\text{algebraic boundary points of } M\}$ and $\widehat{p} : \widehat{M} \to \mathbb{P}^1$ be the holomorphic map extending $p : M \to \mathbb{C}$ which we constructed before.

Let U be a connected open set in $\mathbb{C}$ and f a holomorphic function on U. We assume that (U, f) defines a germ $f_a \in M$ for some $a \in U$ (and hence for all $a \in U$, since U is connected).

The set $E = \widehat{M} \setminus M$ is a discrete set in $\widehat{M}$, and we have our universal function h on M. Let X_f be the union of M with those points $P \in E$, which are not essential singularities of h (i.e., points where h has either a holomorphic extension or a pole). Thus, there is a meromorphic function h_f on X_f with $h_f|_M = h$. Let $p_f : X_f \to \mathbb{P}^1$ be the restriction of $\widehat{p}$.

The triple (X_f, p_f, h_f) is the Riemann surface of the function f on U.

If $P \in X_f \setminus M$, and $p_f(P) = a$, then near P, the map p_f is equivalent to the map $z \mapsto a + z^n$ ($n \geq 1$) or $z \mapsto z^{-n}$. Thus, if z is a local coordinate at P on X_f and w denotes a local coordinate at a on $\mathbb{P}^1$, the local description of the maps p_f, h_f can be written:

$$\begin{cases} w = z^n \,, \\ h_f = \sum_{-k}^{\infty} a_\nu w^{\nu/n} \quad \left(= \sum_{-k}^{\infty} a_\nu z^\nu\right) . \end{cases}$$

4.4.2 Algebraic Functions on Riemann Surfaces

Algebraic curves over $\mathbb{C}$: Algebraic Riemann surfaces. This is a classical example that generalizes the previous one. If we glue together a finite number of planes with a finite number of slits we obtain an algebraic curve spread over $\mathbb{C}$. It is bi-holomorphic to a compact Riemann surface (not necessarily the Riemann sphere as before) minus a finite number of points (those at ∞ and those corresponding to finite ramification points). If the projection map $\pi : \mathcal{S} \to \mathbb{C}$ is taken as the z-variable then the surface can be identified with an algebraic curve over $\mathbb{C}$ given by an algebraic equation

$$P(w, z) = 0$$

where P is a polynomial. These Riemann surfaces are named algebraic Riemann surfaces. Classically algebraic curves are defined by using Riemann surfaces constructed with segment cuts with disjoint endpoints as well.

This Riemann surface description is the classical point of view that mathematicians had of algebraic curves in the XIXth century after B. Riemann's celebrated memoir on abelian integrals [158], [1, pp. 145–211]. The abstract modern definition of Riemann surfaces is due to H. Weyl and T. Radó [188]. The literature of the subject is so vast that some suggestions as to further reading (in English) may be helpful. Classical references on the XIXth century theory of algebraic curves are [8, 92–94, 143–145] or [178]. For a historical survey see [87].

In this section, we put Theorem 4.3.15 to work in showing how analytic functions arise through the solution of algebraic equations.

Definition 4.4.1. Let $P(z, w) \in \mathbb{C}[z, w]$ be a nontrivial polynomial in two variables. The *(unbranched) Riemann surface* X_P of the equation $P = 0$ is the largest open subset of the sheaf $\mathcal{O}_\mathbb{C}$ formed by the germs φ_z of holomorphic functions satisfying the equation $P\left(\zeta, \varphi_z(\zeta)\right) = 0$ in a neighborhood of z.

Given a polynomial $P \in \mathbb{C}[z, w]$, one can write it as

$$P = P_1^{n_1} \dots P_N^{n_N}$$

with P_j irreducible, i.e., prime in the factorial ring $\mathbb{C}[z, w]$.

We claim that the surface X_P is the disjoint union of the surfaces X_{P_j}. It is clear that if $\varphi_z \in X_{P_j}$ then $P\left(\zeta, \varphi_z(\zeta)\right) = 0$ in a neighborhood of z.

On the other hand, the set of points in $\mathbb{C}^2$ defined by

$$P_i(z,w) = P_j(z,w) = 0, i \neq j,$$

is finite. Therefore, a germ φ_z cannot belong to $X_{P_i} \cap X_{P_j}$. This shows that $X_P \supseteq \bigcup_{1 \leq j \leq N} X_{P_j}$. Finally, let $\varphi_z \in X_P$ be defined in $B(z,r)$ and let $z_0 \in B(z,r)$ be such that there is no w_0 and a pair $i \neq j$ with $P_i(z_0, w_0) = P_j(z_0, w_0) = 0$. Since $P(z_0, \varphi_z(z_0)) = 0$, there is only one index, say i_0, for which $P_{i_0}(z_0, \varphi_z(z_0)) = 0$. In a neighborhood of z_0 we have therefore $P_j(\zeta, \varphi_z(\zeta)) \neq 0$ if $j \neq i_0$. Hence $P_{i_0}(\zeta, \varphi_z(\zeta)) \equiv 0$ for all points in a neighborhood of z_0 and we can conclude that $P_{i_0}(\zeta, \varphi_z(\zeta)) \equiv 0$ in $B(z,r)$. In other words, $\varphi \in X_{P_{t_0}}$ and, hence, $X_P = \dot{\bigcup}_j X_{P_j}$.

Therefore we can assume from now on that P is irreducible and write

$$P(z,w) = a_0(z)w^n + a_1(z)w^{n-1} + \cdots + a_n(z),$$

with $a_i \in \mathbb{C}[z]$ having no common factor and $a_0 \not\equiv 0$.

Except for the values z which are zeros of a_0, the equation (in w) $P(z,w) = 0$ has n roots. Except for a finite number of additional values of z, these n roots are distinct. We denote by $\Delta(P)$ the set of those exceptional values of z; this is called the *set of critical points* of P. For simplicity we add ∞ to $\Delta(P)$. In this way, $\Delta(P)$ is the union of ∞, the roots of a_0, and the $z \in \mathbb{C}$ such that the two polynomials in $w, P(z,\cdot)$ and $\frac{\partial P}{\partial w}(z,\cdot)$, have a common root. (the $\left(P, \frac{\partial P}{\partial w}\right)$ have only finitely many common zeros, thanks to the irreducibility of P.)

Lemma 4.4.2. *Let $z_0 \in S^2 \backslash \Delta'(P)$ and $w_1, \ldots, w_n$ be the distinct roots of $w \mapsto P(z_0, w)$. There are germs $\varphi_1, \ldots, \varphi_n \in \mathcal{O}_{z_0}$ such that $\varphi_j(z_0) = w_j$ and*

$$a_0(z) \prod_{j=1}^{n} (w - \varphi_j(z)) = P(z,w)$$

for every z in a neighborhood of z_0 and $w \in \mathbb{C}$.

Proof. Let $\varepsilon > 0$ be such that the closed disks $\bar{B}(w_j, \varepsilon)$ are disjoint. Let $r > 0$ be such that $P(z,w) \neq 0$ when $|z - z_0| \leq r$ and $|w - w_j| = \varepsilon$ for some j. By Rouche's theorem, for each fixed $z, |z - z_0| < r$, there is only one zero of $P(z,w) = 0$ in $B(w_j, \varepsilon)$, and it is given by

$$\varphi_j(z) = \frac{1}{2\pi i} \int_{\partial B(w_j,\varepsilon)} w \frac{\partial P}{\partial w}(z,w) \frac{dw}{P(z,w)}$$

The functions $\varphi_j \in \mathcal{H}(B(w_j, \varepsilon))$ and clearly satisfy the other conditions of the lemma. □

Let us consider now the situation in a neighborhood of a point $z_0 \in \Delta(P) \backslash \{\infty\}$. Let $\varepsilon > 0$ be such that $B(z_0, \varepsilon) \cap \Delta(P) = \{z_0\}$, and define $V_{z_0} = \pi^{-1}(B(z_0,\varepsilon)\setminus\{z_0\}) \cap X_{\vec{P}}$, where $\pi : \mathcal{O}_{\mathbb{C}} \to \mathbb{C}$ is the canonical projection, $\pi(\varphi_z) = z$. For every open simply connected subset U of $B(z_0,\varepsilon)\setminus\{z_0\}$ we can now construct, thanks to the lemma, n holomorphic function $f_1^U, \ldots, f_n^U$ in U, such that for each $z \in U$ the distinct roots $w_1, \ldots, w_n$ of $P(z,w) = 0$ are given by the values $f_1^U(z), \ldots, f_n^U(z)$. Hence

$$\pi^{-1}(U) \cap X_P = \bigcup_{1\le j\le n} N\left(U, f_j^U\right)$$

and $\pi|_N\left(U, f_j^U\right) : N\left(U, f_j^U\right) \to U$ is a biholomorphism for every j. This shows that

$$\pi : V_{z_0} \to B(z_0,\varepsilon)\setminus\{z_0\}$$

is an n-sheeted covering map.

Let γ be a generator of $\pi_1(B(z_0,\varepsilon)\setminus\{z_0\}, a)$ of index 1 with respect to z_0. The n elements $f_{1,a}, \ldots, f_{n,z} \in \mathcal{O}_a$, which are germs of distinct roots of $P(z,w) = 0$ near $z = a$, can be continued along γ into a new system of roots, which are a permutation σ of the original ones. We denote them by $f_{\sigma(1),a}, \ldots, f_{\sigma(n),a}$. One can decompose σ into disjoint cycles. For instance, up to renumbering, we could have $\sigma(1) = 2, \sigma(2) = 3, \ldots, \sigma(q-1) = q$ and $\sigma(q) = 1$ for some $q \le n$. This means that if we start with $f_{1,a}$ then continuing along γ we obtain $f_{2,a}$, continuing once more we obtain $f_{3,a}$, and after q times we come back to $f_{1,a}$. In this situation, the following property holds.

Lemma 4.4.3. *Let $f_a \in \mathcal{O}_a$ be a germ admitting analytic continuation along any path in $U = B(z_0,\varepsilon)\setminus\{z_0\}$ that starts at $z = a$. Let γ be the generator of $\pi_1(U,a)$ that has index 1 with respect to z_0. Assume that the analytic continuation of f_a along γ^q is equal to f_a. Then there is a Laurent series convergent in $B\left(0,\varepsilon^{1/q}\right)\setminus\{0\}$ such that*

$$f_a\left(z_0 + (\zeta - z_0)^q\right) = \sum_{n\in\mathbb{Z}} c_n (\zeta - z_0)^n$$

Proof. We can assume without loss of generality that $z_0 = 0, \varepsilon = 1$, $U = B(0,1)\setminus\{0\}$. Let $b \in U$ be such that $b^q = a$. The map $\varphi : \zeta \to \zeta^q$ of U into U has the property that the germ g_b of $f_a \circ \varphi$ at $\zeta = b$, when continued along γ, returns to itself. By the monodromy theorem, g_b determines a holomorphic function in U. This function can be expanded in a Laurent series $\sum_{n\in\mathbb{Z}} a_n\zeta^n$ in U, hence

$$f_a(\zeta^q) = \sum_{n\in\mathbb{Z}} c_n\zeta^n$$

as we wanted to show. □

Applying the lemma to $f_{1,a}$ one finds

$$f_{1,a}\left(z_0 + (\zeta - z_0)^q\right) = \sum_{n\in\mathbb{Z}} c_n (\zeta - z_0)^n$$

for a series $f(t) = \sum_{n\in\mathbb{Z}} c_n t^n$ convergent in $B\left(0, \varepsilon^{1/q}\right)\setminus\{0\}$. It is usual to write $\xi = z_0 + (\zeta - z_0)^q$, hence

$$f_{1,a}(\xi) = \sum_{n\in\mathbb{Z}} c_n\,(\xi - z_0)^{n/q}$$

which is called a *Puiseux series representation* for $f_{1,a}$. In terms of f we have $f_{1,a}(\xi) = f\left((\xi - z_0)^{1/q}\right)$. With this notation the different branches $f_{1,a}, \ldots, f_{q,a}$ can also be written as

$$f_{j,a}(\xi) = f\left((\xi - z_0)^{1/q}\right), \quad 1 \le j \le q$$

where one just takes into account the different choices of the root. If $(\xi - z_0)^{1/q}$ is the choice of $f_{1.a}$, the others are $e^{2\pi i(j-1)/q}\,(\xi - z_0)^{1/q}$, i.e.,

$$f_{j,a}(\xi) = f\left(e^{2\pi i(j-1)/q}\,(\xi - z_0)^{1/q}\right) = \sum_{n\in\mathbb{Z}} c_n \left[e^{2\pi i(j-1)/q}\,(\xi - z_0)^{1/q}\right]^n$$

or

$$f_{j,a}\left(z_0 + (\zeta - z_0)^q\right) = \sum_{n\in\mathbb{Z}} c_n \left[e^{2\pi i(j-1)/q}\,(\zeta - z_0)\right]^n .$$

If $q < n$, then one can consider the different cycles of $f_{q+1,a}, \ldots, f_{n,a}$, and find corresponding Puiseux series expansions for the different roots.

If we are in the situation where $a_0(z_0) \neq 0$, then the solutions w of $P(z, w) = 0$ remain in a bounded set when z is near z_0. In fact, the functions $a_j(z)/a_0(z)$ are bounded in a neighborhood of z_0, and from the expression

$$w^n = -\left(\frac{a_1(z)}{a_0(z)}w^{n-1} + \cdots + \frac{a_n(z)}{a_0(z)}\right)$$

we see that either $|w| \le 1$ or

$$\begin{aligned}|w| &= \left|\frac{a_1(z)}{a_0(z)} + \frac{a_2(z)}{a_0(z)}\frac{1}{w} + \cdots + \frac{a_n(z)}{a_0(z)}\frac{1}{w^{n-1}}\right| \\ &\le \left|\frac{a_1(z)}{a_0(z)}\right| + \left|\frac{a_2(z)}{a_0(z)}\right| + \cdots + \left|\frac{a_n(z)}{a_0(z)}\right| .\end{aligned}$$

In other words, if $\alpha > 0$ is sufficiently small so that $a_0(z) \neq 0$ in $|z - z_0| \le \alpha$, we have that any root w of $P(z, w) = 0$ satisfies

$$|w| \le \max\left\{1, \max_{|z-z_0|\le\alpha} \frac{|a_1(z) + \cdots + |a_n(z)|}{|a_0(z)|}\right\} \le M < \infty .$$

Therefore, the different branches $f_{j,a}$ of the first cycle, or if one prefers, $f_{1,a}\left(z_0 + (\zeta - z_0)^q\right)$, remain bounded when $\zeta \to z_0$. This implies that f is holomorphic even at $t = 0$, so that

$$f_{j,a}(\xi) = \sum_{n\ge 0} c_n\,(\xi - z_0)^{n/q}, \quad 1 \le j \le q .$$

The same happens for every cycle of the permutation σ. What does this mean for the covering map $\pi : V_{z_0} \to B(z_0, \varepsilon) \setminus \{z_0\}$?

If $q = 1$, there is only one branch in the cycle, which is then the restriction to $B(z_0, \varepsilon) \setminus \{z_0\}$ of a holomorphic function in $B(z_0, \varepsilon)$. This corresponds to a connected component of V_{z_0}, which has only one sheet.

If $q \geq 2$, the cycle corresponds to a connected component of V_{z_0} of exactly q sheets.

If $a_0(z_0) = 0$ with multiplicity $k \geq 1$, then, by the same reasoning as earlier, if w is a root of $P(z, w) = 0, z \in B(z_0, z) \setminus \{z_0\}$, we have

$$|z - z_0|^k |w| \leq \max \left\{ 1, \frac{|a_1(z)| + \cdots + |a_n(z)|}{|a_0(z)| / |z - z_0|^k} \right\} \leq M < \infty$$

as long as $0 < |z - z_0| \leq \alpha, \alpha$ sufficiently small. In terms of the function f associated to the first cycle this means that

$$t^k f(t) = \sum_{n \in \mathbb{Z}} c_n t^{n+k}, \quad 0 < |t| < \varepsilon^{1/q}$$

remains bounded as $t \to 0$. So f has at most a pole of order k and

$$f_{j,a}(\xi) = \sum_{n \geq -k} c_n (\xi - z_0)^{n/q} .$$

The description of the connected components of the covering

$$\pi : V_{z_0} \to B(z_0, \varepsilon) \setminus \{z_0\}$$

and their number of sheets remains the same.

It remains to consider the point $z_0 = \infty \in \Delta(P)$.

Let $d = \max\{\deg a_j, 0 \leq j \leq n\}$. Then we can consider the polynomial equation

$$z^d P\left(\frac{1}{z}, w\right) = 0$$

about $z = 0$ and the local situation is the same as earlier.

Now we are ready to construct the complete Riemann surface of the irreducible equation $P = 0$, proceeding in the same fashion as we did for $\sqrt{f(z)}$ in Subsection 4.4.4, i.e., for $P(z, w) = w^2 - \prod_{1 \leq j \leq n} (z - a_j)$.

For that purpose, let us denote by $z_1, \ldots, z_N$ the points in $\Delta(P)$. For each j, denote by $1 \leq k \leq m_j$ the different connected components of the coverings $\pi : V_j \to U_j \setminus \{z_j\}$, where the U_j are open disks with disjoint closures in S^2 (for $z_N = \infty$ we mean $U_N = \{z \in \mathbb{C} : |z| > r\} \cup \{\infty\}$), and $V_j = \pi^{-1}(U_j \setminus \{z_j\}) \cap X_P$. The set V_j has connected components $V_{j,k}$ of multiplicity q_{jk}. We have $\sum_{1 \leq k \leq m_j} q_{jk} = n$.

Consider now the topological space $\widetilde{X}$, which is the disjoint union of X_P and $\sum_{\leq j \leq N} m_j$ disks $\widetilde{V}_{j,k}$. One glues V_{jk} to $\widetilde{V}_{jk}$ by introducing isomorphisms σ_{jk} between the covering maps

$$\pi|_{V_{jk}} : V_{jk} \to U_j \{z_j\}$$

and

$$p_{jk} : \widetilde{V}_{jk} \backslash \{c_{jk}\} \to U_j \backslash \{z_j\} ,$$

where c_{jk} is the center of $\widetilde{V}_{jk}$ and p_{jk} is a covering map with q_{jk} sheets given by the q_{jk} th power.

To construct σ_{jk}, choose $\xi_0 \in X_{jk}, \zeta_0 \in \widetilde{V}_{jk} \backslash \{c_{jk}\}$ such that $\pi(\xi_0) = p_{jk}(\zeta_0)$, and if ξ is the endpoint of a path γ in V_{jk} starting at ζ_0, let γ_1 be the lifting to $\widetilde{V}_{jk} \backslash \{c_{jk}\}$ of $\pi \circ \gamma$, starting at ζ_0. We define $\alpha_{jk}(\xi) = \gamma_1(1)$. One verifies without difficulty that σ_{jk} is a biholomorphic map from V_{jk} onto $\widetilde{V}_{jk} \backslash \{c_{jk}\}$ such that $p_{jk} \circ \sigma_{jk} = \pi|_{V_{jk}}$.

Introduce in $\widetilde{X}$ the equivalence relation $\mathcal{R}$ which identifies the points that correspond to each other via some σ_{jk}. The quotient space $\widetilde{X}_P = \widetilde{X}/\mathcal{R}$ is now a Riemann surface in which X_P appears as a dense open set. (In fact, $\widetilde{X}_P \backslash X_P$ is a finite collection of points.)

The covering map $\pi : X_P \to \mathbb{C}$ can be extended to a holomorphic map $\pi : \widetilde{X}_P \to S^2$ setting $\pi(\widetilde{c}_{jk}) = z_j$, where $\widetilde{c}_{jk}$ is the class of c_{jk} modulo $\mathcal{R}$. It is a branched covering map.

The global holomorphic function $F : X_P \to \mathbb{C}$ defined by $F(\varphi_z) = \varphi_z(z)$ can be extended by continuity to the points $\widetilde{c}_{jk}$ to a finite or ∞ value according to whether the corresponding Laurent series $\sum_{n \in \mathbb{Z}} a_n t^n$ is holomorphic or has a pole at $t = 0$. The meromorphic function $F : \widetilde{X}_P \to S^2$ thus obtained verifies $P(\pi(\varphi), f(\varphi)) = 0$ for every $\varphi \in X_P$.

Definition 4.4.4. The triple $(\widetilde{X}_P, \pi, F)$ associated by the preceding construction to the irreducible polynomial $P \in \mathbb{C}[z, w]$ is called the *(complete) Riemann surface* of the equation $P = 0$. The function F is the algebraic function associated to the equation $P = 0$.

If P is not irreducible then we denote by $\widetilde{X}_P$ the (disjoint) union of the $\widetilde{X}_{P_j}$ for the different irreducible factors of P.

Proposition 4.4.5. *If P is irreducible, then $\widetilde{X}_P$ is connected.*

Proof. It is enough to prove that X_P is connected. For that, it is enough to show that for a fixed $z_0 \in \Omega = S^2 \backslash \Delta(P)$, the n germs $\varphi_1, \ldots, \varphi_n$ which are solutions of $P(z, w) = 0$ in a neighborhood of z_0, are in the connected component of the germ φ_1 in $\mathcal{O}_\Omega$. We argue by contradiction.

Let us assume that $\varphi_1, \ldots, \varphi_k$ are in the same component and $\varphi_{k+1}, \ldots, \varphi_n$ in other components. Consider the symmetric functions σ_i of the $\varphi_j, 1 \leq j \leq k$, defined by

$$\prod_{1 \leq j \leq k} (w - \varphi_j) = w^k - \sigma_1 w^{k-1} + \cdots + (-1)^k \sigma_k .$$

If γ is a loop in $\pi_1(\Omega, z_0)$, the analytic continuation along γ preserves the set $\{\varphi_j : 1 \leq j \leq k\}$. Hence the σ_j are well-defined holomorphic functions in Ω.

The only possible singularities of the σ_j are poles at the points of $\Delta(P)$. The reason is that, from the Puiseux series expansions of the solutions of $P = 0$ near a point $c \in \Delta(P)$, one sees that all of them are bounded by $|z - c|^{-M}$ for some fixed power M if $c \neq \infty$, and by $|z|^M$ if $c = \infty$. Therefore the $\sigma_j \in \mathcal{M}\left(S^2\right)$. Let $Q(z)$ be the least-common multiple of the denominators of the rational function σ_j. Then

$$R(z, w) := Q(z) \prod_{1 \leq j \leq k} (w - \varphi_j)$$

is a polynomial in (z, w) of degree k in w. It follows that R and P are relatively prime, since P is irreducible and cannot divide R. Therefore there are only finitely many common roots $(z, w) \in \mathbb{C}^2$ of the equations

$$R(z, w) = P(z, w) = 0 .$$

This contradicts the fact that $R(z, \varphi_1(z)) \equiv 0$ and $P(z, \varphi_1(z)) \equiv 0$ in a neighborhood of z_0.

We conclude that $k < n$ is impossible, hence $\widetilde{X}_P$ is connected. □

Proposition 4.4.6. *$\widetilde{X}_P$ is compact.*

We can assume P is irreducible. $\widetilde{X}_P$ is the union of finitely many relatively compact regions. Namely, they are the images under the equivalence $\mathcal{R}$ of the $\widetilde{V}_{jk}$ and of $X_P\left(\bigcup_{j,k} V_{jk}\right)$.

Remark 4.4.7. It follows that $\pi : \widetilde{X}_P \to S^2$ is surjective, since π is an open map and $\widetilde{X}_P$ is compact.

One can prove that the converse of Proposition 4.4.6 is also true. Any compact Riemann surface can be defined as $\widetilde{X}_P$ for a convenient polynomial $P \in \mathbb{C}[z, w]$ (see Section 4.15).

We have only scratched the surface of the subject of algebraic functions, or equivalently, compact Riemann surfaces. The reader should consult [9, 46, 50, 76, 91, 165], and [38] for this beautiful and rich theory.

Another interesting question is to decide in an effective way, which are possible Puiseux expansions of algebraic function around the critical points.

Theorem 4.4.8 (Picard's theorem). *Let $P(u, v)$ be an irreducible polynomial with complex coefficients in two variables u and v. If there exist nonconstant entire functions $f(z)$ and $g(z)$ satisfying $P(f(z), g(z)) = 0$ for all complex numbers z, then the Riemann surface associated with the algebraic equation $P(u, v) = 0$ has genus 0.*

4.4.3 Riemann Surface of Algebraic Function (Classical Approach)

Spheres with Slits Glued Together. A collection of spheres with slits, endowed with their conformal structure, can be glued isometrically along the slits via

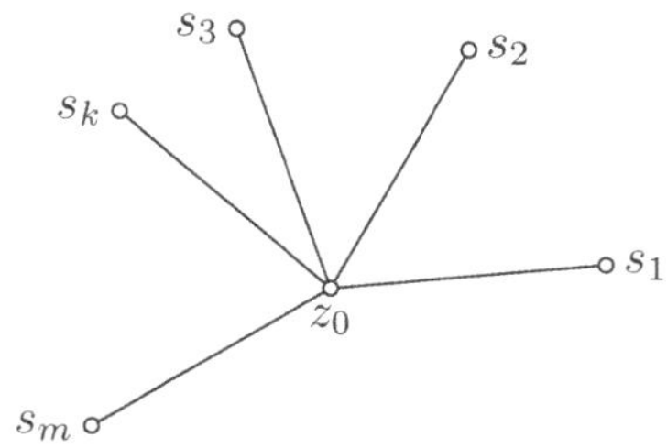

FIGURE 4.5: Cuts along segments.

the identity map. The resulting surface inherits a canonical Riemann surface structure. Conversely, every Riemann surface may be obtained in this way. Such a construction yields a Riemann surface equipped with a distinguished system of charts.

We are given an algebraic equation

$$f(z, w) = a_0(z)w^n + a_1(z)w^{n-1} + \cdots + a_n(z) = 0 \tag{4.11}$$

of degree n with respect to w. As a first, rough approximation to the Riemann surface of the function w, we shall look at the set $\widetilde{S}$ of all solutions (z, w) of $f(z, w) = 0$. On this set, w is tautologically the function that takes on the value w_0 at (z_0, w_0). As we have seen, (4.11) defines an n-valued analytic function $w(z)$. Take a value $z = z_0$ such that all the roots of equation (4.11) at $z = z_0$ are distinct, and otherwise arbitrary. Let these be

$$w_1, w_2, \ldots, w_n\,.$$

Take one of them, for example $w = w_1$. Consider the plane, or better, the Riemann sphere of the complex variable z and assign to each of its points z the value obtained by analytic continuation (see Subsection 4.3.2), starting from the value $w = w_1$ at $z = z_0$. However, such a definition generally leads to an ambiguous result, since the function $w(z)$ has critical points, near which it is multivalued. To avoid this ambiguity, let us mark on the sphere the roots of the discriminant $D(z)$ of equation (4.11) and connect them to the point $z = z_0$ by line segments. At the same time, we impose the additional requirement that these segments must not overlap each other (Fig. 4.5). If we imagine cuts along these segments, they will not destroy the connectedness of the sphere; that is, any point of the sphere can still be joined to z_0 by a curve (even a polygonal line) that does not cross any of the cuts.

If we agree to understand the value $w(z)$ at each point z, not lying on any of the cuts, as the value obtained from $w(z_0) = w_1$ by analytic continuation along a path that does not cross any cut, then this definition of $w(z)$ will be single-valued. Indeed, if we take two such paths, their union forms a closed curve containing no critical points inside. On the other hand, if analytic continuation along a closed curve does not return us to the original value, then the region bounded by this curve must necessarily contain critical points (see Section 1.3).

As for points lying on the cuts, we can define values of $w(z)$ on them by approaching from one or the other side of the cut. In this way, we obtain for the same point two values, which in general will not coincide, but will be different roots of equation (4.11) corresponding to the same value of z. If it turned out that on all cuts both values coincided, this would mean that the function $w(z)$ was single-valued on the entire sphere and therefore had no critical points there, and thus, by Theorem 5.8.2, would be a rational function of z. In that case, the left-hand side of equation (4.11) would have a rational factor linear in w.

Now take n spheres, construct on each the system of cuts described, and assign to each point z on the i-th sphere the value $w(z)$ obtained by analytic continuation of $w(z)$, starting from the value $w = w_i$ at $z = z_0$ $(i = 1, 2, \ldots, n)$. In this way, for each value of z we obtain n values of w. If the point z is not critical, all these values are distinct, since otherwise, by analytic continuation in the reverse direction along a path not crossing the cuts, we would arrive at two different values of w at $z = z_0$, which is impossible. Hence the values of the function $w(z)$ for the same value of z across all n spheres exhaust all the roots of equation (4.11) for that value of z.

Take a point $z = z_1$ lying on one of our cuts. Denote by

$$w_1^+, w_2^+, \ldots, w_n^+ \tag{4.12}$$

the values $w(z_1)$ on the 1-st, 2-nd, …, n-th spheres, defined from the same side of the cut, and by

$$\overline{w}_1, \overline{w}_2, \ldots, \overline{w}_n \tag{4.13}$$

the same values, defined from the opposite side. Both sets (4.12) and (4.13) exhaust all roots of equation (4.11) at $z = z_1$, and therefore coincide. Thus

$$w_1^+ = \overline{w}_{\alpha_1}, \quad w_2^+ = \overline{w}_{\alpha_2}, \quad \ldots, \quad w_n^+ = \overline{w}_{\alpha_n},$$

where

$$S = \begin{pmatrix} 1 & 2 & \cdots & n \\ \alpha_1 & \alpha_2 & \cdots & \alpha_n \end{pmatrix} \tag{4.14}$$

is a certain permutation of the numbers $1, 2, \ldots, n$. This permutation is the same along the entire cut. Indeed, as we move along one side of the cut, the value w_i^+ varies continuously, and likewise for $\overline{w}_{\alpha_i}$. Since at the point $z = z_1$ we have

$$w_i^+ = \overline{w}_{\alpha_i},$$

and everywhere along the cut except one endpoint all values in (4.12) and (4.13) are distinct, relation (4.14) must hold along the whole cut.

To determine the permutation (4.14) corresponding to crossing the ν-th cut, consider points on the cut near its endpoint, i.e., near the critical point

$z = a_\nu$. We have seen [see Section 4.5] that all expansions of $w(z)$ near $z = a_\nu$ split into cycles of orders

$$b_1, b_2, \ldots, b_k, \quad b_1 + b_2 + \cdots + b_k = n\,.$$

The expansions of the μ-th cycle are series in integer powers of $(z - a_\nu)^{1/b_\mu}$:

$$\begin{aligned} w_{\mu,0} &= \psi\left[(z - a_\nu)^{1/b_\mu}\right], \\ w_{\mu,1} &= \psi\left[\varepsilon_\mu (z - a_\nu)^{1/b_\mu}\right], \\ &\vdots \\ w_{\mu,b_\mu-1} &= \psi\left[\varepsilon_\mu^{b_\mu-1}(z - a_\nu)^{1/b_\mu}\right], \\ \varepsilon_\mu &= e^{2\pi i/b_\mu}, \quad (\mu = 1, 2, \ldots, k)\,. \end{aligned} \tag{4.15}$$

Let $z = z_2$ be near $z = a_\nu$. Write

$$z_2 - a_\nu = \rho e^{i\theta_2}\,.$$

Keeping ρ fixed, increase θ from θ_2 to $\theta_2 + 2\pi$. Then z describes a circle of radius ρ centered at a_ν. From (4.15) it follows that

$$w_{\mu,0} \to w_{\mu,1} \to \cdots \to w_{\mu,b_\mu-1} \to w_{\mu,0}\,.$$

Denote by $+$ $(-)$ the side of the ν-th cut corresponding to $\theta = \theta_2$ $(\theta = \theta_2 - 2\pi)$. Then

$$\overline{w}_{\mu,0} = w^+_{\mu,1}, \quad \overline{w}_{\mu,1} = w^+_{\mu,2}, \quad \ldots, \quad \overline{w}_{\mu,b_\mu-1} = w^+_{\mu,0}\,.$$

Replacing the numbering in (4.15) with that in (4.12) and (4.13), we obtain the desired permutation $S = S_\nu$.

Let $a_1, a_2, \ldots, a_\nu$ be the complete set of critical points of $w(z)$, numbered so that a ray from z_0, rotated counterclockwise, meets the 1st, 2th, …, ν-th cuts in order. Then a counterclockwise loop around z_0 induces the sequence of substitutions

$$S_1, S_2, \ldots, S_\nu\,.$$

Thus the full loop corresponds to the product

$$S_1, S_2 \cdots S_\nu\,.$$

Since z_0 is noncritical, continuation around it must return to the initial values, hence

$$S_1 \cdot S_2 \ldots S_\nu = \mathcal{E}\,.$$

To obtain a single surface from the n spheres, we glue, in each ν-th cut, the $-$ side of the i-th sphere to the $-$ side of the α_i-th sphere, since the values $w(z)$ coincide along these sides. Thus

$$S_1 \cdot S_2 \cdots S_\nu = \mathcal{B}\,.$$

To construct a single connected Riemann surface from n separate sheets, we glue along each branch cut by identifying the two banks of the cut: the edge of the $(-i)$-th sheet with the corresponding edge α_i of the n-th sheet. Along these identifications, the values of $w(z)$ match, so the glued surface defines a well-behaved analytic continuation of the function.

Instead of an abstract "identification," Riemann and his followers (Neumann and others) put into it a rather material and visual sense. They imagined n spheres concentrically nested one inside the other, and the same values of $\boldsymbol{z}$ had to correspond to points located on a common radius.

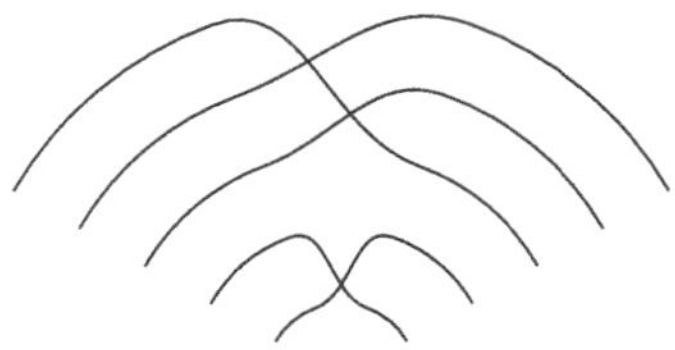

FIGURE 4.6: n spheres concentrically nested one inside the other.

Riemann and his followers (Neumann, etc.) gave this identification a concrete visual meaning: they imagined the n spheres nested concentrically, with points corresponding to the same z lying on a common radius (see Fig. 4.6). The cuts were taken along cones with vertex at the center. The sides of the cuts were glued together according to the permutations describing the behavior of $w(z)$. The resulting system of n glued spheres is called the *Riemann surface*. Each point of this surface corresponds to a unique value of $w(z)$, and analytic continuation along paths on the surface is independent of the path.

4.4.4 Riemann Surfaces of Hyperelliptic Curves

A particular case occurs when we only use two plane sheets. An important special case is that of a *hyperelliptic curve*, defined as the Riemann surface given by an equation of the form

$$Q(w, z) = w^2 - P_d(z) = 0\,,$$

where $P_d(z)$ is a polynomial of degree d. In this case the corresponding multivalued function can be expressed as

$$w = \sqrt{P_d(z)}\,,$$

where the square root symbol is to be interpreted in its multivalued sense.

We call the totality of all solutions of the form $w = w(z)$ to the equation $f(w, z) = 0$ a *multivalued algebraic function*, and we refer to the surface (i.e., complex curve) defined by $f(w, z) = 0$ as the *graph* or *Riemann surface* of this multivalued function. This provides a simplified variant of the usual definition of a Riemann surface (see, for example, [170]). Our definition is equivalent to the standard one in the case where the surface $f(w, z) = 0$ has no singular points or self-intersections.

An *affine hyperelliptic curve* is a curve of the form

$$X_{P_d} = \left\{ (z, w) \in \mathbb{C}^2 \mid w^2 - P_d(z) = 0 \right\} , \tag{4.16}$$

where $P_d(z)$ has d distinct roots. This is an equation that we certainly cannot solve directly if $P_d(z)$ is sufficiently general.

In fact, the curve X_{P_d} is called an *elliptic curve* for $d = 3, 4$, and it is called *hyperelliptic* for $d > 4$. These surfaces are two-sheeted coverings of the z-plane.

Lemma 4.4.9. *The surface X_{P_d} is free of singularities if and only if the polynomial $P_d(z)$ has no multiple roots.*

Proof. Calculating the gradient of the function $f = w^2 - P_d(z)$, we obtain

$$\operatorname{grad}_{\mathbb{C}} f = \left(\frac{\partial f}{\partial w}, \frac{\partial f}{\partial z} \right) = \left(2w, -\frac{dP_d(z)}{dz} \right)$$

Thus a point (w_0, z_0) is a singular point of the curve if and only if

$$2w_0 = 0, \qquad \left. \frac{dP_d(z)}{dz} \right|_{z_0} = 0, \quad w_0^2 - P_d\left(z_0\right) = 0$$

i.e., if and only if $w_0 = 0$, and z_0 is a root both of the polynomial $P_d(z)$ and its derivative. The lemma then follows from the well-known equivalence of this condition on z_0, with the condition that z_0 be a multiple root of $P_d(z)$. $\square$

The complex implicit function theorem allows us to introduce the local coordinate z on the surface (4.16), at (i.e., in neighborhoods of) those points where $\partial f / \partial w = 2w \neq 0$, i.e., where $P_d(z) \neq 0$. At those points z for which $P_d(z) = 0$, we may take w as local co-ordinate, provided that we do not also have $\partial f / \partial z = -dP_d(z)/dz = 0$, i.e., provided that z is not a multiple root of $P_d(z)$.

We first examine cases of low degree d, and from these infer a general result. [It turns out that the points at infinity on such a surface are singular points, so that they may not all appear on the manifolds which we are about to construct (as realizations of such surfaces), while those which do appear should strictly speaking be removed from the manifolds.]

A sphere with one handle is a torus, a sphere with two handles is a double torus, with three handles a triple torus, and so on.

Example 4.4.10. The algebraic function defined by $f(z, w) = w^2 - z = 0$ is not single-valued in the extended z-plane. For, using polar coordinates $z = re^{i\theta}$, we have $w(z) = \pm\sqrt{r}e^{i\theta/2}$. Starting at some point $r_0 e^{i\theta_0}, r_0 \neq 0$, and continuing $w(z)$ along a closed path that winds once around the origin so that θ increases by $2\pi, w(z)$ comes to the value

$w_1(z) = \sqrt{r_0}e^{i\frac{1}{2}(\theta_0+2\pi)} = -\sqrt{r_0}e^{\frac{1}{2}i\theta_0} = w_2(z)$, which is just the negative of its original value. Continuation around this path once again leads back to the original value of $w(z)$. We thus obtain a permutation.

$$\begin{pmatrix} w_1 & w_2 \\ w_2 & w_1 \end{pmatrix} .$$

We join the points $z = 0, z = \infty$ in the domain of the (extended) multiple-valued function $w = \sqrt{z}$ by a line segment α (or to put it more vividly, we "cut" the Riemann sphere S^2, diffeomorphic to $\mathbb{CP}^1$, to obtain the sphere with the segment α removed, depicted in Fig. 4.7. It is not difficult to see intuitively that the restriction of the (extended) surface $f(z, w) = 0$ to those z of S^2 off the slit α, consists of two disjoint connected components, each of which can be seen, by projecting onto the extended z-plane, to be diffeomorphic to the sphere S^2 with the line segment α removed (see Fig. 4.7). (These connected pieces are called the *"branches" of the multi-valued function.*) At the points 0 and ∞ the (extended) function $w = \sqrt{z}$ is single-valued. The desired surface is obtained by identifying the boundary segment α_1 of the component denoted I in Fig. 4.7), with the boundary segment β_2 of the region II, and the boundary segment β_1 of region I with the boundary segment α_2 of region II of the surface (as indicated in Fig. 4.7). It is intuitively plausible that as a result of this cutting and pasting, we obtain a manifold diffeomorphic to S^2.

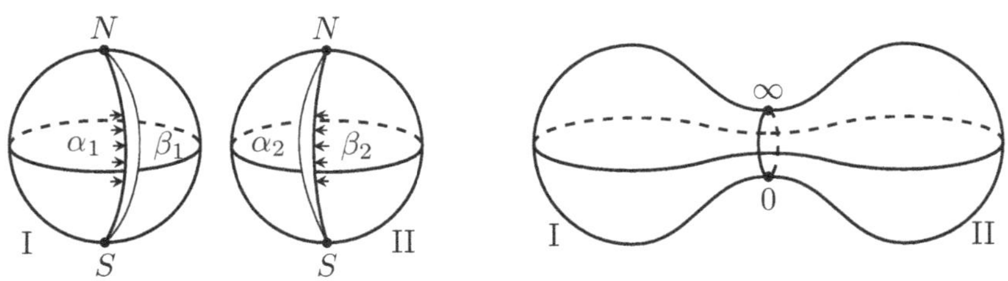

FIGURE 4.7: Two copies of the complex unit sphere Σ are cut along the shortest arc of the great circle on Σ joining the south to north poles glued along S, N.

A branch point of a multivalued function is a point such that if the function is n-valued (has n values) at that point, all of its neighborhoods contain a point that has more than n values. Multi-valued functions are rigorously studied using Riemann surfaces, and the formal definition of branch points employs this concept.

Branch points fall into three broad categories: algebraic branch points, transcendental branch points, and logarithmic branch points. Algebraic branch points most commonly arise from functions in which there is an ambiguity in the extraction of a root, such as solving the equation $w^2 = z$ for w as a function of z. Here the branch point is the origin, because the analytic continuation of any solution around a closed loop containing the origin will result in a different function.

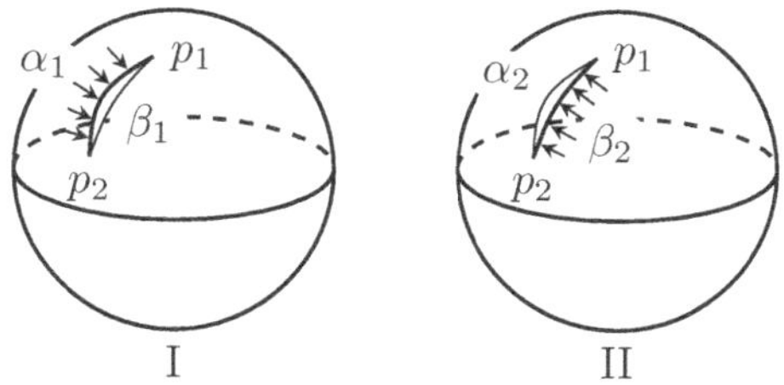

FIGURE 4.8: The points z_1 and z_2 in $\mathbb{C}$ are corresponding to the points p_1 and p_2 on the unit sphere Σ, respectively. Two copies of the sphere are cut along the shortest arc of the great circle on Σ; these are joining p_1 and p_2.

Example 4.4.11. We next consider the case $f(z, w) = w^2 - P_2(z)$, where $P_2(z)$ is a polynomial of degree 2 with simple roots $z = z_1, z = z_2, z_1 \neq z_2$.

To each z there correspond two values of w, one the negative of the other. We go from one to the other by continuing $w(z)$ over any closed path winding once around one of the roots z_1, z_2. For $w = \sqrt{a}\sqrt{z - z_1}\sqrt{z - z_2}$, and the factor $\sqrt{z - z_i}$ changes sign when $\arg(z - z_i)$ changes by 2π.

If we cut the z-plane from z_1 to z_2 by a straight line segment in the z-plane, we cannot wind around either z_1 or z_2 alone without crossing the cut. However, we could choose a path which winds around both z_1 and z_2. But now both $\arg(z - z_1)$ and $\arg(z - z_2)$ change by 2π, both the factors $\sqrt{z - z_1}$ and $\sqrt{z - z_2}$ change sign, and there is no change in w. The other "branch" of the graph is obtained by a global change of sign.

Thus either branch of $w(z)$ is single-valued in the cut plane.

In this case $z = \infty$ is not a branch point. Indeed, using the transformation $z = 1/t$, $t \in \mathbb{C}$, which maps $z = \infty$ to $t = 0$ we have

$$w = \pm\sqrt{\left(\frac{1}{t} - z_1\right)\left(\frac{1}{t} - z_2\right)} = \frac{\pm\sqrt{(1 - tz_1)(1 - tz_2)}}{t} .$$

In a small neighborhood of $t = 0$, the numerator is close to 1 and the denominator is a holomorphic function; it follows that $z = \infty$ is not a branch point.

For z outside that line segment the surface $f(z, w) = 0$ falls into two disjoint connected parts. If we adjoin a point at infinity to each of these connected parts, they will be as shown in Example 4.4.10, with the difference that here $z_1 \neq \infty$ (see Fig. 4.8). As in that example, on identifying the appropriate boundary segments ($\alpha_1 \sim \beta_2$ and $\beta_1 \sim \alpha_2$), we see that the Riemannian manifold is in this case also diffeomorphic to $\widehat{\mathbb{C}}$ (with two points removed).

Example 4.4.12. Consider $f(z, w) = w^2 - P_3(z)$ where $P_3(z)$ is a polynomial of degree 3 with distinct roots z_1, z_2, z_3.

Again, to each z there correspond two values of w, one the negative of the other. We go from one to the other by continuing $w(z)$ over any closed path winding once around one of the roots z_1, z_2, z_3. For $w = \sqrt{a}\sqrt{z-z_1}\sqrt{z-z_2}\sqrt{z-z_3}$, and the factor $\sqrt{z-z_i}$ changes sign when $\arg(z-z_i)$ changes by 2π. If we cut the z-plane from z_1 to z_2, we cannot wind around either z_1 or z_2 alone without crossing the cut.

In this case $z=\infty$ is a branch point. Indeed, using the transformation $z = 1/t$, $t \in \mathbb{C}$, which maps $z=\infty$ to $t=0$, we have

$$w = \pm\sqrt{\left(\frac{1}{t}-z_1\right)\left(\frac{1}{t}-z_2\right)\left(\frac{1}{t}-z_3\right)} = \frac{\pm\sqrt{(1-tz_1)(1-tz_2)(1-tz_3)}}{t\sqrt{t}} .$$

In a small neighborhood of $t=0$ the numerator is close to 1 but the denominator $t\sqrt{t}$ has a branch point in $t=0$. Hence $z=\infty$ is a branch point.

However, we could choose a path which winds around both z_1 and z_2. But now both $\arg(z-z_1)$ and $\arg(z-z_2)$ change by 2π, both the factors $\sqrt{z-z_1}$ and $\sqrt{z-z_2}$ change sign, and there is no change in w. We next cut the z-plane from z_3 to ∞. This prevents us from winding around all three of the roots z_1, z_2, and z_3. Thus either branch of $w(z)$ is single-valued in the cut plane.

Make cuts in $\widehat{\mathbb{C}} = \mathbb{C} \cup \{\infty\}$ from z_1 to z_2 and z_3 to ∞; for z off these slits the extended surface $f(z,w)=0$ again falls into two disjoint connected pieces, as shown in Fig. 4.9. On identifying the appropriate edges of the slits on these two pieces (α_1 with β_2, α_2 with β_1, γ_1 with δ_2, γ_2 with δ_1, as indicated in Fig. 4.9), we obtain the 2-dimensional torus (or "sphere-with-one-handle") with one point removed.

It is easy to see that the torus cannot be mapped onto a sphere topologically. For on the sphere, any closed curve can be deformed to a point and this property is preserved under topological mappings of the surface. This is not the case on the torus.

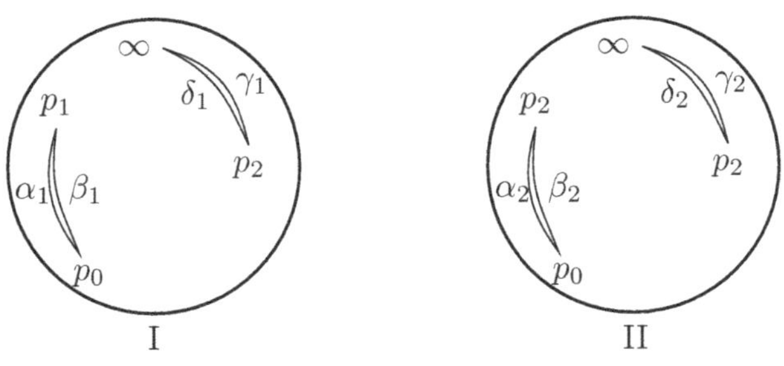

FIGURE 4.9: The points z_0, z_1 and z_2 in $\mathbb{C}$ are corresponding to the points p_0, p_1, and p_2 on the unit sphere Σ, respectively. Two copies of sphere cut along the shortest arc of the great circles on Σ are joining p_0 and p_1, p_1 and p_2, and p_2 and ∞, respectively.

Example 4.4.13. As a final example, consider $f(z,w) = w^2 - P_4(z)$, where $P_4(z)$ is a polynomial of degree 4 with distinct roots z_0, z_1, z_2, z_3. By cutting and pasting as in Example 4.4.12 (with z_3 playing the role of ∞), we again obtain the 2-dimensional torus (with two points removed).

After studying Examples 4.4.10–4.4.13, let us consider the important special case of a *hyperelliptic curve*, given by an equation of the form $w^2 - P_d(z) = 0$ where $P_d(z)$ is a polynomial of degree $d > 4$.

Once again we start by describing the topology.

Proposition 4.4.14. *The Riemann surface of a function of the form $w = \sqrt{P_n(z)}$, where $P_n(z)$ is a polynomial of degree n without multiple roots, is diffeomorphic to a sphere with g handles where $n = 2g+1$ or $n = 2g+2$ (strictly speaking with certain points removed, namely those corresponding to the points at infinity of the original surface.)*

Proof. Suppose first that n is even, and write $n = 2g+2$. Pair off the roots of $P_d(z)$ arbitrarily, and join the members of each pair by an arc in such a way that no two arcs intersect (see Fig. 4.10).

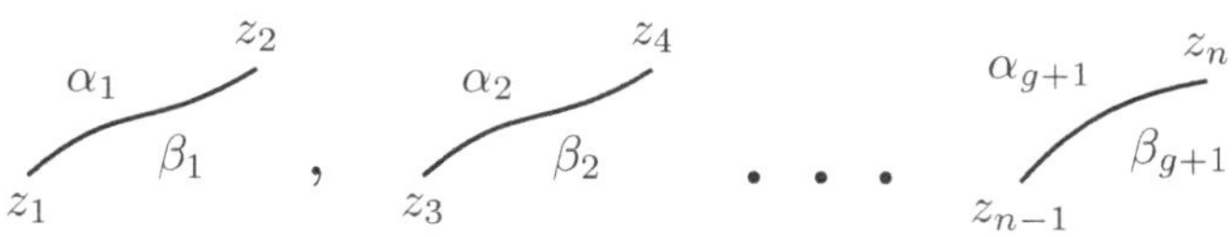

FIGURE 4.10: Join each pair of roots of $P_d(z)$ by an arc α_j in such a way that no two arcs intersect.

If we cut the z-plane along each of these $g+1$ arcs, i.e., if we remove the points on these arcs, then the surface falls into two disjoint connected parts U_1 and U_2. (If we move around any pair of roots on the original surface, we stay on the same branch.) The edges α_i, β_i of the i th cut lie on (or rather are boundary segments of) different connected pieces U_1, U_2. We now glue these edges back together as follows:

$$(U_1, \alpha_i) \sim (U_2, \beta_i) \ , \quad (U_1, \beta_i) \sim (U_2, \alpha_i) \ .$$

[This assertion is justified by the fact that if, on the original surface, we move along the piece U_1 approaching the edge α_i, then on crossing it we pass smoothly over onto the branch U_2 (with corresponding edge β_i), and similarly if we approach on U_1 the edge β_i, we cross over onto U_2 (with corresponding edge α_i).]

For odd n the construction is similar, with $z_{n+1} = \infty$ taken as the $(n+1)$-st branch point. □

Most of the examples are well known (see [34, 44, 170]). The list can be continued, but let us focus on these sources.

One of the most important and remarkable results of linear algebra is the fact that an $n \times n$ matrix over a commutative ring R is invertible if and only if its determinant is invertible in the ring R. This result forms the starting point of several well-known results in analysis: let $U \subset \mathbb{R}^n$ be an open set and $F : U \to \mathbb{R}^n$ a C^1-map. Then for each point p in U one has the local inverse function theorem, asserting that F is locally invertible in a neighborhood of p, i.e., with inverse a C^1-map, if and only if the determinant of the Jacobian matrix at p is non-zero, i.e., $\det JF(p) \neq 0$. We get a similar result if we start with an analytic map $F : U \to \mathbb{C}^n$, where $U \subset \mathbb{C}^n$ is an open neighborhood of a point p in U and assume that $\det JF(p) \neq 0$.

This result will play a crucial role in the study of polynomial/holomorphic functions.

Theorem 4.4.15. *Let $P_n(z)$ be a polynomial of degree n with simple roots. Then the Riemann surface*

$$\Gamma = \left\{(z, w) \in \mathbb{C}^2 \mid F(z, w) = w^q - P_n(z) = 0\right\}$$

is a smooth 2-dimensional manifold in $\mathbb{R}^4$.

Proof. Calculating the gradient of the function F we obtain

$$\operatorname{grad}_{\mathbb{C}} F|_P = \left(\frac{\partial F\,(z, w)}{\partial z}, \frac{\partial F\,(z_0, w_0)}{\partial w}\right) = \left(P_n'(z_0), q w_0^{q-1}\right).$$

Thus a point $P_0 = (w_0, z_0)$ is a singular point of the curve if and only if

$$2w_0 = 0, \quad \left.\frac{dP_n(z)}{dz}\right|_{z_0} = 0, \quad w_0^q - P_n(z_0) = 0$$

i.e., if and only if $w_0 = 0$, and z_0 is a root both of the polynomial $P_n(z)$ and its derivative. The assertion $\operatorname{grad}_{\mathbb{C}}|_P = 0$ follows from the well-known equivalence of this condition on z_0, with the condition that z_0 be a multiple root of $P_n(z)$.

By assumption $P_n(z)$ has only simple roots it follows $\operatorname{grad}_{\mathbb{C}}|_P \neq 0$ for every point $P \in \Gamma$. The complex implicit functions theorem tell us that Γ is a complex manifolds (of complex dimension 1) hence a smooth 2-dimensional manifold in $\mathbb{R}^4$. □

Remark 4.4.16. Let $P(z)$ be a polynomial in one variable, which is not a perfect square. Then the polynomial $Q(z, w) = w^2 - P(z)$ is irreducible. Moreover, if $P(z)$ has distinct roots, then Q is nonsingular, and its locus of roots X is a Riemann surface.

A slight generalization will be useful later. If $P(z, w)$ is an irreducible polynomial, then the points on its locus of roots X where P is singular form a finite set. If we delete these points, then the resulting open subset of X is a Riemann surface, using the same charts as given above. This is referred to as the *smooth part* of the affine plane curve X, and in general, if P is an irreducible polynomial, the smooth part of its zero locus is a Riemann surface.

4.4.5 Global Topological Properties

It is probably true that the most significant example of topological ideas arose in the theory of Riemann Surfaces of algebraic functions. Let us begin, therefore, by describing briefly this example.

If we start from a (non-singular) algebraic curve in the complex projective plane, this defines a compact Riemann surface. Underlying this is a real two-dimensional differentiable manifold, and finally there is the underlying topological space. In other words, we have here a hierarchy of structures

$$\text{algebraic} \rightarrow \text{holomorphic} \rightarrow \text{differentiable} \rightarrow \text{topological.}$$

There are then two basic questions we can ask in this situation.

First, we can ask for invariants of the underlying topological space. Secondly, we can ask for an interpretation of these invariants in terms of the various "superstructures."

One tool that topologists have developed for determining whether or not two topological spaces are homeomorphic is called an *invariant.* An invariant is a function I on the set of all topological spaces of a certain type such that $I(S)$ is equal to a well-defined mathematical object for every appropriate topological space S.

In our particular example there is essentially only one topological invariant of a surface, namely the genus (or number of handles), and the classical results of the theory tell us that this number is also the dimension of the space of holomorphic (or algebraic) differentials. This interprets the genus g in terms of the algebraic or holomorphic structures. The famous theorem of Gauss that the integral of the curvature is equal to $2-2g$ may be regarded as a satisfactory explanation of g in differential terms.

The topological structure of a compact Riemann surface X is determined by its genus g. Topologically, a Riemann surface of genus g is a sphere with g handles or, equivalently, a torus with g holes.

Let us examine the global topological properties of Riemann surfaces. We restrict our attention to the case

$$f(z, w) = w^2 - P_n(z)\,,$$

where $P_n(z)$ is a polynomial without multiple roots. The surface X is defined in $\mathbb{C}^2$ as the graph

$$X = \left\{(z, w) \in \mathbb{C}^2 \;\middle|\; w^2 - P_n(z) = 0\right\}\,.$$

Our goal is to study the behavior of X at infinity.

It is necessary first to clarify what is meant by *"at infinity."* There are several distinct ways to endow this notion with a rigorous meaning. This problem, known as the **compactification problem**, is related to the broader question of how one can transform non-compact open manifolds into compact ones by "attaching a boundary."

In general, there is no universal solution to this problem, since the concept of compactification depends on the specific context. The same open manifold may admit different compactifications (if it is compactifiable at all), depending on the requirements of the problem at hand.

Consider the complex line $\mathbb{C}^1(z)$ and adjoin to it the point at infinity, that is, attach to the real plane $\mathbb{R}^2(x, y)$ the point ∞. In doing so,

$$\mathbb{C}^1 \cup \{\infty\}$$

is transformed into the two-sphere S^2, which we shall refer to as the **extended complex line**.

There exists a canonical way to identify this extended complex line with the one-dimensional complex projective space $\mathbb{CP}^1$. Let us adopt for $\mathbb{CP}^1$ the homogeneous coordinate model $(\lambda z^1, \lambda z^2)$ with $\lambda \neq 0$, and consider the mapping

$$h : (\lambda z^1, \lambda z^2) \longmapsto \frac{\lambda z^1}{\lambda z^2} = \frac{z^1}{z^2} \in S^2 = \mathbb{C}^1(z) \cup \{\infty\}.$$

If

$$\frac{a}{b} = \frac{c}{d} = \rho,$$

we may take $a = c = 1$, hence $b = d = \rho^{-1}$, and therefore $(1, \rho^{-1}) = (1, \rho^{-1})$. Thus, at the "finite" points of the extended plane $\mathbb{R}^2 \cup \{\infty\} = S^2$, the mapping h is a homeomorphism; adjoining the point ∞ preserves this property.

Consequently, we may identify $\mathbb{CP}^1 \cong S^2$. It is worth recalling that this identification of the extended complex plane with the sphere S^2 can be realized by means of stereographic projection.

Consider the case where $P(z, w)$ is a polynomial of degree d in the variables z, w. On making the substitution $z = y^1/y^0, w = y^2/y^0$, we obtain

$$P(z, w) = \frac{1}{(y^0)^d} Q_d\left(y^0, y^1, y^2\right),$$

where Q_d is a homogeneous polynomial in three variables. This furnishes a device for re-realizing our surface $P(z, w) = w^2 - P_d(z, w) = 0$ in $\mathbb{C}^2$, as the surface in the projective space $\mathbb{CP}^2$ given by the equation

$$Q_d\left(y^0, y^1, y^2\right) = 0 \tag{4.17}$$

except that the points of the latter surface for which $y^0 = 0$ correspond to "points at infinity" on the original Riemann surface defined by the equation $w^2 - P_d(z) = 0$. The adjunction of these points at infinity has compactified our surface:

Lemma 4.4.17. *The Riemann surface in complex projective space* $\mathbb{CP}^2$ *defined by equation* (4.17), *is compact.*

Proof. The set of zeros of Q_d is clearly a closed set in $\mathbb{CP}^2$. Since $\mathbb{CP}^2$ is compact, and any closed subset of a compact space is compact, the lemma follows. □

Corollary 4.4.18. *Every natural number is the genus of some algebraic curve.*

In this case we have d spheres, and every two of them connect in a pair of points, so in total we have $\binom{d}{2}$ connections. But $d-1$ of them are needed to glue the d spheres to a connected chain without loops; only the remaining ones then add a handle each. So the genus of C_d is

$$\binom{d}{2} - (d-1) = \binom{d-1}{2} = \frac{d(d-1)}{2}\,.$$

This is commonly called the *degree-genus formula* for complex plane curves.

Notice that the modification in the Examples 4.4.10–4.4.13 is made on the domain of the function, not on the range. Riemann initially thought of his sheets as lying over the complex plane in Euclidean 3-space. Much of the subsequent development of the notion of Riemann surface was done to remove the artificiality and arbitrary character of Riemann's embedding in $\mathbb{R}^3$. In those days of Riemann his surfaces were only a helpful means to represent multivalued functions.

Differentiable Manifolds. If we replace the space $\mathbb{C}^n$ with $\mathbb{R}^n$ and require the transition maps to be differentiable rather than holomorphic, we obtain the definition of differentiable manifolds and their local differentiable coordinate systems. These coordinates make it possible to define differentiable mappings between differentiable manifolds. Further details on these manifolds can be found in the works of Dubrovin, Novikov & Fomenko [44], Hirsch [84], and Narasimhan [138].

Any complex manifold M can also be viewed as a differentiable manifold with the same atlas. To do this, we replace the space $\mathbb{C}^n$ with the corresponding real space $\mathbb{R}^{2n}$ and, in coordinates, replace each complex coordinate

$$z_i = x_i + \sqrt{-1}\, y_i$$

with the two real coordinates x_i and y_i. Since the transition mappings are holomorphic, the transition functions in the corresponding real atlas are automatically differentiable.

Clearly

$$\dim_{\mathbb{R}} M = 2 \dim_{\mathbb{C}} M\,,$$

where $\dim_{\mathbb{R}} M$ is the real dimension of M, that is, the dimension of M regarded as a differentiable manifold. Differentiable manifolds of dimension two are called *surfaces*. A Riemann surface is a surface exactly in this sense, or in a still weaker topological sense.

The Examples 4.4.10–4.4.13 prompt two fundamental questions. First, *what topological forms can a Riemann surface assume?* Second, *how can one determine the topological type of a Riemann surface without relying on ad hoc geometric constructions such as cutting and pasting?*

The answer to the first question—subject to certain qualifications—is that every orientable surface admits a complex structure, and hence can be realized as a Riemann surface (see Theorem 5.1.7).

An answer to the second question, applicable to a broad class of surfaces, is furnished by the Riemann–Hurwitz theorem (Theorem 5.10.33), which expresses the topological invariants of a surface in terms of the algebraic data associated with a holomorphic map.

Moreover, when the surface X_P is completed by adjoining its points at infinity, the resulting surface $\widehat{X}_P$ becomes a compact manifold (see Section 4.6).

Compact oriented topological 2-manifolds are completely characterized; any such manifold is homeomorphic to a sphere with g handles for some positive integer g, and this number is the only homeomorphism invariant. A compact oriented surface with one handle is called a *torus*.

Definition 4.4.19. Let X be a compact oriented topological 2-manifold. The number g of handles of X is called the *genus* of X.

Surfaces arise naturally in many different forms, in branches of mathematics ranging from complex analysis to dynamical systems. The classification theorem, known since the 1860's, asserts that all closed surfaces, despite their diverse origins and seemingly diverse forms, are topologically equivalent to spheres with some number of handles or crosscaps. The proofs found in most modern textbooks follow that of Seifert and Threlfall [163]. Seifert and Threlfall's proof, while satisfyingly constructive, requires that a given surface be brought into a somewhat artificial standard form. In [51] the authors present a completely new proof, discovered by John H. Conway in about 1992, which retains the constructive nature of [163] while eliminating the irrelevancies of the standard form. Conway calls it his zero irrelevancy proof, or "*ZIP proof*," and asks that it always be called by this name, remarking that "otherwise there's a real danger that its origin would be lost, since everyone who hears it immediately regards it as the obvious proof." We trust that Conway's ingenious proof will replace the customary textbook repetition of Seifert–Threlfall in favor of a lighter, fat-free nouvelle cuisine approach that retains all the classical flavor of elementary topology.

What we should learn from the examples above:

Algebraic geometry studies the topological and geometric properties of objects defined by polynomial equations, linking it to topology and differential geometry. Unlike many geometric theories, it naturally accommodates singular objects, and analyzing degenerations from smooth to singular cases often simplifies problems. This connects it to singularity theory. While it may seem limiting to use only polynomial equations, a deep theorem shows that for compact objects, allowing holomorphic equations yields no broader class.

Consequently, algebraic geometry often coincides with complex analytic geometry; for example, any holomorphic map from the Riemann sphere $\widehat{\mathbb{C}}$ to itself is necessarily a rational map.

4.5 Puiseux Series

Locally at each finite critical point ζ the following happens: there is a neighborhood of ζ, say U, so that on U at least one of the zeros w_j ceases to be analytic due to the fact that $P(\cdot, \zeta) = 0$ has at least one multiple root.

The roots of a polynomial, especially multiple roots, depend in a fairly bad way on the coefficients. However, if under a holomorphic transformation of the coefficients the number of geometrically distinct roots does not change, then these roots also holomorphically depend on the parameters.

In a neighborhood of a non-critical point $z = a$ (or $z = \infty$) the elements $w_i(z)$ are holomorphic and may be expanded in power series

$$w_i(t) = c_{0i} + c_{1i}t + \dots, \qquad i = 1, \dots, n,$$

with $t = z - a$ (or $t = 1/z$). They converge in each disc which does not contain a critical point.

In the case of a critical point the situation is more complicated. Let now $z = q$ denote a critical point. We describe three circles c_1, c_2, c_3 through q which do not contain another singular point in their interior, such that the discs d_1, d_2, d_3 bounded by these circles overlap and $d_i \subset \mathbb{D}$ (see Fig. 4.11).

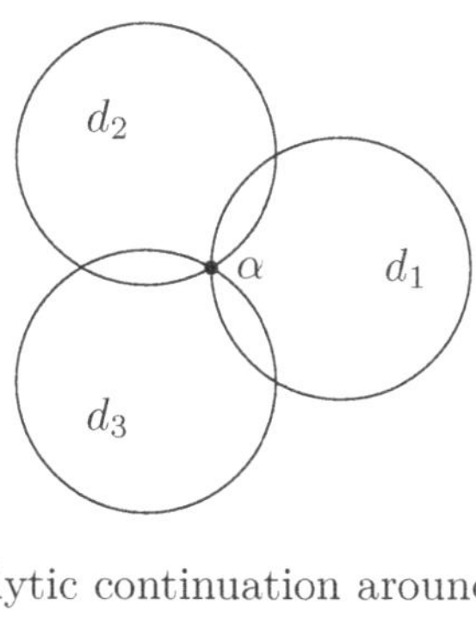

FIGURE 4.11: Analytic continuation around a singular point α.

If we start with a function element $\mathcal{F}_0 = (w_{k_1}, d_1)$ in the first disc we can find an element in the second disc which coincides with $\mathcal{F}_1$ in the common region. Proceeding in this way through d_3 until we return to d_1 then it may occur that we get the same function element $\mathcal{F}_1$, but it is also possible that the process of analytic continuation leads to another element, let us say $\mathcal{F}_2 = (w_{k_2}, d_1)$. Continuing, we may find by this process in d_1 an element $\mathcal{F}_3$, and so on. Now it may happen that for a certain natural number m the function $\mathcal{F}_m = (w_{k_m}, d_1)$ coincides with $\mathcal{F}_0$. Then we shall say that the function elements $\mathcal{F}_0, \ldots, \mathcal{F}_{m-1}$ form a cycle

$$[\mathcal{F}_0, \ldots, F_{m-1}] .$$

These function elements are connected by analytic continuation and form a branch of the given analytic function $\mathcal{F}_1 = (w_1, d_1)$.

Note that in each disc d_i, $i = 1, 2, 3$, there are only n functions (see Fig. 4.11).

What is going on here? We must get back an initial function element in one of the discs after $m \leqq n$ encirclements about the point $z = \alpha$.

Thus we see that the set of n function elements is divided into a certain number of cycles. Suppose that w is defined in D and belongs to a cycle of m elements.

The multi-valued function has, for every point on such a surface that represents its branching, one well-defined value, and can therefore be regarded as a *completely determined function* of position on this surface.

Introducing the uniformizing variable t by $z = \alpha + t^m$, we see that w is holomorphic in a disc about $t = 0$ in the t-plane, except possibly at $t = 0$. If $a_n(\alpha) \neq 0$ then w is also holomorphic at $t = 0$. This may be seen as follows. Let w denote a root of the equation

$$a_0 + a_1 w + \cdots + a_n w^n = 0$$

with $a_n \neq 0$ and let

$$M = \max\left(\frac{|a_0|}{|a_n|}, \ldots, \frac{|a_{n-1}|}{|a_n|}\right) .$$

Then $|w| < 1 + M$. For we have

$$-1 = \frac{a_0}{a_n w^n} + \frac{a_1}{a_n w^{n-1}} + \cdots + \frac{a_{n-1}}{a_n w}$$

and if $|w| \geqq 1 + M$ we should have

$$1 \leqq \left|\frac{a_0}{a_n}\right| \frac{1}{|w|^n} + \cdots + \frac{|a_{n-1}|}{|a_n|} \frac{1}{|w|} \leqq \frac{M}{M+1} < 1 .$$

This is a contradiction.

Hence the functions w_i remain bounded as $t \to 0$ and by Riemann's theorem they are holomorphic at $t = 0$. In this case the point $z = p$ is an ordinary branch point.

There is only the need of a slight modification of the reasoning if $a_n(p) = 0$. Then we consider the function $a_n(z)w$ instead of w. The values of this function remain bounded as $z \to 0$ and it follows that $z = p$ is an algebraic pole.

Summing up we may state: The singularities of an algebraic function are algebraic and they occur at the critical points.

Some frequently occurring singularities will be considered below in somewhat more detail.

Example 4.5.1. Let a punctured disk be represented by

$$\mathbb{D}^* = \{z \in \mathbb{C} \mid 0 < |z| < 1\} ,$$

and consider a fixed point, for instance the point $a = r + i0$, $0 < r < 1$.

Let $\gamma : I \to \mathbb{D}^*$, $\mathbb{D}^* \subset \mathbb{C} \setminus \{0\}$, be any loop in $\mathbb{D}^*$ passing through z_0; then there is an $n \in \mathbb{Z}$ such that $\gamma \sim \gamma_0^n$, where

$$\gamma_0(t) = re^{2\pi it}, \quad 0 \leqslant t \leqslant 1 .$$

Hence

$$\gamma_0^n(t) = r^n e^{2\pi int}, \quad 0 \leqslant t \leqslant 1 .$$

Since $\gamma : I \to \mathbb{C} \setminus \{0\}$, by Theorem 1.8.6 there exists a continuous choice of argument for γ, i.e.,

$$\gamma(t) = |\gamma(t)|\, e^{i\varphi(t)}, \quad t \in [0, 1] ,$$

where $\varphi(t) : I \to [0, 2\pi]$ is a continuous function on $[0, 1]$.

Consider the mapping $\Gamma : I \times I \to \mathbb{D}^*$ defined by the formula

$$\Gamma(s, t) = r^s |\gamma(t)|^{1-s} e^{i\{2\pi nts + \varphi(t)(1-s)\}} .$$

It is continuous and for $s = 0$ coincides with

$$\Gamma(0, t) = |\gamma(t)| e^{i\varphi(t)} = \gamma(t)$$

and for $s = 1$

$$\Gamma(1, t) = re^{2\pi int} = \gamma_0^n(t) .$$

Thus $\Gamma(s, t)$ realizes a homotopy of γ with γ_0^n.

If $\gamma \sim \gamma_0^{n_1} \sim \gamma_0^{n_2}$, then the loop $\delta = \gamma_0^{n_1 - n_2}$ is homotopic to the constant loop $e(t) \equiv z_0$ inside $D^*(a, R)$. Therefore, by Cauchy's theorem,

$$2\pi i\,(n_1 - n_2) = \int_\delta \frac{dz}{z} = 0 ,$$

i.e., $n_1 = n_2$. It follows that there is a least integer n such that $\gamma \sim \gamma_0^n$.

Definition 4.5.2. The *image* of a point z in the domain of a function f is the point $w = f(z)$ in the range of f. For a set S a subset of the domain of f, the set $T = \{w = f(z) \mid z \in S\}$ is the image of S. The inverse image of a point w in the range of the function f is the set of all points z in the domain of f that have w as their image, $f^{-1}(\{w\}) = \{z \in \mathbb{C} \mid f(z) = w\}$. The inverse image of a set T in the range of the function f is the set of values z that have their image in T, $f^{-1}(T) = \{z \in \mathbb{C} \mid f(z) \in T\}$.

The existence of an inverse image is not to be confused with the existence of an inverse function. For example, consider $f(z) = z^4$. The image of $\{1, i, -1, -i\}$ is $f(\{1, i, -1, -i\}) = \{1\}$, and so the inverse image of $\{1\}$ is $f^{-1}(\{1\}) = \{1, i, -1, -i\}$. Notice that f is not a one-to-one function and so has no inverse function.

Example 4.5.3. In the complex plane $\mathbb{C}$ with variable ζ we consider the holomorphic map $\tau : \mathbb{D} \to \mathbb{D}$ given by

$$\tau(\zeta) = \zeta^n . \tag{4.18}$$

If U is an open set in the complex plane $\mathbb{C}$, then a function $h : U \to \mathbb{C}$ is conformal if and only if it is holomorphic and its derivative is everywhere non-zero on U.

The mapping realized by a function $z = \tau(\zeta) = \zeta^n$ is conformal at all points except $\zeta = 0$, since the derivative $\tau'(\zeta) = n\zeta^{n-1}$ exists and is nonzero at all points $\mathbb{C} \setminus \{0\}$.

If we write the variables ζ and z in exponential form, $\zeta = re^{i\varphi}$, $z = \rho e^{i\theta}$, then the equation $w(\zeta) = \zeta^n$ leads to the equalities

$$\rho = r^n, \quad \theta = n\varphi .$$

From this it is clear that the circles $|\zeta| = r$ are transformed into the circles $|z| = r^n$, and the angle of aperture is $\varphi\, (0 < \varphi < \alpha)$, where $\alpha = 2\pi/n$, with the vertex at the origin (for $n = 2$ see Fig. 4.12). This lies in the plane of the variable ζ, and is mapped onto the angle $0 < \theta < n\alpha$ of the plane z.

Let the points ζ_1 and ζ_2 be such that $\zeta_2 = \zeta_1 e^{i2\pi/n}$, $n \geqslant 2$. It is easy to see that $\zeta_1 \neq \zeta_2$, and $\zeta_2^n = \zeta_1^n e^{i2\pi} = \zeta_1^n$. Therefore the mapping (4.18) is not univalent in the entire complex plane $\mathbb{C}$, but is so inside any angle of size $\alpha < 2\pi/n$ with vertex at the origin. Of course, this is why the function $f(z) = z^2$ does not have an inverse.

The mapping $\tau(\zeta) = \zeta^n$ has a simple geometric meaning. Circular sectors are transformed into circular sectors, and their solution increases by a factor of n. If the circular sector is not too large, then the mapping of the intersection of the annulus and the circular sector is conformal.

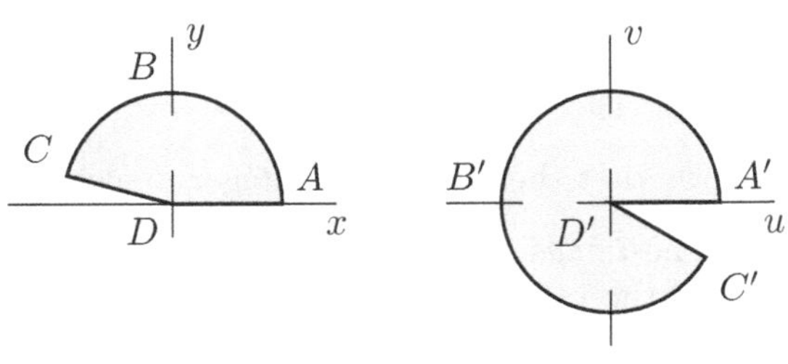

FIGURE 4.12: Geometric meaning of the mapping $\tau(\zeta) = \zeta^2$.

Let

$$V = \{z \in \mathbb{C} \mid 0 < |z - q| < \varepsilon\}$$

and let γ_0 be the circle centered at q passing through $a \in V$. Since function elements $\mathcal{F}_0, \ldots, \mathcal{F}_{m-1}$ form a cycle,

$$[\mathcal{F}_0, \ldots, \mathcal{F}_{m-1}] \,,$$

we have that $\mathcal{F}_k$ is an analytic continuation of $\mathcal{F}_0$ along γ_0^k.

Let $\mathcal{F}$ be a complete analytic function in V that is obtained by analytic continuation from the canonical element, say $\mathcal{F}_0 = (w_1(z-a), U_a)$, $a \in V$, by analytic continuations along all paths starting at the center of the element, along which such a continuation is possible.

Set

$$v = \left\{ \zeta \in \mathbb{C} \;\middle|\; 0 < |\zeta| < \varepsilon^{\frac{1}{m}} \right\} .$$

Let

$$\tau(\zeta) = q + \zeta^m \,.$$

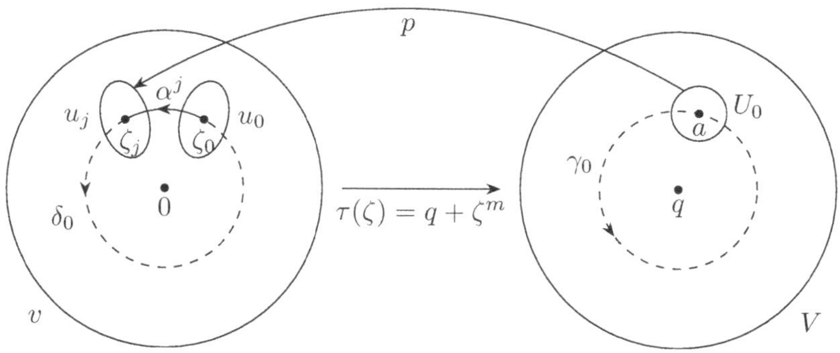

FIGURE 4.13: Geometric meaning of the mapping $\tau(\zeta) = \zeta^m$.

Let ζ_0 be a point in v and its connected neighborhood $u_0 \subset v$ so that the mapping $\tau : u_0 \to U_0$ defines a bijection of u_0 onto the disk $U_0 \subset V$ centered at the point $a = \tau(\zeta_0)$ (see Fig. 4.13).

The preimages u_j, $j = 0, 1, \ldots, m-1$, are obtained from u_0 by the formula $u_j = \alpha^j u_0$, where $\alpha = e^{i2\pi/m}$.

Let δ_0 be the circle centered at $\zeta = 0$ passing through ζ_0. If the point ζ runs through δ_0, then its image $z = \tau(\zeta)$ runs through γ_0^m. The composition of the elements of $\mathcal{F}_j$ with the mapping τ gives

$$G_j := \mathcal{F}_j \circ \tau = (u_0, w_j \circ \tau) =: (u_0, g_j) \,, \quad j = 0, 1, \ldots, m-1 \,.$$

The analytic continuation of G_j along the circle δ_0 corresponds (in the image of the mapping τ) to the analytic continuation of $\mathcal{F}_j$ along γ_0^m, which again leads to the element $\mathcal{F}_j$. Thus, under analytic continuation along δ_0, each element of G_j is mapped to itself. It follows that each element of G_j defines some single-valued holomorphic function $\Phi_j \in \mathcal{O}(v)$.

This shows that the composition $\mathcal{F}\circ\tau$ consists of m single-valued functions $\Phi_0, \Phi_1, \ldots, \Phi_{m-1} \in \mathcal{O}(v)$. All these functions are distinct, since the functions w_j are distinct (see above), and the mapping τ is locally invertible.

Fix any function $\Phi \in \mathcal{O}(v)$ from the set $\{\Phi_0, \Phi_1 \ldots, \Phi_{n-1}\}$. The composition $\Phi \circ \tau^{-1}$ is an analytic function on V obtained by continuing elements of the form $(U_0, \Phi \circ p)$ along all possible paths in V, where $p : U_0 \to V$ (see Fig 4.13) denotes one of the branches of the analytic function

$$\zeta = \tau^{-1}(z) = (z-q)^{1/m} \quad \text{in } V\,.$$

Therefore the analytic function $\Phi \circ \tau^{-1}$ must coincide with $\mathcal{F}$, since it has a common element $(U_0, \Phi \circ p)$ with it. Hence $\Phi \circ \tau^{-1} = \mathcal{F}$.

Any function that is holomorphic in an annulus can be represented there as the sum of a Laurent series. Since $\Phi(\zeta)$ is single-valued and analytic in a punctured neighborhood of the origin, it has a convergent Laurent development of the form

$$\Phi(\zeta) = \sum_{k\in\mathbb{Z}} c_k \zeta^k \quad \text{for } \zeta \in v\,.$$

We formally substitute $\zeta = (z-q)^{1/m}$. We obtain the expansion of w_j in a Puiseux series:

$$w_j(z) = \sum_{k\in\mathbb{Z}} c_{jk}(z-q)^{k/m}\,.$$

If $\Psi \in \mathcal{O}(v)$ is any other function from the set $\{\Phi_0, \Phi_1 \ldots, \Phi_{n-1}\}$, then from the equality $\Phi \circ \tau^{-1} = \Psi \circ \tau^{-1}$ of analytic functions on U (both sides of which are equal to $\mathcal{F}$), it follows that in a neighborhood of the point $z_0 \in V$ the function $\Psi(p(z))$ coincides with one of the elements $\Phi \circ \tau^{-1}$. Therefore, by the above,

$$\Psi(p(z)) = \Phi\left(\alpha^j p(z)\right) \qquad \text{for some } j \in \{1, \ldots, m-1\}\,.$$

Thus we have $\Psi(\zeta) = \Phi\left(\alpha^j \zeta\right)$ for all ζ from some neighborhood of $p(a)$ in v, and hence, by the uniqueness theorem, for all $\zeta \in v$.

We see that locally around a critical point $\zeta \in \mathcal{C}$ there is always a representation of the form (called *Puiseux series*)

$$w_j(z) = \sum_{k=0}^{\infty} c_{jk}(z-\zeta)^{k/\nu_j}$$

with some $\nu_j \geq 1$. If $\nu_j = 1$ then w_j is of course analytic, whereas in all other cases it is analytic as a function of t, which is defined via the uniformizing change of variables $z = \zeta + t^{\nu_j}$.

Riemann proposed replacing the z-plane $\mathbb{C}$ with a surface on which the algebraic function $W(z)$ will be single-valued, while its branches will be the values of $W(z)$ on separate parts of the surface (such surfaces are called *Riemann surfaces*) see Chapter 4. This is the genesis of the idea of a Riemann surface.

It may seem clear from Example 4.5.4 that the number of sheets needs to be precisely n. However, this assertion requires some work as the reader will see below.

Example 4.5.4. For an explicit example of a Puiseux series, consider the irreducible polynomial $P(w,z) = \left(w^2 - 1\right)^2 - z$.

If we solve the above equation for w locally around $(w,z) = (1,0)$ we have

$$w(z) = \sqrt{1+\sqrt{z}} = \sum_{n=0}^{\infty} \binom{1/2}{n} z^{n/2}$$

which converges in $|z| < 1$.

The system $P = 0, \partial_w P = 0$ has solutions

$$(w,z) \in \{(0,1),(1,0),(-1,0)\}.$$

Note that $\partial_w^2 P(w,z) \neq 0$ as well as $\partial_z P(w,z) \neq 0$ at each of these points.

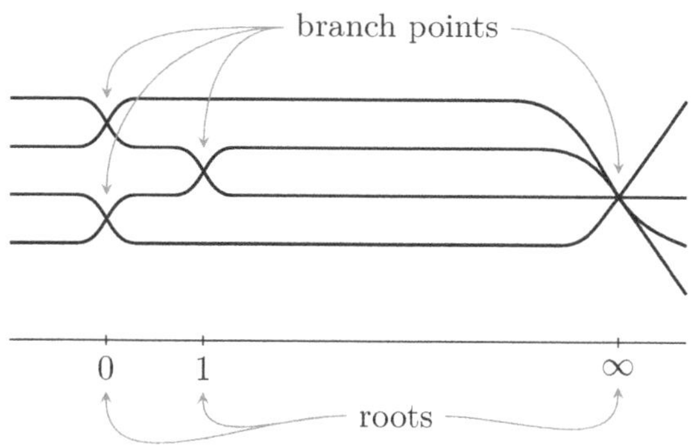

FIGURE 4.14: The sheets of $(w^2-1)^2 - z$.

Thus any branch point associated with them has branching number 1. Moreover, if $z = 0$, then necessarily $w = \pm 1$, whereas $z = 1$ yields $w = 0$ as well as $w = \pm\sqrt{2}$. Finally, $z = \infty$ is a branch point and analytic continuation around a large circle (more precisely, any loop encircling both $z = 0$ and $z = 1$ once) permutes the four sheets cyclically.

For this reason, the sheets look schematically as shown in Fig. 4.14 (the two sheets which branch over $z = 1$ cannot branch again over $z = 0$ since this would contradict the aforementioned cyclic permutation property at $z = \infty$).

Let us now analyze the four sheets and their permutation properties more carefully. First, on the simply connected region

$$\Omega := \mathbb{C} \backslash [-\infty, 0)$$

there exist four branches $f_j(z), 0 \leq j \leq 3$ uniquely determined by the asymptotic equalities

$$f_0(x) \sim x^{\frac{1}{4}}, f_1(x) \sim ix^{\frac{1}{4}}, f_2(x) \sim -x^{\frac{1}{4}}, f_3(x) \sim -ix^{\frac{1}{4}}$$

as $x \to \infty$. The enumeration here has been chosen so that analytic continuation along any circle containing $z = 0, 1$ with positive orientation induces the cyclic permutation

$$f_0 \mapsto f_1 \mapsto f_2 \mapsto f_3 \mapsto f_0 .$$

Next, for all $x > 1$ there are the explicit expressions

$$\begin{aligned} f_0(x) &= \sqrt{1+\sqrt{x}}, \quad & f_2(x) &= -\sqrt{1+\sqrt{x}}\,, \\ f_1(x) &= i\sqrt{\sqrt{x}-1}, \quad & f_3(x) &= -i\sqrt{\sqrt{x}-1}\,. \end{aligned}$$

Analytic continuation to Ω yields

$$f_1(i0+) = f_3(i0-) = f_2(0) = -1, \quad f_1(i0-) = f_3(i0+) = f_0(0) = 1\,.$$

Recall that, by our convention, $\sqrt{z}$ is analytic on Ω with $\sqrt{x} > 0$ if $x > 0$. Analytic continuation around the loop $z(\theta) = 1 + \varepsilon^2 e^{i\theta}, 0 \le \theta \le 2\pi$, with $\varepsilon > 0$ small leaves f_0 and f_2 invariant, whereas f_1 and f_3 are interchanged. Similarly, analytic continuation around $z(\theta) = \varepsilon^2 e^{i\theta}$ (starting with $\theta > 0$ small) takes f_0 into f_3 and f_2 into f_1. This implies that the monodromy group of this Riemann surface is generated by the permutations (12)(03) and (13). The formal definition of the monodromy group is as follows: for any $c \in \pi_1(\mathbb{C}\backslash\{0,1\})$ let $\mu(c) \in S_4$ (the group of permutations on four symbols) be defined as the permutation of the four sheets, which is induced by analytic continuation along the closed loop c. For this we assume, as we may, that c has its base point somewhere in Ω and we pick the four germs defined on each branch over that base point for analytic continuation. Note that μ is well-defined by the monodromy theorem. The map

$$\mu : \pi_1(\mathbb{C}\backslash\{0,1\}) \to S_4$$

is a group homomorphism and the monodromy group is the image $\mu(\pi_1)$. For example, (13)(12)(03) = (0123) which is precisely the cyclic permutation at $z = \infty$. The genus g of this Riemann surface is given by the Riemann–Hurwitz formula as

$$g - 1 = 4(-1) + \frac{1}{2}(1+1+1+3) = -4+3 = -1$$

which implies that $g = 0$.

Now we return to consider the excluded points of the polynomial equation

$$P(w) = w^n + A_{n-1}(z)w^{n-1} + \cdots + A_0(z) = 0$$

with meromorphic coefficients A_i. At a finite excluded point a, we have a punctured disc $0 < |z-a| < R$ which contains no excluded points. Consider a function element θ with center in $0 < |z-a| < R$, which satisfies $P(w) = 0$. If we consider P_{θ_1} we will have m_1 function elements which satisfy $P(w) = 0$. If $m_1 < n$ there will be another function element θ_2 with center in $0 < |z - a| < R$, which satisfies $P(w) = 0$. Finally, we will obtain $P_{\theta_1}, P_{\theta_2}, \ldots, P_{\theta_k}$ with corresponding $m_1, m_2, \ldots, m_k$ and $m_1 + m_2 + \cdots + m_k = n$. Hence the solutions w could be represented by

$$w = \sum_{-\infty}^{\infty} a_\nu^{(j)} (z - a)^{\nu/m_j} \quad , \quad j = 1, 2, \ldots, k\,.$$

But since $w(z)$ can become infinite of at most finite order at $z = a$, we really have

$$w = \sum_{-N_j}^{\infty} a_\nu^{(j)} (z - a)^{\nu/m_j} \quad , \quad j = 1, 2, \ldots, k\,.$$

Furthermore if the complete analytic function S_Q corresponding to the equation $F(w)$ is algebraic, at ∞ we will also have

$$w = \sum_{-N_j}^{\infty} a_\nu^{(j)} \left(\frac{1}{z}\right)^{\nu/m_j} .$$

The series that we have just been writing, power series in a fractional power of the argument with a finite numbers of negative powers is called a "*Puiseux series*."

What appears then to be a natural thing to do would be to adjoin these new "function elements" to our space $\mathfrak{R}$.

A Riemann surface $\mathfrak{R}_f$ can be made a complex manifold exactly as in the case of the square-root function. Each disk D_{z_0} comes with a coordinate: $z - z_0$ or $1/z$ in the case of a regular or polar point, and $(z - z_0)^{1/k}$ or $z^{-1/k}$ in the case of an algebraic branch point. Two such disks are disjoint if the associated functions f disagree on their entire overlap, and are linked if their intersection is non-empty and the associated functions agree on the intersection. The function f is defined, single-valued, and meromorphic on $\mathfrak{R}_f$.

The map $|z, f| \to (z, f(z)) \in \mathbb{S}^2$ identifies the Riemann surface $\mathfrak{R}_f$ with the curve defined by the equation $w = f(z)$.

4.6 Algebraic Curves

Algebraic geometry makes statements about the topological structure of objects defined by polynomial equations. It is therefore related to topology and differential geometry, where similar results are obtained using analytic methods.

Consider the algebraic function $f(z)$ defined implicitly by the equation $P(z, f(z)) = 0$, where

$$P(z, w) = a_n(z) w^n + a_{n-1}(z) w^{n-1} + \cdots + a_0(z) = 0\,.$$

Assume P is irreducible and $q_n \not\equiv 0$. Irreducibility implies that the q_j do not all vanish simultaneously, so $P(z, \cdot)$ is always a polynomial in w of degree $\leq n$.

The relationship of the study of complex algebraic curves with complex analysis arises when one attempts to make sense of "multi-valued holomorphic functions". The foremost characteristic of an algebraic "function" is that it is not a function.

The graph of f is

$$X := \left\{(z, w) \in \mathbb{C}^2 \mid P(z, w) = 0\right\} .$$

This is the right time for a fundamental definition.

Definition 4.6.1. A subset $X \subset \mathbb{C}^2$ is called an *affine algebraic curve* if there exists a polynomial $P \in \mathbb{C}[X_1, X_2]$ such that $\deg f \geq 1$ and

$$X_P = Z(P) = \left\{(x_1, x_1) \in \mathbb{C}^2 : P(x_1, x_2) = 0\right\} .$$

Since

$$Z(P) = Z(\lambda \cdot P) = Z\left(P^k\right)$$

for $\lambda \in \mathbb{C}^*$ and $k \in \mathbb{N}^*$, it is obvious that the polynomial P for a given variety X cannot be uniquely determined. The main result of this chapter will be that this is essentially the only indeterminacy. The situation in the real numbers, on the other hand, is hopeless-as we have already seen in the example of the empty set.

Algebraic plane curves are defined implicitly, as the set of points which satisfy such an equation.

Notice that we call this a curve because it has one complex dimension. However, we will be slightly inconsistent and refer to this occasionally as a surface, especially when we want to emphasize its topological aspects.

The many technical advantages of passing from $\mathbb{R}$ to $\mathbb{C}$ are offset by a loss of direct intuition: from the real point of view, a "curve" in $\mathbb{C}^2$ is a surface in four-dimensional space. We will see latter that this leads to an interesting connection with the Riemann surfaces studied in complex function theory. From the various possibilities of distinguishing a plane in $\mathbb{C}^2 = \mathbb{R}^4$ as "real," the various real curves arise as intersections of a surface in $\mathbb{R}^4$ with different planes.

Definition 4.6.2. Let the complex affine algebraic plane curve $X_P \subset \mathbb{C}^2$ be described by a minimal polynomial $P \in \mathbb{C}[z, w]$. A point p is said to be a smooth, or a *non-singular*, point on X_P if

$$\operatorname{grad}_p P := \left(\frac{\partial P}{\partial z}(p), \frac{\partial P}{\partial w}(p)\right) \neq (0, 0) .$$

Otherwise, p is called a *singular* point on X_P. A curve all of whose points are non-singular is referred to as a non-singular curve.

Thus by definition a singular point is a point on the curve at which the gradient evaluates to the zero vector.

Observe that in that case the affine algebraic plane curve $Z(P)$ is a Riemann surface by the implicit function theorem as at each solution (z, w) we have $\partial P/\partial z \neq 0$ or $\partial P/\partial w \neq 0$.

Complex curves (= compact Riemann surfaces) appear across a whole spectrum of math problems, from Diophantine arithmetic through complex function theory and low dimensional topology to differential equations of math physics.

The study of algebraic curves and Riemann surfaces, involving as it does a rich interplay between algebra, analysis, topology and geometry, with applications in many different areas of mathematics, has been the subject of active research right up to the present day.

The dichotomy Riemann surface/algebraic curve arises from different points of view: analysts think of a real two-dimensional surface with a Riemannian metric which, via isothermal coordinates, defines a holomorphic structure, whereas algebraic geometers think of a complex one-dimensional object.

Remark 4.6.3. Riemann and his followers were investigating complex algebraic curves using complex analysis and topology; other mathematicians began to use purely algebraic methods to obtain the same results. Algebraic curves and surfaces are now reasonably well understood, but the theory of algebraic varieties of dimension greater than two remains very incomplete. (An algebraic variety is, roughly speaking, the set of solutions to finitely many polynomial equations in finitely many variables over $\mathbb{C}$).

If X_P is smooth at p, the line

$$T_pX := \left\{ (z, w) \in \mathbb{C}^2 \;\middle|\; \frac{\partial P}{\partial z}(p) \cdot z + \frac{\partial P}{\partial w}(p) \cdot w = c \right\}$$

is called the *tangent* to X_P at p. Here $c \in \mathbb{C}$ is chosen so that $p \in T_pX_P$.

Just as for differentiable curves, the tangent is the line through p that, near p, passes as close to X_P as possible.

One can also show that if P is irreducible then X_P is connected (see Section 4.7).

The definition has some obvious problems in the case of arbitrary P. Namely, two different polynomials might define the same curve (think of (z, w) and $P^2(z, w)$) and the set X_P might not be connected (for example $P(z, w) = z(z + 1)$). So it is important to use a minimal polynomial because for $P = Q^k$ with $k \geq 2$, X_P would be singular at every point.

Another problem is that the set X_P might not be a smooth subvariety of $\mathbb{C}^2$.

In a neighborhood of a nonsingular point, we can use the implicit function theorem to write x or y as an analytic function of the other variable. So, locally at such a point, X looks like a disk.

By contrast, the *nodal cubic* $y^2 = x^2(x+1)$ looks like a union of two disks touching at $(0,0)$ in a small neighborhood of this point given by $|t \pm 1| < \varepsilon$ in the parametrization

$$x = t^2 - 1$$
$$y = xt\,.$$

The nodal cubic has a singular point at the origin, and it is irreducible meaning that it is not a union of other curves. From an algebraic point of view, this amounts to the irreducibility of the defining polynomial. The singularity is clearly visible from the graph over the reals. Here we see two tangent directions at the origin instead of just one. See Fig. 4.15a for the real picture. The two disks are called *branches* of the singularity.

Singularities may have only one branch, as in the case of the cusp $y^2 = x^3$ (see Fig. 4.15b for the real picture).

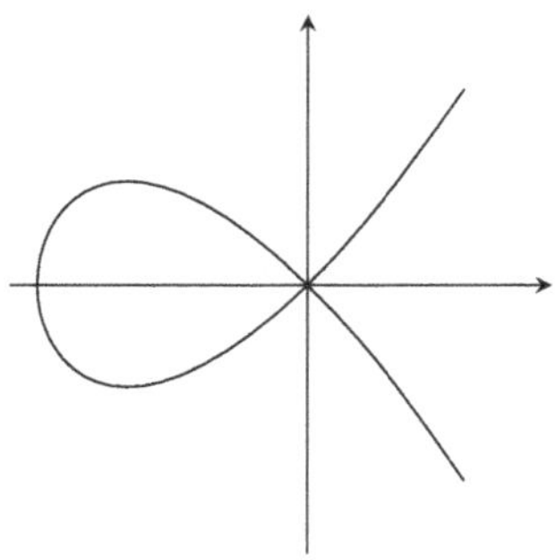

(a) Nodal curve.

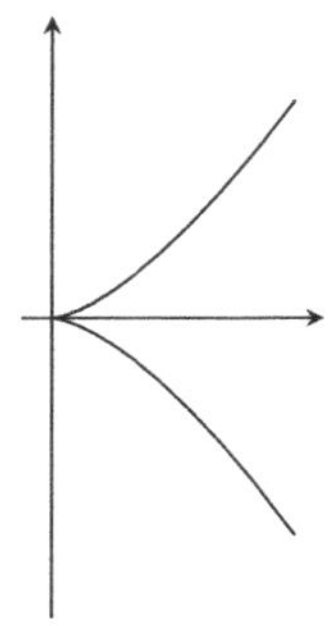

(b) Cuspidal curve.

FIGURE 4.15: Graphs of Nodal and Cuspidal Curves: The Real Case.

In general, every point of intersection of components is singular.

In order to plot the complex graph, we reparameterize it by blowing up the singularity. This means that we separate the lines at the origin, by introducing a new variable $t = y/x$ representing the slope of each line. Plotting this in x, y, z space, with $x = t$, yields the so called *resolution of singularities* of the original curve.

We can find an explicit parameterization by substituting $y = xt$ into

$$y^2 - x^2(x+1) = 0$$

which yields

$$x^2\left(t^2 - x - 1\right) = 0\,.$$

This has two components corresponding to the factorization (see Fig. 4.16). These are given by

$$x = 0, y = 0, z = t\,,$$

and

$$x = t^2 - 1, y = t^3 - t, z = t\,.$$

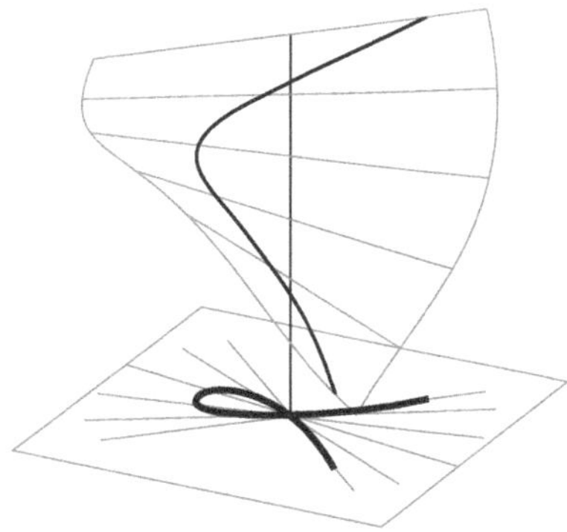

FIGURE 4.16: Resolution of Singularities of a Nodal Curve.

The first two equations provide a rational parametrization of the original curve. As we have seen, the elliptic curve considered above cannot admit such a parametrization. Thus, despite the presence of a singularity, the present example is simpler.

Setting $t = r + is$, with r and s real variables, and letting

$$\begin{aligned} x_1 &= \mathrm{Re}(x) = r^2 - s^2 - 1\,, \\ x_2 &= \mathrm{Im}(x) = 2rs\,, \\ x_3 &= \mathrm{Re}(y) = r^3 - 3rs^2 - r\,, \end{aligned}$$

one obtains a parametrization of the surface.

To visualize the behavior of the function y on this surface, we apply a chessboard pattern corresponding to the preimage of a polar chessboard under y, which reveals its conformal structure.[1]

The resulting image, drawn by Prof. Wegert specifically for this book, is shown in Fig. 4.17. This visualization offers a much clearer insight into the geometric and analytic behavior of the function. Notably, the resulting shape resembles the elliptic curve visualized earlier.

This is because the nodal cubic can be viewed as the limit of elliptic curves

$$y^2 = x(x + \varepsilon)(x + 1)$$

as $\varepsilon \to 0$. In the process, the curve in the previous graph—called a *vanishing cycle* in this context—shrinks to a point. Thus, the global topology changes: it is no longer a torus. In fact, it is not difficult to see from the above parametrization that, after adding a point at infinity, we obtain a sphere with the two points $t = 1$ and $t = -1$ pinched together.

[1] This is explained in detail in Chapter 6 of E. Wegert's book Visual Complex Functions: An Introduction with Phase Portraits [183], which also contains a nice collection of images of special complex functions.

FIGURE 4.17: Graphical representation of the nodal curve using a chessboard phase pattern.

The key notion needed to address the problem of defining an algebraic curve is that of an *irreducible polynomial.*

A (non-constant) polynomial F is said to be irreducible if it cannot be written as $F = Q \cdot R$, where Q and R are non-constant polynomials. Every polynomial can be written in a unique way (up to multiplication by non-zero constants and permutation of factors) as a product of irreducible factors. The following theorem shows that C_F is determined by the irreducible factors of F.

Let us start with the specific elliptic curve

$$y^2 = 4x^3 - x = x(2x-1)(2x+1)\,.$$

When x and y are treated as real variables, the graph of the above equation looks as shown in Fig. 4.18.

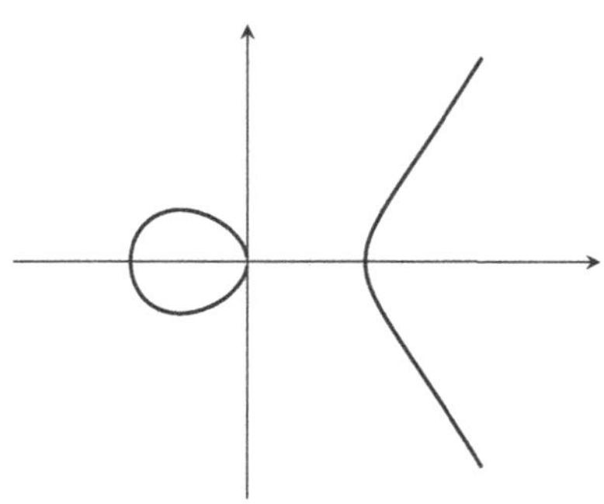

FIGURE 4.18: The real locus of the elliptic curve $y^2 = 4x^3 - x$.

After adding a point at infinity to the curve on the right, we obtain two circles topologically. Now, let us treat the variables x and y as complex. (In fact, historically, significant progress in the study of elliptic integrals was achieved only after the introduction of complex analysis in the 19th century.) With this interpretation, the above equation defines a complex elliptic curve, which is topologically a torus (after adding a point at infinity), as shown in Fig. 4.19.

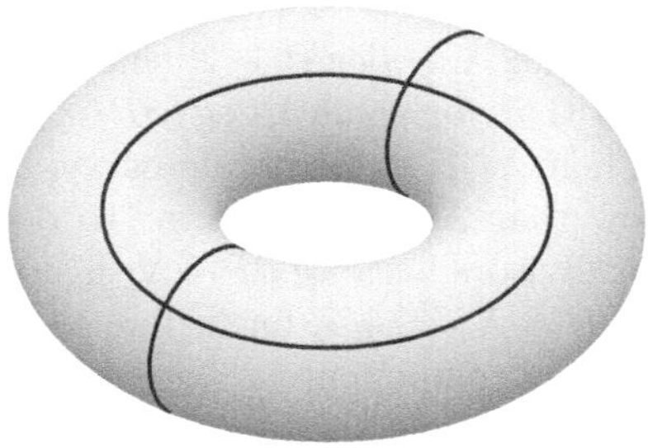

FIGURE 4.19: The complex elliptic curve, topologically a torus.

We will also keep track of the transverse circle in yellow. It may be helpful to think of this as a "purely imaginary" curve, even though this description is not entirely accurate.

The geometric objects studied in algebraic geometry need not be smooth (i.e., they are not necessarily manifolds). Even when our primary interest lies in smooth objects, degenerations into singular ones can greatly simplify a problem (as in Example 4.4.13). This is a key feature distinguishing algebraic geometry from other "geometric" theories, such as differential or symplectic geometry. Naturally, this comes at a cost: the theory must be robust enough to encompass singular objects and to describe how structures vary as we pass from smooth to singular cases. In this respect, algebraic geometry is closely related to singularity theory, which focuses precisely on these questions.

We say that a curve $X_P = \{P = 0\} \subset \mathbb{C}^2$ is irreducible if the polynom P is irreducible. Therefore, if a polynomial is factored into its prime factors as

$$P = P_1^{n_1} \cdots P_k^{n_k} \,,$$

where $n_i \geq 1$, then

$$X_P = X_{P_1 \cdots P_k} \,.$$

We will say that $P_1 \cdots P_k$ is a minimal polynomial. The curves X_{P_i} defined by the irreducible factors of F are the irreducible components of X_P.

Definition 4.6.4. The set $\operatorname{Sing} X := \{p \in X \mid X \text{ is singular at } p\}$ is called the *singular locus* of X.

The affine plane curve X is called *nonsingular* if all its points are nonsingular.

Here is a first nontrivial property of this set.

Proposition 4.6.5. *For an algebraic curve $X_P \subset \mathbb{C}^2$, $\operatorname{Sing} X_P$ is finite.*

Proof. If $X = \{P = 0\}$, set $X_i := \{\partial P/\partial X_i = 0\}$. Then

$$\operatorname{Sing} X = X \cap X_1 \cap X_2 .$$

We may assume that $\deg X = \deg P \geq 2$, since a line is smooth at every point. This also implies that for some $i, \deg(\partial P/\partial X_i) \geq 1$; without loss of generality, let $i = 1$. Then X_1 is an algebraic curve, and $\operatorname{Sing} X \subset X \cap X_1$. Thus it suffices to show that $X \cap X_1$ is finite. To do this, we use Bézout's theorem in the form for divisors because $\partial P/\partial X_1$ is not necessarily a minimal polynomial of X_1 (take $P = X_1 X_2^2 + 1$, for instance). It remains to show that X and X_1 have no common component.

Suppose P and $\partial P/\partial X_1$ have a common prime factor Q. Then

$$P = Q \cdot H \quad \text{and} \quad \frac{\partial P}{\partial X_1} = Q \cdot H_1 = H \frac{\partial Q}{\partial X_1} + Q \frac{\partial H}{\partial X_1} .$$

Hence Q is also a divisor of $H \cdot \partial Q/\partial X_1$. If $\partial Q/\partial X_1 \neq 0$, then Q is a divisor of H, so Q^2 is a divisor of P. But this contradicts the fact that P is a minimal polynomial.

If $\partial Q/\partial X_1 = 0$, then $Q = X_2 - a$. But the coordinates can be chosen at the outset so that X contains no lines parallel to the axes. □

A polynomial is said to be *irreducible* if it cannot be factored into nontrivial polynomials over the same field.

Every polynomial $P \in \mathbb{C}[X_1, X_2]$ of degree ≥ 1 defines a curve

$$Z(P) = \{(X_1, X_2) \in \mathbb{C}^2 \mid P(X_1, X_2) = 0\} \subset \mathbb{C}^2 .$$

If P divides Q, i.e., if $Q = P \cdot H$, then $Z(P) \subset Z(Q)$. Since the ring of polynomials is a unique factorization domain, we have a good understanding of the divisibility properties of polynomials. We would like to use this to draw conclusions about the possible subcurves of a given curve.

The following lemma will help us relate the geometry of the curves to the algebra of the polynomials, and find our way back from the loci of the curves to the polynomials.

For the proof of Lemma 4.6.6 we use a classical geometric method that allows us, by means of a resultant, to reduce the assertion to the fundamental theorem of algebra.

Lemma 4.6.6 (Study's lemma). *Let $P, Q \in \mathbb{C}[X, Y]$. If P is irreducible of degree ≥ 1 and $Z(P) \subset Z(Q)$ then P is a divisor of Q.*

We will give a proof of this lemma to illustrate both a basic technique and the important notion of resultant of two polynomials. The strategy is to reduce the assertion to the fundamental theorem of algebra.

Proof of Lemma 4.6.6. For the proof we use a classical geometric method that allows us, by means of a resultant, to reduce the assertion to the fundamental theorem of algebra. Let

$$P = a_0 X_2^m + a_1 X_2^{m-1} + \cdots + a_m$$
$$G = b_0 X_2^n + b_1 X_2^{n-1} + \cdots + b_n$$

where $a_\mu, b_\nu \in \mathbb{C}[X_1]$ and $a_0, b_0 \neq 0$. Perhaps after switching X_1 and X_2, we may assume that $m \geq 1$. Now consider the sets $Z(P) \subset Z(G)$ and intersect them with the vertical lines $X_1 = x_1$ for all $x_1 \in \mathbb{C}$. We must also have $n > 0$: otherwise choose x_1 such that $a_0(x_1) \neq 0$ and $b_0(x_1) \neq 0$; then $Z(P)$ would intersect the line $X_1 = x_1$ but $Z(G)$ would not. (See Fig. 4.20, 4.21)

Now consider the resultant $\text{Res}(P, G) \in A = \mathbb{C}[X_1]$. Our P was assumed to be irreducible, so if $\text{Res}(P, G) = 0$, then P is a divisor of G by Lemma 11.4.9. Since $\text{Res}(P, G) \in A$ is itself a polynomial, we need only show that $\text{Res}(P, G)(x_1) = 0$ for infinitely many $x_1 \in \mathbb{C}$.

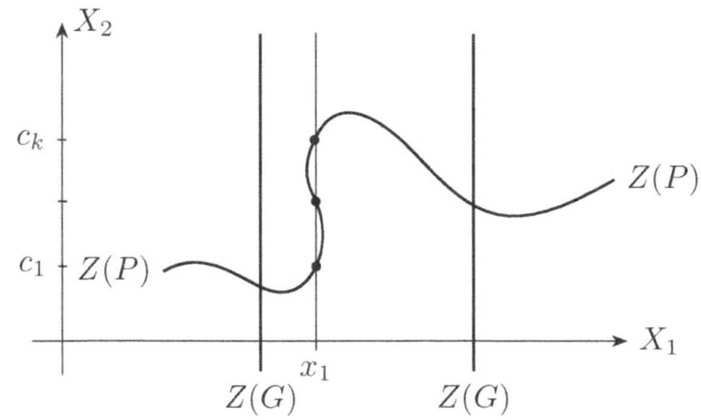

FIGURE 4.20: $\deg(G) = 0$.

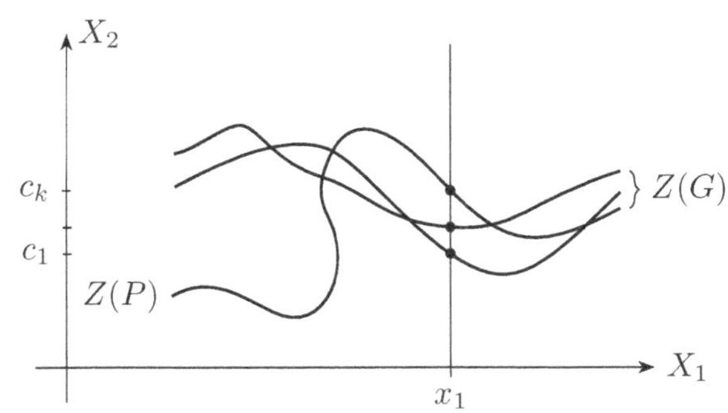

FIGURE 4.21: $\deg(G) > 0$.

Since $a_0, b_0 \neq 0$, we have $a_0(x_1), b_0(x_1) \neq 0$ for almost all $x_1 \in \mathbb{C}$. Setting $X_1 = x_1$ in f and g yields polynomials $P_{x_1}, G_{x_1} \in \mathbb{C}[X_2]$. If P_{x_1} has a zero

$c \in \mathbb{C}$, so does G_{x_1}. Hence

$$X_2 - c$$

is a nonconstant common factor of P_{x_1} and G_{x_1} in $\mathbb{C}[X_2]$. We conclude that $\mathrm{Res}(P, G)(x_1) = \mathrm{Res}(P_{x_1}, G_{x_1}) = 0$ by using the resultant theorem with $A = \mathbb{C}$. □

The analogous statement in the real numbers is obviously false for $P = X_1^2 + X_2^2$ and $G = X_1$, for instance. Lemma 4.6.6 is a precursor of the Hilbert Nullstellensatz [194, vol. II, p. 164]. In the special case of curves, the Nullstellensatz says:

Corollary 4.6.7. *If $P \in \mathbb{C}[X_1, X_2]$ is not a unit, then $\{P = 0\} \neq \emptyset$.*

Proof. Suppose $\{P = 0\} = \emptyset$. If H is an irreducible factor of P, then $\{H = 0\} = \emptyset$. By the lemma, H divides every Q. But this is impossible. □

Remark 4.6.8. By using this method of projecting a curve on a line we can easily see that an algebraic curve $X \subset \mathbb{C}^2$ contains infinitely many points.

Remark 4.6.9. Over $\mathbb{C}$, Hilbert's Nullstellensatz tells us that if an irreducible polynomial f has no zeros, then f must be constant. In other words, an irreducible nonconstant polynomial always defines a nonempty variety.

But over $\mathbb{R}$, the situation changes: an irreducible polynomial can fail to have any real zeros. For example:

$$x^2 + y^2 + a, \quad a > 0$$

is irreducible over $\mathbb{R}$ but has no real solutions, since $x^2 + y^2 \geq 0$ for all reals.

Even quartics like

$$x^4 + y^4 + 1$$

work the same way.

The key reason is that $\mathbb{R}$ is not algebraically closed, so irreducibility doesn't guarantee that the zero set is nonempty in $\mathbb{R}^n$. The "empty curve" happens when the polynomial only has zeros in $\mathbb{C}^n \setminus \mathbb{R}^n$.

This method of projecting a curve onto an axis also easily yields the following result:

Proposition 4.6.10. *An algebraic curve $C \subset \mathbb{C}^2$ contains infinitely many points.*

When do two polynomials of complex variables define the same curve? For complex algebraic curves we have a simple answer to the question when two polynomials define the same curve. The following is an immediate consequence of Lemma 4.6.6.

Lemma 4.6.11. *A curve $X = Z(P) \subset \mathbb{C}^2$ is irreducible if and only if there exists $k \in \mathbb{N}^*$ and an irreducible $G \in \mathbb{C}[x, y]$ such that $P = G^k$.*

The analogous result does not hold for real algebraic curves: there are many irreducible polynomials defining the empty curve.

This lemma and the prime factorization of a defining polynomial implies the following.

Theorem 4.6.12. *Any algebraic curve $X \subset \mathbb{C}^2$ is the union $X = X_1 \cup \cdots \cup X_r$ of irreducible curves. The X_i, $1 \le i \le r$ are called the* components of X, *and they are unique up to a permutation.*

4.7 The Connectedness of Algebraic Curves

An important property of irreducible algebraic curves is the following.

Theorem 4.7.1 (the connectedness of algebraic curves). *C irreducible $\Longrightarrow C \backslash \pi^{-1}(D)$ is connected ($\Longrightarrow C$ connected).*

Proof. We prove this theorem using some results from complex analysis. We are given an affine algebraic curve $C = \{f(x, y) = 0\}$ with:

$$f(x, y) = y^n + a_1(x)y^{n-1} + \cdots + a_n(x) = 0\,. \tag{4.19}$$

The polynomial $f(0, y) = y^n + a_1(0)y^{n-1} + \cdots + a_n(0)$ is a polynomial in y and by the fundamental theorem of algebra can be factored as $(y - y_1(0)) \cdots (y - y_n(0))$, where $y_1(0), \ldots, y_n(0)$ are complex numbers. It follows that $f(0, y) \neq 0$ on the circle $|y| = r$, where $r = 1 + \max\{|y_1(0)|, \ldots, |y_n(0)|\}$. Hence there is, by the continuity of f, a disk $U \ni 0$ in $\mathbb{C}$ such that $f(x, y) \neq 0$ for $x \in U$ and $|y| = r$. The logarithmic residue theorem implies that, for fixed x, the number of zeros of the function $f(x, y)$ in the disk $V = \{y \in \mathbb{C} \mid |y| < r\}$ is

$$n(x) := \left[\frac{1}{2\pi i} \int_{|y|=r} \frac{\partial f(x, y)}{\partial y} / f(x, y)\right] dy\,.$$

This integral obviously depends continuously on x. Since it takes only integer values, $n(x) \equiv n$ in U, i.e., for each fixed $x \in U$ the polynomial $f(x, y)$ has exactly n zeros in V, counted with multiplicities. Suppose that $y_1(x), \ldots, y_n(x)$ are these zeros, each written as many times as its multiplicity (the order is arbitrary). We put

$$P(x, y) := \prod_1^n (y - y_j(x)) =: y^n + c_1(x)\, y^{n-1} + \cdots + c_n(x)$$

and prove that the coefficients c_j holomorphically depend on x. The product $P(x, y)$ is clearly well defined on $\mathbb{C} \backslash D$, since monodromy applied to D simply swaps its factors; hence it is in $\mathcal{O}(\mathbb{C} \backslash D)$.

For this we need two algebraic theorems: Viete's formulas (for coefficients as symmetric functions of the roots) and the fundamental theorem on symmetric polynomials (cf, e.g., [116, §6 Symmetric polynomials, p. 190]), according to which any symmetric polynomial in $\alpha = (y_1, \ldots, y_n)$ can be written as a polynomial in the elementary symmetric functions $\sigma_m(\alpha) = \Sigma_1^n y_j^m$. These theorems imply that $c_f(x) = g_f(\sigma_1(\alpha(x)), \ldots, \sigma_n(\alpha(x)))$, where g_f are polynomials in the variables $\sigma_1, \ldots, \sigma_n$ with coefficients depending only on k and j. Hence, in order to prove holomorphy of the functions c_j it is sufficient to prove holomorphy of the functions $s_m(x) := \Sigma_1^n y_j(x)^m$. Here we use the generalized formula for the logarithmic residue from the theory of functions of one complex variable (see, e.g., [3, p. 153]): if a function h is holomorphic in a disk $|\zeta| < r$, and a function g is holomorphic in it and has no zeros on the circle $|\zeta| = r$, then

$$\frac{1}{2\pi i} \int_{|\zeta|=r} h(\zeta) \frac{g'(\zeta)}{g(\zeta)} d\zeta = \sum_1^n h(y_j),$$

where $y_1, \ldots, y_n$ are the zeros of g in the disk $|\zeta| < r$ (each written as many times as its multiplicity). This theorem can be easily derived from Cauchy's residue theorem. Applying it with $\zeta = y$, $h(\zeta) = y^m$, and $g(\zeta) = f(x, y)$ for fixed $x \in U$, we obtain

$$s_m(x) = \sum_1^n y_j(x)^m = \frac{1}{2\pi i} \int_{\partial U_n} y^m \left(\frac{\partial f(x,y)}{\partial y} / f(x,y) \right) dy$$

and hence $s_m \in \mathcal{O}(U)$. Thus, all c_j are holomorphic in U and hence $P(x, y)$ is a holomorphic function in the domain $U \times \mathbb{C}$.

Finally, because the monic polynomials (in y) f and P have the same roots, it follows that $f(x, y) \equiv P(x, y)$.

We write

$$\begin{aligned} \pi: \quad C &\longrightarrow \mathbb{C} \\ (x, y) &\longmapsto x \end{aligned}$$

for the projection of the curve to the x-axis. Writing $D := \{\mathcal{D}(f)(x) = 0\}$ for the discriminant locus of this projection, we have that for $x \in \mathbb{C} \backslash D$, the fiber $\pi^{-1}(x)$ consists of n distinct points. For some fixed disk $\Delta \subset \mathbb{C} \backslash D$, label these points $\{y_1(x), \ldots, y_n(x)\}$. Notice that $\mathcal{R}\left(f, \frac{\partial f}{\partial y}\right) = \mathcal{D}(f)(x) \neq 0$ implies that $\frac{\partial f}{\partial y} \neq 0$ on $\{f = 0\} \cap \pi^{-1}(\Delta)$, so that the holomorphic implicit function theorem (Theorem 10.1.12) gives $y_i(x) \in \mathcal{O}(\Delta)$. The point here is that the "roots" of (4.19) in y are algebraic—hence multivalued—functions of x, but we can take well-defined holomorphic branches of them over Δ. As we shall see, the multivaluedness will intertwine them outside Δ.

Since P belongs to $\mathbb{C}[x, y]$ and does not factor, then the complete set of "branches" $\{y_i(x)\}$ is acted on transitively by monodromy about D, and one can therefore draw a path on $C \backslash \pi^{-1}(D)$ connecting any two points. □

This theorem is false in the real numbers. The hyperbola, for instance, consists of two connected components but is described by an irreducible polynomial.

4.8 Riemann Surfaces Associated to a Polynomial

As above consider a polynomial

$$P_d(x) = a_d(x)y^d + a_{d-1}(x)y^{d-1} + \cdots + a_1(x)y + a_0(x)\,,$$

and set this expression equal to zero.

As we vary x, we think of there being d solutions for y. Thus, y is an implicitly defined function of x with d possible values that vary holomorphically.

A major motivation for the theory of Riemann surfaces was to provide a proper foundation for such multivalued functions. (They were not developed abstractly as compact complex manifolds of dimension one; rather, they arose as branched covers of $\mathbb{P}^1$ appearing through multivalued functions of this kind.)

We will begin with a complex-analytic discussion. Consider

$$X = \{(x, y) \mid P(x, y) = 0\} \subset \mathbb{C}^2\,,$$

the "complex plane" (an ambiguous term here). Let X^0 denote the smooth locus, so X^0 is a smooth submanifold of $\mathbb{C}^2$. We wish to regard this as a branched cover of $\mathbb{CP}^1$.

We have a projection map $\pi : X \to \mathbb{C}$ that sends a point $p \in X$ to its x-coordinate.

For $p \in X$, there are three possibilities:

1. π is smooth at p. That is, $\frac{\partial P}{\partial y}(p) \neq 0$, so X is smooth at p, and π is a local isomorphism (in either the complex-analytic or étale topology).

2. $\frac{\partial P}{\partial y}(p) = 0$, but $\frac{\partial P}{\partial x}(p) \neq 0$. Here X is smooth at p, but the projection map π is not locally one-to-one. Locally, π is m-to-one and, in local coordinates, looks like $z \mapsto z^m$. This is a branched cover of the disk.

   In this case, we say that p is a branch point of the map, and m is the ramification index.

3. If both derivatives vanish, we have a singular point of the curve (in $X \setminus X^0$).

There is an algebraic explanation, but here we give the complex-analytic one.

However, it does not matter: no matter how bad the singularity is, we can still find a suitable neighborhood of the image point such that, over the punctured disk, the associated map is a covering space. That is, if $q = \pi(p)$, then there exists a small disk Δ containing q such that

$$\pi^{-1}(\Delta^*) \to \Delta^*$$

(for Δ^* the punctured disk) is a covering. This holds because the singular locus is a finite set.

Thus, any covering of a punctured disk Δ^* (say, the unit disk in the complex plane) is of the form

$$\Delta^* \xrightarrow{z \mapsto z^m} \Delta^* .$$

No matter how bad the singularity may appear, if one removes the singular point and its preimage, the result is a disjoint union of punctured disks. Each connected component (a punctured disk) can then be completed by adding one point (turning each punctured disk into a disk). This procedure produces a new complex manifold that projects onto the x-line.

After performing the transformation described above, we obtain a Riemann surface projecting to the x-line. We have resolved the singularities.

Example 4.8.1. Consider the degenerate conic defined by the equation $y^2 - x^2 = 0$, projecting to the x-line. The degenerate conic consisting of two nonparallel lines has a singularity where they meet—at the origin. If we take the preimage over a punctured disk in the x-line, we obtain a union of two punctured disks. Thus the associated Riemann surface has two points above the origin, one for each punctured disk.

Example 4.8.2. Consider the cusp curve defined by $y^2 - x^3 = 0$. See Fig. 4.15b. Take the preimage over the punctured disk $\{x \mid 0 < |x| < \epsilon\}$. Analytically, the curve looks locally like a single punctured disk, so there is one point above the origin in the associated Riemann surface.

Example 4.8.3. Consider $y^3 - x^2 = 0$; here the x-projection is locally three-to-one. The preimage over a small punctured disk in the x-plane is connected, so one point lies above the origin in the Riemann surface.

Thus, to any polynomial equation $P(x, y)$, even a singular one (after possibly adding points), we can associate a branched cover of $\mathbb{C}$ by a Riemann surface X.

Ultimately, we want a compact Riemann surface. We complete $\mathbb{C}$ to $\mathbb{P}^1$ and seek to extend the branched cover $X \to \mathbb{C}$ to a branched cover

$$\overline{X} \to \mathbb{P}^1$$

such that $\overline{X}$ is compact. To do this, take the complement of a large disk,

$$U_R = \{x \mid |x| > R\} .$$

Let $\pi : X \to \mathbb{C}$ be the projection. Then for sufficiently large R,

$$\pi^{-1}(U_R) \to U_R$$

is a covering space. We obtain a covering of the punctured disk U_R. Again, we can complete this to a compact Riemann surface $\overline{X}$ by adding one point to each connected component of $\pi^{-1}(U_R)$.

In this way, we complete X to a compact Riemann surface.

Theorem 4.8.4. *To each equation $P(x, y) = 0$, we can associate a branched cover*

$$\overline{X} \to \mathbb{P}^1$$

where $\overline{X}$ is a compact Riemann surface, by the procedure described above.

This construction also provides a way to compute the number of points in the fiber above ∞: one considers the number of connected components in the preimage of a large disk.

Example 4.8.5. Consider curve $y^2 = x^3$. View this as a two-sheeted cover of the x-line. The branch points occur at the cube roots of unity. Over the complement of a large disk in the x-line, the projection is a covering space. We only need to know whether the preimage of this complement is connected.

It is. To see this, note that as x travels once around a large circle and returns to its starting point, its argument increases by 2π, so the argument of $x^3 - 1$ increases by 6π. Thus, the argument of y increases by 3π, so y changes sign. It follows that the cover is connected, and one point must be added. (The idea is that analytically continuing a germ of a solution for y in terms of x around a circle yields the opposite solution.)

This gives a general recipe: Starting with a polynomial in two variables, one obtains a compact Riemann surface. The same can be done purely algebraically. In algebraic terms, we want to construct a complete variety (the algebraic analogue of "compact") and make it smooth. Typically, one starts with the polynomial $P(x, y)$ and takes the associated curve in $\mathbb{A}^2$. Then we take its closure in $\mathbb{P}^2$, i.e., the hypersurface in $\mathbb{P}^2$ defined by the homogenization of $P(x, y)$. Next, we normalize the projective curve by blowing up to resolve singularities.

If one wishes to proceed algebraically, an alternative is available.

Warning. Do not do this unless necessary. If you can use the analytic approach, do so instead. You may need to blow up a curve multiple times to resolve a singularity (e.g., for a curve like $y^a - x^b$), while the analytic approach is much simpler. Moreover, taking the projective closure is not always well-behaved. There is no reason to expect that the curve naturally belongs in $\mathbb{P}^2$; even the projective closure of a smooth affine curve may be singular. Intuitively examining polynomials and reasoning about their associated Riemann surfaces is often much easier.

We now consider some basic examples of the resolution of singularities by blowing up a point.

Example 4.8.6 (The nodal cubic). Let $C \subset \mathbb{A}^2$ be the cubic given by the equation $y^2 = x^2(x + 1)$; that is, the cubic with equation $y^2z = x^3z + x^2z$

in $\mathbb{CP}^2$ (see Fig. 4.22). This is an irreducible cubic curve with a node at the origin.

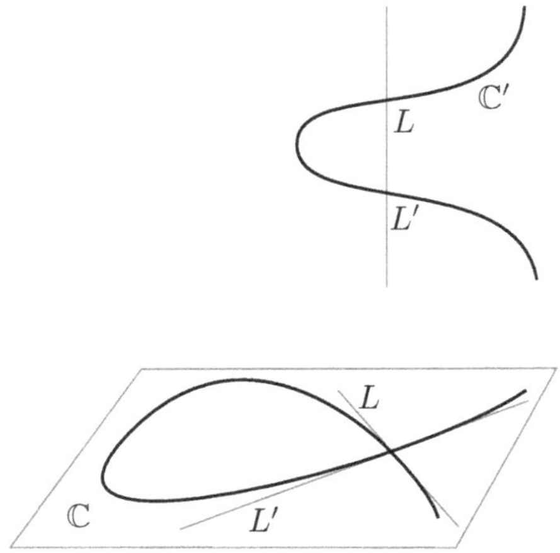

FIGURE 4.22: The nodal cubic.

Blowing up the origin and considering the chart $u \neq 0$, the strict transform $\sigma^{-1}(C)$ is given by

$$\{y^2 = x^2(x+1),\ y = xv\}$$

(where we have set $u = 1$).

Substituting, we obtain

$$(a)\ x^2(v^2 - x - 1) = 0, \quad (b)\ y = xv\,.$$

Equation (a) consists of two components in the coordinates (x, y, v) of $\mathbb{A}^3$. The first is given by $\{x^2 = 0,\ y = 0\}$, which shows that E is counted with multiplicity two in $\sigma^{-1}(C)$. The other component is given by $\{v^2 - x - 1 = 0,\ y = xv\}$, and we denote it by $\widetilde{C}$. This component intersects the exceptional divisor at the points defined by $v^2 = 1$, i.e., $v = \pm 1$, which correspond to the principal tangents of C at O.

Example 4.8.7 (The cusp)**.** Let $C \subset \mathbb{P}^2$ be the cubic given by the equation $zy^2 - x^3 = 0$; that is, the cubic $y^2 - x^3 = 0$ in $\mathbb{A}^2$ under the local chart $z \neq 0$. Examining the blow-up, we find that the exceptional divisor is tangent to $\widetilde{C}$, and the intersection multiplicity is two, although $\widetilde{C}$ itself is smooth. We can improve this situation by performing another blowing up (see Fig. 4.23).

Example 4.8.8 (the tacnode)**.** Let $C \subset \mathbb{P}^2$ be the curve given by the equation $zx^2 - x^4 - y^4 = 0$; that is, the curve $x^2 - x^4 - y^4 = 0$ in $\mathbb{A}^2$ under the local chart $z \neq 0$ (see Fig. 4.24). In this case, a single blowing up does not resolve the singularity, and the multiplicity remains two.

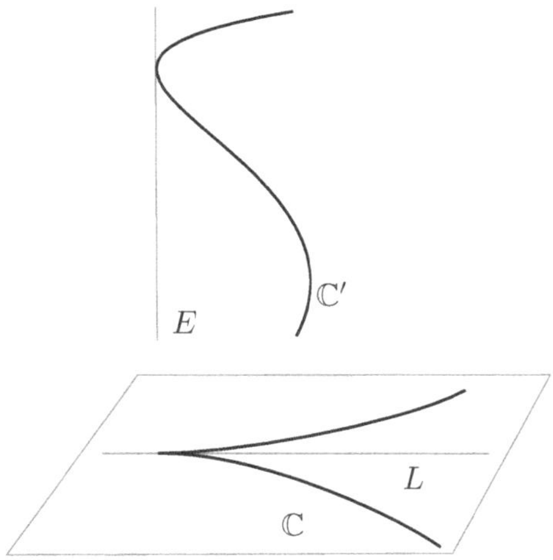

FIGURE 4.23: The cusp.

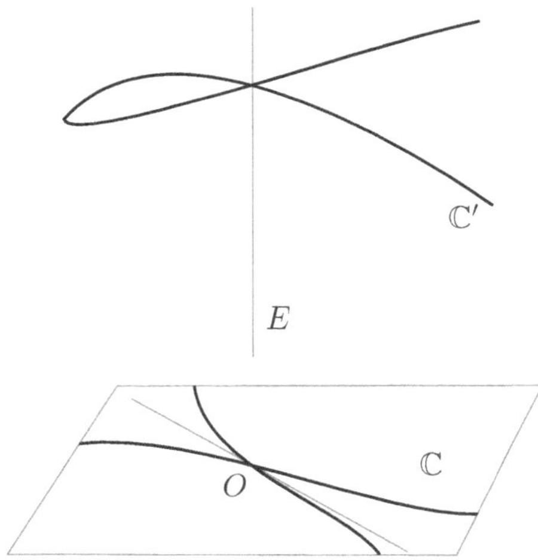

FIGURE 4.24: The tachnode.

4.9 Smooth Projective Plane Curves

In Subsection 4.4.4 we constructed Riemann Surfaces in the complex plane as level sets for a regular value of a polynomial function. In a similar fashion we wish to construct compact Riemann Surfaces as closed subsets of the complex projective plane $\mathbb{CP}^2$. Polynomials in the homogeneous coordinates of the projective plane do not define functions on $\mathbb{CP}^2$.

Example 4.9.1. Consider the polynomial $p(X, Y, Z) = X^2 + Y + Z + 1$. Note that $p(1, 1, 1) = 4 \neq 7 = p(2, 2, 2)$. Since $[1 : 1 : 1] = [2 : 2 : 2]$ are the same point p in $\mathbb{CP}^2$ is attempting to assign two different outputs to the same input, violating the definition of a function.

Affine curves are never compact. Indeed, supposing that $F(x,y)$ depends on x, the function $(x,y) \to x$ is a non-constant holomorphic function on X_F. In order to consider compact surfaces we define projective curves in $\mathbb{CP}^2$. We start with a homogeneous polynomial $F(X,Y,Z)$ defined on $\mathbb{C}^3$.

Definition 4.9.2. A polynomial $P \in \mathbb{C}[X,Y,Z]$ is said to be *homogeneous* of degree d if any of the following equivalent conditions is satisfied:

1. Every monomial of P has degree d;

2. For every $t \in \mathbb{C}$,
$$P(tX,tY,tZ) = t^d P(X,Y,Z)\,;$$

3.
$$X\frac{\partial P}{\partial X} + Y\frac{\partial P}{\partial Y} + Z\frac{\partial P}{\partial Z} = dP$$
(here dP denotes the number d times the polynomial P).

We factor homogeneous polynomials by irreducible homogeneous polynomials and Hilbert's Nullstellensatz is valid for homogeneous polynomials. So a projective curve is defined by a minimal polynomial whose irreducible factors have multiplicity one. We define, as for affine curves, the irreducible components of X_F to be the projective curves defined by the irreducible factors of F.

Definition 4.9.3. Let $Q(X,Y,Z)$ be a homogeneous, nonconstant polynomial of degree d, in the complex variables X,Y and Z with complex coefficients. The locus
$$\Gamma = \big\{[X:Y:Z] \in \mathbb{CP}^2 \;\big|\; Q(X,Y,Z) = 0\big\} \tag{4.20}$$
is the *projective complex curve* defined by the polynomial Q.

Lemma 4.9.4. *The Riemann surface in complex projective space* $\mathbb{CP}^2$ *defined by equation* (4.20), *is compact.*

Proof. The set of zeros of Q is clearly a closed set in $\mathbb{CP}^2$. Since $\mathbb{CP}^2$ is compact, and any closed subset of a compact space is compact, the lemma follows. □

Compact Riemann surfaces can be embedded into projective spaces and are thus, by virtue of Chow's theorem, algebraic curves.

Definition 4.9.5. The curve (4.20) defined by the zeros of the homogeneous polynomial $Q(X,Y,Z)$ is *non-singular* if there are no non zero solutions to the equations
$$Q = \frac{\partial Q}{\partial X} = \frac{\partial Q}{\partial Y} = \frac{\partial Q}{\partial Z} = 0\,.$$
This is the projective curve defined by the polynomial Q.

A *curve* with one or more singular points is called *singular*; a curve with none is called *smooth.*

The singular points are those points on the curve where both partial derivatives vanish.

We recall that the projective space $\mathbb{CP}^n$ is the quotient of $\mathbb{C}^{n+1}\backslash\{0\}$ by the equivalence relation that identifies vectors v and αv in $\mathbb{C}^{n+1}\backslash\{0\}$ with $\alpha \in \mathbb{C}^*$. Namely $\mathbb{CP}^n = \mathbb{C}^{n+1}\backslash\{0\}/\mathbb{C}^*$. The space $\mathbb{CP}^0$ is a single point; note that $\mathbb{CP}^1$ can be thought of as the complex plane $\mathbb{C}$ plus a single point ∞ and it can be identified with the Riemann sphere. We see also that $\mathbb{CP}^2$ can be thought of as $\mathbb{C}^2$ together with a line at infinity, namely a copy of $\mathbb{P}^1$ and so on.

The projective line is the simplest example of a compact Riemann surface. For example, the space $\mathbb{CP}^2$ can be covered with three open sets homeomorphic to $\mathbb{C}^2$:

$$\begin{aligned} U_0 &= \left\{[X:Y:Z] \in \mathbb{CP}^2 \mid x \neq 0\right\} \\ U_1 &= \left\{[X:Y:Z] \in \mathbb{CP}^2 \mid y \neq 0\right\} \\ U_2 &= \left\{[X:Y:Z] \in \mathbb{CP}^2 \mid z \neq 0\right\}. \end{aligned}$$

The homeomorphism on U_0 is given by the map

$$[X:Y:Z] \to (Y/X, Z/X) \in \mathbb{C}^2$$

and similarly for the other open sets U_1 and U_2.

Proposition 4.9.6. *Let $Q(X,Y,Z)$ be a homogeneous polynomial such that the projective plane curve C that is the zero locus of Q in $\mathbb{CP}^2$ is a smooth compact Riemann surface. At every point of C one can take as a local coordinate a ratio of the homogeneous coordinates.*

Proof. Suppose that C is a smooth projective curve. In order to give a complex structure on C let us recall that each C_i is a smooth irreducible affine plane curve and hence a Riemann surface. The coordinate charts are given by the projections. For example for the curve C_0 the coordinate charts are y/x or z/x and the transition functions are as the same as the one obtained for smooth affine plane curves. We have then to check that the complex structures given on each C_i are compatible. Let $P \in C_0 \cap C_1$ where $P = [X:Y:Z]$ and $X \neq 0$ and $Y \neq 0$. Since each affine plane curve is non-singular, we assume without loss of generality that F_X and F_Y are non-zero. Let $\phi_0 : C_0 \to \mathbb{C}$ with $\phi_0(P) = Y/X$ and with inverse $\phi_0^{-1}(Y/X) = [1 : Y/X : h(Y/Y)]$ where h is a holomorphic function. Let $\phi_1 : C_1 \to \mathbb{C}$ with $\phi_1(P) = Z/Z$ with inverse $\phi_1^{-1} = \left[g\left(\frac{Z}{Y}\right), 1, \frac{Z}{Y}\right]$ where $g\left(\frac{Z}{Y}\right)$ is holomorphic for $Y \neq 0$ and non zero since we assume $X \neq 0$. Then $\phi_1 \circ \phi_0^{-1}(Y/X) = Xh(Y/X)/Y$ which is holomorphic because $Y \neq 0, X \neq 0$ and $h(Y/X)$ is holomorphic. In the same way $\phi_0 \circ \phi_1^{-1}(Z/Y) = \frac{1}{g(Z/Y)}$ which is holomorphic because $Y \neq 0$ and g is nonzero. Similar checks can be done with the other coordinate charts. □

Proposition 4.9.7. *An irreducible complex algebraic curve $\{F = 0\} \subset \mathbb{P}^2$ has (if any) finitely many singularities.*

Proof. The complex affine polynomial $f(x, y) = F(1, x, y)$ has multiple roots in y for x in the discriminant locus $\Delta = \{(\mathcal{D}(f))(x) = 0\} \subseteq \mathbb{C}$. We may assume f has positive degree in y, since otherwise $Z(f)$ is just vertical line.

Since f is irreducible in $\mathbb{C}[x, y]$ of positive degree in y, the identical vanishing of $\mathcal{D}(f)$ would imply that $Z(f)$, hence $\bar{Z}(F)$, was reducible. So $\mathcal{D}(f)$ is a nontrivial polynomial, and Δ is finite:

$$\#\{x \in \mathbb{C} \mid \text{ exists } y \text{ such that } f(x, y) = f_y(x, y) = 0\} < \infty .$$

□

Euler's formula. If $F \in \mathbb{C}[X_0, X_1, X_2]$ is homogeneous of degree n, then

$$X_0 \frac{\partial F}{\partial X_0} + X_1 \frac{\partial F}{\partial X_1} + X_2 \frac{\partial F}{\partial X_2} = n \cdot F .$$

This can be checked directly, and is valid for arbitrarily many variables and over any field of characteristic zero. It gives the following result.

Proposition 4.9.8. *Let $C = Z(F) \subset \mathbb{P}_2(\mathbb{C})$ be an algebraic curve, where F is a minimal polynomial, and let $p \in C$. Then*

(a) The curve C is smooth at p if and only if

$$\operatorname{grad}_p F := \left(\frac{\partial F}{\partial X_0}(p), \frac{\partial F}{\partial X_1}(p), \frac{\partial F}{\partial X_2}(p) \right) \neq (0, 0, 0)$$

(b) If C is smooth at p, then the projective tangent T_pC at p is given by the linear equation

$$X_0 \frac{\partial F}{\partial X_0}(p) + X_1 \frac{\partial F}{\partial X_1}(p) + X_2 \frac{\partial F}{\partial X_2}(p) = 0 .$$

Proof. If $p = [p_0 : p_1 : p_2]$, we may assume without loss of generality that $p_0 \neq 0$. An affine part of C that contains p is then described by $f[X_1, X_2] = F[1, X_1, X_2]$, and we have

$$\frac{\partial f}{\partial X_i}[X_1, X_2] = \frac{\partial F}{\partial X_i}[1, X_1, X_2] \quad \text{for } i = 1, 2$$

Thus $\operatorname{grad}_p f \neq 0$ implies that $\operatorname{grad}_p F \neq 0$. Conversely, if $\operatorname{grad}_p f = 0$, then it follows from Euler's formula

$$0 = nF(p) = p_0 \frac{\partial F}{\partial X_0}(p)$$

that $\operatorname{grad}_p F = 0$. This proves (a). The line given in (b) has the right slope, and by Euler it passes through p. □

Our next result follows from this and Proposition 4.6.5.

Corollary 4.9.9. *An algebraic curve $C \subset \mathbb{CP}^2$ has only finitely many singularities.*

4.10 Compact Riemann Surfaces

We establish the first half of the correspondence between compact Riemann surfaces and algebraic functions.

Theorem 4.10.1. *Suppose the Riemann surface R_f is compact. Then $\{(z, f(z))\}$ is an algebraic curve: there exists a polynomial $P(z, w)$ such that*

$$P(z, f(z)) \equiv 0$$

on R_f.

Proof. The poles and branch points of R_f are isolated, hence finite in number. Let $\{z_j\}_{j=1}^d$ denote their base points. Choose $z_0 \in \mathbb{C}$, and let $\{\gamma_j\}$ be pairwise disjoint simple curves joining z_0 to z_j. Let Ω be the complement in $\mathbb{S}$ of $\cup_{j=1}^d \gamma_j$. Then Ω is simply connected.

Suppose f is holomorphic in a neighborhood of $w_0 \in \Omega$. For any $w \in \Omega$, let γ be a curve from w_0 to w. The set of points of γ to which f can be analytically continued is both open and closed in γ; compactness of R_f ensures closedness. Hence f extends to all of Ω. By the monodromy theorem, f is single-valued there.

Thus, for each base point $w_0 \in \Omega$, there are finitely many determinations of f, corresponding to finitely many points of R_f lying over w_0. Let m denote this number. Since it is constant along paths in Ω, R_f consists of m sheets, i.e., m copies of Ω joined along the cuts γ_j.

Number the sheets, and for $z \in \Omega$ let $\{(z, f_j(z))\}_{j=1}^m$ denote the corresponding points of R_f. Define

$$Q(z, w) = \prod_{j=1}^m (w - f_j(z)) = \sum_{k=0}^m (-1)^{m-k} S_{m-k}(z) w^k ,$$

where S_k are the elementary symmetric functions in the f_j:

$$S_0 = 1, \quad S_1 = \sum_{j=1}^m f_j, \quad S_2 = \sum_{j \neq k} f_j f_k, \quad \ldots, \quad S_m = \prod_{j=1}^m f_j .$$

The functions S_k are single-valued on Ω, independent of the choice of z_0 and $\{\gamma_j\}$. They are holomorphic on $\mathbb{S} \setminus \{z_j\}$, and each z_j is an isolated singularity of pole type. Thus each S_k is a rational function.

Let $q(z)$ be the least common multiple of the denominators of $S_0, S_1, \ldots, S_{m-1}$, and define

$$q_k(z) = (-1)^{m-k} q(z) S_{m-k}(z) .$$

Then $q_k(z)$ are polynomials, and

$$P(z,w) = q(z) \prod_{j=1}^{m} (w - f_j(z)) = \sum_{k=0}^{m} q_k(z) w^k$$

is a polynomial in z and w. By construction, $P(z,w) = 0$ whenever $w = f(z)$ for some $(z, f(z)) \in R_f$.

Later we shall show that P is irreducible. The next sections prove the converse: if $P(z,w)$ is irreducible of positive degree in w, then the Riemann surface of a function f satisfying

$$P(z, f(z)) = 0$$

is compact. □

4.11 Projective Algebraic Curves

It is already clear how we define curves in the complex projective plane: as subsets of $\mathbb{CP}^2$ which are the zero sets of homogeneous polynomials

$$F(x_0, x_1, x_2) = \sum_{i+j+k=m} a_{ijk} x_0^i x_1^j x_2^k,$$

with complex coefficients a_{ijk}. The number m is the "order" of the curve. When we want to study these curves, we can view them from several aspects: we can regard such a curve C as a projective algebraic subset of $\mathbb{CP}^2$, i.e., simply as the zero set of a homogeneous polynomial; as a complex analytic subset of the complex manifold $\mathbb{CP}^2$, i.e., a subset which, in a suitable coordinate neighborhood of each point, is the zero set of a complex analytic function; or, finally, as a topological subset of the manifold $\mathbb{CP}^2$.

The first viewpoint is the viewpoint of algebraic geometry, and it is the one we shall mostly adopt later in this book, though not always consistently.

The second viewpoint is that of function theory. In particular, each nonsingular curve C in $\mathbb{CP}^2$ is a 1-dimensional complex manifold by the implicit function theorem, and hence, disregarding the embedding, a one-dimensional compact complex manifold. When the curve has singular points one can associate with it a unique one-dimensional compact complex manifold, its "normalisation," by resolving its singular points. But the one-dimensional compact complex manifolds are precisely the abstract Riemann surfaces studied in complex analysis, and thus from the second viewpoint the study of curves in the complex projective plane is really a part of complex function theory.

There is little difference between the algebro-geometric and function-theoretic starting points. The important theorem of Chow shows that each closed complex analytic subset of $\mathbb{CP}^n$ is a projective algebraic subset.

We begin by considering again the affine curve $C' \subset \mathbb{C}^2$ with the equation $y - x^2 = 0$, and the two-sheeted covering

$$C' \to \mathbb{C}$$

obtained by projection onto the y-axis. When we restrict any entire rational function on $\mathbb{C}^2$, e.g., the function x, to the 1-dimensional complex manifold C', we obtain a well-defined holomorphic function x on C'. (Indeed, we have seen before that one can take x as a coordinate on C'.)

We now consider a point $y \in \mathbb{C}$ and the two points of C' lying over y. Then at these two points the function x has precisely the values

$$x = \pm\sqrt{y}\,.$$

Thus we see the following: on the y-plane a function $\sqrt{y}$ cannot, as is well known, be defined properly, i.e., as a single-valued analytic function of y, because when one travels once around the origin and analytically continues one of the two power series expansions, one returns with the expansion with the opposite sign. But when we pass to the two-sheeted covering $C' \to \mathbb{C}$, the function $x = \sqrt{y}$ becomes a well-defined complex analytic function on C'. If one also admits the value ∞ for y, then one must likewise do so for x; in this way one obtains precisely the branched covering

$$\mathbb{CP}^1 \to \mathbb{CP}^1$$

that we analyzed above, and on the covering space the curve $x = \sqrt{y}$ becomes a well-defined meromorphic function. One calls this branched covering the Riemann surface of the many-valued "function" $\sqrt{y}$.

In general, one can consider a many-valued algebraic "function" $y(x)$ defined by an implicit equation

$$a_m(x)y^m + a_{m-1}(x)y^{m-1} + \cdots + a_0(x) = 0\,,$$

where the $a_i(x)$ are polynomials in x. (In our case the equation was simply $y^2 - x = 0$.) One interprets the $a_i(x)$ as meromorphic functions on $\mathbb{CP}^1$ and y as a many-valued meromorphic "function" on $\mathbb{CP}^1$. One now seeks a 1-dimensional compact complex manifold $\widetilde{C}$ and a holomorphic mapping

$$\widetilde{C} \to \mathbb{CP}^1$$

as a branched covering so that y can be interpreted as a well-defined meromorphic function on $\widetilde{C}$. One expects such a construction to exist, because when one takes an m-sheeted covering, one has in general m points over each point of $\mathbb{CP}^1$, at which the function y can take its m values. Such a $\widetilde{C}$, for which the sheet number of the covering is chosen as small as possible, always exists and is essentially unique. One calls this branched covering $\widetilde{C} \to \mathbb{CP}^1$ the Riemann surface of the many-valued "function" y.

Now this many-valued Riemann surface has a lot to do with algebraic curves in $\mathbb{CP}^2$. Namely, the affine equation

$$a_m(x)y^m + a_{m-1}(x)y^{m-1} + \cdots + a_0(x) = 0$$

also defines a projective algebraic curve C in $\mathbb{CP}^2$, and projection in the direction of the x-axis onto the y-axis defines a mapping

$$C \to \mathbb{CP}^1 .$$

When C is non-singular, as in the previous case of the quadric, then $C \to \mathbb{CP}^1$ is already the Riemann surface of y. When C is singular, one must go to a non-singular normalization $\widetilde{C} \to C$, and the composition

$$\widetilde{C} \to \mathbb{CP}^1$$

is then the desired Riemann surface.

The third, topological, viewpoint is essentially qualitative in comparison with the other two. For that reason it is perhaps especially suitable for obtaining a first impression of the nature of the geometric figures we wish to investigate. In what follows we shall therefore investigate a few simple examples of curves mainly from the topological viewpoint, and they will be curves of orders 1, 2, and 3 in turn—in other words, lines, quadrics, and cubics.

The connection between the algebro-geometric and function-theoretic viewpoints in the investigation of algebraic curves is the following: the concrete Riemann surfaces $\widetilde{C} \to \mathbb{CP}^1$ of many-valued algebraic functions are the branched coverings of $\mathbb{CP}^1$ which result from the composition of a normalization $\widetilde{C} \to C$ and projection $C \to \mathbb{CP}^1$ of an algebraic curve C. In addition, one can show that each abstract Riemann surface $\widetilde{C}$, i.e., each compact 1-dimensional complex manifold, can be viewed as a branched covering $\widetilde{C} \to \mathbb{CP}^1$, i.e., as the concrete Riemann surface of a many-valued algebraic "function."

A good reason to look at algebraic curves in higher dimensional projective spaces is that a general compact connected Riemann surface of genus $g \geq 2$ cannot be realized as a smooth algebraic curve in $\mathbb{CP}^2$. Indeed, the genus of such a Riemann surface is $\frac{1}{2}(n-1)(n-2)$ where n is the degree of the irreducible homogeneous polynomial in $\mathbb{C}[X_0, X_1, X_2]$ which defines it, and clearly, not all positive integers are of this form. Moreover, no hyperelliptic Riemann surface of genus $g \geq 2$ can be embedded in $\mathbb{CP}^2$. However, we have the following theorem: For any compact connected Riemann surface X, there is a holomorphic embedding

$$\varphi : X \to \mathbb{CP}^3 .$$

Of course, we cannot here go precisely into all these things from function theory, and we must refer to textbooks on function theory such as those of Behnke and Sommer [11] or Siegel [165].

4.12 Projective Closure and Affine Restriction

Let us study the global topological properties of Riemann surfaces. Another property we might wish to study is compactness. If we consider plane curves over $\mathbb{R}$, we see that many of them (e.g., a line, or $xy = 1$, or $y = x^2$, or $y^2 = x^3 - x$) have points going "to infinity," and therefore are not compact (in the usual topology on $\mathbb{R}^2$). There are some exceptions, like the circle.

No affine plane curve is compact: as a subset of $\mathbb{C}^2 \simeq \mathbb{R}^4$, it is not a bounded set, since for any fixed z_0 by the fundamental theorem of algebra (or the d'Alembert–Gauss theorem)[2] there will be roots w to the polynomial $f(z_0, w) = 0$. It follows that over $\mathbb{C}$ it turns out that a plane curve always has points going to infinity, and is therefore never compact.

We wish to compactify the picture by adding "points at infinity," and the way to accomplish this is to work not in the affine plane $\mathbb{C}^2$, but in the projective plane $\mathbb{CP}^2$.

We will now extend an affine curve $C \subset \mathbb{C}^2$ to a projective curve $\widehat{C} \subset \mathbb{CP}^2$. To do this, we must decide which points at infinity to adjoin. The complex projective space $\mathbb{CP}^2$, as the quotient of $\mathbb{C}^3$, has a topology, and an obvious possibility is to define $\widehat{C}$ to be the topological closure. The asymptotes of an algebraic curve in the affine plane are the lines that are tangent to the projectivized curve through a point at infinity. A plane curve of degree n intersects its asymptote at most at $n-2$ other points, by Bézout's theorem, as the intersection at infinity is of multiplicity at least two.. Then for every line that is an asymptote of the curve, the corresponding point at infinity is adjoined (see [58, §12.6 Asymptotes], [115, p. 121]). If the curve C is algebraic, then there are only finitely many of them, and the topological construction above can be described by a simple computation with polynomials, as follows.

Given an affine curve $C \subset \mathbb{C}^2$, with equation $f(x, y) = 0$, we may define the projective closure $\widehat{C} \subset \mathbb{CP}^2$ of C. It is defined by the equation $F(x_0, x_1, x_2) = 0$ where $F(x_0, x_1, x_2)$ is constructed by putting $x = \frac{x_1}{x_0}$ and $y = \frac{x_2}{x_0}$ and substituting this in $f(x, y)$, and writing the result as

$$f\left(\frac{x_1}{x_0}, \frac{x_2}{x_0}\right) = \frac{F(x_0, x_1, x_2)}{{x_0}^m}$$

where x_0 does not divide the numerator on the right. Here $F(x_0, x_1, x_2)$ is a homogeneous polynomial with coefficients from $\mathbb{C}$, uniquely determined by $f(x, y)$ as follows: If

$$f(x, y) = \sum_{I=(i_1, i_2)\in\Phi} a_I x^{i_1} y^{i_2}$$

[2] Despite its name, it is not fundamental for modern algebra; it was named when algebra was synonymous with the theory of equations.

where Φ denotes a finite set of tuples of non-negative integers (i_1, i_2), then the degree of f is $d = \max\{i_1 + i_2 \mid (i_1, i_2) \in \Phi\}$, and the projective closure is given by the equation

$$F(x_0, x_1, x_2) = \sum_{I=(i_1,i_2)\in\Phi} a_I x_0^{d-i_1-i_2} x_1^{i_1} x_2^{i_2} = 0\,.$$

Here d is the degree of the original affine curve C as well as of its projective closure C.

Definition 4.12.1. The homogeneous polynomial $F(x_0, x_1, x_2)$ as defined above is denoted by $f^h(x_0, x_1, x_2)$, and referred to as the *homogenization* of the (non-homogeneous) polynomial $f(x, y)$.

The key to understanding the relation between an affine curve and its projective closure lies in the simple and beautiful relation

$$f(a, b) = f^h(1, a, b)$$

which holds for all a and b. Thus if C is the affine curve defined by $f(x, y) = 0$, then the projective closure $\widehat{C}$ of C is defined by the equation $f^h(x_0, x_1, x_2) = 0$. Conversely, if we are given a projective curve $\widehat{C}$ given by the equation $F(x_0, x_1, x_2) = 0$, then we may define its affine restriction to $D_+(x_0)$ as identified with $\mathbb{C}^2$ for the curve given by the equation $F(1, x, y) = 0$. But this affine restriction is not always defined: Namely, if $F(x_0, x_1, x_2) = x_0^d$, then C is the line $L_\infty = V_+(x_0)$, the line at infinity, with multiplicity d. Of course, the affine restriction of this curve to $D_0(x_0)$ is given by the equation $1 = 0$, so we might say that the affine restriction of this curve to $D_+(x_0)$ is empty. On the other hand, if we chose to take the affine restriction to $D_+(x_1)$ instead, and put $x = \frac{x_0}{x_1}$ and $y = \frac{x_2}{x_1}$, then the affine restriction is the curve given by $x^d = 0$, in other words the y-axis counted with multiplicity d.

The mapping

$$[x_0 : x_1 : x_2] \mapsto \left(z = \frac{x_0}{x_2},\ w = \frac{x_1}{x_2}\right)$$

and its inverse

$$(z, w) \mapsto [z : w : 1]$$

establish an isomorphism between the affine part of $\mathbb{CP}^2$ and $\mathbb{C}^2$. The full projective plane is obtained from the affine part $\mathbb{C}^2$ by adding the points at infinity of the form

$$[x_0 : x_1 : 0] \sim \mathbb{CP}^1 \sim S^2\,.$$

The embedding of $C = \left\{(z, w) \in \mathbb{C}^2 \mid P(z, w) = 0\right\}$ into $\mathbb{CP}^2$ is defined as follows: Suppose

$$P\left(\frac{x_0}{x_2}, \frac{x_1}{x_2}\right) = \frac{Q(x_0, x_1, x_2)}{x_2^N}\,,$$

where $Q(x_0, x_1, x_2)$ is a homogeneous polynomial of degree N in x_0, x_1, x_2.

It is assumed that the fraction on the right-hand side is irreducible, i.e., the numerator Q and the denominator x_2^N have no common factors.

This furnishes a device for re-realizing our surface C in $\mathbb{C}^2$ as the surface $\widehat{C}$ in the projective space $\mathbb{CP}^2$ given by the equation

$$Q(x_0, x_1, x_2) = 0 \tag{4.21}$$

except that the points of the latter surface for which $x_2 = 0$ correspond to "points at infinity" on the original Riemann surface

$$C = \{(z, w) \in \mathbb{C}^2 \mid P(z, w) = 0\} .$$

The process of homogenization of a polynomial gives the projective closure of an affine curve. Algebraically, what we did is multiply each monomial of $P(z, w)$ by the appropriate power of x_2 so that we get a homogeneous equation. Having done that, we can also set $x_2 = 0$ to get the points at infinity.

The adjunction of these points at infinity has compactified our surface:

Lemma 4.12.2. *The Riemann surface $\widehat{C}$ in complex projective space $\mathbb{CP}^2$ defined by equation* (4.21) *is compact.*

Proof. In a compact Hausdorff space, a set is compact if and only if it is closed. The set of zeros of Q is clearly a closed set in $\mathbb{CP}^2$. Since $\mathbb{CP}^2$ is compact, and any closed subset of a compact space is compact, the lemma follows. □

Not all the affine plane curves can be compactified in a smooth way by embedding them in $\mathbb{CP}^2$.

Example 4.12.3. A conic curve is a curve defined by a polynomial of degree 2 (also known as a quadratic curve). Consider the conics in $\mathbb{C}^2$

$$C_1 = \{x_2 = x_1^2\}, \quad C_2 = \{x_1 x_2 = 1\} .$$

We extend them to projective curves

$$\widetilde{C}_1 = \{x_0 x_2 = x_1^2\} \ , \quad \widetilde{C}_2 = \{x_1 x_2 = x_0^2\} .$$

Restricting $\widetilde{C}_i$ to the affine chart $\mathbb{C}^2 \subset \mathbb{CP}^2$ recovers C_i for $i = 1, 2$.

Let us first consider $\widetilde{C}_1$. The points added at infinity correspond to $x_0 = 0$. From the defining equation, this implies $x_1 = 0$; the coordinates cannot all vanish, so $x_2 \neq 0$. Thus there is only one point at infinity, namely $[0 : 0 : 1]$. This point represents the *vertical direction* in $\mathbb{C}^2$, which is the direction in which the parabola $x_2 = x_1^2$ extends to infinity (in both directions). See Fig. 4.25a.

For $\widetilde{C}_2$, the added points also satisfy $x_0 = 0$. In this case, $x_1 x_2 = 0$, which yields two points: $[0 : 1 : 0]$ and $[0 : 0 : 1]$ in $\mathbb{CP}^2$. We therefore add two points at infinity—one corresponding to the *horizontal* direction and one to the *vertical* direction in $\mathbb{C}^2$. This agrees with the figure below: the hyperbola $x_1 x_2 = 1$ extends to infinity along both the x_1-axis and the x_2-axis. See Fig. 4.25b.

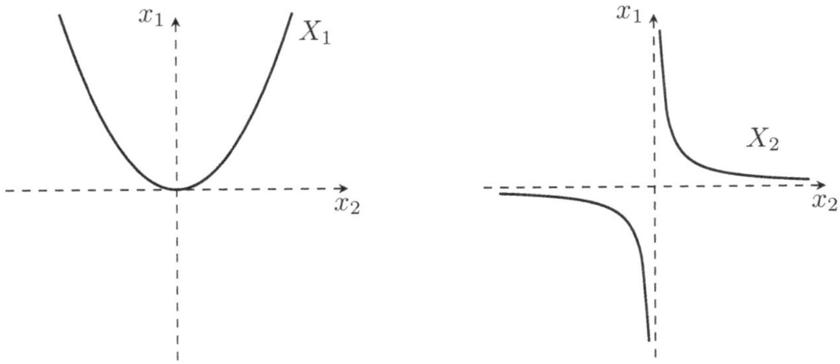

(a) Parabola $x_2 = x_1^2$. (b) Hyperbola $x_1 x_2 = 1$.

FIGURE 4.25: Quadratic curves.

Note that the equations of $\widetilde{C}_1$ and $\widetilde{C}_2$ are exactly the same, up to a permutation of the coordinates. Even if we have not given projective varieties the structure of varieties yet, it should be obvious that $\widetilde{C}_1$ and $\widetilde{C}_2$ will be isomorphic varieties, with the isomorphism being given by exchanging x_0 and x_1. Hence we see that the two distinct types of conics in $\mathbb{C}^2$ become the same in projective space: there is only one projective conic in $\mathbb{CP}^2$ up to isomorphism. The difference in the affine case comes from the fact that some conics "meet infinity" in one point (like C_1), and some in two (like C_2).

4.12.1 Compactification of Affine Plane Curve

Complex affine plane curves $X := \{(z,w) \in \mathbb{C}^2 \mid F(z,w) = 0\}$, where F is a nonsingular polynomial, are noncompact Riemann surfaces. To compactify them one needs to add point(s) $\infty^1, \infty^2, \dots \infty^N$ at infinity and introduce proper local parameters at these points in such a way that

$$\hat{X} = X \cup \infty^1 \cup \infty^2 \cup \cdots \cup \infty^N$$

is a compact Riemann surface. The plane curve X, defined by the polynomial equation $F(z,w) = 0$, can be compactified by embedding it in $\mathbb{CP}^2$. The mappings

$$[X : Y : Z] \to \left(z = \frac{X}{Z}, w = \frac{Y}{Z}\right)$$

and the inverse mapping

$$(z,w) \to [z : w : 1]$$

establish an isomorphism between an affine part of $\mathbb{CP}^2$ and $\mathbb{C}^2$. The whole projective plane is obtained from the affine part $\mathbb{C}^2$ by adding the line at infinity of the form $[X : Y : 0] \simeq \mathbb{CP}^1 \simeq S^2$. An embedding of X in $\mathbb{CP}^2$ is defined as follows. Suppose that

$$F(z,w) = F_k(z,w) + F_{k-1}(z,w) + \cdots + F_0(z,w)$$

where each $F_j(z, w)$ is a homogeneous polynomial of degree j. Then we define the homogeneous polynomial

$$Q(X, Y, Z) = Z^k F\left(\frac{X}{Z'}, \frac{Y}{Z}\right)$$

of degree k. A complex compact curve $\hat{X}$ is given in $\mathbb{CP}^2$ by the homogeneous equation

$$\hat{X} := \left\{[X : Y : Z] \in \mathbb{P}^2 \mid Q(X, Y, Z) = 0\right\} .$$

The affine part of the curve $\hat{X}$ (where $Z \neq 0$) coincides with X. The associated points at infinity have the form

$$Q(X, Y, 0) = 0 .$$

The surface $\hat{X}$ is compact and is thus the desired compactification of the surface X.

Remark 4.12.4. The operation of taking projective closure is not well-behaved. There is nothing to suggest that the curve necessarily wants to live in $\mathbb{CP}^2$: taking the projective closure even of a smooth affine curve may give a non-smooth curve.

If this is the case, the compactification of the smooth affine plane curve as a singular projective curve is not a good compactification.

4.13 Example 4.4.12 (continuation)

Recall that in Example 4.4.12 we consider the elliptic curve

$$X := \left\{(z, w) \in \mathbb{C}^2 \mid w^2 = P_3(z)\right\} ,$$

where $4\,(z - z_1)\,(z - z_2)\,(z - z_3) =: P_3(z)$ and $z_1, z_2, z_3 \in \mathbb{C}$ are pairwise distinct.

It is clear that $P_3'(z_i) \neq 0$ for all $i = 1, 2, 3$. In Example 4.4.12 we constructed the 2-dimensional torus (or "sphere-with-one-handle") with one point removed.

Now we will add this point to that torus and construct a compact Riemann surface on which the function $w = \sqrt{P_3(z)}$ is single-valued.

Let $(z_0, w_0) \in X$ be such that $w_0 \neq 0$. Then $z_0 \neq z_i$ for $i = 1, 2, 3$, and we may take $(z, w) \mapsto z$ as a local coordinate in a neighborhood of (z_0, w_0). At a point $(z_0, 0) \in X$, $z_0 = z_i$ for some i, and hence $P_3'(z_0) \neq 0$. By the implicit function theorem, there is a holomorphic function f defined in a neighborhood of 0 such that $z = f(w)$ near $(z_0, 0)$ and $z_0 = f(0)$. Thus we may take $(z, w) \mapsto w$ as a local coordinate in a neighborhood of $(z_0, 0)$. This defines a complex structure on X, and the projection $\pi_1 : X \to \mathbb{C}$, $\pi_1(z,w) = z$, is holomorphic. The Riemann surface X has two sheets, since each general value of z corresponds to two values of w. (Note that $\pi_2 : X \to \mathbb{C}$ is also holomorphic.)

Let us complete X to a Riemann surface $\widehat{X}$ over $\widehat{\mathbb{C}}$. Let $\mathbb{P}^2$ be the complex projective plane, i.e., $\mathbb{P}^2 = \mathbb{C}^3 \setminus \{0\}/\sim$, where $(z_1, z_2, z_3) \sim (w_1, w_2, w_3)$ if there exists $\lambda \in \mathbb{C}^*$ such that $(z_1, z_2, z_3) = \lambda(w_1, w_2, w_3)$. We denote by $[z_1, z_2, z_3]$ the equivalence class of (z_1, z_2, z_3) and we say that $[z_1, z_2, z_3]$ are homogeneous coordinates. Define

$$\widehat{X} := \{[z, w, t] \in \mathbb{P}^2 : w^2 t = 4(z - z_1 t)(z - z_2 t)(z - z_3 t)\} .$$

We identify X with $\left\{[z, w, t] \in \widehat{X} \;\middle|\; t = 1\right\}$. The complement $\widehat{X} \setminus X$ consists of a single point at infinity with homogeneous coordinates $[0, 1, 0]$ (setting $t = 0$ gives $z^3 = 0$, thus $z = 0$). Let this point be denoted by ∞. In a neighborhood of ∞, we can take $\widehat{X} \ni [z, w, t] \mapsto z/w \in \mathbb{C}$ as a local coordinate. In fact, replacing $[z, w, t]$ by $[z', w', t'] = [z/w, 1, t/w]$ gives

$$t' = 4(z' - z_1 t')(z' - z_2 t')(z' - z_3 t') ,$$

and by the implicit function theorem, t' is a holomorphic function of z' in some neighborhood of $(z', t') = (0, 0)$. This defines a complex structure on $\widehat{X}$ and a holomorphic projection $\widehat{p} : \widehat{X} \to \widehat{\mathbb{C}}$ which coincides with p on X and sends $\infty \in \widehat{X}$ to $\infty \in \widehat{\mathbb{C}}$. The Riemann surface $\widehat{X}$ is the compactification of X, and $\widehat{p} : \widehat{X} \to \widehat{\mathbb{C}}$ is a two-sheeted branched covering with branch points z_1, z_2, z_3, ∞. A geometric model of $\widehat{X}$ is obtained by adding the removed point to the 2-dimensional torus with one point removed obtained in Example 4.4.12.

The Weierstrass $\wp$-function is even and periodic, $\wp(w_i - z) = \wp(z)$ and hence $\wp'(w_i - z) = -\wp'(z)$, in particular, $\wp'(w_i/2) = 0$ for $i = 1, 2$. Similarly, $\wp'((w_1 + w_2)/2) = 0$. The half-periods $w_1/2, w_2/2$, and $(w_1 + w_2)/2$ are precisely the three simple zeros of $\wp'$ which is of order 3. Let us set

$$z_1 = \wp(w_1/2), z_2 = \wp(w_2/2), z_3 = \wp((w_1 + w_2)/2) . \tag{4.22}$$

Proposition 4.13.1. *Let* $\Lambda = \mathbb{Z}w_1 + \mathbb{Z}w_2$, *where* $\operatorname{Im}(w_2/w_1) > 0$, *and assume that* z_1, z_2, z_3 *satisfy* (4.22) *for the associated Weierstrass* $\wp$*-function. Then the mapping* $\varphi : \mathbb{C}/\Lambda \to \widehat{X}$,

$$\varphi(z) := \begin{cases} [\wp(z), \wp'(z), 1], & \text{if } z \neq 0 , \\ [\wp(z)/\wp'(z), 1, 1/\wp'(z)], & \text{if } z = 0 \end{cases}$$

is a biholomorphism.

Proof. We claim that $\wp$ satisfies the differential equation

$$(\wp'(z))^2 = 4(\wp(z) - e_1)(\wp(z) - e_2)(\wp(z) - e_3) . \tag{4.23}$$

Indeed, $f(z) := (\wp(z) - e_1)(\wp(z) - e_2)(\wp(z) - e_3)$ vanishes in the fundamental parallelogram precisely at the points $w_1/2, w_2/2$, and $(w_1 + w_2)/2$ of order 2, respectively. Also $(\wp')^2$ has double zeros at these points. Moreover, f as well as $(\wp')^2$ has poles of order 6 at the points in Λ. It follows that $(\wp')^2/f$ is holomorphic and elliptic, thus constant, which constant equals 4.

Since $\wp$ satisfies the differential equation (4.23), φ maps $\mathbb{C}/\Lambda$ into $\widehat{X}$. We have the following commuting diagram:

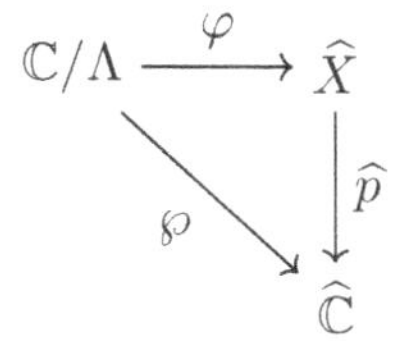

Then φ is surjective, since $\wp$ is surjective. Let us show that φ is injective. Let $z, z' \in \mathbb{C}$ be such that $z' - z \notin \Lambda$ and $\wp(z) = \wp(z')$. Assume first that $2z \notin \Lambda$. We have $z' + z \in \Lambda$, since $\wp(-z) = \wp(z)$ because $\wp$ is even and assumes every value exactly twice. Consequently, $\wp'(z) \neq \wp'(z')$, and hence $\varphi(z) \neq \varphi(z')$, since otherwise $\wp'(z) = \wp'(-z') = -\wp'(z') = -\wp'(z)$, as $\wp'$ is odd, implying $\wp'(z) = 0$, which contradicts $2z \notin \Lambda$. Recall that the zeros of $\wp'$ are $w_1/2$, $w_2/2$, and $(w_1 + w_2)/2$.

In the case that $2z \in \Lambda$, we also have $2z' \in \Lambda$, because $\wp(-z') = \wp(z')$, and $\wp$ assumes every value exactly twice. Thus, $z, z' \in \{w_1/2, w_2/2, (w_1 + w_2)/2\}$ modulo Λ. But $z_1 = \wp(w_1/2)$, $z_2 = \wp(w_2/2)$, and $z_3 = \wp((w_1 + w_2)/2)$ are pairwise distinct, which implies $z' - z \in \Lambda$. This shows that φ is injective.

Thus, $\varphi : \mathbb{C}/\Lambda \to \widehat{X}$ is a holomorphic bijective mapping from a compact space to a Hausdorff space. It follows that φ is a homeomorphism, and hence a biholomorphism. □

The converse is also true: for every Riemann surface $\widehat{X}$ of an equation $w^2 = 4(z - z_1)(z - z_2)(z - z_3)$, where z_1, z_2, z_3 are distinct and satisfy $z_1 + z_2 + z_3 = 0$, there exists a discrete subgroup $\Lambda = \mathbb{Z}w_1 + \mathbb{Z}w_2$, where $\mathrm{Im}(w_2/w_1) > 0$, such that $\widehat{X}$ can be realized as in the proposition.

Example 4.13.2. Let us consider the class of hyperelliptic Riemann surfaces

$$\Gamma = \left\{(z, w) \in \mathbb{C}^2 \mid F(z, w) = w^2 - P_N(z) = 0\right\}$$

where $P_N(z) = \prod_{j=1}^{N}(z - a_j)$, and $a_i \neq a_j$ for $i \neq j$. If we consider the projective curve defined by the zeros of a homogeneous polynomial

$$Q(z, w, t) = w^2 t^{N-2} - t^N P_N(z/t) = 0\,,$$

one can check that the curve is singular at the point $[0 : 1 : 0]$ if $N \geqslant 4$. Therefore, for $N \geqslant 4$, the embedding of Γ in $\mathbb{P}^2$ is not a good compactification.

The above example shows that not all the affine plane curves can be compactified in a smooth way by embedding them in $\mathbb{CP}^2$. Below we are going to illustrate another way of compactifying affine plane curves.

In order to explain this compactification let us define Riemann surfaces with punctures.

We call the Riemann surface with punctures any Riemann surface X such that there exists an open subset $U \subset X$ such that:

(P1) there exists a biholomorphic map from U onto a disjoint finite union of punctured discs $\mathbb{D}^* = \{0 < |z| < 1\}$;

(P2) $X \setminus U$ is compact.

One easily checks that, for any such X, one gets a new Riemann surface $\widehat{X}$ by gluing X and a disjoint finite union of discs $\mathbb{D} = \{z \in \mathbb{C} \mid |z| < 1\}$ along the open set U (see Fig 4.26).

Clearly, $\widehat{X}$ is compact and contains X, and $\widehat{X} \setminus X$ is finite. Moreover, $\widehat{X}$ is characterized by these properties and is therefore well defined.

The surface Γ can be made into a compact Riemann surface $\widehat{\Gamma}$ by joining points $\infty^{(1)}, \ldots, \infty^{(N)}$:

$$\widehat{\Gamma} = \Gamma \cup \left\{\infty^{(1)}\right\} \cup \ldots \cup \left\{\infty^{(N)}\right\}$$

at infinity ($z \to \infty, w \to \infty$), and introducing proper local parameters at this/ese point/s.

Definition 4.13.3. Let X be a Riemann surface such that there exists an open subset U_∞

$$U_\infty^{(1)} \cup \ldots \cup U_\infty^{(N)} = U_\infty \subset X$$

such that $X \backslash U_\infty$ is compact and $U_\infty^{(n)}$ are homeomorphic to punctured discs

$$z_n : U_\infty^{(n)} \to \mathbb{D}\backslash\{0\} = \{z \in \mathbb{C} \mid 0 < |z| < 1\}$$

where homomorphisms z_n are holomorphically compatible with the complex structure of X. Then X is called a *compact Riemann surface with punctures.*

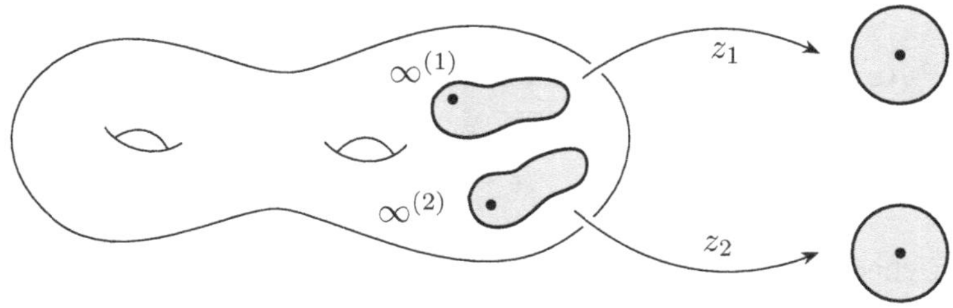

FIGURE 4.26: A compact Riemann surface with punctures.

Let us extend the homeomorphisms z_n to $\mathbb{D}$:

$$z_n : \hat{U}_\infty^{(n)} = U_\infty^{(n)} \cup \infty^{(n)} \to \mathbb{D} = \{z \in \mathbb{C} \mid |z| < 1\} . \tag{4.24}$$

Here we define punctures $\infty^{(n)}$ by the condition $z_n\left(\infty^{(n)}\right) = 0, \;\; n = 1, \ldots, N$. A complex atlas for a new Riemann surface

$$\widehat{\mathcal{R}} = \mathcal{R} \cup \left\{\infty^{(1)}\right\} \cup \ldots \cup \left\{\infty^{(n)}\right\}$$

is defined as a union of a complex atlas $\mathcal{A}$ of $\mathcal{R}$ with the coordinate charts (4.24) compatible with $\mathcal{A}$. We call $\widehat{\mathcal{R}}$ a compactification of $\mathcal{R}$.

4.14 Smooth and Singular Points on Affine and Projective Curves

Consider a projective algebraic curve

$$C = \left\{[x_0 : x_1 : x_2] \in \mathbb{CP}^2 \;\middle|\; F(x_0 : x_1 : x_2) = 0\right\} .$$

Moreover, let $P = (a_0 : a_1 : a_2)$ be a point on C.

Definition 4.14.1. We say that the point P is a *smooth point on* C if

$$\left(\frac{\partial F}{\partial X_0}(a_0, a_1, a_2), \frac{\partial F}{\partial X_1}(a_0, a_1, a_2), \frac{\partial F}{\partial X_2}(a_0, a_1, a_2)\right) \neq (0, 0, 0) .$$

Whenever this condition is not satisfied, the point is referred to as a singular point. Correspondingly, a smooth point is also referred to as a non-singular point.

The set of singular points of X is denoted by $\mathrm{Sing}(X)$. The curve X is called *nonsingular* if $\mathrm{Sing}(X) = \emptyset$.

Earlier we defined the term smooth point for affine curves $C \subset \mathbb{C}^2$. Even if this previous definition is similar to the one we have given here, we need to show that they do not contradict one another. Namely, when we form the projective closure of the affine curve C, we obtain a projective curve $\widehat{C} \subset \mathbb{CP}^2$. A point $p \in C$ should then be smooth as a point of the affine curve C if and only if it is smooth as a point on the projective curve $\widehat{C}$.

This problem is disposed of by means of the following proposition:

Proposition 4.14.2. *With notation as in Sect. 4.12 we have*

$$\left(\frac{\partial f}{\partial x}\right)^h (x_0, x_1, x_2) = \frac{\partial f^h}{\partial x_1}(x_0, x_1, x_2)$$

and

$$\left(\frac{\partial f}{\partial y}\right)^h (x_0, x_1, x_2) = \frac{\partial f^h}{\partial x_2}(x_0, x_1, x_2) .$$

Proof. We put

$$f(x,y) = \sum_{I=(i_1,i_2)\in\Phi} a_I x^{i_1} y^{i_2}\,.$$

Then $F = f^h$ is given by

$$F\,(x_0,x_1,x_2) = \sum_{I=(i_1,i_2)\in\Phi} a_I x_0^{d-i_1-i_2} x_1^{i_1} x_2^{i_2} = 0\,.$$

The verification of the claim is immediate from this. □

Corollary 4.14.3. *Let C be an affine curve, and let $\widehat{C}$ be the projective closure of C, where $V_+\,(X_0)$ is the points at infinity. Then (a,b) is a smooth point on the affine curve C if and only if $(1:a:b)$ is a smooth point on the projective curve $\widehat{C}$.*

Nonsingular complex curves are the most important examples of surfaces with holomorphic atlases.

According to the implicit function theorem, C possesses a relatively simple structure in the neighborhood of a smooth point: in such a neighborhood, C corresponds biholomorphically to an open set in $\mathbb{C}$.

Theorem 4.14.4. *If X is a complex algebraic curve in $\mathbb{C}^2$ defined by a polynomial $P(x,y)$ then $X \setminus \mathrm{Sing}(X)$ has a holomorphic atlas.*

Proof. Suppose that $(a,b) \in C$, i.e., $P(a,b) = 0$, and that

$$\frac{\partial P}{\partial w}(a,b) \neq 0\,.$$

The implicit function theorem tells us that there are open neighborhoods V and W of a and b in $\mathbb{C}$ and a holomorphic function $g : V \to W$ such that if $z \in V$ and $w \in W$ then

$$P(z,w) = 0 \Longleftrightarrow w = g(z)\,.$$

Since $X \setminus \mathrm{Sing}(X)$ is an open subset of X, by choosing V and W small enough we may assume that

$$U = \{(z,w) \in C \mid z \in V, w \in W\}$$

is an open neighborhood of (a,b) in $X \setminus \mathrm{Sing}(X)$. Then the map $\phi : U \to V$ defined by

$$\phi(z,w) = z$$

is a continuous function with continuous inverse

$$z \mapsto (z, g(z))\,.$$

Similarly if $(\partial P/\partial x)(a,b) \neq 0$ there is an open neighborhood U of (a,b) in $X \setminus \mathrm{Sing}(X)$ such that the map $\psi : U \to \mathbb{C}$ defined by

$$\psi(z,w) = w$$

is a homeomorphism onto an open subset V of $\mathbb{C}$ with inverse

$$w \mapsto (h(w), z)\,,$$

where $h(w)$ is a holomorphic function of w.

Thus there is an atlas on $X \setminus \mathrm{Sing}(X)$ such that every chart is of one of the two forms ϕ or ψ. The transition functions are then either the identity or compositions of the form

$$z \mapsto (z, g(z)) \mapsto g(z)$$

or

$$y \mapsto (h(y), y) \mapsto h(y)$$

where g and h are holomorphic. Therefore this atlas is holomorphic. □

Consider a homogeneous polynomial

$$P \in \mathbb{C}\,[\xi, \eta, \zeta]$$

of degree d, that is, $P(t\xi) = P\,(t\xi, \eta, t\zeta) = t^d P\,(\xi, \eta, \zeta)$ for all $t \in \mathbb{C}$. It is clear that

$$\frac{dP(t\xi, t\eta, t\zeta)}{dt} = \frac{dt^d}{dt} P\,(\xi, \eta, \zeta)\,.$$

Then

$$\begin{aligned}\frac{dP(t\xi, t\eta, t\zeta)}{dt} &= \frac{\partial P}{\partial \xi}(t\xi, t\eta, t\zeta)\xi + \frac{\partial P}{\partial \eta}(t\xi, t\eta, t\zeta)\eta + \frac{\partial P}{\partial \zeta}(t\xi, t\eta, t\zeta)\eta \\ &= d \cdot t^{d-1} P\,(\xi, \eta, \zeta)\,.\end{aligned}$$

Let $t = 1$. Then

$$\xi\frac{\partial P}{\partial \xi} + \eta\frac{\partial P}{\partial \eta} + \zeta\frac{\partial P}{\partial \zeta} = d \cdot P\,. \qquad \text{(Euler's formula)}$$

The argument for projective curves is very similar to the proof of Theorem 4.14.4.

Theorem 4.14.5. *If X is a projective curve in $\mathbb{CP}^2$ defined by a homogeneous polynomial $P(\xi, \eta, \zeta)$, then $X \setminus \mathrm{Sing}(X)$ has a holomorphic atlas.*

Proof. Suppose that $[a, b, c] \in C$, i.e., $P(a, b, c) = 0$, and that

$$(\partial P/\partial y)(a, b, c) \neq 0\,.$$

By Euler's formula,

$$a\frac{\partial P}{\partial x}(a,b,c) + b\frac{\partial P}{\partial y}(a,b,c) + c\frac{\partial P}{\partial z}(a,b,c) = 0$$

so

$$a = c = 0 \Longrightarrow a = b = c = 0\,,$$

which is impossible by the definition of $\mathbb{CP}^2$. Hence $a \neq 0$ or $c \neq 0$. Assume that $c \neq 0$. Then by the homogeneity of P

$$\frac{\partial P}{\partial y}(a/c, b/c, 1) = c^{-(d-1)}\frac{\partial P}{\partial y}(a,b,c) \neq 0\,,$$

where d is the degree of P. The implicit function theorem applied to the polynomial $P(\xi, \eta, 1)$ in x and y now tells us that there are open neighborhoods V and W of a/c and b/c in X and a holomorphic function $g : V \to W$ such that if $\xi \in V$ and $\eta \in W$ then

$$P(\xi, \eta, 1) = 0 \Longleftrightarrow \eta = g(\xi)\,.$$

If V and W are chosen small enough then

$$\begin{aligned} U &= \{[\xi, \eta, \zeta] \in C : z \neq 0, \xi/\zeta \in V, \eta/\zeta \in W\} \\ &= \{[\xi, \eta, 1] \in C : \xi \in V, \eta \in W\} \end{aligned}$$

is an open neighborhood of $[a, b, c]$ in $X \setminus \mathrm{Sing}(X)$. The map $\phi : U \to V$ defined by

$$\phi[\xi, \eta, \zeta] = \xi/\zeta$$

is a homeomorphism with inverse

$$w \mapsto [w, g(w), 1]\,.$$

Similarly if $[a, b, c] \in C$ and $(\partial P/\partial \eta)(a, b, c) \neq 0 \neq a$, or if $(\partial P/\partial \xi)(a, b, c) \neq 0$ or $(\partial P/\partial \zeta)(a, b, c) \neq 0$, then we can find a homeomorphism $\phi : U \to V$ from an open neighborhood U of $[a, b, c]$ in $X \setminus \mathrm{Sing}(X)$ onto an open subset V of X such that $\phi[\xi, \eta, \zeta]$ is one of the following:

$$\zeta/\xi\ ,\ \eta/\zeta\ ,\ \zeta/\eta\ ,\ \xi/\eta, \eta/\xi\,,$$

and the inverse of ϕ has the form

$$w \mapsto [1, g(w), w]\ ,\ [g(w), w, 1]\ ,\ [g(w), 1, w]\ ,\ [w, 1, g(w)]\ ,\ \text{or}\ [1, w, g(w)]\,,$$

where $g : V \to C$ is holomorphic. Thus we get an atlas on $X \setminus \mathrm{Sing}(X)$ each of whose transition functions is of one of the following forms

$$w \mapsto w\ ,\ 1/w\ ,\ g(w)\ ,\ 1/g(w)\ ,\ w/g(w)\ \text{ or } g(w)/w$$

such that g is holomorphic and the denominator does not vanish on the set where the transition function is defined. Thus this atlas is holomorphic. □

4.15 Chow's Theorem

The fundamental result of the theory of Riemann surfaces, conjectured by Riemann around 1850 and proved over the next five decades, is:

Theorem 4.15.1. *Every compact Riemann surface is algebraic.*

We have an idea of what this means, because we have considered Riemann surfaces defined by polynomial equations

$$P(z, w) = w^n + a_{n-1}(z)w^{n-1} + \cdots + a_1(z)w + a_0(z) = 0$$

and we have seen how to compactify these; indeed, the result implies that every compact Riemann surface arises in this way. But we would now like to do more than just explain the meaning of the theorem, and to survey the basic algebraic tools available for the study of compact Riemann surfaces.

At first glance, it may seem restrictive to allow only algebraic (polynomial) equations to describe our geometric objects. However, a deep theorem shows that for compact objects, we obtain nothing new even if we allow holomorphic equations. In this sense, algebraic geometry is closely related to, and in certain cases coincides with, complex (analytic) geometry. The simplest example of this correspondence is that any holomorphic map from the Riemann sphere $\mathbb{C}_\infty$ to itself must in fact be a rational map (i.e., the quotient of two polynomials).

The affine space $\mathbb{C}^n$ and the projective space $\mathbb{CP}^n$ are, of course, complex manifolds. Moreover, they are both algebraic and analytic varieties as well, since we can simply regard them as the vanishing loci of the zero function.

We already know that every algebraic variety is also an analytic variety. A natural question, then, is whether every analytic variety is algebraic. In general, this is false. There are two simple counterexamples illustrating this, rooted in the fact that the analytic topology is strictly finer than the Zariski topology, and that there are more holomorphic functions than polynomials.

Example 4.15.2. Consider the polydisc in $\mathbb{C}^n$. It is an analytic variety, but it cannot be defined as the zero locus of any set of polynomials. This is because any polynomial that vanishes on the polydisc must vanish on all of $\mathbb{C}^n$. In other words, the polydisc is an open subset in the analytic topology but not Zariski open.

Example 4.15.3. Consider the zero locus of the function $\sin z$ in $\mathbb{C}$, which consists of the countable set of points $\{k\pi \mid k \in \mathbb{Z}\}$. This is not the zero locus of any polynomial, because the zeros of a nonzero polynomial must be finite in number. This example illustrates that holomorphic functions may have infinite discrete zero sets, unlike polynomials.

It is known that complex submanifolds (and, more generally, analytic subsets) of projective space are algebraic varieties. As usual, we use homogeneous coordinates $[z_0, z_1, \dots, z_n]$ on projective space. For a homogeneous polynomial $F \in \mathbb{C}[z_0, z_1, \dots, z_n]$, the condition $F(z)$ is invariant under scaling (since $F(\lambda z) = \lambda^{\deg F} F(z)$ holds); this means that any collection $F_1, \dots, F_k$ of homogeneous polynomials defines a closed subset $Z = Z(F_1, \dots, F_k)$ of $\mathbb{CP}^n$. It is clearly analytic; in fact, its intersection with each of the standard open subsets $U_0, U_1, \dots, U_n$ is defined by polynomial functions.

Definition 4.15.4. An analytic subset $Z \subseteq \mathbb{CP}^n$ is said to be a *projective algebraic variety* if it is of the form $Z(F_1, \dots, F_k)$ for some collection of homogeneous polynomials.

Let $\sigma : \mathbb{C}^{n+1} \backslash \{0\} \to \mathbb{CP}^n$ be the natural projection, the map that takes $(z_1, \dots, z_{n+1})$ to $[z_1 : \dots : z_{n+1}]$ in homogeneous coordinates. We define an algebraic projective variety of $\mathbb{CP}^n$ as a set $\sigma(V)$ where $V \subset \mathbb{C}^{n+1}$ is the common zero set of a finite family of homogeneous holomorphic polynomials. It is not hard to show that $\sigma(V)$ is a projective algebraic variety.

Usually an algebraic projective variety is just called a *projective variety*, partly because of the following theorem.

Theorem 4.15.5 (Chow's theorem). *If $Z \subseteq \mathbb{CP}^n$ is an analytic set, then there exist homogeneous polynomials $F_1, \dots, F_k \in \mathbb{C}[z_0, z_1, \dots, z_n]$ such that $Z = Z(F_1, \dots, F_k)$.*

Before proving Chow's theorem, we record (without proof) an important result concerning complex manifolds due to Remmert and Stein. An interested reader may consult [190] or [75] for the proof.

Theorem 4.15.6 (proper mapping theorem). *If M and N are complex manifolds, $f : M \to N$ is a holomorphic map, and $V \subset M$ is an analytic variety such that $f|_V$ is proper, then $f(V)$ is an analytic subvariety of N.*

Note that the application of the Remmert–Stein theorem is the key point in this proof.

Proof of Chow's Theorem. This proof follows the proof by Cartan, Remmert, and Stein. Suppose that we have a complex analytic variety $X \subseteq \mathbb{CP}^n$. It is not hard to show that $\sigma^{-1}(X)$ is a complex analytic subvariety of $\mathbb{C}^{n+1} \backslash \{0\}$. By the theorem of Remmert–Stein, the set $V = \sigma^{-1}(X) \cup \{0\}$ is a subvariety of $\mathbb{C}^{n+1}$. Furthermore, V is a complex cone, that is, if $z = (z_1, \dots, z_{n+1}) \in V$, then $tz \in V$ for all $t \in \mathbb{C}$.

The final step is to show that if a complex analytic subvariety $V \subset \mathbb{C}^{n+1}$ is a complex cone, then it is given by the vanishing of finitely many homogeneous polynomials. Take a finite set of defining functions of V near the origin. That is, take $f_1, \dots, f_k$ defined in some open ball $B = B(0, \epsilon)$, such that $B \cap V = \{z \in B \mid f_1(z) = \dots = f_k(z) = 0\}$. We can suppose that ϵ is small enough that the power series for f_j converges in B for all j. Expand f_j in a power series near the origin and group together homogeneous terms as

$f_j = \sum_{m=0}^{\infty} f_{jm}$, where f_{jm} is a homogeneous polynomial of degree m. For $t \in \mathbb{C}$ we write

$$f_j(tz) = \sum_{m=0}^{\infty} f_{jm}(tz) = \sum_{m=0}^{\infty} t^m f_{jm}(z)\,.$$

For a fixed $z \in V$ we know that $f_j(tz) = 0$ for all $|t| < 1$, hence we have a power series in one variable that is identically zero, and so all coefficients are zero. Thus f_{jm} vanishes on $V \cap B$ and hence on V. It follows that V is defined by a family of homogeneous polynomials. Since the ring of polynomials is Noetherian, we need only finitely many, and we are done. □

Conversely, a smooth algebraic curve lying in $\mathbb{CP}^n$ is a compact Riemann surface. The fact that it is described by polynomials gives it a holomorphic structure, and it is compact as a closed subset of the compact space $\mathbb{CP}^n$. Combining the results from the embedding theorem and Chow's theorem, as well as this construction, proves

Theorem 4.15.7. *Every compact Riemann surface is isomorphic to a smooth algebraic curve.*

This result is remarkable because we start from spaces in projective space that are locally defined by the vanishing locus of holomorphic functions, but they turn out to be the vanishing locus of polynomials globally.

Chow's theorem has an immediate corollary.

Corollary 4.15.8. *Meromorphic functions on a projective variety are rational.*

Proof. Given a meromorphic function f on a projective variety $X \subset \mathbb{CP}^k$, we can view it as a map from X to $\mathbb{CP}^1$. Let $\Gamma_f \subset X \times \mathbb{CP}^1$ be the closure of the graph of f. Then Γ_f is a subvariety of $X \times \mathbb{CP}^1$, hence a subvariety of $\mathbb{CP}^k \times \mathbb{CP}^1$, which embeds into $\mathbb{CP}^{k+1}$. By Chow's theorem, the graph Γ_f is algebraic. Let p and q be the two projections from $X \times \mathbb{CP}^1$ to X and $\mathbb{CCP}^1$, respectively. Then p is an isomorphism from Γ_f to X, and f can be recovered as $f(x) = q\left(p^{-1}(x)\right)$. Because p and q are algebraic, so is f. □

Conversely, a smooth algebraic curve lying in $\mathbb{CP}^n$ is a compact Riemann surface. The fact that it is described by polynomials gives it a holomorphic structure, and it is compact as a closed subset of the compact space $\mathbb{CP}^n$.

Theorem 4.15.5 reveals a close connection with algebraic geometry: every compact Riemann surface X can be embedded into some projective space $\mathbb{CP}^n$ (in fact, already into $\mathbb{CP}^3$). By Chow's theorem, X is therefore a (projective) algebraic variety, meaning it can be described by finitely many homogeneous equations. This phenomenon is special to complex dimension one. The key point is that one can always construct a nonconstant meromorphic function on a Riemann surface (see Section 5.16). Given such a function, it is not difficult to obtain a projective embedding of a compact Riemann surface X. By contrast, it is easy to construct a compact two-dimensional torus

$$T = \mathbb{C}^2/L$$

for a suitably chosen lattice L, which cannot be embedded into any projective space $\mathbb{CP}^n$.

5

Abstract Riemann Surfaces

In Chapter 4, we discussed the construction of different surfaces to handle multivalued functions. As a result, we constructed complex manifolds and rational functions on them. That discussion was heuristic and relied on the reader's intuition. However, this topic can also be made rigorous.

In Chapter 4 we saw that a Riemann surface of a function is a one-dimensional complex manifold. It has become entirely natural for mathematicians today to define Riemann surfaces as one-dimensional complex manifolds and to consider holomorphic and meromorphic functions on them.

The modern era in the study of Riemann surfaces, and, indeed, of all manifolds, opened in 1913 with the publication of Hermann Weyl's book *The Concept of a Riemann Surface* (see [188]). Here, for the first time, no defining function is assumed. The definition is purely geometric and also independent of any embedding in Euclidean 3-space.

This chapter is a survey of the basic notions and main results of the theory of Riemann surfaces. Attention is centered on the compact case, as it is directly related to the theory of algebraic curves. A detailed exposition, and proofs, can be found in Ahlfors–Sario [6], Forster [50], Springer [170], and Weyl [187].

This chapter is for a person who wants only some idea of what Riemann surfaces are about. We therefore begin with introducing a number of important concepts in a very down-to-earth context. It is our goal to keep the presentation as elementary as possible.

Here is what we are going to talk about. We are going to discuss compact Riemann surfaces, and a compact Riemann surface is the same thing as a smooth projective algebraic curve (over $\mathbb{C}$). That in turn is really the same thing as a smooth projective curve over any algebraically closed field of characteristic zero. By abuse of notation, we will use $\mathbb{C}$ to denote any such field as well.

The fact that these are the same thing—that is, that a compact Riemann surface is an algebraic curve—is nontrivial. It requires work to show that such an object even admits a nontrivial meromorphic function. Note:

Proposition 5.0.1. *There are compact complex manifolds of* $\dim \geq 2$ *that do not admit nonconstant meromorphic functions.*

The miracle in dimension one is that there are nonconstant meromorphic functions, and enough to embed the manifold in projective space.

DOI: 10.1201/9781003713005-5

5.1 Basic Definitions

Currently, Riemann surfaces are most conveniently regarded as special complex analytic manifolds. This section therefore begins with basic definitions from complex analytic geometry. This is also justified by the fact that many important notions and results of the theory of Riemann surfaces are difficult to explain without resorting to some more general complex analytic manifolds.

In the real case, the one-dimensional manifolds (curves) are quite simple to understand, particularly if one requires the manifold to be connected; for instance, all compact connected one-dimensional real manifolds are homeomorphic to the unit circle. However, in the complex case, the connected one-dimensional manifolds—the ones that look locally like subsets of $\mathbb{C}$—are much richer, and are known as Riemann surfaces.

This suffices to establish their general properties, and dispenses with unnecessary embedding information. (Moreover, smoothness is built in, whereas in the embedded case it must be checked). However, the abstract definition is somewhat complicated and less intuitive. One way to motivate this discussion is with the following observation.

Every Riemann surface in $\mathbb{C}^2$ is non-compact. This is clear for a Riemann surface defined as the zero-set of an algebraic equation $P(z, w(z)) = 0$; it projects surjectively to the complex z-plane. Because the image of a compact set under a continuous map is compact, it follows that the solution-set is not compact.

So there is an obstacle to constructing compact Riemann surfaces, such as the torus without punctures, as graphs of multi-valued functions within $\mathbb{C}^2$. On the other hand, it is easy to produce compact topological surfaces with enough analytic structure to be worthy of Riemann's name.

We assume given a connected Hausdorff space X which will also generally be assumed connected. We want to define what we mean by a complex structure on X (or structure of a complex manifold) which will be an obvious generalization of the concept of a Riemann surface.

Weyl's definition of *Riemann surfaces*, in modern language and without unnecessary hypotheses, is the following: A Riemann surface is a one-dimensional complex manifold. That is,

Definition 5.1.1. Let X be a connected Hausdorff topological space together with an open covering $\{U_j\}$ of X and a family of mappings

$$\psi_j : U_j \to \mathbb{C}$$

such that

(a) each $\psi_j : U_j \to \mathbb{C}$ is a homeomorphism of U_j onto an open subset $\psi_j(U_j)$ of $\mathbb{C}$;

(b) if $U_j \cap U_i \neq \varnothing$, then the function

$$\psi_{ji} = \psi_i \circ \psi_j^{-1} : \psi_j (U_j \cap U_i) \to \psi_i (U_j \cap U_i)$$

is biholomorphic (i.e., the function itself as well as its inverse are both holomorphic).

The diagram in Fig. 5.1 illustrates these concepts.

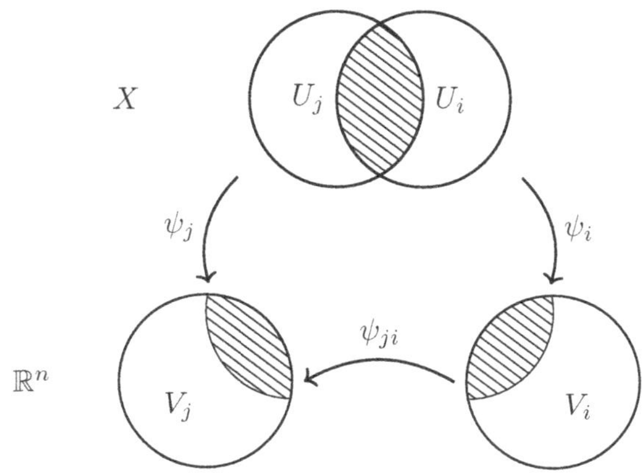

FIGURE 5.1: Charts of a complex manifold. Shaded areas: $U_j \cap U_i$, $\psi_j(U_j \cap U_i)$ and $\psi_i(U_j \cap U_i)$.

Such a pair (U_j, ψ_j) is called a *holomorphic chart* on X and the map ψ_j is called a *local coordinate* at $p \in U_j$. More generally, if f is any holomorphic function defined on a neighborhood of $z = \psi_j(p)$, such that $f'(z) \neq 0$, then $f \circ \psi_j$ defines the same complex structure on a neighborhood of p as ψ_j and is also a local coordinate at p. We call such a (U_j, ψ_j) a *local holomorphic coordinate*, and $\{(U_j, \psi_j)\}$ a *holomorphic coordinate covering.*

A maximal complex atlas is called a *complex structure* on X. An *(abstract) Riemann surface* is a Hausdorff space provided with a complex structure.

We continue with the definitions. Let X be a complex manifold, W an open set in X, and $\{z_j\}$ a coordinate system. Then a mapping $f : W \to \mathbb{C}^l$ is holomorphic (differentiable, and so on) if $f \circ z_j^{-1}$ is holomorphic (differentiable, and so on) for each j where defined. Let N be another complex manifold with coordinates $\{w_\lambda\}$ and $f : W \to N$. Then f is holomorphic (differentiable, and so on) if $w_\lambda \circ f \circ z_j^{-1}$ is holomorphic where defined.

With the notion of a Riemann surface, we run into an entity of a new mathematical nature. It must be rated on a par with such notions as a Riemannian manifold in geometry, or a field in algebra. Just as some metric concepts are defined in a Riemannian manifold, and algebraic operations in a field, so is the notion of analytic function on a Riemann surface. In particular, it is now possible to formulate and prove the theorem stating that a function which is holomorphic on an entire (compact) Riemann surface is constant.

Definition 5.1.2. A *manifold is Hausdorff* if for any two points $p, q \in M$ with $p \neq q$ there exist open sets V_p and V_q such that $p \in V_p, q \in V_q$ and $V_p \cap V_q = \emptyset$.

It is hard to picture spaces that are not Hausdorff but an example exists!

Example 5.1.3. Take two lines $y = 0$ and $y = 1$ in $\mathbb{R}^2$ and identify points for $x < 0$ but not for $x \geq 0$. This construction produces a manifold. However the manifold is not Hausdorff because $(0, 0)$ is not the same point as $(0, 1)$ and yet there are no neighborhoods of $(0, 0)$ and $(0, 1)$ which do not intersect.

Definition 5.1.4. A manifold is *compact* if every atlas contains a finite refinement. That is if every atlas $\{U_j, \psi_j\}_{j \in J}$ contains a subatlas $\{U_j, \psi_j\}_{j \in K}$ which consists of only finitely many U_j but which still satisfies properties (a) and (b) from Definition 5.1.1.

Loosely speaking, compactness corresponds to being closed or finite. The surface of a sphere is a compact space but $\mathbb{R}^n$ is non-compact. However, we have not yet introduced a metric so we do not yet know what "finite" means. Besides, the question of compactness can be subtle. For example, clearly $z_n = n$ has no limit in $\mathbb{C}$ but $z_n = [1 : n] = \left[\frac{1}{n} : 1\right]$ does converge to $[0 : 1]$ in $\mathbb{P}^1$, which is compact.

Another example consists of the surface of revolution obtained by rotating the graph of e^{-x^2} about the x-axis, this yields a surface that is non-compact but which has a finite area.

An arbitrary Riemann surface carries with it a large amount of geometric information. In particular, the Riemann surface of an algebraic function reveals some important characteristics of that function.

Since the transition maps are holomorphic and hence smooth, complex manifolds have a natural smooth structure.

Dimension of a Complex Manifold. The dimension of a chart $\varphi : U \to \mathbb{C}^n$ is the number n, that is, the number of its coordinates. For a connected complex manifold M, this number is independent of the choice of φ on M and is called the *dimension of* M. In what follows, all manifolds will be assumed connected. The dimension of M is denoted by $\dim_{\mathbb{C}} M$ or, simply, $\dim M$.

Orientability of Riemann surfaces. Recall that a surface is orientable if there is a continuous choice of clockwise rotations on it. (A non-orientable surface is the Möbius strip; a compact example without boundary is the Klein bottle.) Orientability of Riemann surfaces will follow from our desire to do complex analysis on them; notice that the complex plane carries a natural orientation, in which multiplication by i is counter-clockwise rotation.

In view of the Cauchy formula, every holomorphic function on a domain in $\mathbb{C}$ is real analytic. Thus every Riemann surface (= manifold of complex dimension one) is a real analytic manifold of real dimension 2. Using the Cauchy–Riemann differential equations, one can see that the real Jacobian matrix of the ψ_{ji} has positive determinant, hence X is orientable as a real surface.

Definition 5.1.5. A manifold M is *orientable* if there exists for M an atlas $\{U_j, \psi_j\}$ such that for every non-empty $U_j \cap U_k$ the Jacobian $\frac{\partial(\psi_j)}{\partial(\psi_k)}$ is positive.

Theorem 5.1.6. *Every Riemann surface is oriented.*

Proof. Let $f = \psi_{ji} := \psi_j \circ \psi_i^{-1}$ be a typical transition map. Write $f = u + iv$. Then the Jacobian determinant of f is

$$\det(Df) = u_x v_y - u_y v_x .$$

On the other hand, since f is holomorphic, we have $v_x = -u_y$ and $v_y = u_x$. Thus

$$\det(Df) = u_x^2 + u_y^2 = \left|\frac{\partial f}{\partial z}\right|^2 > 0 .$$

This completes the proof. □

In fact, since a complex one-dimensional manifold always admits a metric, compactness is equivalent to every sequence of points having a convergent subsequence.

A manifold that is equipped with a metric is called a *Riemannian manifold*, which is not what we have considered up to this point.

Every Riemann surface (like every complex manifold) has Hermitian metrics. Because every 2-form on a real 2-manifold is closed for dimensional reasons, every Hermitian metric on a Riemann surface is Kähler.

Formally, a Riemannian metric (or just a metric) on a smooth manifold is a choice of inner product for each tangent space of the manifold. A Riemannian manifold is a smooth manifold together with a Riemannian metric.

Theorem 5.1.7. *Any orientable two-dimensional Riemannian manifold is a Riemann surface.*

Proof. By definition, Riemannian manifolds admit a positive definite metric. In two dimensions, one can always choose coordinates x, y in a neighborhood of any point such that the metric tensor is of the form

$$ds^2 = \lambda^2(x,y)\left(dx^2 + dy^2\right)$$

for a given real function λ. In complex coordinates $z = x + iy$, this reads

$$ds^2 = \lambda^2(z,\bar{z}) dz d\bar{z} \tag{5.1}$$

In another patch, with complex coordinate $w = u + iv$, one can similarly write

$$ds^2 = \mu^2(w,\bar{w}) dw d\bar{w} , \tag{5.2}$$

for some function μ. If the two patches overlap, one can change coordinates from w to z,

$$dw = \frac{\partial w}{\partial z}dz + \frac{\partial w}{\partial \bar{z}}d\bar{z} \tag{5.3}$$

and complex conjugate. Since the metric is a tensor, hence globally defined, one can equate (5.1) to (5.2) on the overlap. Using (5.3), this leads to the requirement

$$\frac{\partial w}{\partial z}\frac{\partial \bar{w}}{\partial z} = 0$$

Hence the transition function is either holomorphic or anti-holomorphic. If it were antiholomorphic, $\partial w/\partial z = 0$, and the Cauchy–Riemann equations then imply

$$\frac{\partial u}{\partial x} = -\frac{\partial v}{\partial y}, \quad \frac{\partial u}{\partial y} = \frac{\partial v}{\partial x}$$

and hence the Jacobian takes the form

$$J = \begin{pmatrix} \frac{\partial u}{\partial x} & \frac{\partial u}{\partial y} \\ \frac{\partial u}{\partial y} & -\frac{\partial u}{\partial x} \end{pmatrix}.$$

But this means that $\det J < 0$ and so the manifold would not be orientable (by definition). This contradicts the assumption of the theorem, so this means that the transition function must be holomorphic ($\partial w/\partial \bar{z} = 0$), and the manifold is complex. □

Since we can think of $\mathbb{C}^n$ as $\mathbb{R}^{2n}$ we see from the definition that every n-dimensional complex manifold is a $2n$-dimensional real manifold. Therefore Riemann surfaces are orientable surfaces.

It is known from topology that any compact orientable surface (two-dimensional topological manifold) is homeomorphic to a sphere with g handles. The number $g \in \mathbb{N}$ is called the *genus* of the surface (see [126]).

A Riemann surface X, viewed over $\mathbb{R}^2$, is an orientable, compact, connected, smooth 2-manifold. By a theorem in topology, this means that M is homeomorphic to a sphere with g handles, and we say M has genus g, $g = 0, 1, 2, \ldots$.

A compact orientable surface is classified up to homeomorphism by a single number called the *genus*. A sphere with one handle is a torus, a sphere with two handles is a double torus, with three handles a triple torus, and so on.

Compact Riemann surfaces can be embedded into projective spaces and are thus, by virtue of Chow's theorem, algebraic curves.

An (abstract) Riemann surface is a Hausdorff space provided with a complex structure. Every concrete Riemann surface is an abstract Riemann surface. The converse is a deep theorem which requires the construction of nonconstant meromorphic functions.

The assertion that on any compact Riemann surface, there is a nonconstant meromorphic function is the difficult and important point in Theorem 5.16.6.

Let us emphasize that this property is special to one-dimensional compact complex manifolds. (On a compact complex surface, i.e., on a compact two-dimensional complex manifold, there may be no non-constant meromorphic function. For example, a general complex torus of dimension ≥ 2 does not admit any meromorphic function. See Section 5.7.)

This type of existence theorem goes back to Riemann, who gave a non-rigorous construction, based on the Dirichlet principle, of harmonic functions with prescribed singularities on ramified coverings of $\mathbb{CP}^1$. Theorem 5.16.6 was proved rigorously by methods closely related with the ones developed to prove the uniformization theorem, and, at the same time as this last theorem, has been at the origin of many developments in analysis around the term of last century.

5.2 Examples of Riemann Surfaces

In this section, we discuss a number of examples of Riemann surfaces, some of which will play a formative role in the development of the theory.

We begin with an obvious class of examples which serve to illustrate that complex analysis, as we have developed it so far in this book, is really a special case of general Riemann surfaces (and the "local case").

Since Riemann surfaces are locally modeled on $\mathbb{C}$, it should come as no surprise that $\mathbb{C}$ is a Riemann surface.

The most basic examples of Riemann surfaces are open sets of the complex plane $U \subset \mathbb{C}$ with the identity atlas $\{(U, Id)\}$. In particular one has the upper half-plane $\mathbb{H} = \{w \in \mathbb{C} : \operatorname{im}(w) > 0\}$ and the unit disc $\mathbb{D} = \{w \in \mathbb{C} : |w| < 1\}$.

A class of examples of Riemann Surfaces is given by graphs of holomorphic functions.

Example 5.2.1 (graphs of holomorphic functions). Let $U \subset \mathbb{C}$ be a connected open subset of the complex plane, and let g be a holomorphic function defined on all of U. Consider the graph X of g, as a subset of $\mathbb{C}^2$:

$$X = \{(z, g(z)) \mid z \in U\} .$$

Give X the subspace topology, and let $\pi : X \to U$ be the projection on the first variable; note that π is a homeomorphism, whose inverse simply sends the point $z \in U$ to the ordered pair $(z, g(z))$. Thus π is a complex chart on X, whose domain covers all of X. Hence we have a complex atlas on X, composed of a single chart; this gives X the structure of a Riemann surface.

This example can be immediately generalized to any finite collection of holomorphic functions $g_1, \ldots, g_n$ on V; simply take X to be the graph in $\mathbb{C}^{n+1}$:

$$X = \{(z, g_1(z), \ldots, g_n(z)) \mid z \in U\} .$$

This is a further generalization of the graph construction introduced above. We would like to consider a locus $X \subseteq \mathbb{C}^2$ which is locally a graph, but perhaps not globally.

Example 5.2.2 (complex tori)**.** Let $\omega_1, \omega_2 \in \mathbb{C}$ be $\mathbb{R}$-independent. Consider the lattice

$$L := \mathbb{Z}\omega_1 + \mathbb{Z}\omega_2 = \{n\omega_1 + m\omega_2; n, m \in \mathbb{Z}\}$$

The set L is a discrete subgroup of the additive group $\mathbb{C}$, so the quotient space

$$X = X_L := \mathbb{C}/L$$

is well defined. We denote the projection by $\pi : \mathbb{C} \to X$. We therefore have a set (actually, an additive group) X, but in order to endow it with the structure of a Riemann surface, we must first give it the structure of a topological space.

We define a set $U \subset X$ to be open if its inverse image $\pi^{-1}(U) \subset \mathbb{C}$ is open. Then π is continuous in this topology, and in fact any topology that makes π continuous must be contained in this topology.

The map π is an open mapping. Indeed, if $U \subset \mathbb{C}$ is open, then $\pi(U)$ is open, because

$$\pi^{-1}(\pi(U)) = \bigcup_{\omega \in L} U + \omega\,.$$

Observe that X is compact. Indeed, the image of the compact set

$$P_z := \{z + t\omega_1 + s\omega_2; s, t \in [0, 1]\}$$

under the continuous map π covers X. Next we will build charts on X. To this end, observe that since $L \subset \mathbb{C}$ is discrete, there is a positive number ε such that the only point of L that is within 2ε of the origin is the origin itself:

$$L \cap \{z \in \mathbb{C}; |z| < 2\varepsilon\} = \{0\}\,.$$

By definition, for any such ε, any $\omega \in L \setminus \{0\}$, and any $z \in \mathbb{C}$, we find that $(\mathbb{D}(z, \varepsilon) + \omega) \cap \mathbb{D}(z, \varepsilon) = \emptyset$. We define for every $z \in \mathbb{C}$ the map

$$\pi_z := \pi|_{\mathbb{D}(z,\varepsilon)} : \mathbb{D}(z, \varepsilon) \to \pi(\mathbb{D}(z, \varepsilon))\,.$$

Other surfaces that can be given a Riemann surface structure are the unit sphere $\mathbb{S}^2 = \big\{(x, y, z) \in \mathbb{R}^3 : x^2 + y^2 + z^2 = 1\big\}$, the complex extended plane (or Riemann sphere) $\widehat{\mathbb{C}} = \mathbb{C} \cup \{\infty\}$ and the complex projective line $\mathbb{P}^1 := \mathbb{P}^1(\mathbb{C})$.

Example 5.2.3 (Riemann sphere)**.** The name Riemann sphere, or one-point compactification of the complex numbers, is usually given to the Riemann surface $\widehat{\mathbb{C}}$ defined as follows. Add an additional point to the complex plane $\mathbb{C}$, and denote it as ∞ (the notation indicates what the topology at this additional point is going to be: one gets close to ∞ by escaping from every compact set in the plane). A collection of fundamental neighborhoods of ∞ is provided by the family of sets $D(\infty, R) = \{z \in \mathbb{C}, |z| > R\} \cup \{\infty\}$.

Write $\widehat{\mathbb{C}} = \mathbb{C} \cup \{\infty\}$. Obviously $\widehat{\mathbb{C}}$ is a compact, connected Hausdorff topological space. Now consider the covering $\{U_0, U_1\}$:

$$U_0 = \widehat{\mathbb{C}} \backslash \{\infty\} = \mathbb{C}, \quad U_1 = \widehat{\mathbb{C}} \backslash \{0\}$$

and the mappings

$$\begin{aligned} z_0 : U_0 &\to \mathbb{C}\,, \\ z &\mapsto z\,; \\ z_1 : U_1 &\to \mathbb{C}\,, \\ z &\mapsto \begin{cases} 0, & z = \infty\,, \\ 1/z, & z \neq \infty \end{cases} \end{aligned}$$

Clearly

$$\begin{aligned} z_1 \circ z_0^{-1} : \mathbb{C} \backslash \{0\} &\to \mathbb{C} \backslash \{0\}\,, \\ z &\mapsto 1/z \end{aligned}$$

is biholomorphic, and likewise $z_0 \circ z_1^{-1}$. In this way $\widehat{\mathbb{C}}$ becomes a compact Riemann surface.

Since we can identify the unit sphere S with $\widehat{\mathbb{C}}$ through stereographic projection, $\widehat{\mathbb{C}}$ therefore also naturally becomes a Riemann surface, called the *Riemann sphere.*

The precise coordinate representation is as follows: let $N = (0, 0, 1)$ denote the north pole and $P = (0, 0, -1)$ denote the south pole. Consider the covering $\{U_0, U_1\}$ of S and the mappings Φ_0, Φ_1:

$$\begin{aligned} U_0 &= S \backslash \{P\}\,, \quad \Phi_0 : U_0 \to \mathbb{C}\,; \\ U_1 &= S \backslash \{N\}\,, \quad \Phi_1 : U_1 \to \mathbb{C}\,. \end{aligned}$$

Here Φ_0 represents the stereographic projection from the south pole P to $\mathbb{C}$. Taking its conjugate we have

$$\Phi_0(\xi, \eta, \zeta) = \frac{\xi - i\eta}{1 + \zeta}$$

We see that Φ_1 represents the stereographic projection from the north pole N to $\mathbb{C}$, and

$$\Phi_1(\xi, \eta, \zeta) = \frac{\xi + i\eta}{1 - \zeta}\,.$$

Here ξ, η, ζ denote linear coordinates in $\mathbb{R}^3$. We have

$$\begin{aligned} \Phi_1 \circ \Phi_0^{-1} : \mathbb{C} \backslash \{0\} &\to \mathbb{C} \backslash \{0\}\,, \\ z &\mapsto 1/z \\ \Phi_0 \circ \Phi_1^{-1} : \mathbb{C} \backslash \{0\} &\to \mathbb{C} \backslash \{0\}\,, \\ z &\mapsto 1/z\,. \end{aligned}$$

It is clear that $\Phi_1 \circ \Phi_0^{-1}$ and $\Phi_0 \circ \Phi_1^{-1}$ are both biholomorphic.

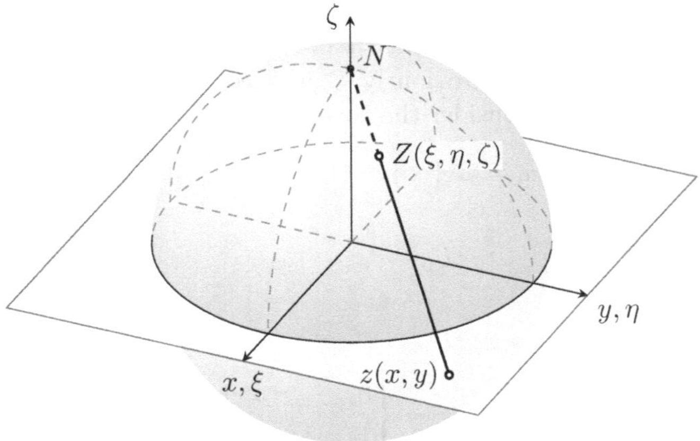

FIGURE 5.2: The stereographic projection.

The Riemann sphere is a compact Riemann surface when it has been constructed by either of the two above procedures, that is, either from the extended complex plane $\widehat{\mathbb{C}} = \mathbb{C} \cup \{\infty\}$, or from the unit sphere $\widehat{\mathbb{C}}$ (Fig. 5.2) obtained by stereographic projection.

The construction of $\widehat{\mathbb{C}}$ can be generalized to any locally compact space X: the open subsets of $\widehat{X} = X \cup \{\infty\}$ containing the point ∞ are set to be the complements of the compact subsets of X. Thus X becomes the complement of a point in the compact space $\widehat{X}$, the Alexandrov one-point compactification of X.

Example 5.2.4 (the complex projective line)**.** The complex projective line $\mathbb{CP}^1$ is covered by two affine charts $U_0 = U_{z_0}$, $U_1 = U_{z_1}$, that is, by two affine lines defined by the equations $z_0 = 1$ and $z_1 = 1$ (see Fig. 5.3).

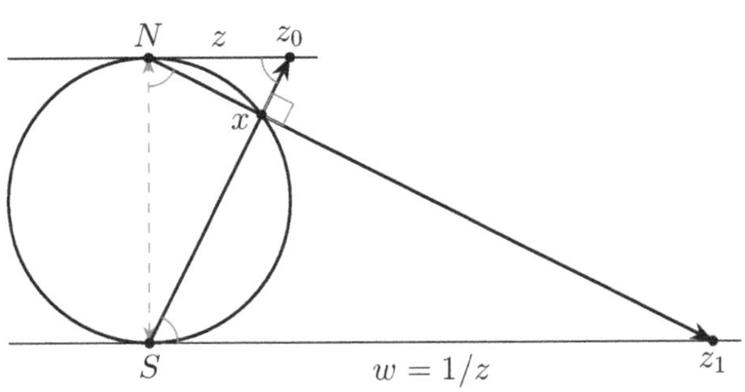

FIGURE 5.3: Standard charts on $\mathbb{P}_1$.

The chart U_0 contains all points except for one point $(0 : 1)$ represented by the vertical coordinate axis. This is the only infinite point for the chart

U_0. Every other point $(z_0 : z_1)$ with $z_0 \neq 0$ is visible in U_1 as $(1 : z_1/z_0)$. The function $t = z_1|_{U_0} = z_1/z_0$ can be used as a local affine coordinate within U_0. The chart U_1 consists of points $(z_0 : z_1) = (z_0/z_1 : 1)$ with $z_1 \neq 0$, and the function $s = z_0|_{U_1} = z_0/z_1$ can be used as a local affine coordinate within U_1. The only infinite point for U_1 is $(1 : 0)$, represented by the horizontal coordinate axis. Local affine coordinates s, t of the same point $(z_0 : z_1) \in \mathbb{P}_1$ lying within $U_0 \cap U_1$ are related by $s = 1/t$.

Thus $\mathbb{CP}^1$ consists of two copies of the affine line $\mathbb{C}^1$ (one line with coordinate s, another with coordinate t) glued together along the complements to their origins by the following rule: point s on the first line is glued with point $t = 1/s$ on the second.

Two copies of $\mathbb{C}$ glued together by the rule $s \leftrightarrow 1/s$ produce a sphere of diameter 1 consisting of two tangent planes drawn through the south and north poles and mapped onto the sphere by the central projections from opposite poles (see Fig. 5.4).

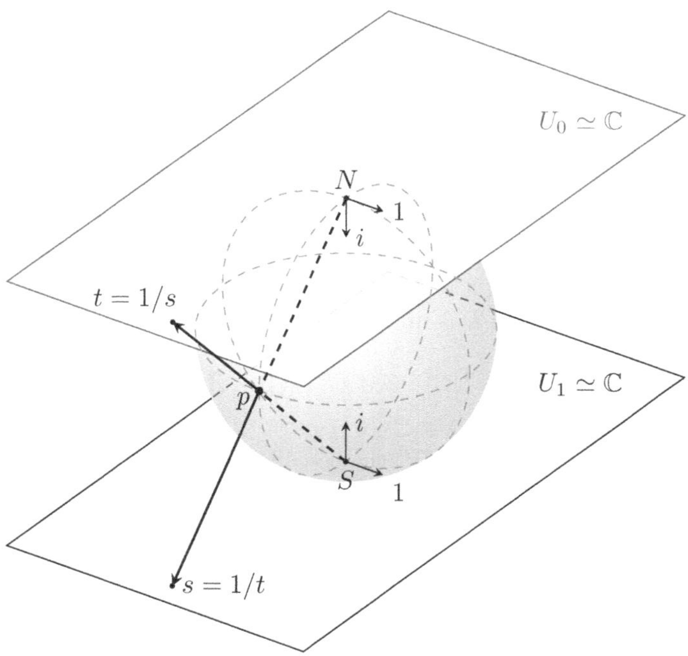

FIGURE 5.4: $\mathbb{CP}^1(\mathbb{C}) \simeq S^2$.

Algebraic varieties in $\mathbb{CP}^1$ are easy to describe.

Proposition 5.2.5. *A subset $V \subseteq \mathbb{CP}^1$ is an algebraic variety if and only if it is either finite or all of $\mathbb{CP}^1$.*

Proof. Suppose $V = \{a_1, \ldots, a_k\}$ is a finite subset of $\mathbb{CP}^1$. If $V = \varnothing$, then it is the variety determined by the polynomial $f \equiv 1$. Otherwise, for each j, we can write $a_j = [a_j^0, a_j^1]$ for some point $(a_j^0, a_j^1) \in \mathbb{C}^2$, so $\{a_j\}$ is the variety determined by the polynomial $f_j(w^0, w^1) = a_j^1 w^0 - a_j^0 w^1$. It follows that V is the variety determined by the product $f_1 \cdots f_k$. On the other hand, $\mathbb{CP}^1$ itself is the variety determined by the zero polynomial.

Conversely, if V is the variety determined by the homogeneous polynomials $f_1, \ldots, f_k$, and V is not equal to $\mathbb{CP}^1$, then at least one of the polynomials, say f_j, is not identically zero. The points of V in $\mathbb{CP}^1 \backslash \{[0,1]\}$ can all be written in the form $[1, z]$ where $f_j(1, z) = 0$. The polynomial function $z \mapsto f_j(1, z)$ has only finitely many zeros by the fundamental theorem of algebra. □

For the case $n = 1$, the complex projective line $\mathbb{CP}^1$, the extended complex plane $\overline{\mathbb{C}} = \mathbb{C} \cup \{\infty\}$, as well as the Riemann sphere S are all isomorphic, and may be regarded as the same object. The space $\mathbb{CP}^1$ turns out to be topologically equivalent to the unit sphere in $\mathbb{R}^3$, and so is called the *Riemann sphere.* It is obtained by taking the complex affine line $\mathbb{C}$, which is topologically identical to the real plane $\mathbb{R}^2$, and adding a point to the entire "boundary" of the plane.

The space $\mathbb{CP}^1$ is an example of a Riemann surface of genus 0; that is, topologically, it is the Riemann sphere. In fact, it is the only example. This follows from an important theorem in algebraic geometry, the Riemann–Roch theorem [129, p. 185] which guarantees that a genus 0 compact Riemann surface admits a degree 1 meromorphic function. One then verifies that such a function defines an isomorphism with $\mathbb{CP}^1$.

Example 5.2.6 (complex projective space)**.** We begin by introducing complex projective space $\mathbb{CP}^d$; it is the space of complex one-dimensional subspaces of $\mathbb{C}^{d+1}$. Such a subspace L is uniquely determined by any one $Z \in L \backslash \{0\}$, so that

$$\mathbb{CP}^d := \{[Z] : Z \in \mathbb{C}^{d+1} \backslash \{0\}, Z \sim \lambda Z\},$$

i.e., we identify Z and λZ ($\lambda \in \mathbb{C} \backslash \{0\}$).

One usually considers on $\mathbb{CP}^d$ the so-called *homogeneous coordinates*

$$Z = [Z_0, \ldots, Z_d];$$

here, the Z_i should not all vanish, and Z is identified with $\lambda Z = (\lambda Z_0, \ldots, \lambda Z_d)$ for all $\lambda \in \mathbb{C} \backslash \{0\}$. Hence, on the subset $U_i := \{[Z_0, \ldots, Z_d] : Z_i \neq 0\}$ of $\mathbb{CP}^d$, we may divide by Z_i and obtain a bijection

$$u_i : U_i \to \mathbb{C}^d$$

$$u_i\left(\frac{Z_0}{Z_i}, \ldots, \frac{Z_{i-1}}{Z_i}, 1, \frac{Z_{i+1}}{Z_i}, \ldots, \frac{Z_d}{Z_i}\right) := \left(\frac{Z_0}{Z_i}, \ldots, \frac{Z_{i-1}}{Z_i}, \frac{Z_{i+1}}{Z_i}, \ldots, \frac{Z_d}{Z_i}\right).$$

The coordinates defined by the u_i are called *Euclidean coordinates.* The coordinate transformations $u_j \circ u_i^{-1}$ are holomorphic on their domains of definition $u_i(U_i \cap U_j)$ (which are open sets in $\mathbb{C}^d$). Hence $\mathbb{CP}^d$ can be made into a complex manifold.

It can also be seen that $\mathbb{CP}^1$ is just the Riemann sphere. Namely, for $d = 1$, we have the coordinate charts

$$u_0 : U_0 = \{(Z_0, Z_1) : Z_0 \neq 0\} \to \mathbb{C}$$
$$(Z_0, Z_1) \mapsto \frac{Z_1}{Z_0} = z$$

and

$$u_1 : U_1 = \{(Z_0, Z_1) : Z_1 \neq 0\} \to \mathbb{C}$$
$$(Z_0, Z_1) \mapsto \frac{Z_0}{Z_1},$$

and the transformation between them is given by

$$u_1 \circ u_0^{-1} : z \mapsto \frac{1}{z}$$

and so we recover the representation of S^2 obtained in an earlier example.

So $\mathbb{CP}^1$ is just $\mathbb{C}$ with one point added. This additional point $(0 : 1)$ can be thought of as a "point at infinity," as you can see from the fact that its affine coordinate is formally $\frac{1}{0}$.

Remark 5.2.7. There is a completely analogous description of $\mathbb{CP}^n$ as $\mathbb{C}^n$ with some points added "at infinity": let $P = [a_0 : \cdots : a_n] \in \mathbb{CP}^n$ be a point. Then we have one of the following cases:

(i) $a_0 \neq 0$. Then $P = [1 : \alpha_1 : \cdots : \alpha_n]$ with $\alpha_i = \frac{a_i}{a_0}$ for all i. The α_i are the affine coordinates of P; they are uniquely determined by P and are obtained by setting $a_0 = 1$. So the set of all P with $a_0 \neq 0$ is just $\mathbb{C}^n$.

(ii) $a_0 = 0$, i.e., $P = (0 : a_1 : \cdots : a_n)$, with the a_i still defined only up to a common scalar.

Obviously, the set of such points is $\mathbb{CP}^{n-1}$; the set of all one-dimensional linear subspaces of $\mathbb{C}^n$. We think of these points as points at infinity; the new twist compared to $\mathbb{CP}^1$ is just that we have a point at infinity for every one-dimensional linear subspace of $\mathbb{C}^n$, i.e., for every "direction" in $\mathbb{C}^n$. So, for example, two lines in $\mathbb{C}^n$ will meet at infinity (when compactified in $\mathbb{CP}^n$) if and only if they are parallel, i.e., point in the same direction. (This is good as it implies that two distinct lines always intersect in exactly one point.)

Usually it is more helpful to think of the projective space $\mathbb{CP}^n$ as the affine space $\mathbb{C}^n$ compactified by adding some points (parametrized by $\mathbb{CP}^{n-1}$) at infinity, rather than as the set of lines in $\mathbb{C}^{n+1}$.

Remark 5.2.8. We claim that $\mathbb{CP}^n$ is a compact space (in the classical topology). In fact, let

$$S^{2n+1} = \left\{(a_0, \ldots, a_n) \in \mathbb{C}^{n+1}; |a_0|^2 + \cdots + |a_n|^2 = 1\right\}$$

be the unit sphere in $\mathbb{C}^{n+1} = \mathbb{R}^{2n+2}$. This is a compact space as it is closed and bounded, and there is an obvious surjective map

$$S^{2n+1} \to \mathbb{P}^n, (a_0, \ldots, a_n) \mapsto (a_0 : \ldots : a_n) .$$

As images of compact sets under continuous maps are compact, it follows that $\mathbb{P}^n$ is also compact.

The spaces $\mathbb{CP}^n$, $n \geqslant 0$, are all compact. Therefore any submanifold of $\mathbb{CP}^n$ is also compact.

An embedding of the affine plane is given by

$$\mathbb{C}^2 \to \mathbb{CP}^2, (x_1, x_2) \to (1, x_1, x_2) .$$

The line at infinity of $\mathbb{C}^2$ is then $\mathbb{P}^2 \backslash \mathbb{C}^2 = \{x_0 = 0\}$. Instead of this line, any other line $L \subset \mathbb{P}^2$ can play the role of the line at infinity of the affine plane $\mathbb{P}^2 \backslash L$.

Remark 5.2.9. In complete analogy to affine algebraic sets, we now want to define projective algebraic sets to be subsets of $\mathbb{P}^n$ that can be described as the zero locus of some polynomials in the homogeneous coordinates. Note however that if $f \in k[x_0, \ldots, x_n]$ is an arbitrary polynomial, it does not make sense to write down a definition like

$$Z(f) = \{(a_0 : \cdots : a_n) ; f(a_0, \ldots, a_n) = 0\}$$

because the a_i are only defined up to a common scalar. For example, if $f(x_0, x_1) = x_1^2 - x_0$ then $f(1,1) = 0$ but $f(-1,-1) \neq 0$, although $(1 : 1)$ and $(-1 : -1)$ are the same point in $\mathbb{P}^1$. To get rid of this problem we have to require that f be homogeneous, i.e., that all of its monomials have the same (total) degree d. This is equivalent to the requirement

$$f(\lambda x_0, \ldots, \lambda x_n) = \lambda^d f(x_0, \ldots, x_n) \quad \text{for all } \lambda$$

so in particular we see that

$$f(\lambda x_0, \ldots, \lambda x_n) = 0 \Longleftrightarrow f(x_0, \ldots, x_n) = 0 ,$$

i.e., the condition that a homogeneous polynomial in the homogeneous coordinates vanishes is indeed well defined.

Definition 5.2.10. A subset $C \subset \mathbb{P}^2$ is called a *projective plane algebraic curve* if there exists a homogeneous polynomial $F \in \mathbb{C}[x_0, x_1, x_2]$ of degree ≥ 1 such that

$$\overline{C} = V(F) = \left\{[x_0 : x_1 : x_2] \in \mathbb{P}^2 \;\middle|\; F(x_0, x_1, x_2) = 0\right\} .$$

A projective algebraic curve $\overline{C} \subset \mathbb{CP}^2$ is a closed subset and hence compact. Why? We need to check that $\pi^{-1}(\overline{C})$ is closed in $\mathbb{C}^3 \backslash \{0\}$. This inverse image is simply

$$\left\{(X, Y, Z) \in \mathbb{C}^3 \backslash \{0\} \;\middle|\; P(X, Y, Z) = 0\right\}$$

for all continuous P. A closed subset of a compact set is compact.

Example 5.2.11. The plane affine curve $C = \{(x, y) \in \mathbb{C}^2 \mid f = x^3 - y^2\}$ has projective closure $\overline{C}$ given by $F = x_1^3 - x_0^2 x_2$. Then $\overline{C}$ has an inflection at $[1 : 0 : 0]$ and a cusp at $[0 : 0 : 1]$. This last point lies in the affine part $x_2 = 1$ with coordinates (x_0, x_1).

Remark 5.2.12. The projective closure $\overline{C}$ of C coincides with the topological closure of C in $\mathbb{P}^2$.

The notions introduced above for affine curves have analogues for projective curves: irreducibility, decomposition into components, degree, etc.

First, note that $f \in \mathbb{C}[x, y]$ is irreducible if and only if its homogenization $F \in \mathbb{C}[x_0, x_1, x_2]$ is irreducible. Then, the degree of the projective plane algebraic curve C is the degree of a minimal defining polynomial F in $\mathbb{C}[x_0, x_1, x_2]$. The irreducible components of C are then defined by the irreducible factors of a minimal F.

Remark 5.2.13. The group $\mathrm{PGL}_2(\mathbb{C})$ of projective transformations acts on $\mathbb{P}^2$. Since the homogeneous coordinates are transformed linearly, we can easily see that the degree and the irreducibility property are independent of projective transformations. Such numbers or properties are called *(projective) invariants.*

Example 5.2.14 (smooth algebraic plane curves)**.** Let $F(z_1, z_2, z_3)$ be a complex polynomial in three variables which is homogeneous of some degree $d \geq 1$, thus

$$F(\lambda z_1, \lambda z_2, \lambda z_3) = \lambda^d F(z_1, z_2, z_3) . \tag{5.4}$$

Then

$$\left\{ F = \frac{\partial F}{\partial x_0} = \frac{\partial F}{\partial x_1} = \frac{\partial F}{\partial x_2} \right\} \subseteq \mathbb{P}^2 \text{ is empty} .$$

(This looks like four constraints, but is in fact essentially just three, due to the Euler identity

$$\sum_{j=1}^{3} z_j \frac{\partial}{\partial z_j} F(z_1, z_2, z_3) = dF(z_1, z_2, z_3)$$

that arises from differentiating (5.4) in λ. The fact that nonsingularity implies irreducibility is another consequence of Bézout's theorem, which is not proven here.) Then the zero locus $C = \{F = 0\} \subseteq \mathbb{P}^2$ is a Riemann surface. This can be checked easily on the standard affine charts of $\mathbb{R}^2$.

Remark 5.2.15. It turns out that the curve $C = \{F = 0\}$ is also connected, but we will prove this later.

For all $d \in \mathbb{N}$, $\mathbb{CP}^d$ is compact, since we have a continuous map of the unit sphere in $\mathbb{C}^{d+1}$ onto $\mathbb{CP}^d$.

Classically, a compact Riemann surface is called *closed*; while a noncompact surface is called *open*. The differences between the compact case and the open case are serious; for instance, a compact Riemann surface does not admit nonconstant holomorphic functions, but an open Riemann surface admits plenty of them.

A more detailed study of Riemann surface (and more generally, complex manifolds) can be found for instance in Griffiths and Harris's "Principles of Algebraic Geometry" [75].

5.3 Genus of Compact Riemann Surfaces

The coarsest classification of a Riemann surface is topological. The one-dimensional complex manifolds are usually called *Riemann surfaces*. Any complex manifold M can be regarded as a differentiable manifold with the same atlas. It suffices to replace the space $\mathbb{C}^n$ by the corresponding real space $\mathbb{R}^{2n}$ and, from the coordinate point of view, each complex coordinate $z_i = x_i + \sqrt{-1}y_i$ by the two real ones, x_i and y_i. Since the transition mappings are holomorphic, those of the corresponding real atlas are differentiable. Clearly,

$$\dim_{\mathbb{R}} M = 2 \dim_{\mathbb{C}} M ,$$

where $\dim_{\mathbb{R}} M$ is the real dimension of M, that is, the dimension of M regarded as a differentiable manifold. Differentiable manifolds of dimension two are called *surfaces*. A Riemann surface is a surface exactly in this sense, or in a still weaker topological sense. It has a canonical orientation: if we identify the real tangent space at any point with the complex tangent space, then for any nonzero vector v, we declare the ordered basis (v, iv) to be positively oriented.

Let us now forget the complex structure and consider the purely topological problem of classifying these surfaces up to homeomorphism.

There is a remarkably simple topological classification of compact Riemann surfaces.

Theorem 5.3.1. *Let X be a compact Riemann surface. Then, as a topological space, X is homeomorphic either to the Riemann sphere $\widehat{\mathbb{C}}$ or to a closed orientable surface of genus $g \geq 1$; that is, a surface obtained by attaching g handles to a sphere.*

The proof of this theorem is relatively straightforward, but it lies outside the scope of this book and will therefore be omitted.

Topologically, a Riemann surface of genus g is a sphere with g handles or, equivalently, a torus with g holes.

There exists a very close link with algebraic geometry: every compact Riemann surface X can be embedded into some projective space $\mathbb{CP}^n$. By Chow's theorem, X is then a (projective) algebraic variety, that is, it can be described by finitely many homogeneous equations. It should be noted that such a phenomenon is special to complex dimension 1. The crucial point is that one can always construct a nonconstant meromorphic function on a Riemann surface (e.g., by Dirichlet's principle). Given such a function, it is not difficult to find a projective embedding of a compact Riemann surface X. On the other hand, it is easy to construct a compact two-dimensional torus $T = \mathrm{C}^2/L$ for some suitably chosen lattice L, which cannot be embedded into any projective space $\mathbb{CP}^n$.

In this book, the expressions compact Riemann surface and (projective) algebraic curve are both used interchangeably. The choice depends on which expression is more commonly used in the part of the theory which is discussed in the relevant section.

An arbitrary Riemann surface carries with it a large amount of geometric information.

We shall deal only with compact Riemann surfaces. By definition, this is a closed (compact) surface S glued from a finite number of disks $U_1, \ldots, U_m$ in the complex plane: for any two disks, U_i and U_j, some domains, $V_{ij} \subset U_i$ and $V_{ji} \subset U_j$, are identified by means of a correspondence $\varphi_{ij} : V_{ij} \to V_{ji}$, which is one-to-one and analytic.

In other words, a Riemann surface is a union of sets $U_1, \ldots, U_N$, each of which is endowed with a coordinate function z_j $(j = 1, \ldots, N)$. This is a one-to-one mapping of U_2 onto a disk in the complex plane. Further, in an intersection $V_{ij} = U_i \cap U_j$, the coordinate z_j is expressed in terms of z_i as an analytic function.

In particular, the Riemann surface of an algebraic function reveals some important characteristics of that function. Since the transition functions $\varphi_{i,j}$ are conformal, and hence orientation-preserving, any Riemann surface is orientable. So, from a topological point of view it has a unique invariant: the genus.

5.4 Triangulations of Riemann Surfaces

A triangle T on X is a set homeomorphic to the standard triangle Δ defined above. The edges of T are the images of the edges of Δ (under the given homeomorphism). Similarly the images of the vertices of Δ (the points $(0,0), (1,0)$ and $(1,1)$ in $\mathbb{R}^2$) are called the *vertices* of T. Observe that in this definition a triangle T is a subset of X, and thus the homeomorphism between T and Δ does not play an important role as in the case of simplices or paths.

Definition 5.4.1. A *triangulation* of a compact surface X is a finite collection of triangles, $\mathcal{T} = \{T_1, \ldots, T_n\}$, satisfying the following properties:

(a) $X = \cup T_i$;

(b) any two triangles either do not intersect, or intersect at one common vertex, or intersect at one common edge;

(c) if the family $\{T_i\}$ is not finite, then its local finiteness is required, which is equivalent to the fact that there are only a finite number of triangles with a common vertex and their union defines a neighborhood of this vertex,.

It is a natural question to ask at this point whether every compact Riemann surface admits a triangulation. The answer is yes.

Theorem 5.4.2 (Rado, [150]). *Every Riemann surface is triangulable.*

For a short proof that compact 2-manifolds can be triangulated see [43]. Every two-dimensional surface has a piecewise linear structure therefore it is triangulable.

The above theorem is obvious in the compact case (see [95] for a modern proof). Any triangulation of a compact Riemann surface is finite.

The construction of Riemann surfaces using conformal structures admits the following variant. Let X be an oriented compact connected C^∞ real surface and let $\mathcal{T}$ be a C^∞ triangulation of X (see Fig 5.5).

FIGURE 5.5: A C^∞ triangulation of a Riemann surface.

A Euclidean structure on $\mathcal{T}$ is the data of a flat Riemannian metric on every triangle Δ of $\mathcal{T}$ such that:

(i) equipped with this metric, Δ is isometric with a triangle (in the usual sense) in the Euclidean plane $\mathbb{R}^2$;

(ii) on any edge E of $\mathcal{T}$, the metrics induced on E by the two triangles which contain E coincide.

A C^∞ triangulation equipped with a Euclidean structure is called a *Euclidean triangulation.* An *equilateral triangulation* is a Euclidean triangulation all of whose triangles are equilateral.

Let $V_{\mathcal{T}}$ be the set of vertices of a Euclidean triangulation $\mathcal{T}$ of X and let $E_{\mathcal{T}}$ be the union of the edges of $\mathcal{T}$. It is easily checked that the flat Riemannian metric in $X \setminus E_{\mathcal{T}}$ given by the Euclidean structure of $\mathcal{T}$ extends to a flat C^∞

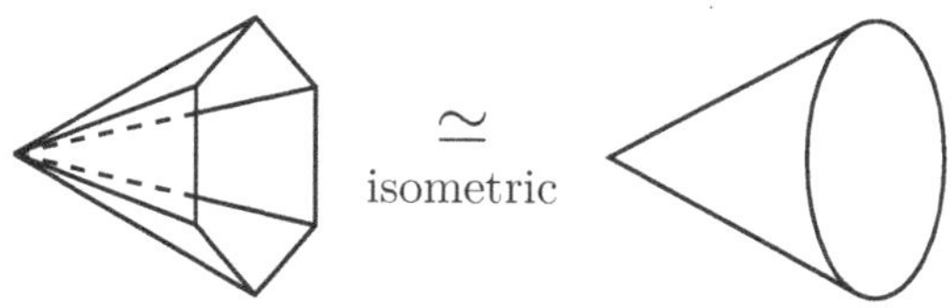

FIGURE 5.6: The singularities of the metric defined through a Euclidean triangulation are conical.

Riemannian metric on $X \setminus V_{\mathcal{T}}$ and that, near a vertex of $\mathcal{T}$, this metric has at worst a conical singularity (see Fig 5.6).

This implies that $X \setminus V_{\mathcal{T}}$ equipped with the holomorphic structure defined by the conformal class of this metric is a compact Riemann surface with punctures (the key point of the proof is pictorially explained on Fig 5.7).

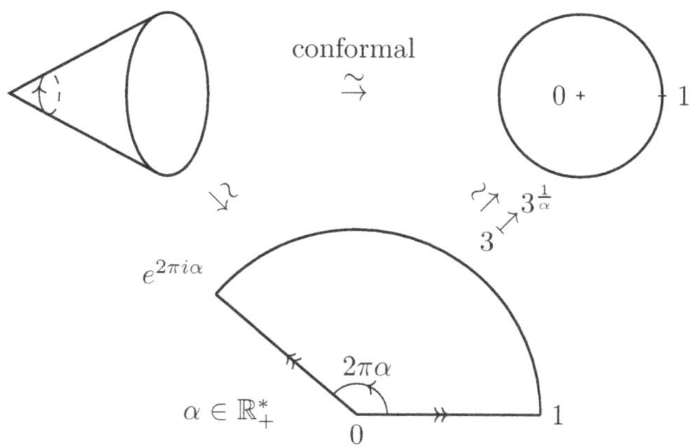

FIGURE 5.7: A punctured cone is conformally equivalent to a punctured disc.

In other words, the holomorphic structure on $X \setminus V_{\mathcal{T}}$ extends to a holomorphic structure on X. To summarize, a Euclidean triangulation of X defines a holomorphic structure on X, hence a compact connected Riemann surface.

Conversely, the holomorphic structure of any compact connected Riemann surface M may be recovered (up to isomorphism) from some Euclidean triangulation $\mathcal{T}$ of M:

- This is true if $M = \mathbb{CP}^1$, using for instance the tetrahedral subdivision of $\mathbb{CP}^1$ (see Fig 5.8; the Riemann surface obtained from this Euclidean triangulation is isomorphic to $\mathbb{CP}^1$ because it is compact and simply connected).

  Moreover, if $a_1, \ldots, a_r$ are given points in $\mathbb{CP}^1$, we can assume, by refining the triangulation $\mathcal{T}$, that $a_1, \ldots, a_r$ are vertices of $\mathcal{T}$ (see Fig 5.9).

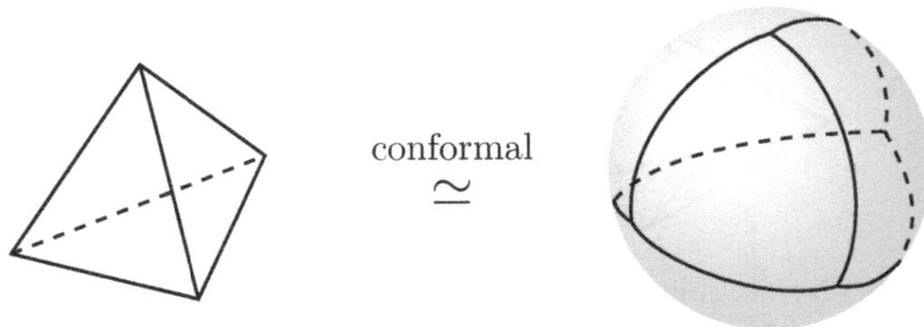

FIGURE 5.8: The Riemann surface $\mathbb{CP}^1$ can be obtained through a Euclidean triangulation.

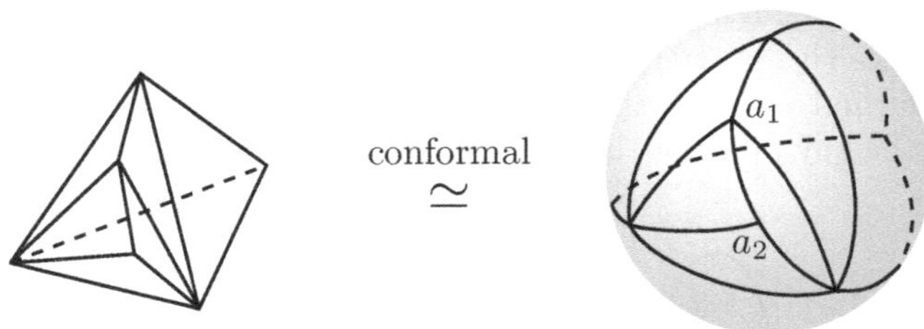

FIGURE 5.9: A refinement of the subdivision shown in Fig. 5.8.

- If M is any compact connected Riemann surface, there exists a ramified covering $\pi : M \to \mathbb{CP}^1$. If $\mathcal{T}$ is a C^∞ triangulation of $\mathbb{CP}^1$ such that $V_\mathcal{T}$ contains the ramification points of π, then there exists a unique C^∞ triangulation $\mathcal{T}'$ of M such that $V_{\mathcal{T}'} = \pi^{-1}(V_\mathcal{T})$ and $E_{\mathcal{T}'} = \pi^{-1}(E_\mathcal{T})$. If $\mathring{\Delta}$ is any open triangle of $\mathcal{T}$, then $\pi^{-1}(\mathring{\Delta})$ is a disjoint union $\dot{\Delta}_1 \cup \cdots \cup \mathring{\Delta}_r$ of open triangles of $\mathcal{T}'$. Moreover, there is a unique Euclidean structure on $\mathcal{T}'$ such that the maps $\pi : \mathring{\Delta}_i \xrightarrow{\sim} \mathring{\Delta}$ are isometries. Finally, the Euclidean triangulation $\mathcal{T}'$ defines the original holomorphic structure on M. Indeed, it defines this holomorphic structure on $M \backslash V_{\mathcal{T}'}$, since $\pi : M \backslash V_{\mathcal{T}'} \to \mathbb{CP}^1 \backslash V_\mathcal{T}$ is locally isometric, hence conformal.

For example, Fig 5.10 shows three valid decompositions of the complex curve $\mathbb{CP}^1$, which is topologically a sphere.

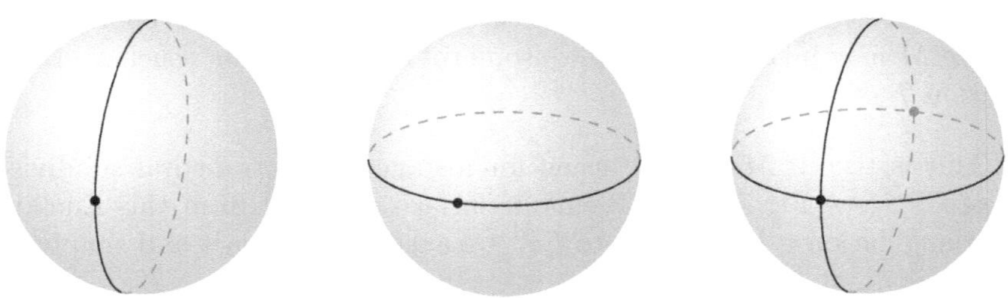

FIGURE 5.10: Decomposition of $\mathbb{CP}^1$.

In the left two illustrations of Fig. 5.10, we have one point, one line, and two discs (the two halves of the sphere), whereas in the illustration on the right we have two points, four lines, and four discs. Of course, there are many possible cell decompositions of X. However, there is an important quantity that does not depend on the chosen decomposition:

Lemma 5.4.3 (Euler characteristic). *Let X be a compact 2-dimensional manifold. Consider a triangulation of X, consisting of σ_0 points, σ_1 lines, and σ_2 discs. Then the number*

$$\chi := \sigma_0 - \sigma_1 + \sigma_2 = \#\ \textit{vertices} - \#\ \textit{edges} + \#\ \textit{faces}$$

depends only on X, and not on the chosen triangulation. We call it the (topological) Euler characteristic *of X. More generally, the number χ is invariant for any surface, and it equals $2 - 2g$, where g is the genus.*

Proof Sketch. Let us first consider the case when we move from one triangulation to a finer one, i.e., if we add points or lines to the triangulation. Such a process is always obtained by performing the following steps a finite number of times:

1. Adding another point on a line: in this case we raise σ_0 and σ_1 by 1 as in the Fig. 5.11, hence the alternating sum $\sigma_0 - \sigma_1 + \sigma_2$ does not change.

2. Adding another line in a disc: in this case we raise σ_1 and σ_2 by 1, so again $\sigma_0 - \sigma_1 + \sigma_2$ remains invariant.

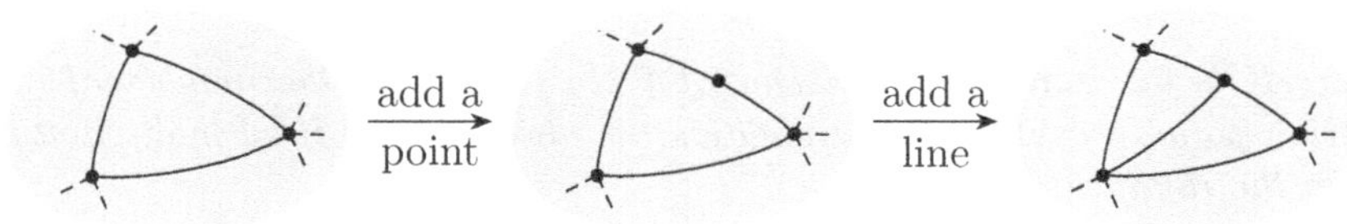

FIGURE 5.11: Move from one triangulation to a finer one.

We conclude that the alternating sum $\sigma_0 - \sigma_1 + \sigma_2$ does not change under refinements. But any two triangulations have a common refinement—which is essentially given by taking all the points and lines in both triangulations, and maybe adding more points where two such lines intersect. For example, in the cell decomposition depicted on Fig. 5.10 the triangulation in the picture on the right is a common refinement of the other two. Hence the Euler characteristic is independent of the chosen triangulation. □

Since we now know that the Euler characteristic of a surface S does not depend on the choice of triangulation, we can write $\chi(S)$ for the Euler characteristic of S.

In particular, the Euler characteristic is even and $\chi(S) \leq 2$. Topological manifolds of dimension ≤ 3 are always triangulable [120, 130], but there are non-triangulable manifolds for dimension n arbitrary but greater than three [101, pp. 299–306], [7, pp. 63–79] Further, differentiable manifolds always admit triangulations [189].

Proposition 5.4.4. *The Euler characteristic of the orientable surface of genus g is $2 - 2g$.*

Proof. Introduce a conformal metric $e^u dz d\bar{z}$ on a Riemann surface. The Gauss–Bonnet theorem provides us with the following formula for the Euler characteristic

$$\chi(\mathcal{R}) = \frac{1}{2\pi} \int_{\mathcal{R}} K \tag{5.5}$$

where

$$K = -2u_{z\bar{z}} e^{-u}$$

is the curvature of the metric. The right-hand side in (5.5) is independent of the triangulation, the left-hand side is independent of the metric we introduced on $\mathcal{R}$. This proves that the Euler characteristics is a topological invariant of $\mathcal{R}$. □

The Hurwitz formulae. With the Euler characteristic we can control the behavior of the genus under mappings of Riemann surfaces.

Theorem 5.4.5 (Hurwitz formula for the Euler characteristic). *Let $f : X \to Y$ be a nonconstant mapping of compact Riemann surfaces. Then*

$$\chi(X) = \deg f \cdot \chi(Y) - \deg R$$

where R is the ramification divisor of f. In particular, the number of ramification points—counting multiplicities—is always even, if by multiplicity we mean the ramification index.

To prove this, we choose on Y a sufficiently fine triangulation $\{T_i\}$, whose vertices include the images of all the ramification points. On X we take the triangulation $f^{-1}\{T_2\}$. Then the inverse image of any single triangle consists of $\deg f$ triangles. Counting the numbers of vertices, edges, and triangles, leads to the required relation (see [75]).

Lemma 5.4.3 allows us to rewrite these relations in terms of the genus.

Theorem 5.4.6 (Hurwitz's genus formula).

$$g(S_1) = \deg f \cdot g(S_2) + \frac{1}{2} \deg R - \deg f + 1 .$$

Corollary 5.4.7. $g(S_1) \geq g(S_2)$.

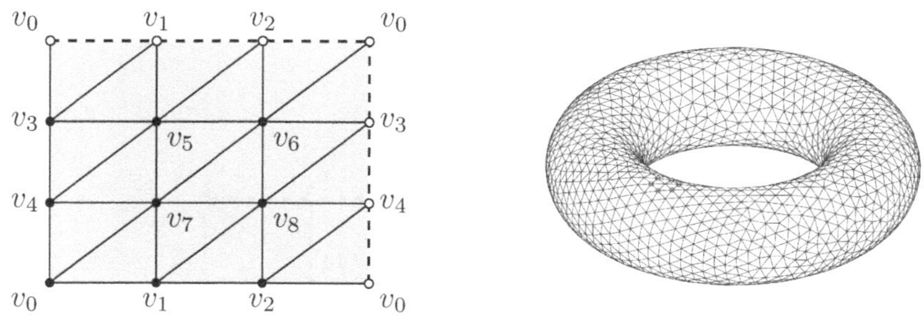

FIGURE 5.12: The most efficient way to triangulate the (2-dimensional) torus is depicted on the left picture. We obtain a triangulation of the torus by identifying the top edge with the bottom edge, and the left edge with the right edge, without any half-twists.

Example 5.4.8 (a genus 0 example)**.** Consider the triangulation of $\mathbb{CP}^1$ as shown on the Fig. 5.10. It is straightforward to show that the Euler characteristic of $\mathbb{CP}^1$ is equal to 2, i.e., $\chi(\mathbb{CP}^1) = 2 = 2 - 2g$. It follows that $\mathbb{CP}^1$ has genus 0.

Example 5.4.9 (a genus 1 example)**.** Let X be a torus, and consider the triangulation of X as shown on the Fig. 5.12.

It has $\sigma_0 = 9$ points, $\sigma_1 = 27$ lines, and $\sigma_1 = 18$ discs, and hence we conclude that the Euler characteristic of X is

$$\chi = \sigma_0 - \sigma_1 + \sigma_2 = 9 - 27 + 18 = 0 = 2 - 2g\,.$$

From this, we deduce that the genus g of the torus satisfies $g = 1$.

Example 5.4.10 (a genus 2 example)**.** Let X be a double torus, and consider the triangulation of X as shown on the Fig. 5.13. It has $\sigma_0 = 2g + 2$ points, $\sigma_1 = 4g + 4$ lines, and 4 discs, and hence we conclude that the Euler characteristic of X is

$$\chi = \sigma_0 - \sigma_1 + \sigma_2 = 2 - 2g\,.$$

In other words, the genus is given in terms of the Euler characteristic as $g = 1 - \frac{\chi}{2}$.

It has $\sigma_0 = 14$ points, $\sigma_1 = 48$ lines, and $\sigma_1 = 32$ discs, and hence we conclude that the Euler characteristic of X is

$$\chi = \sigma_0 - \sigma_1 + \sigma_2 = 14 - 48 + 32 = -2 = 2 - 2g\,.$$

From this, we deduce that the genus g of the double torus satisfies $g = 2$.

Proposition 5.4.11 (degree-genus formula)**.** *A smooth curve of degree d in $\mathbb{CP}^2$ has genus $\binom{d-1}{2}$, where $\binom{d-1}{2} = (d-1)!/[2!(d-2)!]$ is the binomial coefficient.*

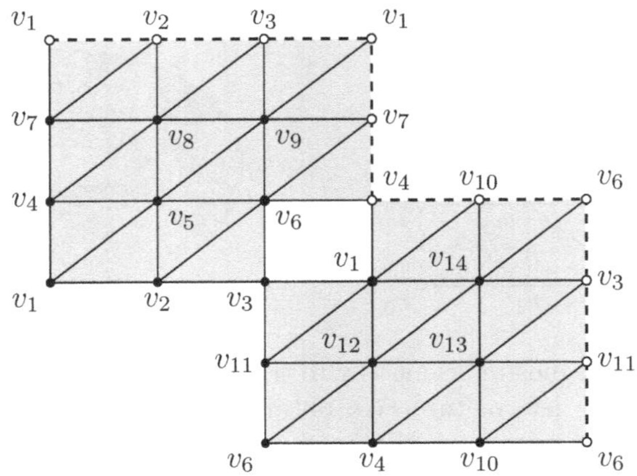

FIGURE 5.13: Triangulation of the double torus. We obtain a triangulation of the double torus by identifying the corresponding edges without introducing any half-twists.

Proof Sketch. For a classical proof see the book of Arbarello, Cornalba, Griffiths and Harris [9, Appendix A] or Grffiths and Harris [75, Chapter 2, section 1].

A singularity of order r decreases the genus by $\frac{1}{2}r(r-1)$. For the proof, see [164, pp. 53–54]

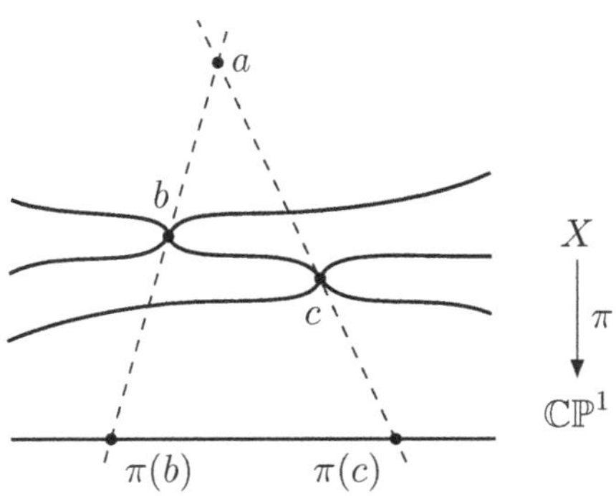

FIGURE 5.14: Branch point.

Let $I(X) = (f)$ for a homogeneous polynomial f of degree d. By a linear change of coordinates we can assume that $a := (1 : 0 : 0) \notin X$. Then the projection

$$\pi : X \to \mathbb{CP}^1, \quad (x_0 : x_1 : x_2) \mapsto (x_1 : x_2)$$

from a as in the Fig. 5.14 is a well-defined morphism on X. Let us study the inverse images of a fixed point $(x_1 : x_2) \in \mathbb{CP}^1$. Of course, they are given by the values of x_0 such that $f(x_0, x_1, x_2) = 0$, so that there are exactly d such points unless the polynomial $f(\cdot, x_1, x_2)$ has a multiple zero in x_0 at a point in the inverse image, which happens if and only if f and $\frac{\partial f}{\partial x_0}$ are simultaneously zero.

If we choose our original linear change of coordinates general enough, exactly two of the zeroes of $f(\cdot, x_1, x_2)$ will coincide at these points in the common zero locus of f and $\frac{\partial f}{\partial x_0}$, so that $\pi^{-1}(x_1 : x_2)$ then consists of $d-1$ instead of d points. These points, such as b and c in the picture above, are usually called the *ramification points* of π. Note that the picture might be a bit misleading since it suggests that X is singular at b and c, which is not the case. The correct topological picture of the map is impossible to draw however since it would require a real 4-dimensional space, namely an affine chart of $\mathbb{CP}^2$. Let us now pick a sufficiently fine triangulation of $\mathbb{CP}^1$, containing all images of the ramification points as points of the triangulation. If $\sigma_0, \sigma_1, \sigma_2$ denote the number of points, lines, and discs in this triangulation, respectively, we have $\sigma_0 - \sigma_1 + \sigma_2 = 2$ by Example 5.4.10 since $\mathbb{CP}^1$ is topologically a sphere, i.e., of genus 0. Now lift this cell triangulation to a triangulation of X by taking all inverse images of the cells of $\mathbb{CP}^1$. By our above argument, all cells will have exactly d inverse images—except for the images of the ramification points, which have one inverse image less. As the number of ramification points is $\left|V_P\left(f, \frac{\partial f}{\partial x_0}\right)\right| = \deg f \cdot \deg \frac{\partial f}{\partial x_0} = d(d-1)$ by Bézout's theorem, the resulting triangulation of X has $d\sigma_0 - d(d-1)$ points, $d\sigma_1$ lines, and $d\sigma_2$ discs. Hence by Lemma 5.4.3 the Euler characteristic of X is

$$\chi = d\sigma_0 - d(d-1) - d\sigma_1 + d\sigma_2 = 2d - d(d-1) = 3d - d^2$$

which means by Example 5.4.10 that its genus is

$$g = 1 - \frac{\chi}{2} = \frac{1}{2}\left(d^2 - 3d + 2\right) = \binom{d-1}{2}$$

□

The degree-genus formula tells us that only Riemann surfaces of genera $0, 1, 3, 6, 10, 15, \ldots$ (numbers expressible as $\frac{(d-1)(d-2)}{2}$, $d \in \mathbb{N}$) can ever be embedded as smooth curves in $\mathbb{P}^2$.

5.5 A Short Proof that Compact 2-Manifolds Can Be Triangulated

The result mentioned in the title of this section was first proved by Rado [150]; a proof can also be found in [6]. The idea for the present proof is due to the first-named author, who discovered it while investigating the possibilities of engulfing in low dimensions. It is shorter than previous proofs and is presented in the interest of economy.

We begin by listing a few familiar facts from geometric topology:

The Jordan–Schoenflies Theorem. A simple closed curve J in $\mathbb{R}^2$ separates $\mathbb{R}^2$ into two regions. There exists a self-homeomorphism of $\mathbb{R}^2$ under which J is mapped onto a circle.

Thickening an Arc. Each arc in the interior of a 2-manifold lies in the interior of a 2-cell (by definition, a 2-*cell* is a rectangle, see [160, p. 39]). This 2-cell can be chosen to be disjoint from any preassigned compact set in the complement of the arc.

Cellularity and Quotients. For our purposes, a *cellular set* K is one that can be written as the intersection of a sequence of 2-cells:

$$K = \bigcap_{i=1}^{\infty} E_i \,, \quad \text{where} \quad E_i \subset \operatorname{Int} E_{i-1} \quad (i = 2, 3, \ldots) \,.$$

If K is a cellular subset of a 2-manifold M, then M/K is homeomorphic to M.

We shall also need the following:

Lemma 5.5.1. *Let M be a closed 2-manifold, and let C be a connected subset of M which is the union of n simple closed curves,*

$$C = \bigcup_{i=1}^{n} C_i \,.$$

Let A be a compact, totally disconnected subset of C. Then A lies in the interior of a closed 2-cell in M. (A totally disconnected set is characterized by the property that each of its connected subsets consists of at most one point.)

Proof. Since A is compact and totally disconnected, some subarc S of C_1 contains no points of A. Thicken $C_1 - S$ to obtain a closed 2-cell D_1 whose interior contains $A \cap C_1$.

Next, suppose that a 2-cell D_k has been constructed to satisfy

$$A \cap \bigcup_{i=1}^{k} C_i \subset \operatorname{Int} D_k \,.$$

Then $C_{k+1} - D_k$ is the union of a countable collection of mutually disjoint open arcs X_μ whose endpoints lie on $\operatorname{Bd} D_k$. Each X_μ contains some subarc S_μ for which $A \cap S_\mu = \emptyset$, so that the two arcs comprising $X_\mu - S_\mu$ can be thickened to form closed 2-cells $E_{\mu 1}$ and $E_{\mu 2}$ which (a) avoid each other, (b) avoid X_λ for $\lambda \neq \mu$, and (c) meet D_k in closed 2-cells. Then

$$D_{k+1} = D_k \cup \bigcup_{\mu} (E_{\mu 1} \cup E_{\mu 2})$$

is a closed 2-cell whose interior contains

$$A \cap \bigcup_{i=1}^{k+1} C_i .$$

□

Theorem 5.5.2. *Any closed* 2*-manifold M can be triangulated.*

Proof. Cover M irreducibly by a finite collection of closed disks $\{B_1, B_2, \ldots, B_n\}$, and put $C_i = \operatorname{Bd} B_i$. Let A be the singular set of

$$C = \bigcup_{i=1}^{n} C_i ,$$

i.e., A consists of those points of C which do not have 1-dimensional Euclidean neighborhoods in C. A is compact and totally disconnected, and, since the cover $\{B_i\}$ was irreducible, no proper subset of $\{B_i\}$ covers M; thus C is connected. Hence, by the lemma, A lies in the interior of a closed 2-cell D in M. Note that $M - C$ is the union of a countable collection of mutually disjoint open 2-cells, each of whose closures in M is a closed 2-cell; this follows easily from the Jordan–Schoenflies theorem. Note also that $C - D$ consists of countably many mutually disjoint open arcs whose endpoints all lie on $\operatorname{Bd} D$.

Now $D \subset U \subset M$, where U is an open 2-cell, as can be seen by thickening $\operatorname{Bd} D$ slightly; hence D is cellular (again by the Jordan–Schoenflies theorem), and so $M_1 = M/D$ is topologically equivalent to M. M_1 is the copy of M with which we shall work from this point on.

Let $R \subset M_1$ be the image of $\overline{C - D}$ under the quotient map; thus R is the one-point union of a countable collection of simple closed curves and is locally Euclidean except at one point p (the image of D under the quotient map). Moreover, $M_1 - R$ is topologically equivalent to $M - (C \cup D)$. Furthermore, any 2-cell neighborhood V of p will contain all but finitely many of the simple closed curves comprising R; this follows from the compactness of $\overline{C - D}$, hence of R. If each of the simple closed curves lying within V is spanned by the disk it bounds, a cellular set T, consisting of a one-point union of closed disks, is obtained. (To see that T is cellular, let N be any neighborhood of T. Then N contains a 2-cell neighborhood V' of p, which must in turn contain all but finitely many of the simple closed curves comprising R. Thickening the arcs of R that lie outside V' and appending the resulting cells to $T \cup V'$ yields a cell Q, where $T \subset \operatorname{Int} Q \subset Q \subset N$.) Therefore, passing to M_1/T if necessary, we lose no generality in assuming that R is an r-leafed rose whose complement in M_1 consists of finitely many components that are open 2-cells.

Finally, enclose p in a small closed 2-cell E meeting each simple closed curve in R in two points of its boundary, $\operatorname{Bd} E$. Then $E \cup R$ is a disk with a finite number of mutually disjoint closed arcs $A_1, A_2, \ldots, A_r$ meeting $\operatorname{Bd} E$ in their endpoints. Each A_i may be enclosed (except for its endpoints) in the interior of a closed disk meeting E in a pair of arcs on its boundary. By selecting these disks to be pairwise disjoint, one obtains a triangulable 2-manifold N with boundary. The triangulation of N can now be extended to the closure of each component of $M_1 - N$. □

Corollary 1. Any compact 2-manifold can be triangulated.

Proof. Include the boundary curves in the set C of the proof of the theorem; minor modifications of the above argument convert it into a proof of the corollary. □

5.5.1 Homology

A sphere is distinguished from a torus by the fact that on the latter a closed curve can be drawn without forming a boundary, while this is not true for the former. In fact, two curves (α, β in Fig. 5.15) can be drawn so that their

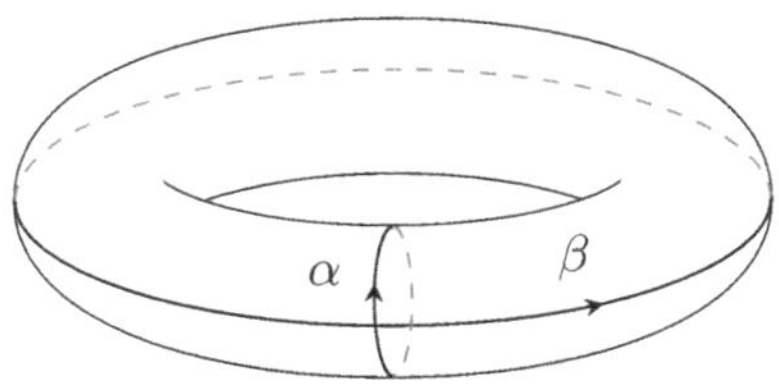

FIGURE 5.15: Torus.

union does not form a boundary; that is, when the surface is cut along them it remains in one piece. On more complicated surfaces (cf. Fig. 5.16) more than two cuts can be made without breaking the surface into separate pieces, and the maximum number of such cuts is a topological property of the surface. Algebraically, this means considering the maximum number of one cycles (one

FIGURE 5.16: Three-dimensional torus.

chains with zero boundary) on the surface such that no linear combination is a boundary, and this number is the same as that of a set of generators of the kernel of the boundary operator ∂, which is linearly independent modulo the image of ∂.

Thus, to give an algebraic account of the question of closed surfaces and boundaries in a space, the appropriate object of study is the quotient group of the kernel of d_p modulo the image of d_{p+1}, for each p. The necessary definitions will now be set up. Let E be a topological space and $\mathcal{G}$ the coefficient group.

A differential group C consists of an abelian group C and an endomorphism $\partial : C \to C$ such that $\partial\partial = 0$. The endomorphism ∂ is called the differential, or *boundary operator* of C. There is a category whose objects are differential groups and whose morphisms are homomorphisms commuting with the differentials.

For a differential group C there is a subgroup of cycles $Z(C) = \ker \partial$ and a subgroup of boundaries $B(C) = \operatorname{im} \partial$. Because $\partial\partial = 0$, $B(C) \subset Z(C)$. The homology group $H(C)$ is defined to be the quotient group

$$H(C) = Z(C)/B(C).$$

The elements of $H(C)$ are called *homology classes.* If z is a cycle, its homology class in $H(C)$ is denoted by $\{z\}$. Two cycles z_1 and z_2 are homologous, denoted by $z_1 \sim z_2$, if their difference is a boundary, that is, if $\{z_1\} = \{z_2\}$.

If $\tau : C \to C'$ is a homomorphism of differential groups commuting with the differentials, then τ maps cycles of C to cycles of C' and boundaries of C to boundaries of C'. Therefore τ induces a homomorphism

$$\tau_* : H(C) \to H(C')$$

such that $\tau_*\{z\} = \{\tau(z)\}$ for $z \in \mathbb{Z}(C)$. Because $(\tau_1\tau_2)_* = \tau_{1*}\tau_{2*}$, there is a covariant functor from the category of differential groups to the category of groups which assigns to a differential group C its homology group $H(C)$ and to a homomorphism τ its induced homomorphism τ_*.

The principal drawback of these easily defined invariants is that they are extremely difficult to compute. Even today, not all the homotopy groups of the 2-sphere—i.e., the space $x^2 + y^2 + z^2 = 1$ in $\mathbb{R}^3$—have been determined. Nonetheless, much is now known about the general properties of homotopy groups, largely thanks to the powerful algebraic techniques afforded by the "cohomological extension" of the component concept, as well as the numerous relations between homotopy and cohomology discovered over the years.

This cohomological extension begins with the dual point of view: a component is characterized by the property that every locally constant function on it is globally constant. Such a component is sometimes called a *connected component*, to distinguish it from a *path component.* Thus, if we define $H^0(X)$ to be the vector space of real-valued locally constant functions on X, then $\dim H^0(X)$ gives the number of connected components of X. Note that for reasonable spaces, where path components and connected components coincide, we obtain the formula

$$\operatorname{card} \pi_0(X) = \dim H^0(X).$$

Still the two concepts are dual to each other, the first using maps of the unit interval into X to test for connectedness and the second using maps of X into $\mathbb{R}$ for the same purpose. One further difference is that the cohomology group $H^0(X)$ has, by fiat, a natural $\mathbb{R}$-module structure.

Suppose X is a topological space and $x_0 \in X$ is a base point of X. A loop based at x_0 is a continuous map $\gamma : [0,1] \to X$ such that $\gamma(0) = \gamma(1) = x_0$. Suppose γ_0 and γ_1 are two loops based at x_0. We say γ_0 is homotopic to γ_1 among all based loops if there is a continuous map $H : [0,1] \times [0,1] \to X$ such that: (i) For $i = 0, 1$ and any $t \in [0,1]$, we have $H(i,t) = \gamma_i(t)$; Definition of composition of loops is compatible with respect to the notion of homotopy.

That is to say, if the based loops γ_0 and γ_0' are homotopic to each other and γ_1 and γ_1' are two other homotopic based loops, then $\gamma_0 * \gamma_1$ and $\gamma_0' * \gamma_1'$ are homotopic to each other among loops based at x_0. Therefore, the composition of loops induces a notion of multiplication on the elements of $\pi_1(X, x_0)$. It is straightforward to show that this multiplication map is associative, admits an identity element and any element of $\pi_1(X, x_0)$ has a multiplicative inverse.

These are some of the basic properties of fundamental groups:

1. If x_0 and x_1 belong to the same connected components of X, then $\pi_1(X, x_0)$ and $\pi_1(X, x_1)$ are isomorphic to each other.

2. Let $f : X \to Y$ be a continuous map which sends $x_0 \in X$ to $y_0 \in Y$. Then f induces a group homomorphism $f_* : \pi_1(X, x_0) \to \pi_1(Y, y_0)$ which maps a loop γ in X based at x_0 to $f \circ \gamma$. If f is a homeomorphism, then f_* is an isomorphism of fundamental groups. More generally, the same claim holds if f is a homotopy equivalence. Recall that f is a homotopy equivalence if there a continuous map $g : Y \to X$ such that $g \circ f : X \to X$ and $f \circ g : Y \to Y$ are both homotopic to the identity maps.

3. If X and Y are two topological spaces with base points x_0 and y_0, then $\pi_1(X \times Y, (x_0, y_0)) = \pi_1(X, x_0) \times \pi_1(Y, y_0)$.

4. If X is a topological space that consists of only one element x_0, then $\pi_1(X, x_0) = \{0\}$. More generally, if X is a contractible topological space, then $\pi_1(X, x_0) = \{0\}$. This is true because the constant map from X to the topological space with one element is a homotopy equivalence.

5. Identify S^1 with the quotient space $\mathbb{R}/\mathbf{Z}$ and let 0 be the base point of S^1. Then $\pi_1(S^1, 0) \cong \mathbb{Z}$. A generator of this topological space is given by the based loop $\gamma(t) = [t]$.

Usually we drop x_0 from our notation $\pi_1(X, x_0)$ for the fundamental groups. This is partly justified because of the first property: if X is a path connected space, then the isomorphism class of $\pi_1(X, x_0)$ is independent of the choice of the base point x_0.

In our class, we are mainly interested in $\pi_1(X(K))$ where K is a knot in S^3. In this case, the fundamental group is called the *knot group* of K. Of course, we can similarly consider fundamental groups of link exteriors.

In this section, we define the homology of a Riemann surface. Given a triangulation of a Riemann surface we define the vertex as a 0-simplex, the edges as a 1-simplex and the triangles as a 2-simplex. The orientation on the manifold induces an orientation on the triangles that can be used to orient the edges bounding each triangle.

Definition 5.5.3. Given a triangulation of a Riemann surface, a simplicial 0, 1 and 2 *chain* is a formal sum of vertices, P_j, edges e_j or triangles t_j:

$$c_0 = \sum_j n_j P_j, \quad c_1 = \sum_j m_j e_j, \quad c_2 = \sum_j k_j t_j \quad n_j, m_j, k_j \in \mathbb{Z}.$$

The set of n-chain C_n has a natural structure of additive abelian group. The element $-c_1$ is the edge with opposite orientation and $-c_2$ is the triangle with opposite orientation. The vertices $P_1, P_2, P_3, \dots$ can be used to identify edges and triangles. For example $\langle P_1 P_2 \rangle$ is the oriented edge from P_1 to P_2 and $\langle P_1, P_2, P_3 \rangle$ is the oriented triangle with sides the oriented edges $\langle P_1 P_2 \rangle, \langle P_2 P_3 \rangle$ and $\langle P_3 P_1 \rangle$. With this notation we define the boundary operator ∂:

Definition 5.5.4. The *boundary operator* $\partial : C_n \to C_{n-1}$ with $n = 0, 1, 2$ is defined as follows:

$$\begin{aligned} &\partial c_0 = 0, \quad c_0 \in C_0 \\ &\partial \langle P_1 P_2 \rangle = P_2 - P_1 \\ &\partial \langle P_1, P_2, P_3 \rangle = \langle P_1 P_2 \rangle + \langle P_2 P_3 \rangle + \langle P_3 P_1 \rangle \end{aligned}$$

The above relation defines ∂ on a 1 and 2-simplex and it can be extend to a 1 and 2-chain by linearity.

With the above definition it is immediate to observe that $\partial^2 = 0$. Furthermore, ∂ is a group homomorphism from C_n to C_{n-1}. We define

$$\begin{aligned} Z_n &= \{c_n \in C_n \mid \partial c_n = 0\}, \\ B_n &= \{c_n \in C_n \mid \exists c_{n+1} \in C_{n+1}, \quad c_n = \partial c_{n+1}\}. \end{aligned}$$

From the above definition it is clear that

$$B_n \subseteq Z_n \subseteq C_n .$$

Since all the groups are abelian, B_n is a normal subgroup. We are interested in the following quotient groups.

Definition 5.5.5. The *n-homology group* of X is defined as

$$H_n(X, \mathbb{Z}) = \frac{Z_n}{B_n}, \quad n = 0, 1, 2 .$$

We remark that the homology groups are independent of the triangulation. A basis for $H_n(X,\mathbb{Z})$ is the set of elements of $H_n(X,\mathbb{Z})$ such that any other element can be written as a linear integer combination of elements of the basis. It is clear from the definition that $\dim H_0(X,\mathbb{Z}) = 1$. Regarding $H_2(X,\mathbb{Z})$, with a little thought it is easy to see that for a compact Riemann surface $\dim H_2(X,\mathbb{Z}) = 1$. The only nontrivial group is $H_1(X,\mathbb{Z})$. We have the following result.

Theorem 5.5.6. *The first homology group $H_1(X,\mathbb{Z})$ is isomorphic to the abelianization of the first homotopy group $\pi(X)$. The group $H_1(X,\mathbb{Z})$ is a free abelian group with $2g$ generators. Any cycle can be written as a sum of generators.*

The theorem above gives the following. Let X be a compact Riemann surface of genus g and let $[\gamma_1],\ldots,[\gamma_{2g}]$ be the set of generators of $\pi_1(X)$. Then any element $[\gamma]\in\pi_1(X)$ can be uniquely written as

$$[\gamma]_{\pi_1} = [\gamma_{k_1}]_{\pi_1}^{j_1}\circ[\gamma_{k_2}]_{\pi_1}^{j_2}\circ\ldots[\gamma_{k_n}]_{\pi_1}^{j_n}\,,\quad k_1,\ldots k_n\in\{1,2,\ldots,2g\}$$

with $j_1,\ldots,j_n\in\mathbb{Z}$ and where we put a subscript of π_1 to denote an element of the fundamental group. Then the corresponding element $[\gamma]_{H_1}$ in the homology is obtained as

$$[\gamma]_{H_1} = j_1\,[\gamma_{k_1}]_{H_1} + j_2\,[\gamma_{k_2}]_{H_1} + \cdots + j_n\,[\gamma_{k_n}]_{H_1}\,,\quad k_1,\ldots k_n\in\{1,2,\ldots,2g\}\,.$$

In the rest of this section we simply denote as γ an element in the homology basis. Let $a_1,\ldots,a_g,b_1,\ldots,b_g$ be a basis in $H_1(X,\mathbb{Z})$. Then any cycle γ is homologous to a linear combination of the basis with integer coefficients:

$$\gamma\simeq\sum_{i=1}^{g} m_i a_i + \sum_{i=1}^{g} n_i b_i,\quad m_i,n_i\in\mathbb{Z}\,.$$

5.6 Holomorphic and Meromorphic Functions

A Riemann surface, locally, is just an open set in $\mathbb{C}$. Many concepts and objects of study in ordinary complex function theory (e.g., holomorphic and meromorphic functions) can be defined as a Riemann surface through local coordinates, as long as the object of study is invariant under coordinate changes.

Definition 5.6.1. Suppose X is a Riemann surface and $\{(U_i,z_i)\}$ is its holomorphic coordinate covering. A *meromorphic function* (resp. *holomorphic function*) f on X is by definition a family of mappings $f_i : U_i\to\mathbb{CP}^1$ satisfying the following conditions:

(a) $f_i = f_j$ on $U_i\cap U_j$ for $U_i\cap U_j\neq\varnothing$

(b) for all i, $f_i\circ z_i^{-1}$ are all meromorphic (resp. holomorphic) functions.

The meromorphic function field on X, denoted by $\mathcal{M}(X)$, is the set of all meromorphic functions on the Riemann surface X with the usual definitions of addition and multiplication. The holomorphic functions on X together with the usual definitions of addition, multiplication, and scalar multiplication form an algebra over $\mathbb{C}$, to be denoted by $\mathcal{O}(X)$.

One consequence of Definition 5.6.1 is that the space $\mathcal{O}(X)$ of holomorphic functions is a commutative complex algebra (a complex vector space closed under pointwise multiplication), while the space $\mathcal{M}(X)$ of meromorphic functions is a complex field (a commutative complex algebra where every non-zero element has an inverse).

A consequence is analytic continuation: if two holomorphic or meromorphic functions on X agree on a non-empty open set, then they agree everywhere.

A third consequence is that one can define the notion of a zero of given order k, or a pole of order k, for a holomorphic or meromorphic function, by composing with a chart map and using the usual complex analysis notions there, noting (from the holomorphicity of transition maps and their inverses) that this does not depend on the choice of chart. However, one cannot similarly define the residue of a meromorphic function on X this way, as the residue turns out to be chart-dependent thanks to the chain rule. (Residues should instead be applied to meromorphic 1-forms, a concept we will introduce later.)

In Section 1.2 we introduced the notions of poles and essential singularities for holomorphic functions defined on punctured neighborhoods in the complex plane. These concepts are preserved under local biholomorphic maps, and therefore extend naturally to Riemann surfaces.

Definition 5.6.2. Let f be holomorphic in a punctured neighborhood of a point $p \in X$. We say that f has a *removable singularity* (resp. a *pole of order n*, an *essential singularity*, or a *zero of order n*) at p if there exists a complex chart $\varphi : U \to V$ with $p \in U$ such that the function $f \circ \varphi^{-1}$ has a removable singularity (resp. a pole of order n, an essential singularity, or a zero of order n) at $\varphi(p)$.

Definition 5.6.3. Let X be a Riemann surface and let $W \subset X$ be open. A function f on W is said to be *meromorphic at $p \in W$* if f is holomorphic in a punctured neighborhood of p and has at p either a pole or a removable singularity. The function f is called *meromorphic on W* if it is meromorphic at every point of W.

Definition 5.6.4. Let X be a Riemann surface and let $W \subset X$ be open. The space of meromorphic functions on W is defined by

$$\mathcal{M}_X(W) := \{f : W \to \mathbb{C} \mid f \text{ is meromorphic}\} .$$

When X is clear from the context, we abbreviate this by $\mathcal{M}(W)$.

A meromorphic function on X is an analytic map $f : X \to \widehat{\mathbb{C}}$ that is not identically ∞. Equivalently, $f(z)$ is meromorphic if it can be locally written as a quotient of two holomorphic functions, not both zero.

The sheaf of meromorphic functions is denoted $\mathcal{M}$. Its construction from $\mathcal{O}$ is one of the most basic examples of sheafification.

Since X is connected, the ring $\mathcal{M}(X)$ (also denoted $\mathbb{C}(X)$) of all meromorphic functions on X forms a field, and the ring $\mathcal{O}(X)$ of all analytic functions forms an integral domain.

We will see that $\mathcal{M}(\widehat{\mathbb{C}}) = \mathbb{C}(z)$, while $\mathcal{O}(\widehat{\mathbb{C}}) = \mathbb{C}$, so not every meromorphic function is a quotient of holomorphic functions. We will see that $\mathcal{O}(\mathbb{C})$ is much wilder (it contains $\exp(\exp z)$, etc.), and yet $\mathcal{M}(\mathbb{C})$ is the field of fractions of $\mathcal{O}(\mathbb{C})$.

Thus we have a well-defined notion of a meromorphic function on a Riemann surface, as well as a well-defined notion of the *order* of a meromorphic function at a point. For $f \in \mathcal{M}(W)$ and $p \in W$, we denote the order of f at p by

$$\operatorname{ord}_p(f)\,.$$

Theorem 5.6.5. *Let X be a connected compact complex manifold, and let $f : X \to \mathbb{C}$ be holomorphic. Then f is constant.*

Proof. Since X is compact, $|f|$ attains a maximum at some $x_0 \in X$ with $f(x_0) = a$. In local coordinates near x_0, f becomes a holomorphic map F on an open subset of $\mathbb{C}^n$ with $|F|$ maximal at 0. Restricting F to complex lines through 0 and applying the Maximum Principle shows F is locally constant, hence constant on X by connectedness. □

Theorem 5.6.6 (Riemann's theorem on removable singularities)**.** *Let U be an open subset of a Riemann surface. Let $a \in U$. If $f \in \mathcal{O}(U\backslash\{a\})$ is bounded in a neighborhood of a, then there is $F \in \mathcal{O}(U)$ such that $F|_{U\backslash\{a\}} = f$.*

Proof. This follows immediately from Riemann's theorem on removable singularities in the complex plane. □

This result is somewhat surprising and even a bit disappointing: the compactness of X, which usually simplifies the study of manifolds, in fact rules out nontrivial holomorphic functions, since all of them must be constant.

Oddly enough, it is not at all easy to prove that there is a non-constant meromorphic function on a Riemann surface X (see Section 5.16)!

The space $X = \mathbb{C}/\Lambda$ is a Riemann surface see Example 5.2.2. First we shows that the complex torus admits a non-constant meromorphic function. Such functions are called *elliptic functions.*

5.7 Elliptic Functions

Any complex torus is of the form $X = \mathbb{C}^g/\Gamma$, where Γ is a lattice in $\mathbb{C}^g$. A complex torus is a complex manifold of dimension g. A meromorphic function on

$\mathbb{C}^g$, periodic with respect to Γ, may be considered as a function on X. In fact, a general complex torus of dimension ≥ 2 does not admit any meromorphic function (see [118]).

A lattice in $\mathbb{C}$ is generated by two complex numbers λ_1 and λ_2, which are linearly independent over $\mathbb{R}$. So we have the following picture (see Fig. 5.17):

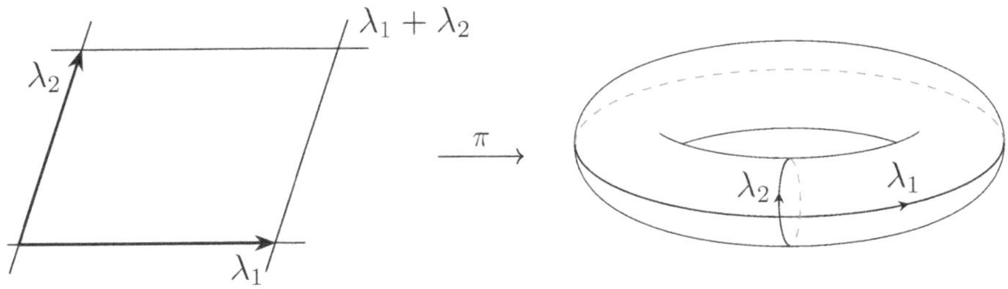

FIGURE 5.17: Construction of a torus by pasting.

Identifying opposite sides of the parallelogram $0, \lambda_1, \lambda_1 + \lambda_2, \lambda_2$, we obtain the complex torus X. The images of the lines $\overline{0\lambda_1}$ and $\overline{0\lambda_2}$ are cycles on X, also denoted by λ_1, and λ_2. Obviously λ_1 and λ_2 generate the group $H_1(X, \mathbb{Z})$. A 1-dimensional complex torus is called an *elliptic curve*. Since Λ is abelian, $\pi_1(X)$ is canonically isomorphic to $H_1(X, \mathbb{Z})$.

Consider the 1-dimensional torus

$$X = \mathbb{C}/\Lambda, \quad \Lambda = \mathbb{Z}\lambda_1 + \mathbb{Z}\lambda_2 .$$

This means we identify points in $\mathbb{C}$ that differ by an element of Λ.

The space X is compact, so any holomorphic function on X is constant. Equivalently, any holomorphic function

$$F : \mathbb{C} \to \mathbb{C}, \quad F(z) = F(z + \lambda_1) = F(z + \lambda_2),$$

is constant. Thus, to obtain nontrivial functions on X, we must allow singularities.

Definition 5.7.1 (elliptic function)**.** Let Λ be a lattice. An *elliptic function* (with respect to Λ) is a meromorphic function $F : \mathbb{C} \to \mathbb{C}$ such that

$$F(z + \lambda) = F(z), \quad \forall z \in \mathbb{C},\ \lambda \in \Lambda .$$

Thus there is a bijective correspondence between Λ-periodic meromorphic functions on $\mathbb{C}$ and meromorphic functions on

$$X = \mathbb{C}/\Lambda .$$

Let $F = G/H$ be a meromorphic function on $V \subseteq \mathbb{C}$. For $a \in V$, the convergent expansions are

$$G(z) = (z - a)^k \left(b_0 + b_1(z - a) + \cdots\right), \qquad k \geq 0,\ b_0 \neq 0,$$
$$H(z) = (z - a)^l \left(c_0 + c_1(z - a) + \cdots\right), \qquad l \geq 0,\ c_0 \neq 0.$$

Hence

$$F(z) = (z-a)^n \left(d_0 + d_1(z-a) + \cdots\right), \quad n = k - l,\ d_0 = \tfrac{b_0}{c_0} \neq 0\,,$$

or equivalently,

$$F(z) = \sum_{m=n}^{\infty} a_m (z-a)^m, \quad a_n = d_0\,.$$

Definition 5.7.2 (order and residue). Let

$$f(z) = \sum_{m=n}^{\infty} a_m (z-a)^m, \quad a_n \neq 0\,.$$

The *order* of f at a, denoted $\operatorname{ord}_a(f)$, is n.

- If $n > 0$, then f has a zero of order n at a.
- If $n < 0$, then f has a pole of order $|n|$ at a.

The *residue* of f at a is

$$\operatorname{Res}_a(f) := a_{-1}\,.$$

Remark 5.7.3. If f is elliptic, then

$$\operatorname{Res}_a(f) = \operatorname{Res}_{a+\lambda}(f), \quad \operatorname{ord}_a(f) = \operatorname{ord}_{a+\lambda}(f), \quad \forall \lambda \in \Lambda\,.$$

Since the zeros and poles of a meromorphic function are isolated and X is compact, every meromorphic function on X has only finitely many zeros and poles. Equivalently, an elliptic function has only finitely many poles in $\overline{\Pi}$.

Define the fundamental parallelogram

$$\Pi := \{s\lambda_1 + t\lambda_2 \in \mathbb{C} \mid 0 \leq s, t \leq 1\}\,.$$

Then the projection $\pi : \Pi \to X$ is bijective. Similarly, for any $\alpha \in \mathbb{C}$,

$$\pi : \alpha + \Pi \to X$$

is also a bijection.

Proposition 5.7.4. *Let F be an elliptic function, and let $\alpha \in \mathbb{C}$ be such that F has no poles and no zeros on*

$$C := \partial(\alpha + \Pi)\,.$$

Then:

(1) $\displaystyle\sum_{a \in \alpha + \Pi} \operatorname{Res}_a(F) = 0.$

(2) $\sum_{a\in\alpha+\Pi} \operatorname{ord}_a(F) = 0.$

Proof. By Cauchy's residue theorem,

$$\sum_{a\in\alpha+\Pi} \operatorname{Res}_a(F) = \frac{1}{2\pi i}\int_C F(z)\,dz = \frac{1}{2\pi i}\sum_{i=1}^{4}\int_{S_i} F(z)\,dz\,,$$

where each S_i is a side of C.

For instance,

$$S_1 : z = \alpha + tw_1, \quad t \in [0,1], \qquad S_3 : z = \alpha + tw_1 + w_2, \quad t \in [0,1]\,.$$

Since F is elliptic,

$$F(\alpha + tw_1 + w_2) = F(\alpha + tw_1), \quad t \in [0,1]\,,$$

so F takes the same values on S_1 and S_3. But these segments are traversed in opposite directions, hence

$$\int_{S_1} F(z)\,dz + \int_{S_3} F(z)\,dz = 0\,.$$

The same reasoning applies to S_2 and S_4. Thus,

$$\int_C F(z)\,dz = 0\,,$$

which proves (1).

By the argument principle,

$$\sum_{a\in\alpha+\Pi} \operatorname{ord}_a(F) = \frac{1}{2\pi i}\int_C \frac{F'(z)}{F(z)}\,dz\,.$$

Since both F' and $\frac{F'}{F}$ are elliptic with the same periods as F, part (1) applies, and the claim follows. □

Corollary 5.7.5. *There is no meromorphic function on X with a single simple pole (a pole of order 1).*

Proof. Suppose such an f exists. Then the corresponding elliptic function

$$F = f \circ \pi$$

has a unique simple pole at some $a \in \Pi$. Thus,

$$F(z) = \frac{a_{-1}}{z-a} + \sum_{m=0}^{\infty} a_m (z-a)^m, \quad a_{-1} \neq 0\,.$$

But then, for suitable α,

$$\sum_{a\in\alpha+\Pi} \operatorname{Res}_a(F) = a_{-1}\,,$$

contradicting Proposition 5.7.4 (1). □

Define the *Weierstrass $\wp$-function* by

$$\wp(z) = \frac{1}{z^2} + \sum_{\omega \in \Lambda \setminus \{0\}} \left(\frac{1}{(z-\omega)^2} - \frac{1}{\omega^2} \right).$$

This series converges normally on compact subsets of $\mathbb{C} \setminus \Lambda$, so $\wp$ is meromorphic on $\mathbb{C}$ with double poles at the lattice points.

It satisfies the periodicity relation

$$\wp(z+\omega) = \wp(z) \quad \text{for all } \omega \in \Lambda.$$

Hence, $\wp$ descends to a meromorphic function on $X = \mathbb{C}/\Lambda$. Since $\wp(z) = z^{-2} + O(z^2)$ near $z = 0$, it is not constant. Thus X admits a non-constant meromorphic function.

Proposition 5.7.6. *There exists a natural bijection between the set of meromorphic functions on a Riemann surface X, which is not necessarily compact, and the set of holomorphic functions (maps) from X to $\widehat{\mathbb{C}}$.*

Proof. If f is our meromorphic function on X, and S is the set of its poles, we can define a map $\bar{f} : X \to \overline{\mathbb{C}}$ by the following rule: $\bar{f}(x) = \infty$, if $x \in S$, and $\bar{f}(x) = f(x)$, otherwise. It is easy to see that $\bar{f}$ is holomorphic: for $x \notin S$, there is no contradiction because f is already holomorphic at x. If $x \in S$, then the composition of f with the chart $\phi_1 : U_1 \to \mathbb{C}$ is $z \mapsto \frac{1}{f(z)}$, and if f has a pole at x, then $\frac{1}{f(z)}$ is bounded in the neighborhood of x and can be made holomorphic in accordance with the removable singularity theorem. Thus, $\bar{f}$ is also holomorphic (since its local representation in the neighborhood of x is.) The converse is also true: if we have a holomorphic map $\bar{f}$, we can easily build a meromorphic function f by restriction of $\bar{f}$ to pre-images of $\mathbb{C}$. Using same ideas we can see that f has poles in the points at the pre-images of ∞. □

In the next section, however, we will see that meromorphic functions on Riemann surfaces (which are compact complex manifolds) yield more interesting results.

5.8 Rational Functions on Riemann Surfaces

Let us introduce the concept of a rational function (meromorphic function) on a complex manifold. More precisely:

Definition 5.8.1. Suppose X is a complex manifold, and $\{(U_i, g_i, h_i)\}$ satisfies the following conditions:

(a) $\{U_i\}$ is an open covering of X;

(b) $g_i, h_i \in \mathcal{O}(U_i), \quad h_i \not\equiv 0;$

(c) compatibility: for $U_i \cap U_j \neq \varnothing$, we have $g_i h_j \equiv g_j h_i$ on $U_i \cap U_j$.

Then the formal expression

$$\varphi = g_i/h_i \quad (\text{on } U_i)$$

becomes well defined on X, and is called the *rational function* (meromorphic function) given by $\{(U_i, g_i, h_i)\}$.

On $\mathbb{C}$, there are of course many holomorphic functions and meromorphic functions; for instance any power series with an infinite radius of convergence will give a holomorphic function, and the quotient of any two such functions (with non-zero denominator) will give a meromorphic function. Furthermore, we have extremely wide latitude in how to specify the zeroes of the holomorphic function, or the zeroes and poles of the meromorphic function, thanks to tools such as the Weierstrass factorization theorem or the Mittag-Leffler theorem (covered in Section 1.4).

With the insight gained from his view of algebraic curves as surfaces covering the plane $\mathbb{C}$ or the sphere $\mathbb{C} \cup \{\infty\} = \mathbb{CP}^1$, Riemann was able to generalize Cauchy's theory of integration on $\mathbb{C}$ to integration on arbitrary algebraic curves.

The fundamental result is Cauchy's theorem, according to which

$$\int_C f(z)dz = 0$$

where f is differentiable or holomorphic (and hence not infinite) on and inside the closed curve C in the complex plane $\mathbb{C}$. It follows that $\int_a^b f(z)dz$ does not depend on the path chosen between a and b, as long as the paths lie in a simply connected region of $\mathbb{C}$ where f does not become infinite.

Of course, it is a different story for functions that do become infinite, such as $f(z) = 1/z$ at $z = 0$. We find, for example, that

$$\int_C \frac{dz}{z} = 2\pi i$$

where C is a counterclockwise path around the unit circle. Nevertheless, the value of the integral of $1/z$ around a closed path C depends only slightly on C—it is the same around any path C' to which C can be deformed without crossing the point $z = 0$.

This example is generalized in Cauchy's integral formula and Cauchy's residue theorem. The integral formula says that

$$f(a) = \frac{1}{2\pi i} \int_C \frac{f(z)dz}{z - a}$$

where C is a circle with center a in a region where f is differentiable, and it has the consequence that

$$f^{(n)}(a) = \frac{n!}{2\pi i} \int_C \frac{f(z)dz}{(z-a)^{n+1}}$$

so a differentiable complex function f in fact has derivatives of all orders. It also follows that f has a Taylor series expansion in the neighborhood of a point where f' exists, and from this we can conclude that if $f(a_k) = 0$ for a convergent sequence of points a_k then f is constantly zero. Another important consequence is Liouville's theorem: if f is differentiable and bounded on $\mathbb{C}$ then f is constant.

The residue theorem generalizes the property of the function $1/z$ to an arbitrary meromorphic function f—one that is differentiable except at isolated poles $z = a$, in the neighborhood of which

$$f(z) = \frac{c_{-h}}{(z-a)^h} + \cdots + \frac{c_{-1}}{(z-a)} + c_0 + c_1(z-a) + c_2(z-a)^2 + \cdots .$$

The integer h is called the *multiplicity* or order of the pole $z = a$, and the coefficient c_{-1} is called the *residue* of f at $z = a$. For such a function,

$$\int_C f(z)dz = 2\pi i c_{-1}$$

where C is a counterclockwise path around $z = a$, small enough not to touch or enclose any other pole of f. Thus the value of the integral depends only on the residue. More generally, if C is a path running counterclockwise around poles of f at which the residues are $r_1, \ldots, r_m$, then

$$\int_C f(z)dz = 2\pi i \left(r_1 + \cdots + r_m\right) .$$

A holomorphic function that takes values in $\mathbb{CP}^1$ is called *meromorphic* if it reaches the value ∞ (it has poles), and holomorphic otherwise.

Every polynomial $f(z)$ defines a meromorphic function on $\mathbb{CP}^1$, which has a unique pole at ∞, provided the degree of the polynomial is ≥ 1. In fact, every rational function $f(z)$ defines a meromorphic function on $\mathbb{CP}^1$, which fails to have a pole at ∞ precisely if $f(z)$ is a proper fraction. Conversely, we can prove the following result.

Theorem 5.8.2. *All meromorphic functions on $\mathbb{CP}^1$ are rational functions.*

Proof. Recall that for a holomorphic function, zeros are isolated unless the function is identically zero (this is the identity theorem). Since $\mathbb{CP}^1$ is compact, it follows that a meromorphic function $f : \mathbb{CP}^1 \to \mathbb{CP}^1$ can have only finitely many zeros $z_1, \ldots, z_n \in \mathbb{C}$ (unless f is identically zero) and only finitely many poles $p_1, \ldots, p_m \in \mathbb{C}$ (since those are isolated zeros of $1/f$). Let $a_1, \ldots, a_n$ and $b_1, \ldots, b_m$ be the orders of the zeros and of the poles. Let

$$g(z) = \prod (z - z_j)^{a_j} / \Pi (z - p_k)^{b_k} .$$

Then f/g is meromorphic, and it no longer has zeros or poles in $\mathbb{C}$ (they are removable singularities). If at infinity it also does not have a pole, then it is a holomorphic map $f/g : \mathbb{CP}^1 \to \mathbb{C}$ and hence is constant. If it has a pole at infinity, then consider the reciprocal $g/f : \mathbb{CP}^1 \to \mathbb{C}$, which again must be constant (here we used that f/g has no zeros in $\mathbb{C}$ so g/f has no poles in $\mathbb{C}$), so actually there was no pole. Thus $f =$ constant $\cdot g$, so f is a quotient of polynomials. □

In this way we also show the existence and uniqueness of the expansion of a proper fraction into partial fractions in the complex sense. Thus $\mathcal{M}\left(\mathbb{CP}^1\right) \simeq \mathbb{C}(z)$, where $\mathbb{C}(z)$ is the field of rational functions of one complex variable z. Hence meromorphic functions constitute a natural generalization of rational functions. Furthermore, these notions coincide for any compact Riemann surface, once the rationality of a function is suitably defined.

Now a rational function $f(z) = p(z)/q(z)$ is determined, up to a constant multiple, by its zeros and poles and their orders. Assuming any common factors of $p(z)$ and $q(z)$ have been canceled, the zeros of $p(z)$ give zeros of $f(z)$ and the zeros of $q(z)$ give poles of $f(z)$.

In addition, if $\deg(p) \neq \deg(q)$, there will be a zero or pole at ∞ according as $\deg(p) < \deg(q)$ or $\deg(p) > \deg(q)$, making the total order of zeros equal to the total order of poles.

To summarize: *a meromorphic function on $\mathbb{C} \cup \{\infty\}$ is determined, up to a constant multiple, by a finite set of zeros and poles, with associated orders. A function with given zeros and poles actually exists (namely, a rational function) provided that the total order of zeros equals the total order of poles.*

Thus Riemann's program of "defining a function by its discontinuities" is easily carried out on $\mathbb{C} \cup \{\infty\}$, a Riemann surface of genus zero. We need only understand the "discontinuities" to be the zeros and poles. Extending this result to all surfaces of genus zero is already a significant problem. It was solved by Riemann (1851) via the Riemann mapping theorem, a difficult result that he proved by appealing to what he called the *Dirichlet principle.*

The Dirichlet principle is a powerful method for proving the existence of functions with given properties (such as specified zeros and poles) on surfaces, but it stems from physical intuition about the flow of electricity and was not rigorously proved in a form suitable for Riemann's applications until after 1900.

It should be stressed that this remarkably simple view of meromorphic functions on $\mathbb{C} \cup \{\infty\}$ is made possible by the point ∞. We would now say that the role of ∞ is to make the surface compact. Compactness ensures that any infinite set of points has a limit, which makes possible the above argument that a meromorphic function on $\mathbb{C} \cup \{\infty\}$ is rational. The meromorphic functions on $\mathbb{C}$ are a much less manageable class, since they include functions that are not even algebraic, such as e^z.

Definition 5.8.3. A *morphism* between two Riemann surfaces X and X' is a continuous mapping $f : X \longrightarrow X'$ such that $\varphi' \circ f \circ \varphi^{-1}$ is a holomorphic function for every choice of coordinates φ in X and φ' in X' for which the composition makes sense. We will denote by $\mathrm{Mor}\,(X, X')$ the set of morphisms from X to X'.

Bijective morphisms are called *isomorphisms* and isomorphisms from a surface to itself are called *automorphisms*. The set of automorphisms of a given Riemann surface X forms a group. We shall denote it by $\mathrm{Aut}(X)$.

Remark 5.8.4. Let $f : X \longrightarrow X'$ be a non-constant morphism between connected compact Riemann surfaces. Since non-constant holomorphic maps are open maps, the image $f(X)$ must be simultaneously open and closed in X', and therefore equal to the whole of X'. Thus f is surjective.

Let X be a Riemann surface.

Proposition 5.8.5. *Let us denote by $c_P \in \mathrm{Mor}(X, \widehat{\mathbb{C}})$ the constant morphism $c_P(x) = P$. Then*

$$\mathcal{M}(X) \equiv \mathrm{Mor}(X, \widehat{\mathbb{C}}) \backslash \{c_\infty\}$$

Proof. A morphism $f : X \to \widehat{\mathbb{C}}$ is clearly a holomorphic mapping outside the set $\{f^{-1}(\infty)\}$. At a point $P \in f^{-1}(\infty)$, choose a suitable coordinate function φ in X and let $\varphi_2(z) = 1/z$ be the chart around $\infty \in \widehat{\mathbb{C}}$. Then $\varphi_2 \circ f \circ \varphi^{-1}(z) = \frac{1}{f\circ\varphi^{-1}(z)} := h(z)$ for some holomorphic function h. Thus $f \circ \varphi^{-1}(z) = \frac{1}{h(z)}$, i.e., f is meromorphic.

Conversely, a meromorphic function f in X defines a continuous mapping $f : X \longrightarrow \widehat{\mathbb{C}}$ by simply sending the poles to $\infty \in \widehat{\mathbb{C}}$. By Riemann's removable singularity theorem †, the corresponding local expression of f is holomorphic in a small neighborhood around each pole. Thus $f : X \longrightarrow \widehat{\mathbb{C}}$ is a morphism of Riemann surfaces. □

Let S be a Riemann surface.

The Riemann (1857) view of an algebraic curve as a surface covering the sphere led him to an equally simple view of the meromorphic functions on such a curve—one that would be the starting point of the Dedekind–Weber theory—they are algebraic functions, of degree bounded by the degree of the curve.

To see why, suppose that f is a meromorphic function on a Riemann surface $\mathcal{S}$, so f has only finitely many poles on $\mathcal{S}$. If $\mathcal{S}$ is of degree n then there are n points $x_1, \dots, x_n$ (not necessarily distinct) over each $x \in \mathbb{C} \cup \{\infty\}$. Consider the values $f(x_1), \dots, f(x_n)$ of f at these points. Then the "multi-valued function" $y = f(x)$ satisfies

$$(y - f(x_1)) \cdots (y - f(x_n)) = 0\,.$$

Expanding the left side, we get

$$y^n + a_{n-1}y^{n-1} + \cdots + a_1 y + a_0 = 0\,, \tag{5.6}$$

where

$$a_0 = (-1)^n f(x_1) \cdots f(x_n)\,,$$

$$\vdots$$

$$a_{n-1} = -\left(f(x_1) + \cdots + f(x_n)\right)$$

are the elementary symmetric functions of $f(x_1), \ldots, f(x_n)$. Because of their symmetry, $a_0, \ldots a_{n-1}$ are well-defined meromorphic functions of $x \in \mathbb{C} \cup \{\infty\}$. So, $y = f(x)$ satisfies a polynomial equation (5.6) of degree n whose coefficients are rational functions of x, and this means that $f(x)$ is an algebraic function of x, of degree at most n. This is essentially the argument of Riemann (1857), [156, Section 5, p. 123].

On a surface of genus $g > 0$ it could likewise be proved that a meromorphic function is determined, up to a constant multiple, by its zeros and poles, and that the total order of zeros equals the total order of poles. However, it could also be proved that there are further constraints on existence, arising from closed paths that do not bound a piece of the surface and the corresponding periods of integrals. For example, on a surface of genus 1 there is no function with one pole of order 1 and one zero of order 1. This raises the problem of finding functions, with given zeros and poles, that satisfy the known constraints.

For genus $g = 1$, constraints were found by Abel (1827), and he was also able to prove that any finite set of zeros and poles satisfying these constraints could be realized by a meromorphic function, that is, by an elliptic function. He did this with the help of "expressions for these functions," which he and Jacobi had already raised to a fine art. "Expressions" for functions on surfaces of higher genus were much less developed in Riemann's time, so he actually made a virtue of a necessity by appealing to Dirichlet's principle in order to prove their existence. The trouble was, Weierstrass (1870) showed that the Dirichlet principle fails in certain cases, so Riemann's methods were under suspicion for some decades.

5.9 The Argument Principle on Riemann Surfaces

Let Γ be a smooth embedded curve in a Riemann surface X (that is, the image of a smooth embedding of $[0,1]$). Let $\{\varphi_j : U_j \to V_j\}_{j=1}^N$ be finitely many coordinate charts covering Γ. For each j, choose relatively compact subsets $W_j \Subset U_j$ such that $\Gamma \subset \bigcup_{j=1}^N W_j$. Let ψ_j be a smooth function with

$\psi_j \equiv 1$ on W_j and $\mathrm{supp}(\psi_j) \subset U_j$. Define a partition of unity along Γ by

$$\chi_j := \left(\sum_{k=1}^{N} \psi_k \right)^{-1} \psi_j \,.$$

Definition 5.9.1. The *logarithmic integral* of a meromorphic function f along Γ is defined by

$$\int_\Gamma \frac{df}{f} := \sum_{j=1}^{N} \int_{\varphi_j(\Gamma \cap U_j)} \chi_j(\varphi_j^{-1}(z)) \, \frac{(f \circ \varphi_j^{-1})'(z)}{f \circ \varphi_j^{-1}(z)} \, dz \,.$$

It is a standard exercise in multivariable calculus to check that this definition is independent of the chosen partition of unity $\{\chi_j\}$. Furthermore, the definition does not depend on the choice of coordinate charts: if $\tilde{\varphi} : \tilde{U} \to \tilde{V}$ is another chart with coordinate $\tilde{z}$, then on $V \cap \tilde{V}$ one verifies using the chain rule that

$$\frac{(f \circ \tilde{\varphi}^{-1})'(\tilde{z})}{f \circ \tilde{\varphi}^{-1}(\tilde{z})} \, d\tilde{z} = \frac{(f \circ \varphi^{-1})'(z)}{f \circ \varphi^{-1}(z)} \, dz \,.$$

Thus the logarithmic integral is well defined on Riemann surfaces, and the argument principle carries over verbatim from the classical setting in the complex plane.

Theorem 5.9.2 (argument principle on Riemann surfaces). *Let f be meromorphic on a domain $W \subset X$ and $\Gamma \subset W$ a closed smooth curve avoiding the zeros and poles of f. Then*

$$\frac{1}{2\pi i} \int_\Gamma \frac{df}{f} = \sum_{p \in \mathrm{int}(\Gamma)} \mathrm{ord}_p(f) \,,$$

where $\mathrm{int}(\Gamma)$ *denotes the set of points inside* Γ *(with orientation induced from* X*).*

Proof. Cover Γ by finitely many charts $\{\varphi_j : U_j \to V_j\}$ and use the partition of unity $\{\chi_j\}$ constructed above. By definition,

$$\int_\Gamma \frac{df}{f} = \sum_{j=1}^{N} \int_{\varphi_j(\Gamma \cap U_j)} \chi_j(\varphi_j^{-1}(z)) \frac{(f \circ \varphi_j^{-1})'(z)}{f \circ \varphi_j^{-1}(z)} \, dz \,.$$

Each integrand is supported in $\varphi_j(W_j)$ and hence the global integral is the finite sum of integrals computed in local complex coordinates.

The partition-of-unity construction ensures that overlaps are handled smoothly; on overlaps the chain rule shows the local differential forms $\frac{(f \circ \varphi_j^{-1})'}{f \circ \varphi_j^{-1}} \, dz$ agree as forms, so the sum of local integrals equals the integral of a globally well-defined 1-form along Γ. Therefore the global integral equals the sum of the usual plane integrals of $\frac{g_j'(z)}{g_j(z)}$ for the local representatives $g_j := f \circ \varphi_j^{-1}$.

Fix a chart φ_j and let $g_j = f \circ \varphi_j^{-1}$ on $\varphi_j(U_j)$. Choose a small closed curve in $\varphi_j(U_j)$ corresponding to the portion of Γ in U_j and enclose the finitely many zeros and poles of g_j that lie in the corresponding interior. The classical argument principle in the plane gives

$$\frac{1}{2\pi i} \int_{\text{local piece}} \frac{g_j'(z)}{g_j(z)}\, dz = \sum_{p \text{ inside}} \operatorname{ord}_{\varphi_j^{-1}(p)}(f)\,.$$

Summing over j and using that the local contributions from overlaps cancel or combine consistently yields the stated global equality.

Combining the local equalities gives

$$\frac{1}{2\pi i} \int_{\Gamma} \frac{df}{f} = \sum_{p \in \operatorname{int}(\Gamma)} \operatorname{ord}_p(f)\,,$$

as required. □

Corollary 5.9.3. *Let X be a compact Riemann surface. If $f \in \mathcal{M}(X)$, then*

$$\sum_{p \in X} \operatorname{ord}_p(f) = 0$$

Proof. Let $p \in X$ such that $f(p) \neq 0$ or ∞. Let U be a small neighborhood of p such that

(i) f has no zeros or poles in U and

(ii) U is the image, under the inverse of some complex chart φ, of a disk in the plane with center $\varphi(p)$.

Since f has no zeros or poles in U, Theorem 5.9.2 shows that

$$\oint_{\partial D} \frac{df}{f} = 0 \quad \text{and} \quad \sum_{p \in X} \operatorname{ord}_p(f) = \sum_{p \in X - U} \operatorname{ord}_p(f) = \oint_{\partial(X-U)} \frac{df}{f}\,.$$

But

$$\oint_{\partial(X-U)} \frac{df}{f} = -\oint_{\partial D} \frac{df}{f}\,.$$

This completes the proof. □

Remark 5.9.4. The same reduction-to-charts argument shows that any local characterization of zeros/poles (for instance via the Laurent expansion of a local representative) is intrinsic: if $\varphi : U \to V$ is a chart at p and $f \circ \varphi^{-1}$ has Laurent expansion

$$f \circ \varphi^{-1}(z) = \sum_{k=n}^{\infty} a_k (z - \varphi(p))^k, \qquad a_n \neq 0\,,$$

then $\operatorname{ord}_p(f) = n$, and this integer does not depend on the chosen chart.

Equivalently, if $\varphi : U \to V$ is a local chart containing p and $f \circ \varphi^{-1}$ admits a Laurent expansion around $\varphi(p)$, then there exists a largest integer n (depending only on p, not on φ) such that all terms of order less than n vanish. This integer n is precisely the order $\mathrm{ord}_p(f)$.

5.10 Construction Conceived by Riemann

Set $X = X \cup \{\text{algebraic boundary points of} X\}$. We can extend the complex structure on X to one on X by taking as a chart containing an algebraic boundary point $P \in \hat{X} \setminus X$ the pair (Ω, φ) constructed above. Let $\hat{p} = \tilde{p}|_{\hat{X}}$. The pair $(\hat{X}, \hat{p})$ will be called the (algebraic) completion of (X, p). The map $\hat{p} : X \to \mathbb{P}^1$ is holomorphic, but does not have to be a local homeomorphism. With the notation above, if P is an algebraic boundary point and Ω is an n-sheeted covering of $D_\varepsilon \setminus \{a\}$ with $n > 1$, then $\hat{p}$ is not a local homeomorphism at P.

This construction can be used to obtain the Riemann surface of a holomorphic function as conceived by Riemann.

5.10.1 Holomorphic Maps Between Riemann Surfaces

Once Riemann surfaces have been defined, it is necessary to specify the holomorphic maps between them.

The space of Riemann surfaces forms a category with the analytic maps $f : X \to Y$ as morphisms. A map is analytic if it gives a locally analytic map on $\mathbb{C}$ in all charts; equivalently, f is analytic if f is continuous and $f^*(\mathcal{O}_Y)$ is a subsheaf of $\mathcal{O}_X$.

Definition 5.10.1. A *biholomorphism* between two open sets $X, Y \subset \mathbb{C}$ is a holomorphic map $f : X \to Y$ with holomorphic inverse.

We continue with the definitions. Let M be a complex manifold, W an open set in M, and $\{z_j\}$ a coordinate system. Then a mapping $f : W \to \mathbb{C}^l$ is holomorphic (differentiable, and so on) if $f \circ z_j^{-1}$ is holomorphic (differentiable, and so on) for each j where defined. Let N be another complex manifold with coordinates $\{w_\lambda\}$ and $f : W \to N$. Then f is holomorphic (differentiable, and so on) if $w_\lambda \circ f \circ z_j^{-1}$ is holomorphic where defined.

Definition 5.10.2. Let X and Y be two Riemann surfaces. A map $f \colon X \to Y$ is a continuous map such that if (U_i, g_i) are an atlas for X and (V_j, h_j) are an atlas for Y, then the resulting *restriction* of f

$$f^{-1}(V_j) \cap U_i \xrightarrow{f} V_j$$

is holomorphic for all i, j.

Loosely speaking, a map of Riemann surfaces is just a map which is holomorphic when restricted to each of the open sets.

Remark 5.10.3. Unless otherwise stated, we will assume f is nonconstant.

Theorem 5.10.4 (uniqueness theorem). *Let $f : X \to Y$ and $g : X \to Y$ be holomorphic maps. Then either $f \equiv g$ or $\{p \in X : f(p) = g(p)\}$ is discrete in X.*

Proof. Define the sets

$$A := \{p \in X : \text{ locally at } p, f \equiv g\}$$
$$B := \{p \in X : \text{ locally at } p, f = g \text{ only on a discrete set }\}$$

It is clear that both A and B are open subsets of X. Hence if we show $X = A \cup B$ we shall be done as X is connected. If $p \in X$ is such that $f(p) \neq g(p)$ then $p \in B$, since if $f = g$ on more than a discrete set near p, i.e., a set containing a nonisolated point, then then the usual uniqueness theorem applied in charts would show $f = g$ locally, so $f(p) = g(p)$. This is clearly absurd. Similarly, if $f(p) = g(p)$ then the uniqueness theorem applied in charts shows $f = g$ locally whenever $\{p \in X : f(p) = g(p)\}$ is not discrete. Thus we indeed have $X = A \cup B$ and these sets are open and disjoint. Applying connectedness, the claimed result follows. □

Theorem 5.10.5 (open mapping theorem). *Let $f : X \to Y$ be holomorphic map. If f is not constant, then f maps open subsets in X to open subsets of Y and $f(X)$ is open.*

Proof. By the uniqueness theorem if f is locally constant around any point, then f is globally constant. Hence applying the usual open mapping theorem in every chart allows us to conclude that f is constant. □

Corollary 5.10.6. *Let X be compact and $f : X \to Y$ holomorphic and nonconstant. Then f is surjective and Y is compact.*

Proof. Since $f(X)$ is both closed (as a compact subset of a Hausorff space) and open by the open mapping theorem, we conclude $f(X) = Y$. □

Corollary 5.10.7. *Let X be a Riemann surface. The following properties are true:*

1. *If X is compact then every holomorphic function $f : X \to \mathbb{C}$ is constant.*
2. *If X is compact then every meromorphic function $f : X \to \widehat{\mathbb{C}}$ is surjective.*

5.10.2 Local Structure of Maps: The Multiplicity of a Map

A holomorphic map defined on a neighborhood of a point in the complex plane is determined by the coefficients (an infinite sequence of numbers) of its power series expansion. In contrast, a holomorphic map defined on a Riemann surface does not, in general, have a meaningful power series associated with it.

Indeed, up to a change of chart, it always has the same local form, depending only on a natural number described in Lemma 5.10.8 below.

An important feature of holomorphic maps between Riemann surfaces is that they can have many different local expressions near a point $x \in X$, depending on the choice of charts and local coordinates around x and $f(x)$. Given a local coordinate function φ_x, post-composing with any biholomorphism h of $\mathbb{C}$ produces a new local coordinate function.

For example, choosing

$$h(z) = e^{(\pi/4)i} z$$

rotates the coordinates by 45°, while choosing

$$h(z) = z + (2 + i)$$

yields translated coordinates.

In fact, h need not be biholomorphic on all of $\mathbb{C}$; it suffices that h be biholomorphic in a neighborhood of the point $\varphi_x(x)$. In that case, one can use $h \circ \varphi_x$ to obtain new local coordinates around x.

A key fact is that for any holomorphic map f, there exist choices of local coordinates in which the local expression of f near a given point $x \in X$ takes a particularly simple form: specifically,

$$z \mapsto z^k,$$

for some uniquely determined integer $k \geq 1$. This result has remarkably deep implications in the theory of functions on Riemann surfaces.

We say that a *chart* (U_x, φ_x) for a Riemann surface X is *centered at* x if $\varphi_x(x) = 0$.

Lemma 5.10.8 (key lemma). *Let $f : X \to Y$ be a non-constant holomorphic map of Riemann surfaces. For any $x \in X$ there are charts centered at $x, f(x)$, such that the local expression of f using these charts is $z \mapsto z^k$ for some integer $k \geq 1$.*

Proof. Start with any charts φ, ψ centered at $x, f(x)$, yielding a local expression denoted by $F := \psi \circ f \circ \varphi^{-1}$. Consider the Taylor expansion of F at 0 and let k be the smallest positive integer such that the coefficient of z^k does not vanish. Since $F(0) = 0, k \geq 1$ and:

$$F(z) = z^k \left(\sum_{n=0}^{\infty} a_{k+n} z^n \right) \tag{5.7}$$

Denote by $G(z) = a_k + a_{k+1}z + \cdots$ the second factor in (5.7). The function $G(z)$ is holomorphic at 0 and $G(0) = a_k \neq 0$. Thus we may make a choice of branch so that the map $\sqrt[k]{G(z)}$ is well defined and holomorphic around $0 = \varphi(x)$.

Defining $h(z) = z\sqrt[k]{G(z)}$, we have that $F = h^k$. The function h is holomorphic in a neighborhood U of $0 = \varphi(x), h(0) = 0$ and $h'(0) \neq 0$. The Inverse

Function Theorem implies that h is bi-holomorphic on a neighborhood $U' \subset U$ of $\varphi(x)$, and therefore the composition $\widetilde{\varphi} = h \circ \varphi$ gives a local chart for X centered at x. The local coordinate $\widetilde{z}$ coming from $\widetilde{\varphi}$ is related to z via

$$\widetilde{z} = h(z)$$

The local expression for f around x is now obtained by changing coordinates from z to $\widetilde{z}$ in F:

$$\widetilde{F}(\widetilde{z}) = F(z(\widetilde{z})) = h(z)^k = \widetilde{z}^k\,.$$

□

Observe that in the case $n = 1$ the map f is a local biholomorphism at $f(x) \in Y$.

We now introduce the concept of the multiplicity of a map between Riemann surfaces. Informally, it refers to the number of preimages that a "generic" point has under the map.

Definition 5.10.9. For $f : X \to Y$ define the *multiplicity* of f at x to be

$$\mathrm{mult}_x(f) = \begin{cases} d & \text{if } f \text{ locally at } x \text{ looks like } z \mapsto z^d \\ \infty & \text{if } f \text{ is constant in some neighborhood of } x \end{cases}$$

Lemma 5.10.10. *If $f : X \to Y$ is a map of Riemann surfaces, either f is constant or* $\mathrm{mult}_x(f) < \infty$ *for all $x \in X$.*

Proof. Note that the set of points where $\mathrm{mult}_x(f) < \infty$ is open and the set where $\mathrm{mult}_x(f) = \infty$ is open, using Lemma 5.10.8. Since X is connected, only one of these two conditions may hold. □

Definition 5.10.11. A continuous map $p : X \to Y$, where X and Y are locally compact (Hausdorff) spaces, is called *proper* if, for any compact set $K \subset Y$, the inverse image $p^{-1}(K)$ is compact in X.

Our next goal is to show that proper maps are covering maps away from their ramification points. To this end, we now define covering maps.

Definition 5.10.12. A map of topological spaces $f : X \to Y$ is a *covering map* if, for each $y \in Y$, there exists an open neighborhood V of y such that $f^{-1}(V)$ is homeomorphic to a disjoint union of open sets, each homeomorphic to V.

Definition 5.10.13. A map $f : X \to Y$ is a *local homeomorphism* if, for every $x \in X$, there exist open sets $U \subset X$ and $V \subset Y$ with $x \in U$ such that the restriction $f|_U : U \to V$ is a homeomorphism.

Lemma 5.10.14. *If X and Y are locally compact, a proper map $p : X \to Y$ is necessarily closed; that is, it takes closed sets in X to closed sets in Y.*

Proof. Let $A \subset X$ be closed, and let $y_0 \in Y$. Let K be a compact neighborhood of y_0 in Y. Then $p(A) \cap K = p\left(A \cap p^{-1}(K)\right)$ is compact (since A is closed and $p^{-1}(K)$ is compact), and therefore closed in K. □

Corollary 5.10.15. *If $f : X \to Y$ is a proper (nonconstant) map of Riemann surfaces, it is surjective.*

Proof. $f(X)$ is open by the open mapping theorem and closed by Lemma 5.10.14. Hence, $f(X) = Y$, since Y is connected. □

Lemma 5.10.16. *If $f : X \to Y$ is proper, then $f^{-1}(q)$ is finite for all $q \in Y$.*

Proof. The set $f^{-1}(q)$ is compact by properness. It is also discrete by Lemma 5.10.8. □

Proposition 5.10.17. *If $f : X \to Y$ is a proper local homeomorphism, then it is a covering map.*

Proof. Choose $y \in Y$. By Lemma 5.10.16, $f^{-1}(y)$ is finite. Let $x_1, \ldots, x_n$ be the preimages. Since f is a local homeomorphism, we may take open sets U_i containing x_i such that $f : U_i \to V_i$ is a homeomorphism, where $V_i = f(U_i)$. By shrinking the U_i, we may assume that their pairwise intersections are empty. Define $V = \bigcap_i V_i$. Note that $f^{-1}(V) \subset \bigcup_i U_i$, and $f^{-1}(V) \cap U_i \to V$ is a homeomorphism, as it is the restriction of the homeomorphism $U_i \to V_i$. Thus, f is a covering map. □

Proposition 5.10.18. *Let $f : X \to Y$ be a proper, nonconstant morphism. Then the function $d(q) := \sum_{f(p)=q} \operatorname{mult}_p(f)$ is finite and constant (independent of the point q).*

Proof. It suffices to show that $d(q)$ is locally constant, since X is connected. For a fixed q, we can find open sets U_i containing x_i, where $\{x_1, \ldots, x_k\} = f^{-1}(q)$, such that $R(f) \cap U_i \subset \{x_i\}$ (i.e., there are no ramification points in U_i other than possibly x_i). If there are no branch points among the x_i, the result follows from Proposition 5.10.17. Using this, it suffices to show that the degree is locally constant on each U_i. That is, we have reduced to the case where $f : U_i \to Y$ is given by $z \mapsto z^d$. In this case, we have $d(q) = d$ for every point $q \in U_i$. □

Assuming Proposition 5.10.18, we can make the following definition.

Definition 5.10.19. For a proper map $f : X \to Y$ of Riemann surfaces, the constant value

$$d(q) := \sum_{f(p)=q} \operatorname{mult}_p(f)$$

is called the *degree* of f.

5.10.3 Ramification Loci of Maps

We next introduce the notions of branch loci and ramification loci. These are just the places at which the map does not look like $z \mapsto z$.

Definition 5.10.20. Let $f : X \to Y$ be a non-constant holomorphic map of Riemann surfaces.

1. Given a point $x \in X$, the integer k_x, such that there exists a local expression centered at p of the form $F(z) = z^{k_x}$, is called the *ramification index* of f at x.

2. The quantity $v_x = k_x - 1$ is called the *differential length* of f at x.

3. If a point x has ramification index $k_x = 1$, then we say f is *unramified* at x.

4. A point x such that $k_x \geq 2$ is called a *ramification point.* The *ramification locus* R is the subset of X consisting of all ramification points.

5. If x is a ramification point, then $f(x) \in Y$ is called a *branch point.* The *branch locus* B is the subset of Y consisting of all branch points—i.e., the image via f of the ramification locus.

Definition 5.10.21. The *ramification locus*, denoted by $R(f)$, of a map $f : X \to Y$ is the set of points at which f is locally of the form $z \mapsto z^d$ for $d > 1$. The branch locus, notated $B(f)$, is $f(R(f))$, the image of the ramification locus.

Warning! The branch locus is the image of the ramification locus, but the ramification locus is not necessarily the inverse image of the branch locus.

Proposition 5.10.22. *Let $f : X \to Y$ be a non-constant holomorphic map between Riemann surfaces, and let $x_0 \in X$. Then f is unramified at a point $x \in X$ (i.e., $k_x = 1$) if and only if, for any local expression F of f around x (not necessarily centered at x), one has*

$$F'(\varphi(x)) \neq 0\,,$$

that is, f is locally invertible at x. In particular, there exists a neighborhood U_0 of x_0 such that $k_x = 1$ for all $x \in U_0 \setminus \{x_0\}$.

Proof. Left to the reader. □

Corollary 5.10.23. *The ramification locus R is a discrete subset of X, i.e., there exist open sets $U_i \subset X$ such that each U_i contains exactly one $x \in R$.*

Example 5.10.24. The map $f : \hat{\mathbb{C}} \to \hat{\mathbb{C}}$ given by $z \mapsto z + \frac{1}{z}$ is a proper map of degree 2. The branch points of this map are $\{-2, 2\}$. To see this, note that if the map has a branch point at x, then the equation $z + \frac{1}{z} = x$ has only one solution in z. Multiplying by z, we obtain $z^2 - xz + 1 = 0$, which has a double root, meaning it factors as $(z-a)^2$. Therefore, $a^2 = 1$, so $a = \pm 1$, and in that case $x = 2a = \pm 2$.

A discrete subset of a compact topological space is finite. This is the contrapositive of the statement "an infinite subset of a compact topological space has a limit point," which follows directly from the definition of compactness. As a consequence, we obtain the following lemma.

Lemma 5.10.25. *If X is a compact Riemann surface and $f : X \to Y$ is a nonconstant holomorphic map between Riemann surfaces, then the ramification locus is a finite set. Since the branch locus is the image of the ramification locus under f, it follows that the branch locus is also finite.*

5.10.4 Riemann's Existence Theorem

In a previous subsection we saw that holomorphic maps of compact Riemann surfaces are covering spaces away from a finite set of points. Conversely, we now show that any topological cover of a punctured Riemann surface gives rise to a unique holomorphic map of compact Riemann surfaces.

Definition 5.10.26. A continuous map between compact topological surfaces $p : X \to Y$ is called a *ramified cover* if there is a finite set of points $B \subset Y$ such that:

1. $p^{-1}(B) \subset X$ is finite;

2. $p : X \backslash p^{-1}(B) \to Y \backslash B$ is a covering.

We can now succinctly state that maps of Riemann surfaces are ramified coverings. The following classical theorem provides the converse.

Theorem 5.10.27 (Riemanns existence theorem). *Let Y be a compact Riemann surface and $\widetilde{X}$ a topological surface. Suppose there exist finitely many points $b_1, \ldots, b_n \in Y$ and a map $\widetilde{f} : \widetilde{X} \to Y \setminus \{b_1, \ldots, b_n\}$ which is a finite-degree topological covering. Then there exists a unique (up to isomorphism) compact Riemann surface X containing $\widetilde{X}$ as a dense open subset (indeed, X is obtained from $\widetilde{X}$ by adjoining finitely many points), such that $\widetilde{f}$ extends to a holomorphic map $f : X \to Y$.*

Sketch of Proof. The proof has two main steps. The first is purely topological: we complete $\widetilde{X}$ to a compact surface and extend $\widetilde{f}$ to a continuous map. The second consists in endowing X with a complex structure so that f becomes holomorphic.

For the first step, fix one of the special points $b_1, \ldots, b_n \in Y$; to simplify notation, call it b. Take a coordinate chart φ around b and set $\Delta := \varphi^{-1}(\{|w| < 1\})$. Then Δ is an open neighborhood of b, homeomorphic to a disk. By assumption, the restriction

$$\widetilde{f} : \widetilde{f}^{-1}(\Delta \setminus \{b\}) \to \Delta \setminus \{b\}$$

is a covering of finite degree d. Let $\widetilde{U}_1, \ldots, \widetilde{U}_m$ be the connected components of $\widetilde{f}^{-1}(\Delta \setminus \{b\})$. Since $\Delta \setminus \{b\}$ is homotopy equivalent to a circle, its fundamental group is $\mathbb{Z}$. Just as in the case of the circle, connected finite covers of a punctured disk are themselves homeomorphic to punctured disks. Thus there exist positive integers $k_1, \ldots, k_m$ and homeomorphisms

$$\widetilde{\phi}_i : \widetilde{U}_i \to \{0 < |z| < 1\} \subset \mathbb{C}$$

such that

$$\varphi \circ \widetilde{f} \circ \widetilde{\phi}_i^{-1}(z) = z^{k_i}\,.$$

For each $i = 1, \ldots, m$, add a new point x_i to $\widetilde{X}$ so that $\widetilde{\phi}_i$ extends to a homeomorphism

$$\phi_i : \widetilde{U}_i \cup \{x_i\} \to \{|z| < 1\} \subset \mathbb{C}\,,$$

with $\phi_i(x_i) = 0$.

Repeating this construction for all $b_1, \ldots, b_n$, we adjoin finitely many points to $\widetilde{X}$ and obtain a new space X. This is a surface: local Euclidean neighborhoods around the added points are provided by the homeomorphisms ϕ_i. Compactness of X follows because every infinite subset has a limit point (limit point compactness), which is equivalent to compactness for metrizable spaces[1]. The map $\widetilde{f}$ extends naturally to a continuous map $f : X \to Y$ by sending each x_i to the corresponding $b \in Y$.

For the second step, we construct a complex structure on X so that f is holomorphic. For each $x \in \widetilde{X}$, choose an open neighborhood $U_x \subset \widetilde{X}$ such that $\widetilde{f}|_{U_x}$ is a homeomorphism onto its image, which lies inside some coordinate chart φ_x of Y. Then

$$\varphi_x \circ \widetilde{f}|_{U_x}$$

defines a chart around x. In these coordinates, f is locally the identity map, hence holomorphic.

For each new point x_i, we use ϕ_i as the local chart. As observed, the local expression for f in these coordinates is $w = z^{k_i}$, which is holomorphic. One checks that the charts just defined are compatible, so they define a complex structure on X. Finally, any other atlas on X making f holomorphic is equivalent to this one, proving uniqueness. □

5.10.5 Applications

Lemma 5.10.28. *Any map $f : X \to Y$ is discrete, meaning for $q \in Y, f^{-1}(q)$ is a discrete set.*

Proof. Let $x \in f^{-1}(q)$. We wish to show there is an open neighborhood around x on which where are no other points mapping to q. This follows as we can choose an open neighborhood on which the map is of the form $z \mapsto z^d$, and the map is discrete when restricted to such an open set. □

[1]Surfaces are metrizable by the Urysohn metrization theorem Munkres, [136, p. 215].

Holomorphic maps $f : X \to Y$ between two connected compact Riemann surfaces enjoy many nice properties:

(1) Any holomorphic map $f : X \to Y$ is either constant or surjective.

Proof. If f in not constant then it is open. Moreover f is closed because X is compact. Hence $f(X)$ is both open and closed and as Y is connoted it follows that $f(X) = Y$ □

(2) Any nonconstant holomorphic map $f : X \to Y$ has finite fibers.

Proof. From the local form of on holomorphic function, we can see then the fibers $f^{-1}(p)$ are discrete. Since Y is compact, $f^{-1}(p)$ must be compact and discrete, hence finite. □

(3) Any nonconstant holomorphic map $f : X \to Y$ has finitely many ramification points.

Proof. The proof is similar to that of statement (2). □

Now we can define the most important invariant of a holomorphic map between compact Riemann surfaces.

Proposition 5.10.29. *Let $f : X \to Y$ be a nonconstant holomorphic map between compact connected Riemann surfaces. Then for any $q \in Y$ the number of points in the fiber, counted with multiplicity, is constant:*

$$d(q) = \sum_{p \in f^{-1}(q)} \mathrm{mult}(f, p)\,.$$

This proposition motivates our next definition.

Definition 5.10.30. Let $f : X \to Y$ be a nonconstant holomorphic map between compact Riemann surfaces. The *degree* of f, denoted $\deg(f)$, is the integer $d_q(f)$ for arbitrary $q \in Y$.

Proof of Proposition 5.10.29. We will show that d is locally constant. Let $q \in Y$ and let $f^{-1}(q) = \{p_1, \ldots, p_n\}$. Then we confined charts $U_i \subseteq X$ around each of the p_i and $V \subseteq Y$ around g such that in the corresponding local coordinates, f looks like $f(\dot{z}) = z^{e_i}$ around each p_i. Now, since X, Y are both compact, by shrinking V if needed, we can assume that $f^{-1}(V) \subseteq U_1 \cup \ldots U_n$ (exercise in set topology).

Now we show that the function is constant on V. Let $q \in V$ then $f^{-1}(q) \subseteq U_1 \cup \cdots \cup U_n$. Then we can simply count the points of the fiber in each U_i: on each of these U_i the map looks like

$$f(z) = z^{e_i}$$

and it is now clear that

$$f^{-1}(a) = \begin{cases} \text{one point of multiple } e_i\,, & \text{if } a = p_i \\ e_i \text{ distinct points}\,, & \text{if } a \neq p_i \end{cases}.$$

Hence $d\,(q') = \sum e_i$ for all $q' \in V$. □

This gives us the following corollary.

Corollary 5.10.31. *A holomorphic map between compact Riemann surfaces is an isomorphism if and only if it has degree one.*

Proof. The multiplicity of a holomorphic map is nonnegative, so a degree one map must associate to each $q \in Y$ precisely one preimage $p \in X$. This is precisely what it means to be injective. Surjectivity follows by Corollary 5.10.6. □

Definition 5.10.32 (ramification point)**.** A point $p \in X$ is called a *critical point*, or *ramification point*, of a holomorphic map f if $\deg_p f > 1$, i.e., $(w_a \circ f_i \circ z_i^{-1})'\,(z_\alpha(p)) = 0$.

A point $q \in Y$ is called a *critical value* of a holomorphic map f if the preimage $f^{-1}(q)$ has at least one critical point.

Furthermore, given a holomorphic map between two compact Riemann surfaces we can relate the two genera using information about the map.

Theorem 5.10.33 (Riemann–Hurwitz formula)**.** *Let $f : X \to Y$ be a nonconstant holomorphic map from a Riemann surface X of genus $\widetilde{g}$ to a Riemann surface Y of genus g. Then*

$$\widetilde{g} = (\deg f)(g-1) + \frac{1}{2}\sum_{p\in P}\left(\deg_p f - 1\right) + 1\,.$$

Proof. Consider a triangulation of the surface Y whose vertices include all critical values of f. Assume that it has F faces, E edges, and V vertices. The preimage of this triangulation is a triangulation of the surface X. It consists of $(\deg f)F$ triangles and $(\deg f)E$ edges. The number of vertices in this triangulation is equal to $(\deg f)V - \sum_{p\in X}\left(\deg_p f - 1\right)$, since every triangle with vertex $q \in Y$ produces exactly $\deg_p f$ triangles having a vertex at $p \in f^{-1}(q)$. Calculating the Euler characteristics yields $2-2g = F-E+V$ and

$$2 - 2\widetilde{g} = (\deg f)F - (\deg f)E + (\deg f)V - \sum_{p\in P}\left(\deg_p f - 1\right)\,,$$

whence $\deg f(2-2g) - (2-2\widetilde{g}) = \sum_{p\in P}\left(\deg_p f - 1\right)$. □

Remark 5.10.34. Riemann viewed an algebraic curve as a surface covering the sphere $\mathbb{C} \cup \{\infty\}$. For a curve of degree n the covering is of n "sheets": that is, above each point of $\mathbb{C} \cup \{\infty\}$ there are n different points of the curve, with the exception of finitely many ramification points where two or more sheets come together.

Riemann (1857) showed that the genus can be computed from the degree n and numbers e_P that give the number of sheets fused together at a ramification point P. Because of a generalization made by Hurwitz (1891), the method is now called the *Riemann–Hurwitz formula.*

The Riemann–Hurwitz formula is most easily explained in terms of the oldest manifestation of the genus concept, the Euler characteristic, which goes back to Euler (1752). Euler observed that, for any decomposition of the sphere into F faces, E edges, and V vertices one has $V - E + F = 2$. More generally, the number $V - E + F$ is invariant for any surface, and it equals $2 - 2g$, where g is the genus. The invariance of $V - E + F$ follows by observing that any two partitions of a surface have a common subdivision, obtained by superimposing one on the other. (This argument is merely plausible, since an edge in one partition may intersect edges of the other partition infinitely often, but Riemann used it in Riemann (1851), [156, Section 6, p. 8].) And any subdivision may be obtained by a series of elementary subdivisions of the following two types:

1. Subdividing an edge by a new vertex.

2. Subdividing a face by a new edge connecting two of its vertices.

The first increases both V and E by 1, the second increases both E and F by 1, so neither changes $V - E + F$.

Thus $V - E + F$ is invariant and, by computing it for a standard subdivision of the genus g surface (such as the one shown for the pretzel with three handles in Fig 5.5), one finds that $V - E + F = 2 - 2g$. This quantity is now called the *Euler characteristic*, $\chi(\mathcal{S})$, of the surface $\mathcal{S}$ of genus g. It enables us to compute the genus from a decomposition of the surface into vertices, edges, and faces, which is particularly convenient when $\mathcal{S}$ is realized as a ramified covering of the sphere.

If $\mathcal{S}$ were simply an n-sheeted covering of the sphere then we should have $\chi(\mathcal{S}) = 2n$, since 2 is the Euler characteristic of the sphere and $\mathcal{S}$ has n vertices over each vertex on the sphere, n edges over each edge, and n faces over each face.

However, when $n > 1$, the covering has ramification points, so we have to adjust the count of vertices. We assume that the ramification points are included among the vertices on the sphere, in which case the count of vertices on $\mathcal{S}$ must be reduced by $e_P - 1$ at each ramification point P. On the other hand, the numbers of edges and faces on $\mathcal{S}$ are still n times the corresponding numbers on the sphere, so we have

$$\chi(\mathcal{S}) = 2n - \sum_P (e_P - 1) \ .$$

Thus the genus g of $\mathcal{S}$ depends only on the degree n of the curve and its ramification number e_p.

Since $\chi(\mathcal{S}) = V - E + F = 2 - 2g$, the above formula for $\chi(\mathcal{S})$ may be written as the following formula for genus:

$$g = \frac{1}{2}w - n + 1,$$

where $w = \sum_P (e_P - 1)$ is the sum of the ramification numbers. This formula is used as the definition of genus in the Dedekind–Weber paper [41, §24].

The determination of genus by degree and ramification numbers is crucial to Dedekind and Weber's program of deriving Riemann's results by algebraic methods. Their algebraic definition of a "Riemann surface" $\mathcal{S}$ lacks geometric structure, so it is not meaningful to decompose $\mathcal{S}$ by vertices, edges, and faces (nor can one make "cuts" in it between the ramification points). However, it is possible to show that there are n distinct points of $\mathcal{S}$ over each point of the sphere $\mathbb{C} \cup \{\infty\}$, with the exception of finitely many ramification points P, and the ramification number e_P at each such point can be defined algebraically. Thus they can define genus in a way that agrees with Riemann's concept of genus, by means of the Riemann–Hurwitz formula.

5.11 The Riemann–Hurwitz Formula—Applications

The Riemann–Hurwitz formula allows us to determine the genus of a compact, connected Riemann surface X without using explicit triangulation. The background of this formula is that holomorphic functions on Riemann surfaces are ramified coverings.

If we view f as a holomorphic map

$$f : \mathbb{C} \to \mathbb{CP}^1,$$

this expresses the well-known topological fact that the degree of the divisor $f^{-1}(q)$ is independent of $q \in \mathbb{CP}^1$, and agrees with the degree (also called *sheet number*) of f.

Example 5.11.1 (coverings of the Riemann sphere). For an n-sheeted holomorphic covering $\pi : X \to \widehat{\mathbb{C}}$ of the Riemann sphere we obtain

$$g = \frac{b}{2} - n + 1.$$

In particular, for a double covering of $\widehat{\mathbb{C}}, b$ is the number of branch points and

$$g = \frac{b}{2} - 1. \tag{5.8}$$

A compact Riemann surface of genus > 1 which admits a double covering of $\widehat{\mathbb{C}}$ is called *hyperelliptic.*

Example 5.11.2. Let $P(z) = (z - a_1) \cdots (z - a_k)$ be a polynomial of degree k with distinct roots a_j. Let $p : X \to \widehat{\mathbb{C}}$ be the Riemann surface of $\sqrt{P(z)}$. Then X is branched over ∞ precisely if k is odd. The total branching order b is k or $k+1$, depending on whether k is even or odd. By (5.8), we have

$$g = \left\lfloor \frac{k-1}{2} \right\rfloor$$

An explicit basis $\omega_1, \ldots, \omega_g$ for $\mathcal{O}^1(X)$ is given by

$$\omega_j := \frac{z^{j-1} dz}{\sqrt{P(z)}}, \quad j = 1, \ldots, g,$$

where z denotes the function $p : X \to \widehat{\mathbb{C}}$. Using local coordinates one shows that the ω_j are holomorphic on all of X.

Example 5.11.3. Consider a holomorphic map $f : X \to Y$ between compact Riemann surfaces. Show that $\deg f = \sum_{p \in f^{-1}(q)} \deg_p f$ for every point $q \in Y$.

Example 5.11.4. Let $f : \mathbb{CP}^1 \to \mathbb{CP}^1$ be a polynomial of degree d. We have $f(\mathbb{CP}^1) = 0$, so Riemann–Hurwitz gives

$$2d = 2 + \sum_{p \in P} \left(\deg_p f - 1\right)$$

or $\sum_{p \in P} \left(\deg_p f - 1\right) = 2(d-1)$.

To see why that is, we pull the following theorem out of our algebraic hat.

Theorem 5.11.5. *Let $f : \mathbb{C} \to \mathbb{C}$ be a polynomial mapping of degree d; then the total branching index over the finite branch points is exactly $(d-1)$.*

This means that the equation $f(x) = y$ has multiple roots for precisely $(d-1)$ values of y, counting y a total of k times when the equation $f(x) - y$ has only $d - k$ distinct roots.

Remark 5.11.6. This is quite clear in examples such as $f(x) = x^n$, or for any polynomial of degree 2. In general, f has multiple roots if and only if the discriminant of f vanishes; and $\operatorname{disc}(f(x) - \alpha)$ is a polynomial in α of degree $(d-1)$. For generic f, $f - \alpha$ will have a double root for precisely $(d-1)$ values of α. Of course, reversing our arguments here will deduce Theorem 5.11.5 from the Riemann–Hurwitz formula.

But we need another degree $(d-1)$ to agree with Riemann–Hurwitz. This of course comes from the point at ∞.

Lemma 5.11.7. *We have $v_f(\infty) = \deg f$ for a polynomial f.*

Proof. Let

$$f(z) = z^n + a_{n-1}z^{n-1} + \cdots + a_1 z + a_0 .$$

Changing coordinates to $w = 1/z$, we have

$$\frac{1}{f\left(\frac{1}{w}\right)} \to 0 \quad \text{as} \quad w \to 0$$

and

$$\frac{1}{f\left(\frac{1}{w}\right)} = \frac{w^n}{1 + a_{n-1}w + \cdots + a_1 w^{n-1} + a_0 w^n}$$

and this has a zero of order n at 0. □

5.12 Any Two Meromorphic Functions Are Algebraically Related

The existence of a non-constant meromorphic function on any Riemann surface X enables one to describe the stock of all meromorphic functions on X.

If X is the Riemann surface of an algebraic function w given by $f(z, w) = 0$, then both w and z are meromorphic functions on X. Therefore any rational function of w and z is meromorphic. It can easily be proved that this is the way all meromorphic functions on X are obtained.

Furthermore, one can construct along the same lines two meromorphic functions w and z on S, connected by a relation of the form $f(z, w) = 0$, where f is a polynomial, and with the property that S is just the Riemann surface of the algebraic function w defined by the equation $f = 0$. This result is known as "Riemann's existence theorem."

Theorem 5.12.1 (any two meromorphic functions are algebraically related). *Let f and g be two meromorphic functions on a connected compact Riemann surface X. Then, there exists a bivariate polynomial Q such that*

$$Q(f, g) = 0$$

Proof. The proof is nothing but the Lagrange interpolation polynomial. Let us call $d = \deg f$ the *total degree* of f, i.e., the sum of degrees of all its poles. Let $Z \in \mathbb{CP}^1$, then $f^{-1}(Z)$ has generically a cardinal equal to d (see Proposition 5.10.29). Let us define $Q_0 = 1$ and for $k = 1, \ldots, d$:

$$Q_k(Z) = \sum_{I \subset_k f^{-1}(x)} \prod_{p \in I} g(p)$$

where we sum over all subsets of $f^{-1}(Z)$ of cardinal k, and we count preimages with multiplicities when Z is not generic. Q_k is clearly a meromorphic function $\mathbb{CP}^1 \to \mathbb{CP}^1$, therefore it is a rational function $Q_k(Z) \in \mathbb{C}(Z)$. We then define:

$$Q(Z, y) = \sum_{k=0}^{d} (-1)^k Q_k(Z) y^k \quad \in \mathbb{C}(Z)[y]$$

It is also equal to

$$Q(Z, y) = \prod_{p \in f^{-1}(Z)} (y - g(p))$$

Therefore, for any $p \in X$ we have

$$Q(f(p), g(p)) = 0$$

□

This theorem is a generalization of Theorem 5.8.2 saying that a meromorphic function on the Riemann sphere is a rational function of z.

One defines, in an obvious way, global meromorphic functions on a Riemann surface. If (and only if) the surface is compact, then any two meromorphic functions are connected by an algebraic equation (see Theorem 5.12.1), and may be treated as the coordinates of a plane algebraic curve.

Conversely, given a plane algebraic curve, there is a definite and well-known procedure of associating with it a compact Riemann surface such that the coordinates of the curve become meromorphic functions on the surface, and the field of rational functions on the curve becomes identical with the field of meromorphic functions on the surface.

Exercises

1. Consider the Riemann sphere $\widehat{\mathbb{C}}$. Let z be the chart which comes from stereographic projection. Define a meromorphic function f on $\widehat{\mathbb{C}}$ by

$$f(z) = \frac{z^3}{1 - z^2}$$

and $f(\infty) = \infty$. (∞ is the north pole of the sphere.) We can think of this as a holomorphic map of $\widehat{\mathbb{C}}$ to $\widehat{\mathbb{C}}$. Find the degree of this map and find its ramification points and their orders.

5.13 Holomorphic and Meromorphic Differential Form

In this section, we shall discuss differential forms on a Riemann surface X. Riemann's approach to meromorphic functions on a Riemann surface was to analyze the differentials first, and use them to understand the (single-valued) functions. The reason for this may be that the Mittag-Leffler problem for differentials has a simpler answer than for functions.

There is exactly one obstruction to the existence of a meromorphic differential with given principal parts: namely, each rational differential, regular on a punctured neighborhood of a point q, has a well-defined residue at q. A given configuration of principal parts at a finite set of points of a compact connected Riemann surface, i.e., a finite family of local rational differentials, occurs as the principal parts of a global meromorphic differential if and only if the residues add up to zero. Then to solve the Mittag-Leffler problem for functions we just have to analyze which differentials are "exact," i.e., are of form df. This leads to the study of "periods" of differentials.

5.13.1 Differential Forms on Riemann Surfaces

Let X be a (connected) Riemann surface with holomorphic atlas

$$\left\{\psi_i : U_i \xrightarrow{\sim} V_i \subseteq \mathbb{C}\right\}$$

and transition maps $\psi_{ij} := \psi_j \circ \psi_i^{-1} : V_{ij} \to V_j$ defined for all $U_i \cap U_j \neq \emptyset$, where $V_{ij} := \psi_i\left(U_i \cap U_j\right)$.

Definition 5.13.1. A *differential form* ω on X is given by a family of local expressions

$$\omega_i = f_i\left(\psi_i, \bar{\psi}_i\right) d\psi_i + g_i\left(\psi_i, \bar{\psi}_i\right) d\bar{\psi}_i$$

where f_i, g_i are smooth functions which obey the following transformation laws:

$$f_i\left(\varphi_{ij}\left(\psi_j\right), \overline{\varphi_{ij}\left(\psi_j\right)}\right) \frac{d\psi_{ij}\left(\psi_j\right)}{d\psi_j} = f_j\left(\psi_j, \bar{\psi}_j\right),$$

$$g_i\left(\varphi_{ij}\left(\psi_j\right), \overline{\varphi_{ij}\left(\psi_j\right)}\right) \overline{\frac{d\psi_{ij}\left(\psi_j\right)}{d\psi_j}} = g_j\left(\psi_j, \bar{\psi}_j\right).$$

This must hold for all charts (U, ψ_i). (Basically, you just take everything to the affine plane using charts, use definitions you're familiar with, and insist that this works for all charts.)

Holomorphic differential forms are locally defined on each coordinate neighborhood $\psi_j(U_j) \subset \mathbb{C}$ as

$$g_j(z)\, dz\,,$$

where the variable z lies in $\psi_j(U_j)$ and the function $g_j : \psi_j(U_j) \to \mathbb{C}$ is holomorphic. On overlaps $U_j \cap U_i$, they satisfy the expected compatibility condition

$$g_i(\psi_{ij}(z))\, \psi_{ij}'(z) = g_j(z)\,,$$

where $\psi_{ij}(z) = \psi_i \circ \psi_j^{-1}(z)$.

A 1-form is called *holomorphic* if all its components ω_i are holomorphic. It is enough to demand this for an atlas of analytic charts, since the transform of a holomorphic 1-form by a holomorphic transformation is holomorphic again. (Here we make use of the fact that we have a Riemann surface and not only a differentiable surface in the sense of real analysis). We denote by $\Omega^1(X)$ the space of holomorphic 1-forms.

5.13.2 Integration of Forms

The invariance property of forms, which resembles the change of variables used in calculus, allows us to define integrals of forms on surfaces, as we show in this subsection.

Integration of 1*-forms.*

Let X be a Riemann surface and $\omega \in \mathcal{E}^1(X)$. Let $\gamma : [0,1] \to X$ be a piecewise C^1-curve. That means γ is continuous and there is a partition $0 = t_0 < t_1 < \cdots < t_n = 1$ and charts $(U_k, z_k), k = 1, \ldots, n$, such that $\gamma([t_{k-1}, t_k]) \subseteq U_k$, and $z_k \circ \gamma : [t_{k-1}, t_k] \to \mathbb{C}$ are C^1. On U_k we have $\omega = f_k dx_k + g_k dy_k$ for smooth f_k, g_k. We define

$$\int_\gamma \omega := \sum_{k=1}^{n} \int_{t_{k-1}}^{t_k} \left(f_k(\gamma(t))\,(x_k \circ \gamma)'(t) + g_k(\gamma(t))\,(y_k \circ \gamma)'(t)\right) dt\,.$$

This definition is independent of the choice of partition and charts.

Theorem 5.13.2. *Let X be a Riemann surface, $\gamma : [0,1] \to X$ a piecewise C^1-curve, and $f \in \mathcal{E}(X)$. Then*

$$\int_\gamma df = f(\gamma(1)) - f(\gamma(0))$$

Proof. Exercise. □

Primitives.

Let X be a Riemann surface and $\omega \in \mathcal{E}^1(X)$. A function $f \in \mathcal{E}(X)$ is called a *primitive* of ω if $df = \omega$. Two primitives differ by an additive constant.

A 1-form with a primitive must be closed. Locally, a closed 1-form has a primitive and hence is exact. In fact, if $\omega = fdx + gdy$ is a closed 1-form on a disk centered at 0 in $\mathbb{C}$ then

$$F(x,y) := \int_0^1 (f(tx,ty)x + g(tx,ty)y)dt$$

provides a primitive of ω. Globally, a primitive of a closed 1-form exists in general only as a multivalued function:

Theorem 5.13.3. *Let X be a Riemann surface and let $\omega \in \mathcal{E}^1(X)$ be closed. Then there is a covering map $p : \widehat{X} \to X$ with $\widehat{X}$ connected and a primitive $F \in \mathcal{E}(\widehat{X})$ of $p^*\omega$.*

Proof. For open $U \subseteq X$ define $\mathcal{F}(U) := \{f \in \mathcal{E}(U) : df = \omega\}$. This defines a sheaf $\mathcal{F}$ on X which satisfies the identity theorem (indeed, on a connected open set V any two elements of $\mathcal{F}(V)$ differ by a constant). By Theorem 11.3.4, $|\mathcal{F}|$ is Hausdorff. We claim that the natural projection $p : |\mathcal{F}| \to X$ is a covering map. Each $a \in X$ has a connected open neighborhood U and a primitive $f \in \mathcal{F}(U)$ of ω. So

$$p^{-1}(U) = \bigcup_{c\in\mathbb{C}} (U, f + c)$$

where the sets $(U, f + c)$ are pairwise disjoint and $p|_{(U,f+c)} : (U, f + c) \to U$ are homeomorphisms.

If $\widehat{X}$ is a connected component of $|\mathcal{F}|$, then $p : \widehat{X} \to X$ is also a covering map. The function $F : \widehat{X} \to \mathbb{C}$ defined by $F(\varphi) := \varphi(p(\varphi))$ is a primitive of $p^*\omega$. □

Corollary 5.13.4. *Let X be a Riemann surface, $\pi : \widetilde{X} \to X$ its universal covering, and let $\omega \in \mathcal{E}^1(X)$ be closed. Then there is a primitive $F \in \mathcal{E}(\widetilde{X})$ of $\pi^*\omega$.*

Proof. Let $p : \widehat{X} \to X$ be the covering provided by Theorem 5.13.3. There is a holomorphic fiber-preserving map $\tau : \widetilde{X} \to \widehat{X}$, by the universal property of π. Then τ^*F is a primitive of $\pi^*\omega = \tau^*(p^*\omega)$. □

Corollary 5.13.5. *On a simply connected Riemann surface every closed 1-form is exact.*

Theorem 5.13.6. *Let X be a Riemann surface, $\pi : \tilde{X} \to X$ its universal covering, and let $\omega \in \mathcal{E}^1(X)$ be closed. Let $F \in \mathcal{E}(\tilde{X})$ be a primitive of $\pi^*\omega$. If $\gamma : [0,1] \to X$ is a piecewise C^1-curve and $\tilde{\gamma}$ is a lifting of γ, then*

$$\int_\gamma \omega = F(\tilde{\gamma}(1)) - F(\tilde{\gamma}(0)) \tag{5.9}$$

Proof. This follows from Theorem 5.13.2 and

$$\int_{\tilde{\gamma}} \pi^*\omega = \int_{\pi\circ\tilde{\gamma}} \omega = \int_\gamma \omega$$

We may take (5.9) as a definition of the integral of a closed 1-form along an arbitrary continuous curve $\gamma : [0,1] \to X$. Clearly, this definition is independent of the choice of the primitive F of $\pi^*\omega$. It is also independent of the lifting of the curve γ. Indeed, let $\tilde{\gamma}_1, \tilde{\gamma}_2$ be two liftings. Since the covering π is normal, by Theorem 11.1.18, there is a deck transformation σ such that $\tilde{\gamma}_1 = \sigma \circ \tilde{\gamma}_2$. Since $\pi \circ \sigma = \pi$, we have $\sigma^*\pi^*\omega = \pi^*\omega$. It follows that σ^*F is a primitive of $\pi^*\omega$ and hence $\sigma^*F - F = \text{const}$. Then

$$F(\tilde{\gamma}_1(1)) - F(\tilde{\gamma}_1(0)) = \sigma^*F(\tilde{\gamma}_2(1)) - \sigma^*F(\tilde{\gamma}_2(0)) = F(\tilde{\gamma}_2(1)) - F(\tilde{\gamma}_2(0))$$

□

Integration of 2-forms.

The integration of a 2-form requires a little more work. First, observe that one cannot speak of the value of a form at a point. However, since changes of coordinates have non-zero Jacobians, it makes sense to say whether a form vanishes or not at a given point. More precisely, if $\alpha \in \mathcal{S}^2(X)$, we define its *support* by

$$\mathrm{supp}(\alpha) = \overline{\{p \in X \; ; \; \alpha(p) \neq 0\}}\,.$$

We have that $\mathrm{supp}(\alpha)$ is a well-defined set. Assume that for a certain form α, the set $\mathrm{supp}(\alpha)$ is compact and contained in a single coordinate patch, say $(U, z = x + iy)$. We define the integral of α by

$$\int_X \alpha = \int_{z(U)} (f \circ z^{-1})\, ds\, dt\,,$$

where (s,t) are coordinates on $\mathbb{R}^2$. As in the case of 1-forms, the invariance property can be used to prove that this integral is well defined (its value does not depend on the local coordinate z).

More generally, if α is a 2-form with compact support, we cover $\mathrm{supp}(\alpha)$ by finitely many coordinate patches (U_j, z_j), $j = 1, \ldots, n$. Let $\{\phi_1, \ldots, \phi_n\}$ be a partition of unity subordinate to this covering (see 1.1.23). We then set

$$\int_X \alpha = \sum_{j=1}^{n} \int_X \phi_j \alpha\,,$$

where α is given by $f_j\, dx_j \wedge dy_j$ on U_j. Observe that the forms $\phi_j \alpha$ have compact support contained in U_j; their integrals are defined by the previous expression. One can check that this is independent of the choices made to compute it.

Let $U \subseteq \mathbb{C}$ be a domain. A 2-form $\omega \in \mathcal{E}^2(U)$ is of the form

$$\omega = f dx \wedge dy = -\frac{1}{2i} f dz \wedge d\bar{z} \quad \text{for } f \in \mathcal{E}(U)\,.$$

If f is integrable on U, we define

$$\int_U \omega := \iint_U f(x,y) dx dy\,.$$

Let $\varphi : V \to U$ be a biholomorphism. The Cauchy–Riemann equations imply that the Jacobian determinant of φ equals $|\varphi'|^2$. On the other hand

$$\varphi^* \omega = -\frac{1}{2i}(f \circ \varphi) d\varphi \wedge d\bar{\varphi} = -\frac{1}{2i}(f \circ \varphi) \left|\varphi'\right|^2 dz \wedge d\bar{z} = (f \circ \varphi) \left|\varphi'\right|^2 dx \wedge dy\,.$$

The transformation formula implies

$$\int_{\varphi(V)} \omega = \int_V \varphi^* \omega\,.$$

Let $\omega \in \mathcal{E}^2(X)$ be a 2-form on a Riemann surface with compact support. There are finitely many charts $\varphi_k : U_k \to V_k, k = 1, \ldots, n$, such that the support of ω is contained in the union of the U_k. Let (f_k) be a smooth partition of unity subordinate to (U_k). We define

$$\int_X \omega := \sum_{k=1}^{n} \int_X f_k \omega := \sum_{k=1}^{n} \int_{V_k} \left(\varphi_k^{-1}\right)^* (f_k \omega) \, .$$

Using the transformation formula, one checks that this definition is independent of the choice of charts and of the partition of unity.

Another obvious property of holomorphic transformations is that they preserve the types $A^{0,1}$ and $A^{0,1}$. This also follows from the Cauchy–Riemann differential equations. Hence we can define $A^{0,1}(X)$ and $A^{1,0}(X)$ for a Riemann surface componentwise. The operations d and ∂ commute with holomorphic transformations. Hence we can define operators

$$d : \mathcal{C}^\infty(X) \longrightarrow A^1(X)\, , \quad d : A^1(X) \longrightarrow A^2(X)\, ,$$
$$\bar{\partial} : \mathcal{C}^\infty(X) \longrightarrow A^{0,1}(X)\, , \quad \bar{\partial} : A^{1,0}(X) \longrightarrow A^2(X)$$

also componentwise (for example $d(\omega)_\varphi := d\left(\omega_\varphi\right)$.) Also the exterior product generalizes to Riemann surfaces via charts.

Theorem 5.13.7 (Stokes's theorem for differential forms)**.** *Suppose C is a Riemann surface, Ω is an open set in C, and λ is a differential 1-form defined on an open set containing $\overline{\Omega}$. Then*

$$\int_{\partial\Omega} \lambda = \iint_\Omega d\lambda \, .$$

The proof of this theorem can be found in any text containing a discussion of differentiable manifolds.

Corollary 5.13.8. *If λ is a differential 1-form on a compact Riemann surface X, then*

$$\iint_X d\lambda = 0 \, .$$

See [168] for proofs of the next two theorems.

Theorem 5.13.9 (Stokes)**.** *Let ω be a 1-form on a Riemann surface X. Let U be an open subset of X with compact closure and smooth boundary ∂U (that is, ∂U consists of a finite collection of smooth curves). Then*

$$\int_U d\omega = \int_{\partial U} \omega \, .$$

Corollary 5.13.10. *Let X be a compact Riemann surface, and $\omega \in \mathcal{S}^1(X)$. Then*

$$\int_X d\omega = 0 \, .$$

The integration over the boundary of U must be done with a certain orientation. We will use Stokes's theorem in two types of situations: in one case, we consider two disjoint domains U_1 and U_2 with a common boundary given by a smooth curve c, $\partial U_1 = \partial U_2 = c$. Then $\int_{\partial U_1} \omega = -\int_{\partial U_2} \omega$.

In the other situation, the domain of integration will be a disc $\mathbb{D}_r = \{z \in \mathbb{C} ; |z| < r\}$ with the standard orientation of $\mathbb{C}$; its boundary, the circle of center 0 and radius R, is oriented clockwise (opposite to the standard orientation).

For more details on Stokes' theorem and its applications, see the references in [169].

We will need two more theorems from differential geometry. The first one tells us that the set of "bad" (critical) values of a smooth map has measure zero. More precisely, let $f : X \to \mathbb{R}$ be a smooth proper map defined on a smooth surface X. Recall that f is called *proper* if $f^{-1}(K)$ is compact for any compact $K \subset \mathbb{R}$. Let p be a point of X, and $z = x + iy$ a local coordinate near p. We say that p is a *critical point* of f if $f_x(p) = f_y(p) = 0$. Critical points are well defined because changes of coordinates have non-vanishing derivatives. A *critical value* is the image of a critical point under f.

Theorem 5.13.11 (Sard). *Let $f : X \to \mathbb{R}$ be a smooth proper mapping defined on a smooth surface. Let C be the set of critical values of f. Then C has measure zero in $\mathbb{R}$.*

The next result classifies all smooth curves (1-dimensional manifolds) [128].

Theorem 5.13.12. *Any smooth, connected 1-dimensional manifold is diffeomorphic either to the circle S^1 or to some interval of real numbers.*

(An interval is a connected subset of R which is not a point. It may be finite or infinite; closed, open, or half-open.)

Since any interval is diffeomorphic either to $[0, 1]$, $(0, 1]$, or $(0, 1)$, it follows that there are only four distinct connected 1-manifolds.

Theorem 5.13.13. *Any compact, smooth 1-dimensional manifold is diffeomorphic to a circle.*

5.13.3 Meromorphic 1-Forms

A 1-form ω is called *meromorphic* if it can be written locally as $\omega = f(z)dz$, where f is a meromorphic function. The set of poles of the functions f are also known as the poles of ω. They are well defined (do not depend on the function f chosen to represent the form) and form a discrete set of points of X. The set of meromorphic forms on X will be denoted by $\mathcal{M}^1(X)$. In a local coordinate (U, z) the function f has a power series expansion of the form

$$f(p) = \sum_{j=N}^{\infty} a_j (z(p))^j, \quad p \in U$$

which we simplify as $f = \sum_{j=N}^{\infty} a_j z^j$. The coefficient a_{-1} is called the *residue* of the form ω at the point p, and we will denote it by $\mathrm{res}_p(\omega)$.

Proposition 5.13.14. *The residue of a meromorphic form is well defined.*

Proof. We need to show that a_{-1} is independent of f and the coordinate z. Assume then that $z(p) = 0$ as above. Let r and ϵ be positive numbers such that $\mathbb{D}_{r+\epsilon}$, the disc of center 0 and radius $r + \epsilon$ is contained in $z(U)$. Since the poles of ω form a discrete set we can further assume that the only singularity of the form ω in $z^{-1}(\mathbb{D}(r+\epsilon)) = \{q \in U \mid |z(q)| < r + \epsilon\}$ lies at the point p. Consider now the smooth curve $c : [0, 1] \to U$ given by $c(\theta) = z^{-1}\left(re^{i\theta}\right)$. One easily sees that

$$a_{-1} = \frac{1}{2\pi i} \int_c \omega$$

Since this formula gives an way of computing a_{-1} in terms of the integral of ω over a path we see that the residue does not depend of local coordinates (since path integrals are coordinate-independent). □

The following result is a consequence of Stokes's theorem.

Theorem 5.13.15 (residue theorem). *Let ω be a meromorphic form on a compact surface X. Then*

$$\sum_{p \in X} res_p(\omega) = 0 .$$

Proof. First of all, since X is compact and the set of poles of ω is discrete we have that the above sum is finite. So let us assume that these poles are at points $p_1, \ldots, p_n$. It is possible to find a simply connected set with smooth boundary, say U, containing all the poles, as follows: let c' be a smooth curve passing though the poles (once by each pole) and take a small, smooth neighborhood of c', (which can be done in local coordinates). Let c denote the boundary of U.

We have that c is homotopic to a collection of small "circles" around the poles p_j, similar to the curves used in the proof of the previous result. Thus

$$\int_c \omega = \sum_{j=1}^{n} \mathrm{res}_{p_j}\, \omega .$$

On the other hand the form ω has no poles on $X \backslash U$ so it is holomorphic and therefore closed. Applying Stokes' theorem (Theorem 6.3.10) we get

$$0 = \int_{X \backslash U} \mathrm{d}\omega = \int_c \omega ,$$

which completes the proof. □

Theorem 5.13.16. *Let X be a Riemann surface and let $\omega \in \mathcal{E}^1(X)$ have compact support. Then $\int_X dw = 0$.*

Proof. In the plane this follows immediately from Stoke's theorem. In general one may use a partition of unity to decompose ω into 1-forms each of which with has compact support in one chart. □

5.14 Harmonic Differential Forms

In this section, we introduce and study harmonic 1-forms on a compact Riemann surface X. We will see that every smooth closed 1-form on X can be uniquely written as the sum of a harmonic and an exact 1-form. So the first de Rham cohomology group of X is isomorphic to the vector space of harmonic 1-forms on X. This will imply that the genus is a topological invariant.

5.14.1 The $*$-Operator and Harmonic 1-Forms

Let X be a Riemann surface. Let $\omega \in \mathcal{E}^1(X)$. Locally, $\omega = \sum f_j dg_j$ for smooth function f_j, g_j, and we may consider the complex conjugate $\overline{\omega} = \sum \overline{f}_j d\overline{g}_j$. This defines the complex conjugate $\overline{\omega} \in \mathcal{E}^1(X)$ of ω. The real part of ω is defined by $\mathrm{Re}(\omega) = (\omega + \overline{\omega})/2$. We say that ω is real if $\omega = \mathrm{Re}(\omega)$. We have $\mathrm{Re}\left(\int_\gamma \omega\right) = \int_\gamma \mathrm{Re}(\omega)$, because $\overline{\int_\gamma \omega} = \int_\gamma \overline{\omega}$.

If $\omega \in \mathcal{O}^1(X)$ is holomorphic, then $\overline{\omega}$ is called *antiholomorphic*. The vector space of all antiholomorphic 1-forms on X is denoted by $\overline{\mathcal{O}}^1(X)$.

Let $\omega \in \mathcal{E}^1(X)$. There is a unique decomposition

$$\omega = \omega_1 + \omega_2, \qquad \text{where } \omega_1 \in \mathcal{E}^{1,0}(X), \omega_2 \in \mathcal{E}^{0,1}(X).$$

Defining

$$*\omega := i\,(\overline{\omega}_1 - \overline{\omega}_2)$$

we obtain an $\mathbb{R}$-linear automorphism

$$* : \mathcal{E}^1(X) \to \mathcal{E}^1(X)$$

with $*\mathcal{E}^{1,0}(X) = \mathcal{E}^{0,1}(X)$ and $*\mathcal{E}^{0,1}(X) = \mathcal{E}^{1,0}(X)$.

Lemma 5.14.1. *We have the following properties.*

(1) $**\omega = -\omega$ *and* $*\omega = *\overline{\omega}$ *for all* $\omega \in \mathcal{E}^1(X)$.

(2) $d * (\omega_1 + \omega_2) = i\partial\overline{\omega}_1 - i\overline{\partial}\overline{\omega}_2$ *for all* $\omega_1 \in \mathcal{E}^{1,0}(X), \omega_2 \in \mathcal{E}^{0,1}(X)$.

(3) $*\partial f = i\overline{\partial f}, *\overline{\partial} f = -i\overline{\partial f}$, *and* $d * df = 2i\partial\overline{\partial f}$ *for all* $f \in \mathcal{E}(X)$.

Proof. (1) is obvious. For (2),

$$d * (\omega_1 + \omega_2) = (\partial + \overline{\partial})\,(i\,(\overline{\omega}_1 - \overline{\omega}_2)) = i\partial\overline{\omega}_1 - i\overline{\partial}\overline{\omega}_2\,.$$

By (2),

$$d * df = d * (\partial f + \overline{\partial} f) = i\partial\overline{\overline{\partial} f} - i\overline{\partial}\overline{\partial f} = 2i\partial\overline{\partial f}\,.$$

The rest follows easily from the definition. □

Proposition 5.14.2. *Let $\omega \in \mathcal{E}^1(X)$. The following conditions are equivalent.*

(1) $d\omega = d * \omega = 0$.

(2) $\partial\omega = \overline{\partial}\omega = 0$.

(3) $\omega = \omega_1 + \omega_2$, *where* $\omega_1 \in \mathcal{O}^1(X)$ *and* $\omega_2 \in \overline{\mathcal{O}}^1(X)$.

(4) ω *is locally the exterior derivative of a harmonic function.*

Proof. The equivalence of the first three items follows from Lemma 5.14.1. (1) $\Rightarrow$ (4) Since $d\omega = 0$, locally $\omega = df$ for some smooth function f. Since $0 = d * \omega = d * df = 2i\partial\overline{\partial f}$, f is harmonic. (4) $\Rightarrow$ (1) Assume that $\omega = df$ and f is harmonic, then $d\omega = ddf = 0$ and $d * \omega = d * df = 2i\partial\overline{\partial f} = 0$. □

A 1-form $\omega \in \mathcal{E}^1(X)$ on a Riemann surface X which satisfies the equivalent conditions in the proposition is called *harmonic*. We denote the vector space of all harmonic 1-forms on X by $\mathcal{E}^1_{\text{har}}(X)$. By Proposition 5.14.2,

$$\mathcal{E}^1_{\text{har}}(X) = \mathcal{O}^1(X) \oplus \overline{\mathcal{O}}^1(X)\,.$$

If X is a compact Riemann surface of genus g, then

$$\dim \mathcal{E}^1_{\text{har}}(X) = 2g\,.$$

Theorem 5.14.3. *Let $\sigma \in \mathcal{E}^1_{har}(X)$ be a real harmonic 1-form. There exists a unique $\omega \in \mathcal{O}^1(X)$ such that $\sigma = \mathrm{Re}(\omega)$.*

Proof. We may write $\sigma = \omega_1 + \overline{\omega}_2$ for $\omega_1, \omega_2 \in \mathcal{O}^1(X)$. Since σ is real, $\omega_1 + \overline{\omega}_2 = \sigma = \overline{\sigma} = \overline{\omega}_1 + \omega_2$, and so $\omega_1 = \omega_2$. Then $\sigma = \mathrm{Re}\,(2\omega_1)$.

If $\sigma = \mathrm{Re}(\omega) = \mathrm{Re}\,(\omega')$ for $\omega, \omega' \in \mathcal{O}^1(X)$, then $\tau := \omega - \omega' \in \mathcal{O}^1(X)$ and $\mathrm{Re}(\tau) = 0$. Locally, there is a holomorphic function f such that $\tau = df$. Thus, f has constant real part and hence is constant itself. Therefore, $\tau = 0$. □

5.14.2 A Scalar Product on $\mathcal{E}^1(X)$

Let X be a compact Riemann surface. For $\omega_1, \omega_2 \in \mathcal{E}^1(X)$ let

$$\langle \omega_1, \omega_2 \rangle := \int_X \omega_1 \wedge *\omega_2\,.$$

This defines a (sesquilinear) scalar product on $\mathcal{E}^1(X)$. Let us check that $\langle\cdot,\cdot\rangle$ is positive definite. In a local chart $\omega \in \mathcal{E}^1(X)$ has the form $\omega = f dz + g d\bar{z}$, thus $*\omega = i(\bar{f} d\bar{z} - \bar{g} dz)$ and

$$\omega \wedge *\omega = i\left(|f|^2 + |g|^2\right) dz \wedge d\bar{z} = 2\left(|f|^2 + |g|^2\right) dx \wedge dy\,.$$

Consequently, $\langle \omega,\omega\rangle \geq 0$ and $\langle \omega,\omega\rangle = 0$ if and only if $\omega = 0$. Thus $(\mathcal{E}^1(X), \langle\cdot,\cdot\rangle)$ is a unitary space, which however is not complete.

Lemma 5.14.4. *Let X be a compact Riemann surface.*

(1) $\partial\mathcal{E}(X), \bar{\partial}\mathcal{E}(X), \mathcal{O}^1(X)$, and $\overline{\mathcal{O}}^1(X)$ are pairwise orthogonal subspaces of $\mathcal{E}^1(X)$.

(2) $d\mathcal{E}(X)$ and $*d\mathcal{E}(X)$ are orthogonal subspaces of $\mathcal{E}^1(X)$ and

$$d\mathcal{E}(X) \oplus *d\mathcal{E}(X) = \partial\mathcal{E}(X) \oplus \bar{\partial}\mathcal{E}(X)\,. \tag{5.10}$$

Proof. (1) Since $\mathcal{E}^{1,0}(X)$ and $\mathcal{E}^{0,1}(X)$ are clearly orthogonal also $\partial\mathcal{E}(X) \perp \bar{\partial}\mathcal{E}(X)$, $\bar{\partial}\mathcal{E}(X) \perp \mathcal{O}^1(X)$, and $\partial\mathcal{E}(X) \perp \overline{\mathcal{O}}^1(X)$.

Let us prove $\partial\mathcal{E}(X) \perp \mathcal{O}^1(X)$. Let $f \in \mathcal{E}(X)$ and $\omega \in \mathcal{O}^1(X)$. By Lemma 5.14.1 and Proposition 5.14.2,

$$\omega \wedge *\partial f = i\omega \wedge \bar{\partial}\bar{f} = i\omega \wedge d\bar{f} = -id(\bar{f}\omega)$$

and by Theorem 5.13.16,

$$\int_X \omega \wedge *\partial f = -i \int_X d(\bar{f}\omega) = 0$$

For $\bar{\partial}\mathcal{E}(X) \perp \overline{\mathcal{O}}^1(X)$, use $\bar{\omega} \wedge *\bar{\partial} f = -i\omega \wedge \partial\bar{f} = -i\omega \wedge d\bar{f} = id(\bar{f}\omega)$.

(2) Let $f, g \in \mathcal{E}(X)$ of the line Then, by Lemma 5.14.1,

$$df \wedge *(*dg) = -df \wedge dg = -d(f dg)$$

Again by Theorem 5.13.16, $\langle df, *dg\rangle = 0$. For the identity (5.10) observe that, by Lemma 5.14.1,

$$df + *dg = \partial f + \bar{\partial} f + *(\partial g + \bar{\partial} g) = \partial(f - i\bar{g}) + \bar{\partial}(f + i\bar{g})$$

□

Corollary 5.14.5. *Let X be a compact Riemann surface. Every exact harmonic 1-form on X vanishes. Every harmonic function on X is constant.*

Proof. By Lemma 5.14.4, $d\mathcal{E}(X)$ is orthogonal to $\mathcal{E}^1_{\text{har}}(X) = \mathcal{O}^1(X) \oplus \overline{\mathcal{O}}^1(X)$. □

Corollary 5.14.6. *Let X be a compact Riemann surface.*

(1) *Let* $\sigma \in \mathcal{E}^1_{har}(X)$. *If* $\int_\gamma \sigma = 0$ *for every closed curve* γ *in* X, *then* $\sigma = 0$.

(2) *Let* $\omega \in \mathcal{O}^1(X)$. *If* $\mathrm{Re}\int_\gamma \omega = 0$ *for every closed curve* γ *in* X, *then* $\omega = 0$.

Proof. A closed 1-form $\omega \in \mathcal{E}^1(X)$ has a primitive $f \in \mathcal{E}(X)$ if and only if all periods of ω vanish. Since both σ and $\mathrm{Re}\,\omega$ are exact. Then the assertions follow from Theorem 5.14.3 and Corollary 5.14.5. □

5.14.3 The Hodge-deRham Theorem

Theorem 5.14.7 (the Hodge–de Rham theorem). *Let* X *be a compact Riemann surface. We have the orthogonal decompositions*

$$\mathcal{E}^{0,1}(X) = \bar{\partial}\mathcal{E}(X) \oplus \overline{\mathcal{O}}^1(X)$$

and

$$\mathcal{E}^1(X) = *d\mathcal{E}(X) \oplus d\mathcal{E}(X) \oplus \mathcal{E}^1_{har}(X)\,.$$

Moreover,

$$\ker\left(d : \mathcal{E}^1(X) \to \mathcal{E}^2(X)\right) = d\mathcal{E}(X) \oplus \mathcal{E}^1_{har}(X)$$

and

$$H^1(X, \mathbb{C}) \cong \mathrm{Rh}^1(X) \cong \mathcal{E}^1_{har}(X)\,.$$

Corollary 5.14.8. *Let* X *be a compact Riemann surface.*

(1) *Let* $\sigma \in \mathcal{E}^{0,1}(X)$. *Then* $\bar{\partial} f = \sigma$ *has a solution* $f \in \mathcal{E}(X)$ *if and only if* $\sigma \perp \overline{\mathcal{O}}^1(X)$.

(2) *A 1-form* $\sigma \in \mathcal{E}^1(X)$ *is exact if and only if*

$$\int_X \sigma \wedge \omega = 0$$

for every closed $\omega \in \mathcal{E}^1(X)$.

For more of this topic see [75, 89, 179, 185]

5.15 The Dimension of the Space of Holomorphic Differential Forms $\Omega^1(X)$

Holomorphic functions on a compact Riemann surface are necessarily constant. However, the space of holomorphic differential forms on such a surface is a finite-dimensional complex vector space, and its dimension depends only on the topology of the surface. In fact, this dimension equals the genus of the surface. For example, on an elliptic curve, all holomorphic differentials are scalar multiples of the "constant" form dz.

On the other hand, the Riemann sphere admits only the trivial holomorphic differential form, which is identically zero. Let us prove this assertion.

Consider the covering of $\mathbb{CP}^1$ by the two open sets U_1 and U_2 as before. On the overlap, we have $w(z) = 1/z$, and therefore

$$-\frac{1}{z^2}\, g_1\left(\tfrac{1}{z}\right) = g_2(z)\,.$$

That is,

$$g_1\left(\tfrac{1}{z}\right) = -z^2 g_2(z)\,.$$

The only holomorphic solution with g_2 defined on $\mathbb{C}$ and g_1 on $\mathbb{CP}^1 \setminus \{0\}$ (in the w-coordinate, with ∞ corresponding to $w = 0$ and $g_1(w) = c_0 + c_1 w + \cdots + c_k w^k$) is the trivial solution. Hence, the only holomorphic differential form on the Riemann sphere is zero.

Example 5.15.1. Let X be given by an equation $F(z, w) = 0$, where the partial derivatives of F do not vanish simultaneously. Suppose $\frac{\partial F}{\partial w} \neq 0$, so that we may solve locally for z in terms of w. A holomorphic differential can then be written as

$$\frac{\partial F}{\partial z}\, dz$$

in the coordinate z. On the other hand, using the coordinate w we may define

$$-\frac{\partial F}{\partial w}\, dw\,.$$

The relation

$$\frac{\partial F}{\partial z}\, dz + \frac{\partial F}{\partial w}\, dw = 0$$

shows that this differential form is globally well defined on the whole surface.

Moreover, consider

$$\frac{dz}{\frac{\partial F}{\partial w}} \quad \text{and} \quad -\frac{dw}{\frac{\partial F}{\partial z}}$$

defined in the corresponding coordinate charts. Since the partial derivatives never vanish simultaneously, these expressions patch together to define a global holomorphic differential form.

The topological structure of a compact Riemann surface X is determined by its genus g. Topologically, a Riemann surface of genus g is a sphere with g handles, or equivalently, a torus with g holes.

Analytically, the genus can be characterized as the maximal number of linearly independent holomorphic 1-forms on X (see also the section "The Riemann–Roch Theorem and Applications").

We use $\Omega^1(X)$ to denote the set of holomorphic differentials on X. Here is then the main result of this section:

Theorem 5.15.2. *Suppose X is a compact Riemann surface of genus g. Then*

$$\dim \Omega^1(X) = g\,.$$

We prove this theorem in two parts. First we prove:

Proposition 5.15.3. *Hypotheses as in Theorem 5.15.2. Then we have*

$$\dim \Omega^1(X) \geq g\,.$$

Proof. We shall base our proof on the following general fact: for any compact Riemann surface X, there exists a holomorphic mapping π from X into $\mathbb{CP}^2$, such that $X' = \pi(X)$ is an irreducible algebraic curve of degree d in $\mathbb{CP}^2$ possessing δ ordinary double points (denoted $p_1, p_2, \ldots, p_\delta$) and no other singular points (X is in fact the normalization of X').

If $d = 1$ or 2, then $g = 0$ (genus formula), and for these cases Theorem 5.15.2 has already been proven. Therefore we shall assume below that $d \geq 3$.

Now, suppose that as already mentioned above, X is mapped holomorphically into $\mathbb{CP}^2$. After a suitable transformation of coordinates, we may assume that X' is not tangent to the line at infinity L_∞ in $\mathbb{CP}^2, L_\infty$ does not pass through any of the points $p_1, p_2, \ldots, p_\delta$, and X' does not pass through the point $[0, 0, 1]$.

More precisely, we can select a straight line in $\mathbb{CP}^2$ which intersects X' in d distinct points. By making this straight line L_∞, we can guarantee that X' and L_∞ are not tangent for the following reason.

Suppose X' and L_∞ are tangent; then at the point of tangency, the intersection number of X' and L_∞ is at least 2. This then leads to:

$$(X' \cdot L_\infty) \geq d + 1$$

which contradicts Bézout's theorem[2].

By a similar reasoning we can guarantee that L_∞ does not pass through any of the points $p_1, p_2, \ldots, p_\delta$. Finally, select a point which is on L_∞ but not on X' and make this point $[0, 0, 1]$. Then the resulting coordinate system satisfies our requirements. Suppose the affine equation of X' in this coordinate system is $f(x, y) = 0$. We write $\Gamma = p_1 + p_2 + \cdots p_\delta$ and let S^{d-3} represent the set of homogeneous polynomials with complex coefficients of degree $(d-3)$ in three variables. Let

$$S^{d-3}(-\Gamma) = \left\{ G\left(\xi^0, \xi^1, \xi^2\right) \in S^{d-3} \mid G\left(p_i\right) = 0 \right\}\,, \quad \text{for all } i = 1, 2, \ldots, \delta\,.$$

Clearly, both S^{d-3} and $S^{d-3}(-\Gamma)$ are complex vector spaces, and

$$\dim S^{d-3} = \frac{1}{2}(d-1)(d-2)\,.$$

[2]Bézout's theorem describes the number of times two algebraic curves intersect in projective space, up to multiplicity. Bézout's theorem, basic version: Two distinct irreducible plane curves of degrees m, n intersect in at most mn points.

On the other hand, the fact that a polynomial vanishes at a point p_i is equivalent to the fact that the coefficients of this polynomial satisfy a linear equation. Hence

$$\dim S^{d-3}(-\Gamma) \geq \frac{1}{2}(d-1)(d-2) - \delta = g$$

(where we have used the genus formula). We shall construct an injective homomorphism from $S^{d-3}(-\Gamma)$ into $\Omega^1(X)$; this would guarantee that

$$\dim \Omega^1(X) \geq \dim S^{d-3}(-\Gamma)\,,$$

which then completes the proof of the proposition. For any $G \in S^{d-3}(-\Gamma)$, we let

$$g(x,y) = G(1,x,y)\,.$$

We also let $U_0 = \left\{[1,x,y] \in \mathbb{CP}^2\right\}$ and $U_1 = \left\{[u,1,v] \in \mathbb{CP}^2\right\}$ and define

$$\alpha = \left\{ \begin{array}{ll} \left.\dfrac{g(x,y)dx}{\frac{\partial f(x,y)}{\partial y}}\right|_{C'}, \text{ on } C' \cap U_0\,, \text{ when } \left.\dfrac{\partial f(x,y)}{\partial y}\right|_{C'} \neq 0\,, \\ -\left.\dfrac{g(x,y)dy}{\frac{\partial f(x,y)}{\partial x}}\right|', \text{ on } C' \cap U_0\,, \text{ when } \left.\dfrac{\partial f(x,y)}{\partial y}\right|_{C'} = 0\,; \\ -\left.\dfrac{g\left(\frac{1}{u},\frac{v}{u}\right)\frac{du}{u^2}}{\frac{\partial f}{\partial y}\left(\frac{1}{u},\frac{v}{u}\right)}\right|_{C'}, \text{ on } C' \cap U_1 \end{array} \right\}$$

(here $(\partial f/\partial y)(1/u, v/u)$ denotes the value of $\partial f/\partial y$ at the point $(1/u, v/u)$).

From the fact that X' does not pass through the point $[0,0,1], \alpha$ is defined over the whole of X'. We shall now see that it is uniquely defined.

Note that inside U_0, the affine equation of X' is $f(x,y) = 0$, and therefore on X'

$$\frac{\partial f}{\partial x}dx + \frac{\partial f}{\partial y}dy = 0$$

whence, we have

$$\left.\frac{dx}{\frac{\partial f(x,y)}{\partial y}}\right|_{X'} = -\left.\frac{dy}{\frac{\partial f(x,y)}{\partial x}}\right|_{X'}\,,$$

so in the above definition, the value for α when $(\partial f(x,y)/\partial y)|_{C'} = 0$ is a continuation of that for α when $\partial f(x,y)/\partial y \neq 0$. From this it is easy to verify that $\pi^*(\alpha)$ is a meromorphic differential on X. Thus we have a mapping

$$\begin{aligned} &S^{d-3}(-\Gamma) \to K^1(X)\,, \\ &G \mapsto \omega_G = \pi^*(\alpha)\,. \end{aligned}$$

Obviously, this mapping is a vector space homomorphism over $\mathbb{C}$; it is furthermore injective, since if $\omega_G = 0$ then X' is contained in the curve $g(x,y) = 0$, which is impossible since $\deg G = d-3$. We only need to prove that ω_G is a holomorphic differential on X.

From the above discussion, a point p of ω_G on $\pi^{-1}(C' \cap U_0)$ can become a pole of ω_G only when

$$\left.\frac{\partial f(x,y)}{\partial y}\right|_{p'} = \left.\frac{\partial f(x,y)}{\partial x}\right|_{p'} = 0$$

(where $p' = \pi(p)$). In this case, p' is clearly a double point of X', and hence $p' \in \Gamma$, so that

$$g(x,y)|_{p'} = 0\,.$$

But p' is a second-order zero of F, and therefore a first-order zero of $\partial f/\partial r$, so

$$\alpha|_{p'} = -\left.\frac{g(x,y)dy}{\frac{\partial f(x,y)}{\partial x}}\right|_{p'} \neq \infty\,.$$

This says that ω_G is holomorphic at p.

Consider now the behavior of ω_G on $\pi^{-1}(C' \cap L_\infty)$. We have

$$\alpha = -\left.\frac{u^{d-3}g\left(\frac{1}{u},\frac{v}{u}\right)du}{u^{d-1}\frac{\partial f}{\partial y}\left(\frac{1}{u},\frac{v}{u}\right)}\right|_{C'}\,.$$

Note that the numerator and denominator are polynomials in u and v. Denoting the denominator by $h(u,v)$, we only need to show that $h(u,v) \neq 0$ on $X' \cap L_\infty$. In fact, the equation of X' on U_1 is

$$k(u,v) = u^d f\left(\frac{1}{u},\frac{v}{u}\right) = 0\,.$$

Clearly,

$$\frac{\partial k(u,v)}{\partial v} = u^d\frac{\partial f}{\partial u}\left(\frac{1}{u},\frac{v}{u}\right)\cdot\frac{1}{u} = h(u,v)\,.$$

Suppose $p' \in C' \cap L_\infty$ is such that

$$h(u,v)|_{p'} = 0\,;$$

then since p' is not a double point of X' we necessarily have

$$\left.\frac{\partial k(u,v)}{\partial u}\right|_{p'} \neq 0\,.$$

Since $p' \in L_\infty$, the coordinates of p' are of the form $[0,1,v]$, and we can then write the tangent line to X' at p' as

$$\left.\frac{\partial k(u,v)}{\partial u}\right|_{p'}(u-0) = 0\,,$$

i.e., the tangent line is just L_∞. This contradicts our choice of the coordinate system. □

Next we shall prove that

$$\dim \Omega^1(X) \leq g\,.$$

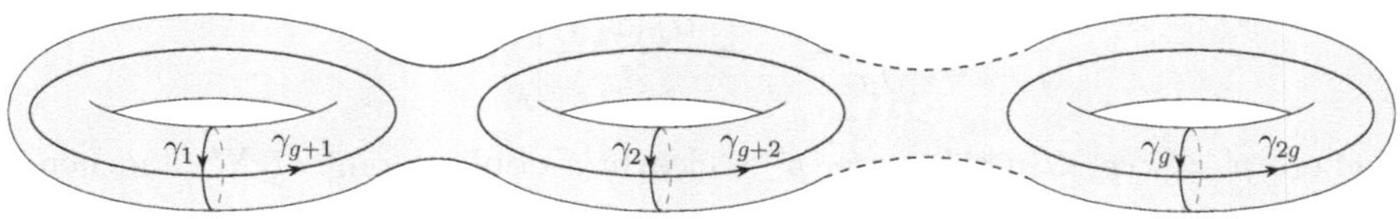

FIGURE 5.18: Riemann surface X of genus g. $\gamma_1, \gamma_2, \ldots, \gamma_{2g-1}, \gamma_{2g}$ forming a $\mathbb{Z}$-basis of the homology group $H_1(X, \mathbb{Z})$ of X.

Suppose X is a compact Riemann surface of genus g (see Fig. 5.18), with $\gamma_1, \gamma_2, \ldots, \gamma_{2g-1}, \gamma_{2g}$ forming a $\mathbb{Z}$-basis of the homology group $H_1(C, \mathbb{Z})$ of X, and suppose γ is a closed curve on X. We then have

$$\gamma = \sum_{i=1}^{2g} n_i \gamma_i + \partial\Omega$$

where $n_i \in \mathbb{Z} \quad (i = 1, 2, \ldots, 2g)$, and Ω is a region on X. Let $[\gamma]$ represent the homology class containing γ.

From Stokes's theorem (Theorem 6.3.10), we obtain easily:

Proposition 5.15.4. *If λ is a closed differential 1-form on X, then the mapping*

$$\eta_\lambda : H_1(X, \mathbb{Z}) \to \mathbb{C}\,,$$
$$[\gamma] \mapsto \int_\gamma \lambda\,,$$

where γ is a closed curve on X, is well defined.

Proof. If γ' is homologous to γ, then $\gamma - \gamma' = \partial\Omega$, where Ω is a region on C. By Stokes' theorem (Theorem 5.13.7), we have

$$\eta_\lambda(\gamma) - \eta_\lambda(\gamma') = \int_{\gamma - \gamma'} \lambda = \int_{\partial\Omega} \lambda = \iint_\Omega d\lambda = 0\,.$$

□

Proposition 5.15.5. *If λ is a closed differential 1-form on X, and if for all i we have*

$$\int_{y_t} \lambda = 0\,,$$

then λ is exact.

Proof. Fix the point p_0 on X. For any point p on X, choose a path γ from p_0 to p, and define

$$f(p) = \int_\gamma \lambda .$$

We must ascertain that the definition of $f(p)$ is independent of the choice of the path γ. Suppose γ' is another path from p_0 to p. Then $\gamma - \gamma'$ is a closed curve on X, and therefore

$$\gamma - \gamma' = \sum_{i=1}^{2g} n_i \gamma_i + \partial\Omega ,$$

where $n_i \in \mathbb{Z}$, and Ω is a region on X. From

$$\int_{\gamma_1} \lambda = 0 \quad (i = 1, 2, \ldots, 2g)$$

as well as

$$\int_{\partial\Omega} \lambda = \iint_\Omega d\lambda = 0$$

we then know that

$$\int_\gamma \lambda = \int_{\gamma'} \lambda ;$$

that is, the definition of $f(p)$ is independent of the choice of the path γ. Thus f is a well-defined X^∞ function on X, and by the fundamental theorem of calculus, we have $df = \lambda$. □

Proposition 5.15.6. *Suppose X is a compact Riemann surface, and $\omega, \varphi \in \Omega^1(X)$. Considering ω and φ as differential 1-forms on X, if*

$$\omega + \overline{\varphi} = df$$

where f is a X^∞ function on X, then

$$\omega = \varphi = 0 .$$

Proof. Suppose $\omega = h(z)dz, \varphi = g(z)dz$. Then

$$\omega \wedge \varphi = 0$$

and

$$\frac{i}{2}\varphi \wedge \bar{\varphi} = |g(z)|^2 \frac{i}{2} dz \wedge d\bar{z} = |g(z)|^2 du \wedge dv .$$

We shall prove by reductio ad absurdum that $\varphi = 0$. Suppose $\varphi \neq 0$, then

$$\frac{i}{2} \iint_C \varphi \wedge \bar{\varphi} = \iint_C |g(z)|^2 du \wedge dv > 0 . \tag{5.11}$$

But

$$\varphi \wedge \bar{\varphi} = \varphi \wedge \omega + \varphi \wedge \bar{\varphi} = \varphi \wedge (\omega + \bar{\varphi}) = \varphi \wedge df$$

and we shall prove below that

$$\iint_C \varphi \wedge df = 0\,.$$

This directly contradicts (5.11), thereby showing $\varphi = 0$. To prove the last claim, first observe that

$$d(f\varphi) = df \wedge \varphi + f \wedge d\varphi\,.$$

Since $\varphi \in \Omega^1(X)$, g is holomorphic and it follows that

$$\frac{\partial g}{\partial \bar{z}} = 0\,.$$

Therefore

$$\begin{aligned} d\varphi &= d(gdz) = dg \wedge dz \\ &= \left(\frac{\partial g}{\partial z}dz + \frac{\partial g}{\partial \bar{z}}d\bar{z}\right) \wedge dz = 0 \end{aligned}$$

whence

$$\varphi \wedge df = -df \wedge \varphi = -d(f\varphi)\,.$$

Noting that $f\varphi$ is a differential 1-form on X, from Corollary 5.13.8, we know then that

$$\iint_X d(f\varphi) = 0\,.$$

So we arrive at the desired conclusion that

$$\iint_X \varphi \wedge df = -\iint_X d(f\varphi) = 0\,.$$

Similarly $\omega = 0$. □

We now give the other part of the proof of Theorem 5.15.2.

Proposition 5.15.7. *Under the hypothesis of Theorem 5.15.2, we have*

$$\dim \Omega^1(X) \le g\,.$$

Proof. Assuming $\dim \Omega^1(X) \ge g+1$, let us suppose that the $(g+1)$ elements $\omega_1, \ldots, \omega_{g+1} \in \Omega^1(X)$ are linearly independent over $\mathbb{C}$, and consider the equations

$$\int_{\gamma_i} \sum_{j=1}^{g+1} (\lambda^j \omega_j + \eta^j \bar{\omega}_j) = 0 \quad (i = 1, 2, \ldots, 2g)\,,$$

where γ_i form a $\mathbb{Z}$-basis for $H_1(C, \mathbb{Z})$. Considering $\lambda^1, \ldots, \lambda^{g+1}, \eta^1, \ldots, \eta^{g+1}$ as unknowns, the above equations are just $2g$ linear equations in these $(2g+2)$

unknowns. This system of equations must have nontrivial solutions, and let us suppose that one is $\lambda_0^1, \ldots, \lambda_0^{g+1}, \eta_0^1, \ldots, \eta_0^{g+1}$. Then, choosing

$$\omega = \sum_{j=1}^{g+1} \lambda_0^j \omega_j \in \Omega^1(X), \quad \varphi = \sum_{j=1}^{g+1} \overline{\eta_0^j} \omega_j \in \Omega^1(X),$$

we get for all $i = 1, 2, \ldots, 2g$,

$$\int_{\gamma_t} (\omega + \bar{\varphi}) = \int_{\gamma_t} \sum_{j=1}^{g+1} \left(\lambda_0^j \omega_j + \eta_0^j \omega_j \right) = 0$$

From Proposition 5.15.5, there exists a X^∞ function f on X such that

$$\omega + \bar{\varphi} = df$$

and from Proposition 5.15.6

$$\omega = \varphi = 0 .$$

Since $\lambda_0^1, \ldots, \lambda_0^{g+1}, \eta_0^1, \ldots, \eta_0^{g+1}$ are not all zero, this contradicts the linear independence of $\omega_1, \ldots, \omega_{g+1}$. □

5.16 Every Riemann Surface Admits a Non-Constant Meromorphic Function

Every non-compact Riemann surface X admits non-constant holomorphic functions with values in $\mathbb{C}$. A theorem of Behnke and Stein [12] asserts that X is a Stein manifold.

In contrast, on a compact Riemann surface X, every holomorphic function with values in $\mathbb{C}$ is constant by the maximum principle.

If we look at Riemann's construction of a Riemann surface X as a multi-sheeted branched covering of the extended complex plane $\widehat{\mathbb{C}}$, then the covering mapping $\pi : X \to \widehat{\mathbb{C}}$ is indeed a meromorphic function on X. Riemann asked whether there were any other meromorphic functions on X.

For an arbitrary Riemann surface, the existence of a non-constant meromorphic function is not evident. As observed by Riemann and Klein and rigorously proved by Weyl, every connected Riemann surface carries many such functions, and the compact ones are algebraic curves.

Recall that in Hodge's theorem (see Subsection 5.14.3) we gave a nice decomposition of the closure of the set of 1-forms with compact support.

We shall now derive some consequences from Hodge's theorem. More precisely, we answer the fundamental question: does a general Riemann surface carry a nonconstant meromorphic function? It is essential that we ask for meromorphic functions rather than holomorphic ones: on compact surfaces the latter ones are trivial.

We will use that decomposition to prove a remarkable, non-trivial theorem; namely, that any Riemann surface admits a non-constant meromorphic function to $\mathbb{C}$.

It is important to note that the basis for our proof of this theorem is the answer to the following question: Given $x \in M$, can we find a u harmonic on $M \setminus \{x\}$ that has a $\frac{1}{z}$ or $\log|z|$ singularity in some parametric disc centered at x?

We will answer this question using Hodge's theorem. Let us begin with the following theorem.

Theorem 5.16.1. *Let $\overline{N} \subset M, \overline{N}$ compact with smooth boundary. Fix $p_0 \in N$ and h harmonic on $N \setminus \{p_0\}$ with $h \in C^1(\overline{N})$ and $\frac{\partial h}{\partial n} = 0$ on ∂N where n is some normal vector field on ∂N that never vanishes.*

Then there exists u harmonic in $M\{p_0\}$, $u - h$ harmonic on N, and $u \in \Omega(M \setminus K)$ for any compact neighborhood K of p_0. Moreover, u is unique up to constants.

Proof. We begin by showing existence. Choose a smooth function θ on N which agrees with h on $N \setminus K$ where K is arbitrary but fixed compact neighborhood of p_0. Then we extend θ to M by letting it equal 0 outside N. Since θ is a smooth function with compact support we can use Hodge's theorem to decompose it.

$$d\theta = g + \alpha + \beta, \text{ where } g \in h_2(M; \mathbb{C}), \alpha \in E, \beta \in *E$$

If we take a $\phi \in C^\infty_{\text{comp}}(M)$, then

$$\langle d\theta, d\phi \rangle = \langle \alpha, d\phi \rangle .$$

Note that since $\beta + g \in E^\perp$ and $\alpha \in E$,

$$\langle \beta, d\phi \rangle = \langle \alpha, *d\phi \rangle = 0 .$$

If $\text{supp}(\phi) \subset M \setminus K$. Then from $\frac{\partial h}{\partial n} = 0$ and $d * dh = 0$ on N, we obtain that

$$\langle d\phi, d\theta \rangle = \int_N d\phi \wedge *d\overline{h} = \int_{\partial N} \phi \overline{v^*(*dh)} = 0 .$$

Here $v : \partial N \to M$ is the inclusion map on the boundary and v^* is the pullback. The main point to note here is that $v^*(*dh)$ is proportional to $\frac{\partial h}{\partial n} = 0$. Therefore α is harmonic on $M \setminus K$ since it is perpendicular to both $d\phi$ and $*d\phi$. If $\text{supp}(\phi) \subset N$, then

$$\langle d\theta - \alpha, d\phi \rangle = 0, \quad \langle d\theta - \alpha, *d\phi \rangle = 0 .$$

Thus, $\alpha - d\theta$ is harmonic on N. In particular, α is smooth on M and thus $\alpha = df$ with f smooth. Now we set

$$u = f - \theta + h$$

To show existence, we suppose there exists v that has the same properties as u. Then $u - v$ would be harmonic on M and $d(u - v) \in \Omega_2^1(M; \mathbb{C})$. In conclusion, $d(u - v) \in E \cap h = \{0\}$, so $u - v$ is constant. □

We note that the same exact proof allows for several exceptional points $x_0, x_1, \ldots, x_k \in N$. Keep in mind that we still need to find a harmonic function that has either a $\frac{1}{z}$ or $\log|z|$ singularity. The following corollaries are critical to finding such a harmonic function.

Corollary 5.16.2. *Given $n \geq 1$ and a coordinate chart (D, ϕ) around $x \in M$ with $\phi(x) = 0$ there is u harmonic on $M \setminus \{x\}$ with $u - z^{-n}$ harmonic on U and $du \in \Omega_2^1(M - K)$ for any compact neighborhood K of x.*

Proof. Without loss of generality we can take $\phi(D) = \mathbb{D}$ and have $\phi(x) = 0$ by the Riemann mapping theorem. We define $h(z) = z^{-n} + z^n$ for every $z \in \mathbb{D}$. Take $N = \overline{D}$. Now h is harmonic on N. By the preceding theorem we have that there exists a harmonic ω on $D \setminus \{x\}$ with $\omega - h$ harmonic on D. We see that this implies that $\omega - z^{-n}$ is harmonic on D since $\omega - h$ effectively gets rid of the z^{-n} term. □

Corollary 5.16.3. *Let $x_0, x_1 \in M$ be distinct and suppose z and λ are local coordinates around x_0 and x_1, respectively. Then there exists u harmonic on $M \setminus \{x_0, x_1\}$ with $u - \log|z|$ and $u + \log|\lambda|$ harmonic locally around x_0, x_1 respectively. Moreover, $du \in \Omega_2^1(M \setminus K)$ where K is any compact neighborhood of $\{x_0, x_1\}$.*

Proof. For the existence part, take θ a X^∞ function on N which agrees with h on $N \setminus K$ where K is an arbitrary but fixed (small) compact neighborhood of p_0. Then extend θ to M simply by setting it $= 0$ outside of N and define $d\theta = 0$ on $M \setminus N$ (see Fig. 5.19).

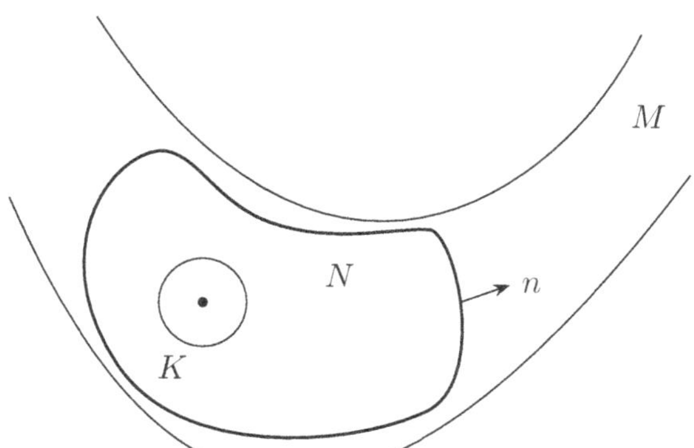

FIGURE 5.19: The sets considered in the proof.

Note that $d\theta \in \Omega_2^1(M; \mathbb{C})$. By Hodge,

$$d\theta = \alpha + \beta, \quad \alpha \in E, \beta \in E^\perp$$

If $\phi \in C^{\infty}_{\mathrm{comp}}(M)$, then

$$\langle d\theta, d\phi\rangle = \langle \alpha, d\phi\rangle, \quad \langle \alpha, *d\phi\rangle = 0$$

First, suppose that $\mathrm{supp}(\phi) \subset M\backslash K$. Then from $\frac{\partial h}{\partial n} = 0$ and $d * dh = 0$ on N, we obtain that

$$\langle d\phi, d\theta\rangle = \int_N d\phi \wedge *d\overline{h} = \int_{\partial N} \phi \overline{i_*(*dh)} = 0$$

where $i : \partial N \hookrightarrow M$ is the inclusion map. The main point to note here is that $i_*(*dh)$ is proportional to $\frac{\partial h}{\partial n} = 0$.

Hence, α is harmonic on $M\backslash K$. On the other hand, if $\mathrm{supp}(\phi) \subset N$, then

$$\langle d\theta - \alpha, d\phi\rangle = 0, \quad \langle d\theta - \alpha, *d\phi\rangle = 0$$

so that $\alpha - d\theta$ is harmonic on N. In particular, α is smooth on M and thus $\alpha = df$ with f smooth. Now set

$$u = f - \theta + h$$

By inspection, u has all the desired properties. Finally, if v had the same properties as u, then $u - v$ would be harmonic on M and $d(u - v) \in \Omega^1_2$. In conclusion, $d(u - v) \in E \cap \mathfrak{h}_2 = \{0\}$, so $u - v = \text{const}$. □

Corollary 5.16.4. *Given a Riemann surface M:*

(1) For $n \geq 1$ and $x_0 \in M$ there exists a meromorphic differential ω with $\omega - \frac{dz}{z^{n+1}}$ holomorphic locally around x_0. Moreover, $\omega \in \Omega^1_2(M \setminus K)$ for every compact neighborhood K of x_0.

(2) Let $x_0, x_1 \in M$.

There exists ω meromorphic on M with $\omega - \frac{dz}{z}$ holomorphic around x_0 and $\omega + \frac{d\lambda}{\lambda}$ holomorphic around x_1, respectively (z, λ are local coordinates). Moreover, $\omega \in \Omega^1_2(M \setminus K)$ for every compact neighborhood K of $\{x_0, x_1\}$.

Proof. With u in Corollary 5.16.3 and 5.16.4 respectively, we set $\alpha = du$. For (1) we take $\omega = \frac{-1}{2n}(\alpha + i * \alpha)$, whereas for (2), we take $\omega = \alpha + i * \alpha$. □

This leads us to the following theorem

Theorem 5.16.5. *Let $\{x_j\}_{j=1}^J \subset M$, $J \geq 2$, and $c_j \in \mathbb{C}$ with $\sum_{j=1}^J c_j = 0$. Then there exists a meromorphic differential ω, holomorphic on*

$$M \setminus \{x_1, x_2, \ldots, x_J\}$$

so that ω has a simple pole at each x_j with residue c_j.

Proof. Pick any other point $x_0 \in X$ and let ω_j be meromorphic with simple poles at x_0, x_j and residues $-c_j, c_j$ respectively. We know that it will have residues of this form by Corollary 5.16.4. The differential $\omega = \sum_{j=1}^{J} \omega_j$ has the desired properties. □

We now have all the tools necessary to prove that every Riemann surface admits a non-constant meromorphic function.

Theorem 5.16.6. *Every compact connected Riemann surface X carries a non-constant meromorphic function.*

Proof. Take three points $x_0, x_1, x_2 \in M$ and let ω_1 be a meromorphic 1-form with simple poles at x_0, x_1 and residues $1, -1$ respectively and holomorphic everywhere else. Similarly, let ω_2 be a meromorphic 1-form with simple poles at x_1, x_2 and residues $-1, 1$ respectively and holomorphic everywhere else. We cant do this by Theorem 5.16.6. Now set $f = \frac{\omega_1}{\omega_2}$ where the division is well-defined in local coordinates and this defines a meromorphic function. Since ω_1 and ω_2 both have poles at x_1 with residues -1 we see that $f(x_1) = 1$. At x_2, ω_1 is fixed, whereas ω_2 has a simple pole. Thus, $f(x_2) = 0$ and f is not constant. □

Equivalently, every compact Riemann surface can be made into a branched cover of $\mathbb{CP}^1$.

Theorem 5.16.6 was proved rigorously by methods closely related with the ones developed to prove the uniformization theorem, and, at the same time as this last theorem, has been at the origin of many developments in analysis around the turn of the last century.

In the proof of the next proposition we will assume that compact Riemann surfaces have non-constant meromorphic functions.

Proposition 5.16.7. *Compact Riemann surfaces are triangulable.*

Proof. Ley $f :\to \widehat{\mathbb{C}}$ be a non-constant meromorphic function and let S denote the set of critical points of f,

$$S = \{p \in X; b_p(f) > 0\} .$$

Since X is compact S is a finite set. Let S' be the set of critical values, $S' = f(S)$. If $\mathcal{T}$ is any triangulation of $\widehat{\mathbb{C}}$ we can subdivide it to obtain another triangulation, say $\mathcal{T}'$, such that all points of S are vertices of $\mathcal{T}'$, and at most one vertex of any triangle of $\mathcal{T}'$ lies in S'. The triangles of $\mathcal{T}'$ lift via f to triangles in X, i.e., if T is a triangle of $\mathcal{T}'$ then $f^{-1}(T)$ is a disjoint union of triangles in X. This follows from the fact that $f : X \backslash f^{-1}(S) \to Y \backslash S'$ is a covering map and the interiors of the triangles are simply connected. Similarly, the (open) edges lift to edges. Thus we can lift $\mathcal{T}'$ to a triangulation on X. □

Exercises

1. Let X be a compact Riemann surface. We know that it has a non-constant meromorphic function f on it. The goal of this problem is to use f to prove that X has a triangulation. The rough idea is to find a triangulation of $\widehat{\mathbb{C}}$ that lifts to X. Since X is compact, it only has a finite number of points with ramification number greater than one. Suppose $U \subset \widehat{\mathbb{C}}$ is open such that f^{-1} does not contain any ramified points. Let n be the degree of f. Then $f^{-1}(U)$ is the disjoint union of n open sets, which are homeomorphic to each other. A triangulation of U lifts to triangulations on these open sets. Show that if we take a triangulation of $\widehat{\mathbb{C}}$ such that every point with ramification number greater than one gets mapped by f to a vertex of the triangulation and the triangles are "small enough," then this triangulation lifts to a triangulation of X.

6

The Riemann–Roch Theorem

A compact Riemann surface is a one-dimensional compact complex manifold; that is, a manifold on which there exist no nonconstant globally defined holomorphic functions, although globally defined meromorphic functions do exist.

This naturally leads to the question of determining the dimension of the space of meromorphic functions on such a surface. The answer to this question is provided by the Riemann–Roch theorem, which establishes a fundamental relationship between the geometry of the surface and the analytical properties of functions defined on it.

In contrast to the non-compact example of a Riemann surface, namely the complex plane, it is not possible, in general, to construct a meromorphic function possessing singularities at arbitrary points of a given Riemann surface. The precise conditions under which this can be achieved are described by Abel's theorem.

This section is by no means intended to be self-contained. Numerous excellent books and online resources present these topics with far greater rigor and completeness. The only difficulty lies in the fact that it is often unclear who should be credited with particular proofs, as many of them appear—sometimes verbatim—across various materials available on the internet.

The Riemann–Roch theorem is an important theorem in mathematics, specifically in complex analysis and algebraic geometry, for the computation of the dimension of the linear subspace of meromorphic functions with prescribed zeros and allowed poles. It relates the complex analysis of a connected compact Riemann surface with the surface's purely topological genus g, in a way that can be carried over into purely algebraic settings.

There are some different ways to prove the Riemann–Roch theorem. One way is to take an analytic approach and study holomorphic and meromorphic functions together with using Serre's duality, which is a transition from analysis to algebraic geometry with an isomorphism. This way to prove it can be found in Rick Miranda's book "Algebraic curves and Riemann surfaces," chapter VI [129]. Then there is another, more modern approach, which includes the concept schemes and more importantly uses sheaf cohomology. This can for example be found in *Algebraic Geometry* by Robin Hartshorne, chapter V [79].

Riemann surfaces are one-dimensional (connected) complex manifolds. It is interesting to understand the class of meromorphic functions defined on them. The Riemann–Roch theorem relates a linear subspace of meromorphic

DOI: 10.1201/9781003713005-6

functions on a Riemann surface to the genus of the surface. As a remarkable consequence we get that every compact Riemann surface of genus 0 is conformally isomorphic only to the Riemann sphere.

Perhaps a brief introduction to the subject, as given by Lefschetz, would be more appropriate here. In a course on Riemann surfaces, Lefschetz began with a rapid sequence of statements, without writing anything on the blackboard:

> "Well, a Riemann surface is a certain kind of Hausdorff space. You know what a Hausdorff space is, don't you? It's also compact, OK. I suppose it is also a manifold. Surely you know what a manifold is. Now let me tell you one nontrivial theorem: the Riemann–Roch Theorem."

6.1 Introduction

It is well known that a compact Riemann surface X admits no nonconstant holomorphic functions. Suppose instead that we allow a single pole at a fixed point $p \in X$, with no other poles. In this case, the only possibility occurs when $X \cong \mathbb{CP}^1$, where the function $(z - z(p))^{-1}$ provides an example. If $g = \text{genus}(X) \geq 1$, then there exists a nonzero holomorphic 1-form ω that does not vanish at p. For any meromorphic function f on X, one has

$$\sum_{q \in X} \text{Res}_q(f\omega) = 0 .$$

If f has a simple pole at p, then $\text{Res}_p(f\omega) \neq 0$, which forces the existence of an additional pole, a contradiction.

If instead one allows a double pole at p (and no other poles), then the situation changes. For $g = 0$ or $g = 1$, there exist nonconstant examples; for instance, the Weierstrass $\wp$-function on an elliptic curve. For $g \geq 2$, the existence of such functions depends in general on the choice of p. More generally, the Riemann–Roch theorem shows that the space of meromorphic functions $f \in \mathcal{M}(X)$ with a unique pole at p of order at most k has dimension at least

$$\max\{1,\, k - g + 1\} .$$

In particular, nonconstant functions exist whenever $k - g + 1 \geq 2$.

In the 1850s, Riemann established a more general inequality in which the single point p (and order k) is replaced by an arbitrary divisor involving multiple points and orders. About a decade later, his student Roch sharpened this to an exact equality, incorporating an additional term related to meromorphic 1-forms. This result subsumes the identity $\dim \Omega^1(X) = g$ and provides a powerful tool for studying embeddings of Riemann surfaces into higher-dimensional projective spaces, among other applications. Its formulation is expressed in terms of spaces of functions and forms associated with divisors; we begin in Section 6.8 by defining these spaces precisely.

6.2 Baby Proof of the Riemann–Roch Theorem

Even the most complicated theorem has very simple considerations behind it. In this subsection we will give an elementary "proof" of the Riemann–Roch theorem due to Harris which almost coincides with the original proof of Riemann. Next we will give a rigorous proof of the Riemann theorem.

Meromorphic functions are almost everywhere holomorphic, but allowed to have isolated poles at certain points.

We know that if no poles are allowed, then there are no non-constant functions on X. What happens if we allow a single pole? Or two? Or three? And so on.

The question of interest is: given a set of points p_i and integers $\alpha_i \in \mathbb{Z}$ (that is, a divisor $D = \sum_i \alpha_i \cdot p_i$), is it possible to find a meromorphic function (respectively, a meromorphic 1-form) with poles at p_i of order at most $-\alpha_i$ if $\alpha_i < 0$, and zeros of order at least α_i if $\alpha_i > 0$? Furthermore, how many linearly independent such functions (respectively, 1-forms) exist?

This leads to the basic question in the Riemann–Roch circle of ideas:

[The Riemann–Roch Problem (First version)]Determine the number of meromorphic functions on X with prescribed poles.

6.2.1 Vector Space $L(D)$

Let us organize our setup. Fix distinct points $p_1, \dots, p_d \in X$ (the assumption of distinctness is for convenience), and for notational purposes define

$$D = p_1 + \cdots + p_d ,$$

which is simply a formal sum of these points.

We want to describe the set of meromorphic functions having at worst simple poles at the p_i. Given a local coordinate z_i around each p_i, a function $f \in \mathcal{L}(D)$ can locally be expressed as

$$f = \frac{a_i}{z_i} + f_i ,$$

where f_i is holomorphic. Since there are no nonzero global holomorphic functions, f is determined up to an additive constant by its principal parts a_i/z_i.

We then define $L(D)$ to be the set of functions that are holomorphic on $X \setminus \{p_1, \dots, p_d\}$ and have at most a simple pole at each p_j. We also include the constant function 0, so that $L(D)$ becomes a complex vector space. Since this is a vector space, the "number" of such functions (as alluded to earlier) should be interpreted as its dimension, denoted by

$$\ell(D) = \dim L(D) .$$

6.2.2 The Riemann–Roch Problem

The Riemann–Roch problem may now be expressed as

> [The Riemann–Roch Problem (Second version)]Determine $\ell(D)$ in terms of data of the points themselves and the topology of X.

Since we feel that the choice of points should not effect the question too much, it seems reasonable to expect their contribution to be in terms of their number d. On the other hand, the topology of X is entirely determined, among topological surfaces, by the non-negative integer g, its genus. Thus we wish to describe $\ell(D)$ in terms of d and g.

Now we will assume that the function f is defined on a Riemann surface, and consider the mapping $f \mapsto (\mathrm{Res}_{x_0}(f), \ldots, \mathrm{Res}_{x_d}(f))$ defines[1] a clearly $\mathbb{C}$-linear map $\mathrm{Res} : L(D) \to \mathbb{C}^d$. Its kernel consists of those functions $f \in L(D)$ whose residues at all the points x_j are zero, i.e., which are holomorphic at all those points. Since a function in $L(D)$ is already holomorphic everywhere else, the kernel of Res is $\mathcal{O}(X) \simeq \mathbb{C}$ (see Corollary 6.3.14).

At this point we have the exact sequence

$$0 \to \mathbb{C} \to L(D) \xrightarrow{\mathrm{Res}} \mathbb{C}^d \,,$$

which however is not necessarily exact on the right—the map Res need not be surjective. That is to say, we have

$$\ell(D) = \dim \mathbb{C} + \dim \mathrm{Res}(L(D)) = 1 + \dim \mathrm{Res}(L(D))\,. \tag{6.1}$$

Now the problem has become determining the image $\mathrm{Res}(L(D))$ inside $\mathbb{C}^d$. We approach this by finding a system of linear constraints that its elements must satisfy.

6.3 The Residue Theorem on Compact Riemann Surfaces

The Residue theorem states that if f is a meromorphic function on a neighborhood of the closure $\overline{U}$ of a relatively compact open set $U \subseteq \mathbb{C}$, with poles $z_1, \ldots, z_n \in U$, then

$$\int_{\partial U} f(z)\, dz \;=\; 2\pi i \sum_{j=1}^{n} \mathrm{Res}_{z_j}(f)\,.$$

[1] We should note that this map depends on the choice of coordinates (z_j, U_j) around each point x_j. There is nothing inherently problematic about this, though it may feel somewhat inelegant. In fact, we will return to this issue later on.

It is natural to ask whether a similar argument applies to a Riemann surface X. One might attempt to decompose X into coordinate charts and observe that the integrals along the overlaps cancel, since each curve appears twice with opposite orientations.

Because X has no boundary, the left-hand side would vanish, suggesting that the sum of the residues is zero. This is exactly the sort of relation we seek: *a linear dependence among residues.*

Remark 6.3.1 (A Problem with This Idea). This line of reasoning is attractive but ultimately flawed: one cannot consistently define the integral of a holomorphic *function* along a curve on a Riemann surface.

The difficulty is the absence of a globally defined differential dz. Such a term must be derived from a local holomorphic coordinate z. In the complex plane $\mathbb{C}$—the setting in which complex analysis is typically introduced—one can fix such a coordinate globally, and thus freely switch between writing $f(z)$ and $f(z)\,dz$.

On a general Riemann surface X, however, no global coordinate exists, and hence no global dz. This prevents a consistent formulation of the construction above.

The resolution is straightforward: instead of working with *functions* $f : X \to \mathbb{C}$ that, in local charts (z, U), appear as meromorphic functions $f(z)$, one should work with meromorphic 1-forms.

6.3.1 Holomorphic and Meromorphic Differentials

In this section, we shall discuss differentials on a Riemann surface X.

Definition 6.3.2. A *holomorphic* 1*-form* on a Riemann surface X is a differential 1-form on X that can be written as $\omega = f dz$, where f is holomorphic and z is a local coordinate. A similar definition applies to *meromorphic* 1*-forms.*

Then a holomorphic 1-form (respectively, meromorphic 1-form) ω is by definition a family $\{(U_i, z_i, \omega_i)\}$ such that:

(a) $\{(U_i, z_i)\}$ is a holomorphic covering of X, and

$$\omega_i = f_i\left(z_i\right) dz_i$$

where $f_i \in \mathcal{O}\left(U_i\right)$ (resp. $\mathcal{M}\left(U_i\right)$);

(b) if $z_i = \varphi_{ij}\left(z_j\right)$ is the coordinate transformation on $U_i \cap U_j (\neq \varnothing)$, then

$$f_i\left(\varphi_{ij}\left(z_j\right)\right) \frac{d\varphi_{ij}\left(z_j\right)}{dz_j} = f_j\left(z_j\right)$$

i.e., the local representation of the differential form changes according to the chain rule

$$f_i\left(\varphi_{ij}\left(z_j\right)\right) d\varphi_{ij}\left(z_j\right) = f_j\left(z_j\right) dz_j.$$

We use $\Omega^1(X)$ to denote the set of holomorphic 1-forms on X, and we will use the notation $\mathcal{M}^1(X)$ for the space of meromorphic 1-form on X.

Thus $\mathcal{M}^1(X)$ is in fact a one-dimensional vector space over the field $\mathcal{M}(X)$ (here we use the fact that non-zero meromorphic 1-forms exist). In particular, the canonical divisor is unique up to linear equivalence.

Example 6.3.3 (holomorphic 1-forms on a complex torus). Take a complex torus $X_\tau = \mathbb{C}/\mathbb{Z} + \tau\mathbb{Z}$ as usual.

Then one can see that this induces a nonzero holomorphic form dz on X. Furthermore this is essentially the only one; indeed, if ω is a differential form on X then it must have the form

$$\omega = f(z)dz$$

with f holomorphic on $\mathbb{C}^1$ and such that $f(z+m+\tau n) = f(z)$, but any such f is constant. Hence

$$\Omega^1(X_\tau) = \mathbb{C} \cdot dz$$

Example 6.3.4 (holomorphic 1-forms on smooth plane curves). Suppose $X = \{f(x,y) = 0\}$ is a smooth plane curve and set $f_x = \frac{\partial f}{\partial x}$, $f_y = \frac{\partial f}{\partial y}$. On the set $U_1 = \{f_x \neq 0\}$ we can define the form

$$w_1 = \frac{1}{f_x} dy\,.$$

Can we extend this to a form also on the chart $U_2 = \{f_y \neq 0\}$? Morally we do it like this: since $f(x,y) = 0$ on X, we have that

$$f_x dx + f_y dy = 0$$

so $1/f_x dy = -1/f_y dx$ and this extends ω_1 to a form ω on the whole of X. As an exercise, make this rigorous.

Example 6.3.5 (Holomorphic 1-forms on $\mathbb{CP}^1$). Take the usual charts U_0, U_1 of $\mathbb{CP}^1$ with affine coordinates w and z respectively, such that $z = 1/w$ on $U_0 \cap U_1$. Then a holomorphic 1-form is

$$\begin{aligned} \omega_0 &= f_0(w)dw \quad (f_0 \text{ holomorphic }) \\ \omega_1 &= f_1(z)dz \quad\;\; (f_1 \text{ holomorphic }) \end{aligned}$$

such that $\omega_0|_{U_0\cap U_1} = \omega_1|U_0 \cap U_1$. On this intersection we see that

$$\begin{aligned} \omega_0|_{U_0\cap U_1} &= f_0(w)dx = f_0\left(\frac{1}{z}\right) d\left(\frac{1}{z}\right) \\ &= -\frac{f_0(1/z)}{z^2}dz\,. \end{aligned}$$

So the condition is that $f_1(z) = -\frac{1}{z^2} \cdot f_0\left(\frac{1}{z}\right)$ However this is impossible because the expression on the right has always a pole at $z = 0$. Hence, on $\mathbb{CP}^1$ the only holomorphic differential form is the zero form. Hence we have $\Omega^1(X) = 0$.

Proposition 6.3.6. *Let ω_1 and ω_2 be two meromorphic 1-forms on a Riemann surface X, with ω_1 not identically zero. Then there is a unique meromorphic function f on X with $\omega_2 = f\omega_1$.*

Proof. Choose a chart $\varphi : U \to V$ on X giving local coordinate z. Write $\omega_i = g_i(z)dz$ for meromorphic functions g_i on V. Let $h = g_2/g_1$ be the ratio of these functions, which is well-defined as a function of the common local coordinate z of g_1 and g_2. In particular, h is holomorphic whenever g_2 and g_1 are with $g_1 \neq 0$. At all other points h may have a pole, and it is easy to see that this is a discrete set as X is compact and the set of poles (and for g_1, zeroes) of g_1 and g_2 is discrete. Thus h is also a meromorphic function on V. Now define $f = h \circ \varphi$, a meromorphic function on U.

It is easy to check that f is well-defined, independent of the choice of coordinate chart. This is the desired function. □

Suppose X is a Riemann surface, with

$$f = \{(U_i, z_i, f_i(z_i))\} \in \mathcal{M}(X).$$

Then the following expression defines a meromorphic differential form $df \in \mathcal{M}^{(1)}(X)$:

$$df = \left\{ \left(U_i, z_i, df_i(z_i) = \frac{df_i(z_i)}{dz_i} dz_i \right) \right\}.$$

Definition 6.3.7. We call df, as defined above, the *differential of the meromorphic function f.*

Definition 6.3.8. Suppose X is a Riemann surface with

$$\omega = \{(U_i, z_i, f_i(z_i)\, dz_i)\} \in \mathcal{M}^1(X), \quad p \in U_i \cap U_j \ .$$

Then

$$\nu_p(f_i) = \nu_p\left(f_i(\varphi_{ij}(z_j)) \frac{d\varphi_{ij}(z_j)}{dz_j} \right) = \nu_p(f_j)$$

and thus we can define

$$\nu_p(\omega) = \nu_p(f_i), \quad p \in U_i \ .$$

If $\nu_p(\omega) > 0$, then p is called a *zero of* ω; if $\nu_q(\omega) < 0$, then q is called a *pole* ω.

Definition 6.3.9. Suppose X is a Riemann surface with

$$\omega = \{(U_i, z_i, f_i(z_j)\, dz_i)\} \in \mathcal{M}^1(X),$$

γ is a piecewise smooth curve on X not containing the poles of ω, and $\gamma = \bigcup_i \gamma_i$ is any partition of γ satisfying $\gamma_i \subset U_i$. We define the integral

$$\int_\gamma \omega = \sum_i \int_{\gamma_1} f_i(z_i)\, dz_i \ .$$

This idea is well defined because suppose $\gamma = \bigcup_i \gamma_i'$ were another partition of γ_i satisfying $\gamma_i' \subset U_i$. Then by the change of variables formula:

$$\int_{\gamma_1 \cap \gamma_j'} f_i(z_i)\, dz_i = \int_{\gamma_i \cap \gamma_j'} f_j(z_j)\, dz_j\,,$$

so that

$$\begin{aligned}\sum_i \int_{\gamma_t} f_i(z_i)\, dz_i &= \sum_i \sum_j \int_{\gamma_1 \cap \gamma_j'} f_i(z_i)\, dz_i \\ &= \sum_j \sum_i \int_{\gamma_i \cap \gamma_j'} f_j(z_j)\, dz_j \\ &= \sum_j \int_{\gamma_j'} f_j(z_j)\, dz_j\,.\end{aligned}$$

Theorem 6.3.10 (Stokes's theorem for holomorphic differential forms)**.** *Suppose X is a Riemann surface with $D \subset X$ an open set, $\overline{D}$ compact, $\partial D = \gamma$ a piecewise smooth curve, and ω a holomorphic differential defined on an open set containing $\overline{D}$. Then*

$$\int_{\partial D} \omega = 0\ .$$

Proof. Suitably subdivide D into a disjoint union $D = \bigcup_i D_i$ such that for each i,

$$D_i \subset U_i$$

and ∂D_i a piecewise smooth curve. By using local coordinate representations and applying Cauchy's theorems, we get

$$\int_{\partial D_t} \omega = 0$$

whence

$$\int_{\partial D} \omega = \sum_i \int_{\partial D_i} \omega = 0$$

since the contributions of the boundaries ∂D_i cancel out other than those arcs contributing to ∂D. □

Definition 6.3.11. Let X be a Riemann surface with $\omega \in \mathcal{M}^1(X)$, $p \in X$, γ_p a small circle around the point p, and ω having no poles other than p on the disc surrounded by γ_p (p itself may or may not be a pole). Then we define the *residue* of the point p of ω to be

$$\mathrm{Res}_p(\omega) = \frac{1}{2\pi i} \oint_{\gamma_p} \omega\,.$$

From Stokes's theorem (Theorem 6.3.10), this definition is independent of the choice of the small circle γ_p. Moreover, for $p \in U_j, \gamma_p \subset U_j$, if $\omega = f_j(z_j)\, dz_j$ in U_j, then

$$\begin{aligned}\operatorname{Res}_p(\omega) &= \frac{1}{2\pi i}\oint_{\gamma_p} \omega = \frac{1}{2\pi i}\oint_{\gamma_p} f_j(z_j)\, dz_j \\ &= \operatorname{Res}_p\left(f_j(z_j)\, dz_j\right) .\end{aligned}$$

Theorem 6.3.12 (residue theorem)**.** *Suppose X is a compact Riemann surface. For $\omega \in \mathcal{M}^1(X)$, we have*

$$\sum_{p \in X} \operatorname{Res}_p(\omega) = 0 \ .$$

Proof. Since X is compact, ω can have only a finite number of poles on X, $p_1, p_2, \ldots, p_m$. Now choose mutually disjoint small discs $\Delta_1, \Delta_2, \ldots, \Delta_m$ so that each contains a distinct p_i and each satisfies the conditions in Definition 6.3.11. Write

$$D = X \backslash \bigcup_i \Delta_i$$

Clearly, for a suitably chosen orientation, we have

$$\partial D = -\bigcup_i \partial \Delta_i$$

and from Stokes' theorem (Theorem 6.3.10), we obtain

$$\begin{aligned}2\pi i \sum_{p \in X} \operatorname{Res}_p(\omega) &= 2\pi i \sum_{i=1}^{m} \operatorname{Res}_{p_i}(\omega) \\ &= \sum_{i=1}^{m} \int_{\partial \Delta_i} \omega \\ &= -\int_{\partial D} \omega = 0 .\end{aligned}$$

□

Theorem 6.3.13. *Let X be a compact Riemann surface. If $f \in \mathcal{M}(X)$ is not a constant function, then*

$$\sum_{p \in X} \nu_p(f) = 0 .$$

This implies in particular that the number of zeroes of f is equal to the number of poles of f (counting multiplicity):

$$\#\{\textit{zeroes of } f\} = \#\{\textit{poles of } f\} .$$

Proof. Choose $\omega = df/f$ and apply the above residue theorem. □

Corollary 6.3.14. *The only holomorphic functions on a compact Riemann surface X are the constant functions. This implies that*

$$\mathcal{O}(X) = \mathbb{C}, \quad \dim \mathcal{O}(X) = 1\,.$$

Proof. Let f a holomorphic function. Let $p_0 \in X$ a given generic point, define the function $g(p) = f(p) - f(p_0)$. This function has no pole and has at least one zero. This would contradict Theorem 6.3.13, unless g is identically vanishing, i.e., f is constant. □

Corollary 6.3.15. *If $f \in \mathcal{M}(X)$ is not constant, then for any $a \in X$, we have*

$$\# f^{-1}(a) = \#\{\textit{poles of } f\}\,.$$

Here, $\# f^{-1}(a)$ is the number of points p such that $f(p) = a$, counting each such point with suitable multiplicity.

This says that any complex number is a functional value of f, and every functional value is assumed the same number of times (counting multiplicity).

Remark 6.3.16. If we think of $f \in \mathcal{M}(X)$ as being a holomorphic mapping from X onto the Riemann sphere S, then the foregoing conclusion can be interpreted as follows: any point on S is an image point of f, and the number of inverse image points of each $p \in X$ does not depend on p (counting multiplicity).

Meromorphic are functions almost everywhere holomorphic, but which are allowed to have isolated poles at certain points.

Corollary 6.3.17 (more than one simple pole)**.** *There is no meromorphic 1-form with only one simple pole. It has either several simple poles, or poles of higher orders, or no pole at all.*

Theorem 6.3.18. *If $g \geq 1$, there is no meromorphic function with only one simple pole.*

Proof. If f is a meromorphic function with 1 simple pole p. Then, consider a holomorphic 1-form ω (it is possible if $g > 0$, we anticipate on the next section). Assume that ω either doesn't vanish at p $(k = 0)$ or has a zero of order k at p. Then $f^{k+1}\omega$ is a meromorphic form, and has only 1 simple pole at p, which is impossible. □

Theorem 6.3.19. *Let ω a meromorphic 1-form not identically vanishing. Then*

$$\deg(\omega) = 2g - 2\,,$$

where g is the genus of X. In other words

$$\#\ \textit{zeros} - \#\ \textit{poles} = 2g - 2\,.$$

Proof. The ratio of 2 meromorphic 1-forms is a meromorphic function, for which # zeros = # poles, therefore # zeros − # poles is the same for all 1-forms.

In particular, let us assume that there exist some non-constant meromorphic function $f : X \to \mathbb{CP}^1$, then $\omega = df$ is an exact meromorphic 1-form.

Let Z be the set of zeros of $\omega = df$, and P the set of poles. Choose an arbitrary cellular (all faces are homeomorphic to discs) graph Γ on $\mathbb{CP}^1$, whose vertices are the points of $f(Z) \cup \{\infty\}$.

Let F the number of faces, E the number of edges and V the number of vertices. Let d the degree of f, i.e., the sum of degrees of all its poles, which is also $d = \# f^{-1}(x)$ for x in a neighborhood of ∞, and thus is the number of preimages of generic x.

The Riemann sphere has Euler characteristic 2:

$$\chi\left(\mathbb{CP}^1\right) = 2 = F - E + V$$

Now consider the graph $\Gamma' = f^{-1}(\Gamma)$ on X. Since the faces of Γ contain no zero of df, then the preimages of each face is homeomorphic to a disc too, so that Γ' is a cellular graph on X. Its Euler characteristic is

$$\chi(X) = 2 - 2g = F' - E' + V'$$

The number of faces $F' = dF$ and edges $E' = dE$ because they are made of generic points. The points of Z and P are by definition not generic. For $x \in f(Z)$ we have

$$\# f^{-1}(x) = d - \sum_{r \in f^{-1}(x)} \operatorname{ord}_r df$$

Their sum is

$$\sum_{x \in f(R)} \# f^{-1}(x) = d(V-1) - \deg(df)_+$$

where $(df)_+$ is the divisor of zeros of df. Similarly for $x = \infty$ we have

$$\# f^{-1}(\infty) = \sum_{p \in P} (-1 - \operatorname{ord}_p df)$$

This implies

$$V' = d(V-1) - d - \deg(df)$$

Putting all together we get

$$\deg(df) = 2g - 2$$

Since $\deg(\omega)$ is the same for all 1-forms, it must be the same as for the exact form df and the theorem is proved.

In fact all what remains to prove is the existence of at least one non-constant meromorphic function. For Riemann surfaces coming from an algebraic equation $P(x, y) = 0$, the function $(x, y) \mapsto x$ can play this role. More generally the existence of non-constant meromorphic functions will be established below (and won't use this theorem). □

One key divisor on a compact Riemann surface is the *canonical divisor*. To define it, one first defines the divisor of a nonzero meromorphic 1-form along the lines above. Since the space of meromorphic 1-forms is a 1-dimensional vector space over the field of meromorphic functions, any two nonzero meromorphic 1-forms yield linearly equivalent divisors. Any divisor in this linear equivalence class is called the *canonical divisor* of X, denoted K_X. The genus g of X can be read from the canonical divisor: namely, K_X has degree $2g - 2$.

The key trichotomy among compact Riemann surfaces X is whether the canonical divisor has negative degree (so X has genus zero), zero degree (genus one), or positive degree (genus at least 2).

For example, this determines whether X has a Kähler metric with positive curvature, zero curvature, or negative curvature. The canonical divisor has negative degree if and only if X is isomorphic to the Riemann sphere $\mathbb{CP}^1$.

6.3.2 Linear Constraints on $\mathrm{Res}(L(D))$

After we have determined the residue of a meromorphic function defined on a Riemann surface (Definition 6.3.11), we will return to the implementation of the program announced in Section 6.3.

Recall that the set $L(D)$ consists of functions that are holomorphic away from the points x_j, and are allowed to have at most a simple pole at each point x_j.

Thus to estimate the image of Res, following the program of Section 6.3, let's fix an arbitrary non-zero holomorphic 1-form ω on X. For any non-zero $f \in L(D)$, the product $f\omega$ is then a meromorphic 1-form on X with poles only possible at the points x_i, and so the residue theorem shows that

$$\mathrm{Res}_{x_1}(f\omega) + \cdots + \mathrm{Res}_{x_d}(f\omega) = 0\,.$$

In holomorphic local coordinates (z_j, U_j) around each point x_j, suppose you write $\omega|_{U_j} = g_j dz_j$ for a holomorphic function $g_j \in \mathcal{O}(U_j)$. Then you have

$$\mathrm{Res}_{x_j}(f\omega) = \mathrm{Res}_{x_j}(f)g_j(x_j)$$

and consequently the above residue theorem relation becomes

$$\sum_{j=1}^{d} \mathrm{Res}_{x_j}(f)g_j(x_j) = 0\,.$$

Thus if $(a_1, \ldots, a_n) \in \mathbb{C}^d$ belongs to the image of the map Res, it must satisfy the linear relation

$$\sum_{j=1}^{d} a_j(f)g_j(x_j) = 0\,.$$

This was for a fixed holomorphic 1-form ω, and since the relation is clearly linear in ω, we get $\dim \Omega^1(X)$-many linear relations on the image of Res.

Here we have used $\Omega^1(X)$ to denote the vector space of holomorphic 1-forms on X. It is a basic truth of Riemann surface theory that $\dim \Omega^1(X) = g$, the genus of X. Indeed, that may (and in algebro-geometric contexts often is) taken as the definition of the genus.

6.4 Dependence Among the Linear Constraints

In the Subsection 6.3.2, we found a system of linear constraints on the elements of $\mathrm{Res}(L(D)) \subseteq \mathbb{C}^2$, one for each element $\omega \in \Omega^1(X)$.

But not all of these constraints will be independent: indeed, if $g_j(x_j) = 0$ for all $1 \le j \le d$, i.e., if ω vanishes at all the points x_j, then the relation imposed by ω is void. Let $\Omega^1_D(X)$ denote the vector subspace of $\Omega^1(X)$ consisting of all holomorphic 1-forms which vanish at all points x_j. Then the non-trivial linear constraints obtained in Subsection 6.3.2 are given by $\Omega^1(X)/\Omega^1_D(X)$.

These linear constraints are also sufficient[2], i.e., they precisely cut out the subset $\mathrm{Res}(L(D)) \subset \mathbb{C}^d$. It follows that

$$\begin{aligned}\dim \mathrm{Res}(L(D)) &= \dim \mathbb{C}^d - \#\text{ constraints}\\ &= d - g + \dim \Omega^1_D(X)\,.\end{aligned}$$

Combining this with the formula (6.1) we got before for $\ell(D)$ in terms of $\mathrm{Res}(L(D))$, we get the formula

$$\ell(D) = 1 + d - g + \dim \Omega^1_D(X)\,. \tag{6.2}$$

This is the Riemann–Roch formula! In many ways, it is the solution to the Riemann–Roch problem, as set out it Subsection 6.2.2—it expresses $\ell(D)$ in terms of d and g. We could perfectly well stop our discussion here.

6.5 Next Steps Toward the Riemann–Roch Theorem

Equality (6.2) is *not* the Riemann–Roch theorem as it appears in the modern literature. Moreover, this equality still contains the undetermined term $\dim \Omega^1_D(X)$.

6.5.1 Riemann's Inequality

Riemann himself had reached only this point in [157]. As in Section 6.4, he did not actually prove that the system of constraints is sufficient; instead, he

[2]This is the one point in this argument where we merely assert something, and not prove it. That is fine though, as shown in [78], it is not all that hard to make this into a real proof.

established the inequality

$$\ell(D) \geq 1 + d - g\,.$$

From the perspective of the first formulation of the Riemann–Roch problem in Subsection 6.2.2, this is quite natural. If one seeks to determine how many poles must be allowed before non-constant meromorphic functions appear, the inequality already provides the answer.

For example, Riemann's goal in [154] was to prove that every compact Riemann surface is algebraic. To this end, he required a sufficiently large supply of linearly independent meromorphic functions, from which to construct an embedding

$$X \longrightarrow \mathbb{CP}^n\,,$$

whose image he then showed to be defined by polynomial equations. The inequality supplied precisely the information he needed: namely, at how many points poles were required in order to produce the necessary functions.

6.5.2 Roch's Contribution

Riemann's student Roch, in [159], refined the inequality into an equality. He first proved that the system of constraints is indeed sufficient, and second, he identified $\Omega^1_D(X)$ with a more concrete object. To explain this identification, it is useful to discuss the generalizations of the Riemann–Roch theorem, through which the relevant ingredients arise naturally.

6.6 Higher Order Poles

Instead of restricting attention to meromorphic functions with simple poles, as in the preceding discussion, we now allow higher-order poles.

Fix points

$$p_1, \ldots, p_k \in X$$

together with integers

$$n_1, \ldots, n_k \in \mathbb{Z}_{\geq 0}\,,$$

and consider meromorphic functions on X that are holomorphic away from the points p_j and have poles of order at most n_j at each p_j ($j = 1, \ldots, k$). Collecting this data into the divisor

$$D = n_1 p_1 + \cdots + n_k p_k\,,$$

the corresponding space of meromorphic functions is the vector space $L(D)$.

The arguments of the previous sections carry over verbatim, with the exception of Section 6.3, which requires substantial modification. If f has a pole of order at most n at a point p, then with respect to a local holomorphic coordinate (z, U) centered at p, the Laurent expansion of f is

$$f|_U = \sum_{j=-n}^{\infty} a_j z^j .$$

The coefficients $a_{-n}, \ldots, a_{-1} \in \mathbb{C}$ encode the contribution of the pole.

In this setting the quantity $\deg D = n_1 + \cdots + n_k$, the degree of the divisor D, replaces the parameter d used earlier. In particular, in place of (6.2) the Riemann–Roch formula reads

$$\ell(D) = 1 + \deg D - g + \dim \Omega_D^1(X) .$$

Although the term $\dim \Omega_D^1(X)$ has not yet been identified with a more explicit object, this reformulation highlights the central role of divisors in the Riemann–Roch theorem. It is therefore natural to examine divisors in greater detail.

6.7 Divisors

Let X be a Riemann surface. A *divisor* on X is a map $D : X \to \mathbb{Z}$ such that for each compact subset $K \subseteq X$, the set of points $x \in K$ with $D(x) \neq 0$ is finite. The set $\operatorname{Div}(X)$ of all divisors on X forms an abelian group under addition. There is a natural partial ordering on $\operatorname{Div}(X)$: for $D_1, D_2 \in \operatorname{Div}(X)$, we set $D_1 \leq D_2$ if $D_1(x) \leq D_2(x)$ for all $x \in X$.

For any meromorphic function $f \in \mathcal{M}(X) \setminus \{0\}$, the map $x \mapsto \operatorname{ord}_x(f)$ is a divisor on X. It is called the *principal divisor* of f and is denoted by (f).

We say that f is a multiple of the divisor D if $(f) \geq D$. Note that f is holomorphic if and only if $(f) \geq 0$.

The divisor (ω) of a meromorphic 1-form $\omega \in \mathcal{M}^1(X) \setminus \{0\}$ is the map $x \mapsto \operatorname{ord}_x(\omega)$, where $\operatorname{ord}_x(\omega) = \operatorname{ord}_x(f)$ if $\omega = f\,dz$ in a local chart (U, z) at x.

For $f, g \in \mathcal{M}(X) \setminus \{0\}$ and $\omega \in \mathcal{M}^1(X) \setminus \{0\}$ we have

$$(fg) = (f) + (g), \quad (1/f) = -(f), \quad (f\omega) = (f) + (\omega) .$$

A divisor $D \in \operatorname{Div}(X)$ is called *principal* if there exists $f \in \mathcal{M}(X) \setminus \{0\}$ such that $D = (f)$. Two divisors D_1, D_2 are said to be *equivalent* if their difference $D_1 - D_2$ is principal.

A divisor $D \in \operatorname{Div}(X)$ is *canonical* if there exists $\omega \in \mathcal{M}^1(X) \setminus \{0\}$ such that $D = (\omega)$. Note that any two canonical divisors are equivalent. Indeed, for any $\omega_1, \omega_2 \in \mathcal{M}^1(X) \setminus \{0\}$ there exists $f \in \mathcal{M}(X) \setminus \{0\}$ with $\omega_1 = f\,\omega_2$, whence $(\omega_1) = (f) + (\omega_2)$.

6.7.1 The Degree of a Divisor

Let X be a compact Riemann surface. Then the degree, i.e., the map

$$\deg : \mathrm{Div}(X) \to \mathbb{Z}, \quad \deg D = \sum_{x \in X} D(x),$$

is well-defined. It is a group homomorphism. For every principal divisor (f) we have $\deg(f) = 0$, since there are as many zeros as poles on a compact Riemann surface. It follows that equivalent divisors have the same degree.

6.7.2 Degree of Principal Divisors

It may seem at first glance at the divisor class group $Cl(X)$ that it should be trivial. Indeed, given any finite collection of points in the complex plane $\mathbb{C}$, one may find a holomorphic function which vanishes there (or has poles, with arbitrary specified degree, etc.).

However, upon passing to a Riemann surface, the situation changes. Indeed, to convince ourselves that many divisors are not principal, note that the degree of any principal divisor is zero.

Definition 6.7.1. A *holomorphic* 1*-form* on a Riemann surface X is a differential 1-form on X that can be written as $\omega = f dz$, where f is holomorphic and z is a local coordinate. A similar definition applies to meromorphic 1-forms.

Definition 6.7.2. If ω is a meromorphic 1-form on a (not necessarily compact) Riemann surface X, and p is a point of X, we define the *order* of ω at p to be $\mathrm{ord}_p(\omega) = \mathrm{ord}_p(f_p)$, where f_p is a local representation of ω at the point p, i.e., ω is represented locally as $f_p dz$ in a neighborhood of p.

In a similar manner, we can associate a divisor to a meromorphic 1-form.

Theorem 6.7.3 (vanishing of the total residue). *For a meromorphic form ω on a compact Riemann surface,*

$$\sum_{p=poles} \mathrm{Res}_p\, \omega = 0\,.$$

Proof. Every Riemann surface X is triangulable (see Theorem 5.4.2). Hence there exists a graph Γ on X whose faces are triangles entirely contained in a single chart, and such that the poles of ω are not on Γ. By homotopic deformation, the sum of the residues inside each triangle equals the integral along its edges. Each edge is traversed twice, once in each direction, so the total sum of the integrals along the edges is zero. □

Corollary 6.7.4 (more than one simple pole). *There is no meromorphic* 1*-form with only one simple pole. It has either several simple poles, or poles of higher orders, or no pole at all.*

Theorem 6.7.5. *If $g \geq 1$, then there is no meromorphic function with only one simple pole.*

Proof. If f is a meromorphic function with 1 simple pole p. Then, consider a holomorphic 1-form ω (it is possible if $g > 0$, we anticipate on the next section). Assume that ω either doesn't vanish at $p(k = 0)$ or has a zero of order k at p. Then $f^{k+1}\omega$ is a meromorphic form, and has only 1 simple pole at p, which is impossible. □

Theorem 6.7.6 (meromorphic functions: # zeros = # poles)**.** *Let f a meromorphic function, not identically vanishing, then the number of poles (with multiplicity) equals the number of zeros:*

$$\# \textit{ zeros} - \# \textit{ poles} = \deg(f) = 0$$

Proof. Recall now the Argument Principle: if f is meromorphic on a (neighborhood of a) relatively compact domain $U \subseteq \mathbb{C}$, so that all its poles and zeros are contained inside U, then

$$\# \text{ zeros of } f - \# \text{ poles of } f = \frac{1}{2\pi i} \int_{\partial U} \frac{f'(z)}{f(z)} dz$$

where the left-hand side must be counted with multiplicities. The integrand on the right-hand side may be rewritten as $d(\log f)$. Covering X by such holomorphic local charts U, we obtain a global result valid on X. But since X does not have any boundary, which is to say, because the contributions from different ∂U cancel out, we conclude that # zeros of f − # poles of $f = 0$ for any non-zero meromorphic function f. Thus $\deg(f) = 0$. □

Theorem 6.7.7 (fundamental theorem of algebra)**.** *Any polynomial $p(z) = a_0 z^n + a_1 z^{n-1} + \cdots + a_n \in \mathbb{C}[z], a_0 \neq 0$, has n roots (counted according multiplicities).*

Proof. The polynomial $p \in \mathcal{M}(\widehat{\mathbb{C}})$ has a pole of order n at ∞. By Theorem 6.7.6, for $p(z)$ the number of poles (with multiplicity) equals the number of zeros. □

Theorem 6.7.8 (holomorphic function are constant)**.** *Any holomorphic function is constant. This implies that*

$$\mathcal{O}(X) = \mathbb{C} \quad , \quad \dim \mathcal{O}(X) = 1$$

Proof. Let f a holomorphic function. Let $p_0 \in X$ a given generic point, define the function $g(p) = f(p) - f(p_0)$. This function has no pole and has at least one zero. This would contradict Theorem 6.7.6, unless g is identically vanishing, i.e., f is constant. □

For a non-zero meromorphic function f on X, we have

$$\deg(f) = \#\text{ zeros of } f - \#\text{ poles of } f$$

counted with multiplicity.

It follows that the notion of the degree $\deg D$ of a divisor D indeed depends only on its divisor class.

It is obvious from the given definitions that if f is a meromorphic function and ω is a meromorphic 1-form, then $(f \cdot \omega) = (f) + (\omega)$. Due to additivity of deg we see that

$$\deg((f\omega)) = \deg((f)) + \deg((\omega)) = \deg((\omega)).$$

Since any meromorphic 1-form can be presented as any other meromorphic 1-form multiplied by some meromorphic function, we see that all canonical divisors have the same degree.

Proposition 6.7.9. *Let X be a compact Riemann surface. Meromorphic functions on X are determined by their divisors, up to a nonzero constant. So we can study them via their divisors.*

Proof. Let $f_1, f_2 \in \mathbb{C}(X)$ meromorphic functions such that $(f_1) = (f_2)$. Then $\left(\frac{f_1}{f_2}\right) = 0$. So $\frac{f_1}{f_2}$ have no poles, so it is holomorphic but since X is compact $\frac{f_1}{f_2} = \lambda \in \mathbb{C}^*$. □

Theorem 6.7.10 (Riemann sphere)**.** *There is no non-identically vanishing holomorphic 1-form on the Riemann sphere:*

$$\mathcal{O}^1\left(\mathbb{CP}^1\right) = \{0\} \quad , \quad \dim \mathcal{O}^1\left(\mathbb{CP}^1\right) = 0.$$

Proof. Consider the differential

$$\omega(z) = f(z)\, dz.$$

If we change coordinates to $z' = 1/z$, then since $dz = -z'^{-2}\, dz'$, we obtain

$$\omega(z) = -z'^{-2} f\left(\tfrac{1}{z'}\right) dz'.$$

For ω to be well-defined globally, it must be holomorphic in both coordinate charts.

Step 1. In the z-chart (the copy of $\mathbb{C}$), the requirement is that $f(z)$ has no poles. Therefore $f(z)$ must be a polynomial.

Step 2. In the z'-chart, we require that the coefficient

$$z'^{-2} f\left(\tfrac{1}{z'}\right)$$

have no pole at $z' = 0$.

Suppose $f(z)$ is a polynomial of degree d. Then

$$f\left(\tfrac{1}{z'}\right) = a_d z'^{-d} + a_{d-1} z'^{-(d-1)} + \cdots + a_0\,,$$

so multiplying by z'^{-2} gives

$$z'^{-2} f\left(\tfrac{1}{z'}\right) = a_d z'^{-(d+2)} + \cdots + a_0 z'^{-2}\,.$$

For this expression to have no pole at $z' = 0$, every exponent of z' must be nonnegative. This requires

$$-(d+2) \geq 0 \quad \Longrightarrow \quad d \leq -2\,.$$

Step 3. But a polynomial cannot have negative degree. The only possibility is the zero polynomial, which implies

$$\omega \equiv 0\,.$$

Therefore, there is no nonzero holomorphic differential of this form on the Riemann sphere. □

In fact this applies to every genus zero curve (but as we shall see later, every genus zero Riemann surface is isomorphic to the Riemann sphere):

Theorem 6.7.11 (genus zero). *There is no non-identically vanishing holomorphic* 1*-form on a curve* X *of genus* 0*:*

$$\mathcal{O}^1(X) = \{0\} \quad , \quad \dim \mathcal{O}^1(X) = 0\,.$$

Proof. Suppose ω is a holomorphic 1-form on X. Fix a base point $o \in X$, and define a function

$$f(p) := \int_o^p \omega, \qquad p \in X\,.$$

Since X is simply connected, the integral is independent of the path from o to p, and hence f is well defined. Moreover, f is holomorphic on X. By Theorem 6.7.8, every holomorphic function on X is constant, so f is constant. Differentiating gives $\omega = df = 0$, and therefore $\mathcal{O}^1(X) = \{0\}$. □

Proposition 6.7.12. *Every divisor of degree 0 on* $\mathbb{CP}^1$ *is principal.*

Proof. Write the divisor as $D = \sum_i n_i[p_i] + n_\infty[\infty]$. As $\deg D = 0$, $n_\infty = -\sum_i n_i$. Let $f = \prod_i (t-a)^{n_i}$. Then since $(t - p_i)$ is a local parameter at p_i and a unit at $b \neq p_i$, $\operatorname{ord}(f; p_i) = n_i$, and since $1/(t-p_i)$ is a local parameter at ∞ for any p_i, $v_\infty(f) = -\sum n_i = n_\infty$. □

Example 6.7.13. Let $X = \mathbb{CP}^1$ and let f be the function $\frac{z-a_1}{(z-a_2)^2}$. This function has a zero of order 1 at a_1 and another zero of order 1 at ∞. It has a pole of order 2 at a_2. The principal divisor (f) associated to this function is $a_1 + \infty - 2 \cdot a_2$.

6.7.3 The Sheaf $\mathcal{L}_D$

Let D be a divisor on a Riemann surface X. For any open subset $U \subseteq X$, let

$$\mathcal{L}_D(U) := \{f \in \mathcal{M}(U) : \operatorname{ord}_x(f) \geq -D(x) \text{ for all } x \in U\}$$

be the set of multiples of the divisor $-D$. Then $\mathcal{L}_D$ (together with the natural restriction maps) forms a sheaf on X. Note that $\mathcal{L}_0 = \mathcal{O}$. For equivalent divisors $D_1, D_2 \in \operatorname{Div}(X)$, the sheaves $\mathcal{L}_{D_1}$ and $\mathcal{L}_{D_2}$ are isomorphic. In fact, if $D_1 - D_2 = (g)$, then the map $\mathcal{L}_{D_1} \ni f \mapsto gf \in \mathcal{L}_{D_2}$ is a sheaf isomorphism.

6.8 The Riemann–Roch Problem in Terms of Divisors

One of the most basic questions about a compact Riemann surface X is: does there exist a nonconstant holomorphic map $f : X \to \widehat{\mathbb{C}}$? But as we have seen already in the discussion of the field $\mathcal{M}(X)$, it is desirable to ask more: for example, do the meromorphic functions on X separate points? Even better, does there exist a holomorphic embedding $X \to \mathbb{P}^n$ for some n?

To answer these questions we aim to determine the dimension of the space of meromorphic functions with controlled zeros and poles. This is the Riemann–Roch problem.

The idea for divisors is to collect information on poles and zeros. We denote the functions with prescribed poles and zero as $L(D)$.

We call a divisor $D = \sum_{P \in X} n_P P \in \operatorname{Div}(X)$ *positive*, or *effective*, if $n_P \geq 0$ for all $P \in X(\mathbb{C})$, and write $D \geq 0$. For every divisor $D \in \operatorname{Div}(X)$ we have an associated Riemann–Roch space

$$L(D) := \{f \in \mathbb{C}(X) \mid f = 0 \text{ or } \operatorname{div}(f) + D \geq 0\}$$

which is a finite-dimensional $\mathbb{C}$-vector space, with $L(0) = \mathbb{C}$ and $L(D) = \{0\}$ for $\deg(D) < 0$. We denote its dimension by

$$\ell(D) := \dim_{\mathbb{C}} L(D)\,.$$

Using the technology of divisors from the last subsection, we can rephrase the definition of the vector space $L(D)$, associated in Subsection 6.2.1 to an positive divisor $D = n_1 x_1 + \cdots n_k x_k$, in terms of it. In fact

$$L(D) = \{f \in \mathcal{M}(X) : \operatorname{div}(f) + D \geq 0\} \cup \{0\}\,.$$

That is, if $D = n_1 x_1 + \cdots + n_k x_k$, a function $f \in L(D)$ is

(1) holomorphic away from the points x_j, or at x_j if $n_j = 0$,

(2) allowed to have a pole of order up to n_j at the point x_j if $n_j > 0$,

(3) must have a zero of order at least n_j at the point x_j if $n_j < 0$.

The Riemann–Roch problem, as always, is to determine $\ell(D)$ in terms of the genus g of X and the degree of the divisor $\deg D = n_1 + \cdots + n_k$. In fact, statement (3) is still valid, so long as we interpret $\Omega^1_D(X)$ as consisting of those meromorphic 1-forms which are

1. holomorphic away from the points x_j, or at x_j if $n_j = 0$,
2. must have a zero of order at least n_j at the point x_j if $n_j > 0$,
3. allowed to have a pole of order up to n_j at the point x_j if $n_j < 0$.

Note that this is precisely opposite to the requirements for a meromorphic function to be in $L(D)$, suggesting that perhaps $\Omega^1_D(X)$ has to do with $-D$ instead of D.

Definition 6.8.1. Given a divisor D, we define the *vector spaces* (over $\mathbb{C}$)

$$\begin{aligned} \mathcal{L}(D) &= \left\{ f \in \mathcal{M}^0(X) \;\middle|\; (f) \geq D \right\}, \qquad & l(D) &= \dim \mathcal{L}(D)\,, \\ \Omega(D) &= \left\{ \omega \in \mathcal{M}^1(X) \;\middle|\; (\omega) \geq D \right\}, \qquad & i(D) &= \dim \Omega(D)\,. \end{aligned}$$

Theorem 6.8.2. *The spaces $l(D)$ and $i(D)$ depend only on the divisor class (modulo principal divisors):*

$$l(D) = l([D])\,, \quad i(D) = i([D])\,.$$

Here $[D]$ is the divisor class of the devisor D.

Proof. If $D_2 - D_1 = (f)$ is a principal divisor, then the maps

$$\begin{aligned} \Omega\left(D_1\right) &\to \Omega\left(D_2\right) \\ \omega_1 &\mapsto f\omega_1 \end{aligned}$$

$$\begin{aligned} \mathcal{L}\left(D_1\right) &\to \mathcal{L}\left(D_2\right) \\ f_1 &\mapsto ff_1 \end{aligned}$$

are isomorphisms (they are obviously invertible: the inverse is dividing by f), therefore $i\left(D_1\right) = i\left(D_2\right) = i\left(\left[D_1\right]\right)$ and $r\left(D_1\right) = r\left(D_2\right) = r\left(\left[D_1\right]\right)$. □

Theorem 6.8.3.

$$i([D]) = l([D] - K)\,.$$

Proof. Let ω an arbitrary meromorphic 1-form. The map

$$\begin{aligned} \Omega(D) &\to \mathcal{L}(D - (\omega)) \\ \omega' &\mapsto \omega'/\omega \end{aligned}$$

is an isomorphism. □

It is easy to see that $f \in L(D)$ means: if for p_i in the divisor D, the coefficient n_i is less than 0, then f must have a zero of multiplicity at least $|n_i|$ at p_i; if $n_i \geq 0$, then f can have a pole of multiplicity at most n_i at p_i.

From the properties of principal divisors (and the fact that non-zero constant functions have principal divisor zero) we see that each $\mathcal{L}(D)$ is a vector space.

Write $D = D_+ - D_-$, where $D_+ \geq 0$ and $D_- \geq 0$. We clearly have

$$\Omega(D_+) \subset \Omega(D) \subset \Omega(-D_-) \Longrightarrow 0 \leq i(D_+) \leq i(D) \leq i(-D_-)$$
$$\mathcal{L}(-D_-) \subset \mathcal{L}(-D) \subset \mathcal{L}(-D_+) \Longrightarrow 0 \leq l(-D_-) \leq l(-D) \leq l(-D_+)\,.$$

We clearly have the nesting properties $\mathcal{L}(D_1) \subset \mathcal{L}(D_2)$ if $D_1 \leq D_2$, and also if $f \in \mathcal{L}(D_1), g \in \mathcal{L}(D_2)$ then $fg \in \mathcal{L}(D_1 + D_2)$.

Theorem 6.8.4. *Let D a divisor. If there exists a negative divisor $\tilde{D} \leq 0$ whose divisor class is $[\tilde{D}] = [D]$ or if $[\tilde{D}] = X - [D]$, then the Riemann–Roch theorem holds.*

The Riemann–Roch problem, as always, is to determine $\ell(D)$ in terms of the genus g of X and the degree of the divisor $\deg D = n_1 + \cdots + n_k$.

In fact, formula

$$\ell(D) = 1 + \deg D - g + \dim \Omega^1_D(X)$$

is still valid, so long as we interpret $\Omega^1_D(X)$ as consisting of those meromorphic 1-forms which are

- holomorphic away from the points x_j, or at x_j if $n_j = 0$,
- must have a zero of order at least n_j at the point x_j if $n_j > 0$,
- allowed to have a pole of order up to n_j at the point x_j if $n_j < 0$.

Note that this is precisely opposite to the requirements for a meromorphic function to be in $L(D)$, suggesting that perhaps $\Omega^1_D(X)$ has to do with $-D$ instead of D.

This theorem is extremely useful and powerful. One reason is that most often one of the two indices $r(-D)$ or $i(D)$ is easy to compute and the other is more difficult, so the theorem gives it without effort.

Also, using positivity $i(D) \geq 0$ (resp. $l(-D) \geq 0$) we easily get a lower bound for $l(-D)$ (resp. $i(D)$), and thus the Riemann–Roch theorem is often used to prove the existence of functions or forms with given poles and degrees.

We can thus easily compute $\mathcal{L}(D)$ for zero or negative divisors:

Lemma 6.8.5. *Let X be a compact Riemann surface. Then $\mathcal{L}(0)$ consists only of the constant functions, and $\mathcal{L}(D)$ consists only of 0 if $D < 0$.*

Proof. If $D \leq 0$ and $f \in \mathcal{L}(D)$, then f is holomorphic on X and hence (by Corollary 6.3.14) constant.

Suppose that $f \in L(D)$ and f is not identically zero. Consider the divisor $E = (f) + D$. Since $f \in L(D)$, $E \geq 0$, so certainly $\deg(E) \geq 0$. However since $\deg((f)) = 0$, we have $\deg(E) = \deg(D) < 0$. This contradiction proves the result. □

Corollary 6.8.6. *$\mathcal{L}(D)$ has dimension 1 when $D = 0$ and 0 when $D < 0$.*

Theorem 6.8.7. *The linear space $I(D)$ is isomorphic to the space $L(K-D)$, where K is a canonical divisor. The linear space $I(K-D)$ is isomorphic to the space $L(D)$.*

Proof. We shall prove that $I(K-D)$ is isomorphic to $L(D)$ first. We assume that $K = (\omega)$ for some meromorphic 1-form ω.

If f is a meromorphic function, then

$$(f) + D = (f\omega) - (\omega) + D = (f\omega) - (K - D)\,.$$

So, $f \in L(D) = \{g \mid (g) + D \geq 0\}$ if and only if

$$f\omega \in I(K - D) = \{\beta \mid (\beta) \geq K - D\}\ .$$

Thus, we have a map:

$$\alpha : L(D) \to I(K - D), \quad \alpha(f) = f\omega\,.$$

Since every meromorphic 1-form in $I(K-D)$ can be presented as $f\omega$ for some meromorphic function, we similarly have an inverse map

$$\alpha^{-1} : I(K - D) \to L(D)\,.$$

Since linearity of α is obvious, we have an isomorphism between $I(K-D)$ and $L(D)$ for any divisor D.

Now, if we substitute $K-D$ instead of D, we will get isomorphism between $I(D)$ and $L(K-D)$. □

Lemma 6.8.8. *The linear space $L(K)$, where K is a canonical divisor, is isomorphic to the linear space of all holomorphic* 1*-forms on the same Riemann surface.*

Proof. From the previous lemma $L(K)$ is isomorphic to

$$I(K - K) = I(0) = \{\omega \mid (\omega) \geq 0\}\ ,$$

which is obviously the space of all holomorphic 1-forms. From the previous lemma and the Theorem 6.9.6 it follows that $l(K) = \dim(L(K)) = g$. □

Proposition 6.8.9. *If* $\deg(D) > 0$ *then* $L(D) = 0$.

Proof. If f is non-zero and contained in $L(D)$ then $\deg(\operatorname{div}(f)) \geq \deg(D) > 0$. But this is impossible since the degree the principal divisor $\operatorname{div}(f)$ is zero. □

Proposition 6.8.10. *If $D \in \operatorname{Div}(X)$ and ω is any non-zero meromorphic differential then*

$$\dim(\Omega(D)) = \dim(L(D/\operatorname{div}(\omega)))\,.$$

Proof. The linear map from $\Omega(D) \to L(D/\operatorname{div}(\omega))$ defined by $\eta \mapsto \eta/\omega$ is clearly one to one and onto. □

A divisor D is called *integral* (or effective) if $D \geq o$ and strictly integral if $D > 0$.

Theorem 6.8.11. *Let X be a compact Riemann surface and let D be a divisor on X. Then the space of functions $L(D)$ is a finite-dimensional complex vector space. Indeed, given the decomposition $D = P - N$, we have the dimensional estimate $l(D) \leq 1 + \deg(P)$.*

Proof. Note that the statement is true for $D = 0$: on a compact Riemann surface, $L(0)$ consists of only the constant functions and therefore has dimension one. We go by induction on the degree of the positive part P of D. If $\deg(P) = 0$, then $P = 0$, so that $l(P) = 1$; since $D \leq P$ we see that $L(D) \subset L(P)$, so that $l(D) \leq l(P) = 1 = 1 + \deg(P)$ as required.

Assume then that the statement is true for divisors whose positive part has degree $k - 1$, and let us prove it for a divisor whose positive part has degree $k \geq 1$. Fix such a divisor D, and write $D = P - N$ as above, with $\deg(P) = k$. Choose a point p in the support of P, so that $P(p) \geq 1$. Consider the divisor $D - p$; its positive part is $P - p$, which has degree $k - 1$. Hence the induction hypothesis applies, and we have that $l(D-p) \leq \deg(P-p)+1 = \deg(P)$. Now we apply the codimension statement above and conclude that $l(D) \leq 1 + l(D-p)$. Hence $l(D) \leq \deg(P) + 1$, as claimed. □

6.9 The Canonical Divisor

Finally, we come to Roch's contribution to the Riemann–Roch story, alluded to in Subsection 6.5.2. First, he introduced a special kind of divisor on X, called the canonical divisor and denoted by K.

Definition 6.9.1 (canonical divisor). Let $\omega = f dz$ be a meromorphic 1-form on X, where f is meromorphic and z is a local coordinate. Let $\{z_1, \ldots, z_k\}$ be zeroes of f and $\{p_1, \ldots, p_l\}$ be its poles, both counted with multiplicities. Then we can define a divisor (ω), called the *canonical divisor* of ω

$$(\omega) := \sum_{i=1}^{k} z_i - \sum_{j=1}^{l} p_j\,.$$

Example 6.9.2. Consider $\mathbb{CP}^1$ with two coordinates z and w around 0 and ∞ respectively, related by $z = \frac{1}{w}$. We can define a form dz around 0, which transforms to $\frac{1}{w^2}dw$. That means this form has no zeroes, and a pole of order 2 at ∞. The canonical divisor (ω) is given by $-2 \cdot \infty$.

Example 6.9.3. Consider a rational function

$$f(z) = c\frac{(z-z_1)\dots(z-z_k)}{(z-p_1)\dots(z-p_l)}$$

for some non-zero complex number c and some complex numbers $z_1, \dots, z_k$, $p_1, \dots, p_l$. This is a meromorphic function on $\mathbb{C}$, and $f(1/z)$ is also meromorphic, so f extends to a meromorphic function on the Riemann sphere $\mathbb{C}\cup\{\infty\}$. It has zeroes at $z_1, \dots, z_k$ and poles at $p_1, \dots, p_l$, and also has a zero of order $l-k$ (or a pole of order $l-k$) at ∞, as can be seen by inspection of $z \mapsto f(1/z)$ near the origin (or the growth of $f(z)$ near infinity), and thus

$$(f) = \sum_{j=1}^{k}(z_j) - \sum_{j=1}^{l}(p_l) + (l-k)\cdot(\infty)\,.$$

In particular, (f) has degree zero.

Using the divisor notation, Theorem 6.3.13 of Section 5.13 can be stated as follows:

Theorem 6.9.4. *If $f \in \mathcal{M}(X)$ is not constant, then*

$$\deg(f) = \sum_{p\in X}\operatorname{ord}_p(f) = 0\,.$$

Theorem 6.9.5. *If K is a canonical divisor on a compact Riemann surface X, then* $\deg(K) = -\chi = 2g-2$.

Proof. Since all canonical divisors have the same degree, consider the meromorphic 1-form $K = df$, where $f : S \to \mathbb{CP}^1$ is a nonconstant meromorphic function whose existence is guaranteed by Theorem 5.16.6.

Suppose there are n poles (counted with multiplicity). The map f is a ramified cover of degree n. We have that, by Riemann–Hurwitz formula (Theorem 5.10.33),

$$2-2g = 2n - \sum_{p\in X}(\operatorname{ord}_p(f)-1)$$

□

Exercises

1. Let ω_1, ω_2 be two meromorphic 1-forms. Let (U, z) be a chart. On its domain we define ω_1/ω_2 to be $f_1(z)/f_2(z)$ where $\omega_i = f_i(z)dz$. Show that this definition is well defined, i.e., it doesn't depend on the choice of chart. So this defines a meromorphic function.

This canonical divisor has the special property that

$$L(K) \simeq \Omega^1(X)$$

We may define it by taking an arbitrary non-zero meromorphic 1-form ω, and associating to a divisor $div(\omega)$ of its zeros minus its poles. This gives rise to the canonical divisor $K = -\operatorname{div}(\omega)$.

The isomorphism

$$L(K) \simeq \Omega^1(X)$$

is now given by sending $f \to f\omega$, for the chosen meromorphic 1-form ω.

Since at the points at which ω has a pole of some order, f must have a zero of at least that order, the obtained 1-form $f\omega$ is indeed holomorphic.

Conversely, for an arbitrary $\omega' \in \Omega^1(X)$, a holomorphic chart (z, U) gives rise to local expressions $\omega|_U = g(z)dz$ and $\omega' = g'(z)dz$ for a holomorphic function g and a meromorphic function g'. Then

$$\left.\frac{\omega'}{\omega}\right|_U = \frac{g'(z)dz}{g(z)dz} = \frac{g'(z)}{g(z)} \in \mathcal{M}(U)$$

and so $f = \frac{\omega'}{\omega}$ is a meromorphic function on X, which must have zeros where ω has poles, and may have poles where ω has zeros, i.e., we have found an element $f \in L(K)$, such that $\omega' = f\omega$, establishing the desired isomorphism.

Theorem 6.9.6. *If the genus of Riemann surface X is equal to g, there exist on X exactly g independent holomorphic 1-forms.*

Definition 6.9.7 (equivalent divisor)**.** Two divisors D_1 and D_2 on a Riemann surface are called *linearly equivalent* if there exists a meromorphic function $f \in \mathcal{M}_*(X)$ defined on the same surface such that D_1 and D_2 differ by its principal divisor (f), i.e., if $D_2 = D_1 + (f)$.

Lemma 6.9.8. *Linearly equivalent divisors have equal degrees.*

Proof. If $D_2 = D_1 + (f)$, then

$$\deg(D_2) = \deg(D_1 + (f)) = \deg(D_1) + \deg((f)) = \deg(D_1) ,$$

since $\deg((f)) = 0$. □

Lemma 6.9.9. *If divisors D_1 and D_2 are linearly equivalent, the space $L(D_1)$ is isomorphic to $L(D_2)$, and the space $I(D_1)$ is isomorphic to $I(D_2)$.*

Proof. Let us assume that $D_2 = D_1 + (f)$ for a certain meromorphic function f. If g is a meromorphic function and $g \in L(D_2)$, then

$$(fg) + D_1 = (f) + (g) + D_1 = (g) + D_1 + (f) = (g) + D_2 .$$

It follows that $(g) + D_2 \geq 0$ if and only if $(fg) + D_1 \geq 0$. Therefore, we have a map $L(D_2) \to L(D_1)$. Since, f is a meromorphic function, its inverse f^{-1} also is a meromorphic function, which clearly provides us with the inverse map $L(D_2) \mapsto L(D_1)$. Since linearity is also obvious we have an isomorphism between $L(D_1)$ and $L(D_2)$. The fact that $I(D_1)$ is isomorphic to $I(D_2)$ is proved similarly. □

6.10 The Riemann–Roch Formula

The argument of the previous section is easy to modify in the presence of a divisor D. It allows us to entirely analogously obtain a vector space isomorphism

$$L(K-D) \simeq \Omega_D^1(X).$$

Using this identification, the Riemann–Roch formula

$$\ell(D) = 1 + \deg D - g + \dim \Omega_D^1(X)$$

may be rewritten in the today-standard form

$$\ell(D) - \ell(K-D) = \deg D + 1 - g.$$

This works for an arbitrary divisor D on X, and *this is the celebrated Riemann–Roch theorem in its definitive form.*

A meromorphic function on a compact Riemann surface X is, up to a constant, uniquely determined by its Laurent principal parts, because if $f_1, f_2 \in K(X)$ and they have the same Laurent principal parts, then $f_1 - f_2$ is a holomorphic function on the compact X and is hence a constant.

This observation allows us to identify the quotient space $\mathcal{L}(D)/\mathbb{C}$ as a subspace of $\mathbb{C}^d$, where $d = \deg D$, as follows. With D fixed, $D = \sum_{i=1}^{k} n_i p_i \geq 0$, let z_i be a coordinate function around p_i as before, with $z_i(p_i) = 0$. For $f \in \mathcal{L}(D)$, let its principal part at p_i be denoted by η_i as above. We have seen that η_i must be of the form above, i.e.,

$$\eta_i = \frac{a_{in_1}}{z_i^{n_t}} + \cdots + \frac{a_{i2}}{z_i^2} + \frac{a_{i1}}{z_i},$$

where $a_{in_i}, \ldots, a_{i1} \in \mathbb{C}$. Now let P_i be the n_i-dimensional vector space over $\mathbb{C}$ spanned by $\{1/z_i^{n_t}, \ldots, 1/z_i^2, 1/z_i\}$. The preceding observation implies that the map

$$\begin{aligned} \mathcal{L}(D)/\mathbb{C} &\to P_1 \oplus \cdots \oplus P_k \\ [f] &\mapsto (\eta_1, \ldots, \eta_k) \end{aligned} \tag{6.3}$$

is well defined, where $[f]$ denotes the coset containing f. This is clearly a vector space homomorphism which is in addition injective. We now identify each P_i with $\mathbb{C}^{n_i}$ in the obvious manner. Then $P_1 \oplus \cdots \oplus P_k$ is identified with $\mathbb{C}^d$, i.e.,

$$\{(\eta_1, \eta_2, \ldots, \eta_k)\} \cong \mathbb{C}^{n_1} \oplus \mathbb{C}^{n_2} \oplus \cdots \oplus \mathbb{C}^{n_k} \cong \mathbb{C}^d \tag{6.4}$$

where $d = n_1 + n_2 + \cdots + n_k =$ degree D. In the sense of (6.3) and (6.4), $\mathcal{L}(D)/\mathbb{C}$ is now identified with a subspace of $\mathbb{C}^d$.

Lemma 6.10.1. *Let X be a Riemann surface, and let D be a divisor on X, and let p be a point of X. Then either $L(D-(p)) = L(D)$ or $L(D-(p))$ has codimension one in $L(D)$.*

Proof. Let $\phi : U \to \mathbb{C}$ be a chart that maps p to the origin, and suppose that D already had order m at p (so that $D-(p)$ had order $m+1$). Then functions $f \in \mathcal{L}(D-(p))$, when composed with the inverse ϕ^{-1} of the chart function have Laurent expansion

$$f\left(\phi^{-1}(z)\right) = \frac{a_{m+1}}{z^{m+1}} + \frac{a_m}{z^m} + \dots$$

for some complex coefficients $a_{m+1}, a_m, \dots$ (which will depend on the choice of chart).

The map $\alpha : f \mapsto a_{m+1}$ is clearly a linear map from $\mathcal{L}(D-(p))$ to $\mathbb{C}$. If α is identically the zero map, then $L(D-(p)) = L(D)$. Otherwise, α is onto, and so $L(D-(p))$ has codimension one in $L(D)$, and the claim follows. □

As a corollary of Lemma 6.10.1 and Lemma 6.8.5, we see that the spaces $\mathcal{L}(D)$ are all finite dimensional, with the dimension $\dim(L(D))$ increasing by zero or one each time one adds an additional pole to D. Here is another simple linear algebra relation between the dimensions of the spaces $\mathcal{L}(D)$:

Lemma 6.10.2. *Let X be a compact Riemann surface, and let D_1, D_2 be divisors. Then*

$$\begin{aligned} &\dim \mathcal{L}(D_1) + \dim \mathcal{L}(D_2) \\ &\quad \le \dim \mathcal{L}(\min(D_1, D_2)) + \dim \mathcal{L}(\max(D_1, D_2)) \end{aligned} \tag{6.5}$$

Proof. From the dimension formula for subspaces we have

$$\dim \mathcal{L}(D_1) + \dim \mathcal{L}(D_2) - \dim\left(L(D_1) \cap \mathcal{L}(D_2)\right) = \dim\left(\mathcal{L}(D_1) + \mathcal{L}(D_2)\right)$$

Since

$$\mathcal{L}(D_1) \cap \mathcal{L}(D_2) = \mathcal{L}(\min(D_1, D_2))$$

and

$$\mathcal{L}(D_1) + \mathcal{L}(D_2) \subset \mathcal{L}(\max(D_1, D_2)) ,$$

the claim follows. □

If D is a divisor and (f) is a principal divisor, then (6.5) gives an isomorphism between $\mathcal{L}(D)$ and $\mathcal{L}(D+(f))$, by mapping $g \in \mathcal{L}(D)$ to $g/f \in \mathcal{L}(D+(f))$.

In particular, the dimensions $\dim(\mathcal{L}(D))$ and $\dim(\mathcal{L}(D+(f)))$ of the linearly equivalent divisors $D, D+(f)$ are the same.

If we define a divisor class to be a coset

$$D + \mathrm{PDiv}(X) = \{D + (f) \mid f \in \mathcal{M}_*(X)\}$$

of the principal divisors in $\mathrm{Div}(X)$ (that is to say, an equivalence class for linear equivalence), then we conclude that the dimension $\dim(\mathcal{L}(D))$ depends only on the divisor class $D + \mathrm{PDiv}(X)$ of D. The space $\mathrm{Div}(X)/\mathrm{PDiv}(X)$ of divisor classes is an abelian group, which is known as the divisor class group. (For nonsingular algebraic curves, this group also coincides with the Picard group, though the situation is more subtle if one allows singularities.)

6.11 The Geometric Riemann–Roch Theorem

Define the vector spaces

$$L(D) = \{\text{meromorphic functions } f : (f) \geq -D\},$$
$$\Omega(D) = \{\text{meromorphic differentials } \omega : (\omega) \geq -D\}.$$

Here are some elementary facts about these spaces.

Proposition 6.11.1. $L(1) = \mathbb{C}$.

Proof. If $(f) \geq 1$, then f is a holomorphic function on X, hence constant. □

Proposition 6.11.2. *If* $\deg(D) > 0$, *then* $L(D) = \{0\}$.

Proof. Suppose $f \in L(D)$ is nonzero. Then

$$\deg((f)) \geq \deg(D) > 0.$$

But the degree of a principal divisor (f) is always zero, a contradiction. □

Proposition 6.11.3. *If* $D \in \mathrm{Div}(X)$ *and* ω *is any nonzero meromorphic differential, then*

$$\dim \Omega(D) = \dim L(D/(\omega)).$$

Proof. The linear map

$$\Omega(D) \longrightarrow L(D/(\omega)), \quad \eta \mapsto \eta/\omega$$

is clearly bijective. □

We now have all the tools we need to prove one of the most important theorems in the study of Riemann surfaces, the Riemann–Roch theorem.

Theorem 6.11.4. *Let* X *be a compact Riemann surface of genus* g *and* $D \in \mathrm{Div}(X)$. *Then*

$$\dim L(-D) = \deg(D) - g + 1 + \dim \Omega(D).$$

Proof of Theorem 6.11.4 when $D = 0$. In this case, $L(-D) = L(0)$ consists of holomorphic functions. These are precisely the constant functions (by Theorem 6.7.8), so $\dim L(0) = 1$.

On the other hand, $\Omega(0)$ is the space of holomorphic differentials, which has dimension g.

Finally, $\deg(0) = 0$. Therefore,

$$1 = 0 - g + 1 + g,$$

and the Riemann–Roch theorem holds. □

Proof of Theorem 6.11.4 when $D > 0$. Let $D = p_1 n_1 + \cdots + p_m n_m$ with $n_j > 0$. Then $\deg(D) = \sum n_j > 0$.

Choose representatives $\{a_1, \ldots, a_g, b_1, \ldots, b_g\}$ for a canonical homology basis. Define

$$B = p_1(n_1 + 1) + \cdots + p_m(n_m + 1),$$

and set

$$\Omega_0(-B) = \left\{ \omega \in \Omega(-B) : \operatorname{res}_{p_j}(\omega) = 0, \int_{a_j} \omega = 0, \, j = 1, \ldots, g \right\}.$$

The point of introducing this space is that the exterior derivative d maps $L(-D)$ into $\Omega_0(-B)$. This follows by considering the action of d locally near the poles. If $f \in L(-D)$, then the poles of df have order one greater than those of f. Moreover, all residues of df vanish, as do its integrals over closed cycles, in particular the a_j cycles. Thus,

$$d : L(-D) \longrightarrow \Omega_0(-B).$$

Hence we may compute

$$\dim L(-D) = \dim(\ker d) + \dim(\operatorname{Im} d).$$

The kernel has dimension 1, since any f with $df = 0$ is constant, and constants lie in $L(-D)$ because $D > 0$. Therefore,

$$\dim L(-D) = 1 + \dim(\operatorname{Im} d). \tag{6.6}$$

Next, we claim that

$$\dim \Omega_0(-B) = \deg(D). \tag{6.7}$$

To prove this, recall that for each $p \in M$ and $n \geq 2$, there exist meromorphic 1-forms $\tau_p^{(n)}$ such that, in a local coordinate vanishing at p, one has $\tau_p^{(n)} = z^{-n} dz$, and $\int_{a_j} \tau_p^{(n)} = 0$ for each j. For $j = 1, \ldots, m$, fix local coordinates vanishing at p_j, and obtain n_j forms

$$\tau_{p_j}^{(2)}, \ldots, \tau_{p_j}^{(n_j+1)} \in \Omega_0(-B).$$

These are clearly linearly independent. Hence

$$\dim \Omega_0(-B) \geq \sum_j n_j = \deg(D).$$

Conversely, expand $\omega \in \Omega_0(-B)$ locally near each p_j:

$$\omega = \sum_{k=-n_j-1}^{\infty} d_{j,k} z^k dz.$$

Define

$$\begin{aligned}S &: \Omega_0(-B) \to \mathbb{C}^{\deg(D)}\,,\\ S(\omega) &= (d_{1,-2},\ldots,d_{1,-n_1-1},\ldots,d_{m,-2},\ldots,d_{m,-n_m-1})\,.\end{aligned}$$

If $S(\omega)=0$, then ω is holomorphic with vanishing integrals over all a_j, hence $\omega=0$. Thus S is injective, and

$$\dim \Omega_0(-B) = \dim \mathrm{Im}(S) \leq \deg(D)\,.$$

This proves (6.7) and shows that the forms $\tau_{p_j}^{(k)}$ with $k=2,\ldots,n_j+1$, $j=1,\ldots,m$, form a basis of $\Omega_0(-B)$.

We now compute $\dim(\mathrm{Im}\, d)$. The image consists of those $\omega \in \Omega_0(-B)$ with $\int_{b_j}\omega = 0$ for $j=1,\ldots,g$. Indeed, since residues and a_j-integrals vanish by definition, these extra conditions ensure that ω has path-independent integrals. Thus, given such ω, define $f=\int_0^z \omega$, with base point 0. Then $df=\omega$, and the poles of f have order one less than those of ω, so $f\in L(-D)$.

Hence $\mathrm{Im}(d)$ arises from $\Omega_0(-B)$ by imposing g linear conditions, so

$$\dim(\mathrm{Im}\, d) \geq \dim \Omega_0(-B) - g = \deg(D) - g\,.$$

This gives the Riemann inequality

$$\dim L(-D) \geq 1 + \deg(D) - g\,.$$

To establish equality, consider the basis

$$\{\tau_{p_j}^k : k=2,\ldots,n_j+1,\ j=1,\ldots,m\}\,.$$

By the bilinear relation,

$$\int_{b_l} \tau_{p_j}^k = 2\pi i \frac{\alpha_{j,k-2}^{(l)}}{k-1}\,,$$

where $\alpha_{j,k-2}^{(l)}$ are the coefficients in the local expansion near p_j of the normalized holomorphic forms $\{\omega_1,\ldots,\omega_g\}$ with $\int_{a_j}\omega_k = \delta_{jk}$. Explicitly,

$$\omega_l = \sum_k \alpha_{j,k}^{(l)} z^k dz \quad \text{near } p_j\,.$$

An element

$$\omega = \sum_{j=1}^{m}\sum_{k=2}^{n_j+1} d_{j,k}\, \tau_{p_j}^k$$

lies in $\mathrm{Im}(d)$ iff, for each $l=1,\ldots,g$,

$$0 = \int_{b_l}\omega = 2\pi i \sum_{j=1}^{m}\sum_{k=2}^{n_j+1} d_{j,k} \frac{\alpha_{j,k-2}^{(l)}}{k-1}\,.$$

This yields the matrix equation

$$T \cdot M \cdot \begin{bmatrix} d_{1,2} \\ \vdots \\ d_{1,n_1+1} \\ \vdots \\ d_{m,2} \\ \vdots \\ d_{m,n_m+1} \end{bmatrix} = 0 \,,$$

where

$$T = \begin{bmatrix} \alpha_{1,0}^{(1)} & \cdots & \alpha_{1,n_1-1}^{(1)} & \cdots & \alpha_{m,0}^{(1)} & \cdots & \alpha_{m,n_m-1}^{(1)} \\ \vdots & \vdots & \vdots & \vdots & \vdots & \vdots & \vdots \\ \alpha_{1,0}^{(g)} & \cdots & \alpha_{1,n_1-1}^{(g)} & \cdots & \alpha_{m,0}^{(g)} & \cdots & \alpha_{m,n_m-1}^{(g)} \end{bmatrix} ,$$

and

$$M = \operatorname{diag}\left(1, 1/2, \ldots, 1/n_1, \ldots, 1, 1/2, \ldots, 1/n_m\right) .$$

Since M is invertible, $\dim(\operatorname{Im} d) = \dim \ker T$.

To interpret this, define $R : \Omega(1) \to \mathbb{C}^{\deg(D)}$ as follows. For $\omega \in \Omega(1)$, expand near p_j:

$$\omega = \sum_{k=0}^{\infty} e_{j,k} z^k dz \,.$$

Set

$$R(\omega) = \left(e_{1,0}, \ldots, e_{1,n_1-1}, \ldots, e_{m,0}, \ldots, e_{m,n_m-1}\right) .$$

Then $\ker R = \Omega(D)$, since $R\omega = 0$ forces ω to vanish to the required orders.

With respect to the basis $\{\omega_1, \ldots, \omega_g\}$, the matrix of R is $T^\top$. Hence

$$\dim \ker(T^\top) + \dim \operatorname{Im}(T^\top) = \dim \Omega(1) = g \,.$$

So

$$\dim \Omega(D) + \dim \operatorname{Im}(T^\top) = g \,.$$

On the other hand,

$$\dim \ker T = \deg(D) - \dim \operatorname{Im}(T^\top) \,.$$

Therefore,

$$\dim \operatorname{Im}(d) = \deg(D) - g + \dim \Omega(D) \,.$$

Combining with (6.6) completes the proof. □

Proof of Theorem 6.11.4 when D is equivalent to an integral divisor. This is immediate, since none of the quantities in the Riemann–Roch equality change when D is replaced by an equivalent divisor. □

Corollary 6.11.5. *Let X be a compact Riemann surface of genus g. Then:*

1. *For each nonzero $\omega \in \Omega^1(X)$ we have*

$$\deg((\omega)) = 2g - 2\,, \qquad \dim L((\omega)) = g\,.$$

2. *For each nonzero $\omega \in \Omega^n(X)$ we have*

$$\deg((\omega)) = 2n(g-1)\,.$$

3. *For each $D \in \mathrm{Div}(X)$ with $\deg(D) > 2g-2$ we have*

$$\dim L(D) = \deg(D) - g + 1\,.$$

Definition 6.11.6. A meromorphic differential $\omega \in \Omega^n(X)$ is *holomorphic* if either $\omega = 0$ or $(\omega) \geq 0$.

Corollary 6.11.7. *Let X be a compact Riemann surface of genus g. Then*

$$\dim_{\mathbb{C}} \Omega^1(X) = g\,.$$

Proof. Let $\omega_1 \in \Omega^1(X)$ be nonzero. Then

$$L\left((\omega_1)\right) = \{f \in \mathbb{C}(X) : f = 0 \text{ or } (f\omega_1) \geq 0\}\,.$$

The isomorphism $\mathbb{C}(X) \xrightarrow{\sim} \Omega^1(X)$ defined by $f \mapsto f\omega_1$ implies

$$L\left((\omega_1)\right) \simeq \Omega^1(X)\,.$$

Therefore,

$$\dim_{\mathbb{C}} \Omega^1(X) = \dim L((\omega_1)) = g\,.$$

□

Lemma 6.11.8. *If $g = 0$ then X is biholomorphic to $\mathbb{C} \cup \{\infty\}$.*

Proof. Pick any $p \in X$. Then

$$\dim L(-p) \geq \deg(p) - g + 1 = 2\,.$$

Since the holomorphic functions (constants) account for only one dimension, there must be a function in $L(-p)$ with a pole of order one at p. Thinking of f as a holomorphic map $X \to \mathbb{C} \cup \{\infty\}$, we see that f has degree one. Hence f is a biholomorphic map. □

Lemma 6.11.9. *Let ω be a meromorphic differential. Then*

$$\deg((\omega)) = 2g - 2\,.$$

Proof. Since all meromorphic differentials are equivalent and therefore have the same degree, we may choose any meromorphic differential to compute this degree.

When $g = 0$ and $X \cong \mathbb{C} \cup \{\infty\}$ we choose $\omega = dz$. If $w = 1/z$ then $dz = w^{-2}dw$, so there is a single pole of order 2 at $z = \infty$. Thus $\deg((\omega)) = -2 = 2g - 2$.

When $g > 0$ there exist holomorphic differentials. Let ω be a holomorphic differential. Because ω is holomorphic, (ω) is integral. Using the Riemann–Roch theorem, we conclude

$$g = \deg((\omega)) - g + 1 + 1\,,$$

which proves the lemma. □

1. If $\deg D \geq g - 1 + n$, then $\dim L(-D) \geq n$.

   Proof.

   $$\dim L(-D) = \deg(D) - g + 1 + \dim \Omega(D) \geq n + \dim \Omega(D) \geq n\,.$$

   □

2. X has a nonconstant meromorphic function.

   Proof. Choose a divisor of degree $g + 1$. □

3. If K is a canonical divisor on a compact Riemann surface X, then $\deg(K) = 2g - 2$.

   Proof. By the Riemann–Roch theorem, for any divisor D we have

   $$\dim L(D) - \dim L(K - D) = \deg(D) - g + 1\,.$$

   Take $D = K$. Then

   $$\dim L(K) - \dim L(0) = \deg(K) - g + 1\,.$$

   Since $\dim L(0) = 1$, we get

   $$\dim L(K) - 1 = \deg(K) - g + 1\,,$$

   or equivalently,

   $$\deg(K) = \dim L(K) + g - 2\,.$$

   By definition, $\dim L(K) = g$, the dimension of holomorphic 1-forms on X. Hence

   $$\deg(K) = g + g - 2 = 2g - 2\,.$$

   □

6.12 Embedding into Projective Space

Example 6.12.1. Let X be a compact Riemann surface. Let

$$f_0, \dots, f_N \in \mathcal{M}(X)$$

be non-trivial meromorphic functions on X. Fix $x \in X$ and let (U, z) be a coordinate neighborhood of x with $z(x) = 0$. If

$$k := \min_j \{\mathrm{ord}_x (f_i), j = 1, \dots, N\},$$

then we have $f_i = z^k g_i$ for all i on U, where the g_i are holomorphic and for some i we have $g_i(x) \neq 0$. Define

$$F(x) := [g_0(x) : \cdots : g_N(x)].$$

This defines a map $F : X \to \mathbb{CP}^N$ which we shall denote by $F = (f_0 : \cdots : f_N)$. Clearly, the definition is independent of the local coordinate chosen. The map F is holomorphic, since if $g_i(x) \neq 0$, then near x we have

$$\varphi_i \circ F = \left(\frac{g_0}{g_i}, \dots, \frac{g_{i-1}}{g_i}, \frac{g_{i+1}}{g_i}, \dots, \frac{g_N}{g_i}\right).$$

Definition 6.12.2. The *complete linear system* of a divisor D, denoted by $|D|$, is the set of all effective divisors linearly equivalent to D:

$$|D| = \{E \in \mathrm{Div}(X) : E \sim D,\ E \geq 0\}.$$

Definition 6.12.3. A complete linear system $|D|$ on a compact Riemann surface X is called *base-point-free* (or simply *free*) if

$$\dim L(D - p) = \dim L(D) - 1$$

for every $p \in \mathrm{supp}(D)$.

Associated to a divisor D on a compact Riemann surface X is the map

$$\varphi_D : X \longrightarrow \mathbb{CP}^n, \qquad p \longmapsto [f_0(p), f_1(p), \dots, f_n(p)],$$

where $f_0, \dots, f_n$ are nontrivial meromorphic functions forming a basis of $L(D)$.

Lemma 6.12.4. *Let X be a compact Riemann surface and let D be a divisor such that $|D|$ is base-point-free. Fix $p \in X$. Then there exists a basis $f_0, f_1, \dots, f_n$ of $L(D)$ with*

$$\mathrm{ord}(f_0, p) = -D(p), \qquad \mathrm{ord}(f_i, p) > -D(p) \ \textit{for}\ i \geq 1.$$

Proof. Consider the codimension-one subspace $L(D-p) \subset L(D)$ and let $f_1, \ldots, f_n$ form a basis of $L(D-p)$. Extend this to a basis of $L(D)$ by including some $f_0 \in L(D) \setminus L(D-p)$. Then for each $i \geq 1$ we have

$$\operatorname{ord}(f_i, p) \geq -D(p) + 1 > -D(p).$$

On the other hand, if $\operatorname{ord}(f_0, p) > -D(p)$, then $f_0 \in L(D-p)$, contradicting the choice of f_0. Thus $\operatorname{ord}(f_0, p) = -D(p)$, as required. □

Lemma 6.12.5. *For distinct points $p, q \in X$, we have $\varphi_D(p) = \varphi_D(q)$ if and only if*

$$L(D-p-q) = L(D-p) = L(D-q).$$

Hence φ_D is injective if and only if

$$\dim L(D-p-q) = \dim L(D) - 2.$$

Proof. It suffices to check whether $\varphi_D(p) = \varphi_D(q)$ using the basis of $L(D)$ provided by Lemma 6.12.4. In this basis, $\varphi_D(p) = [1, 0, \ldots, 0]$. Injectivity fails precisely when $\varphi_D(q) = [1, 0, \ldots, 0]$ as well. By construction, this occurs if and only if

$$\operatorname{ord}(f_0, q) = -D(q), \qquad \operatorname{ord}(f_i, q) > -D(q) \text{ for all } i \geq 1,$$

which means that $\{f_1, \ldots, f_n\}$ is a basis of $L(D-q)$. But these were also chosen as a basis of $L(D-p)$, so the spaces $L(D-p)$ and $L(D-q)$ coincide. Thus $\varphi_D(p) = \varphi_D(q)$ if and only if $L(D-p-q) = L(D-p) = L(D-q)$.

Since $|D|$ is base-point-free, we have

$$\dim L(D-p) = \dim L(D-q) = \dim L(D) - 1.$$

Therefore, $\dim L(D-p-q)$ is either $\dim L(D) - 1$ or $\dim L(D) - 2$. By the argument above, φ_D is injective exactly when $\dim L(D-p-q) = \dim L(D) - 2$, as claimed. □

We have established a criterion for φ_D to be injective. To ensure that φ_D is an embedding, we must additionally require that it is locally an immersion. This happens precisely when, for some $i \geq 1$, the function f_i vanishes at p with order $-D(p)+1$, while f_0 realizes the minimal order $-D(p)$. Equivalently, $f_i \in L(D-p) \setminus L(D-2p)$. A necessary and sufficient condition for this is

$$\dim L(D-p-q) = \dim L(D) - 2 \qquad \text{for all } p, q \in X,$$

the same requirement obtained for injectivity.

Lemma 6.12.6. *Let X be a compact Riemann surface and let $|D|$ be base-point-free. Then φ_D is a holomorphic embedding if and only if*

$$\dim L(D-p-q) = \dim L(D) - 2$$

for all $p, q \in X$.

The Riemann–Roch theorem provides a means of computing $\dim L(D-p-q)$ and verifying that $|D|$ is base-point-free:

$$\dim L(D) - \dim L(K-D) = \deg(D) - g + 1\,.$$

Lemma 6.12.7. *If* $\deg(D) \geq 2g-1$, *then*

$$\dim L(D) = \deg(D) - g + 1\,.$$

Proposition 6.12.8. *The degree of a canonical divisor* K *is* $\deg(K) = 2g-2$. *Hence for* $\deg(D) \geq 2g-1$, *we have* $\deg(K-D) \leq -1$ *and thus* $\dim L(K-D) = 0$.

Theorem 6.12.9 (embedding theorem)**.** *Every compact Riemann surface* X *of genus* g *admits a holomorphic embedding into projective space* $\mathbb{CP}^n$ *for some* n.

Proof. Choose a divisor D on X with $\deg(D) \geq 2g+1$. Then for arbitrary $p, q \in X$ we have

$$\begin{aligned}
\dim L(D) &= \deg(D) - g + 1\,,\\
\dim L(D-p) &= \deg(D-p) - g + 1\,,\\
\dim L(D-p-q) &= \deg(D-p-q) - g + 1\,.
\end{aligned}$$

Since $\deg(D-p-q) = \deg(D) - 2$, it follows that $|D|$ is base-point-free and Lemma 6.12.6 applies. Therefore φ_D defines an embedding of X into projective space. □

Remark 6.12.10. Actually, every compact Riemann surface admits an embedding into $\mathbb{CP}^3$.

6.13 Compact Riemann Surfaces and Algebraic Curves

Theorem 6.13.1. *Every compact Riemann surface is isomorphic to a smooth algebraic curve.*

Proof. Let X be a compact Riemann surface of genus g. Fix an effective divisor D on X of degree

$$\deg D = d \geq 2g + 1\,.$$

By Lemma 6.12.7, for $d \geq 2g-1$ we have

$$\ell(D) = d - g + 1\,.$$

In particular, for $d \geq 2g+1$ we get $\ell(D) \geq g+2 \geq 3$, so the linear system $|D|$ is nontrivial.

1. Base-point freeness.

   Fix a point $p \in X$. Consider the divisor $D - p$, which has degree

   $$\deg(D - p) = d - 1 \geq 2g\,.$$

   Applying Riemann–Roch to $D - p$ gives

   $$\ell(D - p) = (d - 1) - g + 1 = d - g\,.$$

   On the other hand, applying Riemann–Roch to D gives

   $$\ell(D) = d - g + 1\,.$$

   Therefore

   $$\ell(D) - \ell(D - p) = 1\,.$$

   This means that not all sections in $L(D)$ vanish at p. Equivalently, there exists a section

   $$s \in L(D)$$

   such that $s(p) \neq 0$.

   Hence $|D|$ has no base points and defines a holomorphic map

   $$\varphi_D : X \to \mathbb{CP}^{\ell(D)-1}$$

   by sending $x \in X$ to the projective class of evaluations of a basis of $L(D)$ at x.

2. Separation of distinct points.

   For distinct $p, q \in X$ consider $\ell(D - p - q)$. Since $\deg(D - p - q) = d - 2 \geq 2g - 1$, we get $\ell(D - p - q) = d - g - 1$. Hence

   $$\ell(D) - \ell(D - p - q) = 2\,,$$

   so there exists $s \in L(D)$ vanishing at p but not at q. Thus $\varphi_D(p) \neq \varphi_D(q)$, and φ_D is injective on points.

3. Injectivity on tangent directions (immersion).

   To check that φ_D is an immersion, we must separate tangent directions at each point p. Consider $\ell(D - 2p)$. With $\deg(D - 2p) = d - 2 \geq 2g - 1$, again Riemann–Roch gives $\ell(D - 2p) = d - g - 1$. So

   $$\ell(D) - \ell(D - 2p) = 2\,,$$

   which implies there exists a section in $L(D)$ vanishing to order exactly 1 at p while some other section does not. Hence the differential $d\varphi_D(p)$ is nonzero, so φ_D separates tangent vectors and is an immersion.

From Steps 2 and 3 we conclude that φ_D is a holomorphic embedding of X into $\mathbb{CP}^{\ell(D)-1}$. The image $\varphi_D(X)$ is therefore a compact complex submanifold of projective space of complex dimension 1. By Chow's theorem (or the fact that compact complex subvarieties of projective space are algebraic) $\varphi_D(X)$ is a smooth projective algebraic curve.

Thus the holomorphic isomorphism

$$\varphi_D : X \xrightarrow{\simeq} \varphi_D(X)$$

gives the desired identification of X with a smooth algebraic curve over $\mathbb{C}$. □

6.14 Hyperelliptic Integral

A hyperelliptic integral is, by definition, an integral of the form

$$\int_\gamma \frac{dz}{\sqrt{P(z)}},$$

where γ is a path in the complex plane $\mathbb{C}$ with coordinate z, and

$$P(z) = (z - a_1) \cdots (z - a_d)$$

with pairwise distinct constants a_i. If $d = \deg P$ is 1 or 2, explicit integration by elementary functions is well known from calculus. If $d = 3$ or 4, integration is possible using elliptic functions. However, if $d \geq 5$, no explicit integration is known in general.

The reason for this is the following: the differential $\omega = \frac{dz}{\sqrt{fP(z)}}$ is not single-valued when considered as a function on $\mathbb{C}$. Let C denote the compact Riemann surface associated with $\sqrt{P}$. By definition, C is the double covering of the Riemann sphere $\mathbb{CP}_1$, ramified at the points $a_1, \ldots, a_d$, together with ∞ if d is odd. Now ω may be regarded as a holomorphic differential on C. It is essentially the topological structure of C that causes the problem: the more complicated it is, the more difficult it becomes to integrate ω.

At the beginning of the 19th century, the Norwegian mathematician Niels Henrik Abel (1802–1829) and the German mathematician Carl Gustav Jacob Jacobi (1804–1851) found a way to approach this problem. In geometric terms, their method can be described as follows. The idea is to integrate not just ω alone, but simultaneously the entire set of holomorphic differentials

$$\omega_i = z^{i-1} \frac{dz}{\sqrt{P(z)}} \quad \text{for} \quad i = 1, \ldots, g = \left\lfloor \frac{d-1}{2} \right\rfloor$$

on C. For this, fix a point $p_0 \in C$ and consider the map

$$p \longmapsto \left(\int_{p_0}^{p} \omega_1, \ldots, \int_{p_0}^{p} \omega_g \right),$$

defined on a neighborhood U of p_0. Unfortunately, this map cannot be extended to the whole of C, since the integrals depend on the path from p_0 to p. However, if we consider it modulo the values of the integrals along all possible closed paths, we obtain a well-defined map. To be more precise, we consider $H_1(C,\mathbb{Z})$ to be the group of closed paths starting from p_0 modulo homology.

6.14.1 Homology

A 1-chain on a Riemann surface X is a formal finite linear combination of curves $\gamma_j : [0,1] \to X$ with integer coefficients,

$$\gamma = \sum_{j=1}^{k} n_j \gamma_j, \quad n_j \in \mathbb{Z}.$$

The integral along γ of a closed 1-form $\omega \in \mathcal{E}^1(X)$ is defined by

$$\int_\gamma \omega := \sum_{j=1}^{k} n_j \int_{\gamma_j} \omega .$$

Let $C_1(X)$ denote the set of all 1-chains which is an abelian group in a natural way. We define the following boundary operator

$$\partial : C_1(X) \to \operatorname{Div}(X).$$

1. If $\gamma : [0,1] \to X$ is a closed curve, set $\partial\gamma = 0$.
2. If γ is a curve which is not closed, let $\partial\gamma$ be the divisor which is 1 at $\gamma(1)$, -1 at $\gamma(0)$, and 0 everywhere else.
3. For an arbitrary 1-chain $\gamma = \sum_{j=1}^{k} n_j\gamma_j$ set $\partial\gamma := \sum_{j=1}^{k} n_j \partial\gamma_j$.

Then

$$\deg(\partial\gamma) = 0 \quad \text{for all } \gamma \in C_1(X).$$

On a compact Riemann surface X, for any divisor D with $\deg D = 0$, there exists a 1-chain γ such that $\partial\gamma = D$.

Indeed, any divisor D with zero degree can be written as $D = D_1 + \cdots + D_k$, where each D_j is 1 at some point b_j, -1 at some point a_j, and 0 everywhere else. Then it suffices to choose curves γ_j from a_j to b_j and take $\gamma = \gamma_1 + \cdots + \gamma_k$.

The kernel $Z_1(X) := \ker(\partial : C_1(X) \to \operatorname{Div}(X))$ is called the *group of* 1-*cycles*. Two cycles $\gamma_1, \gamma_2 \in Z_1(X)$ are said to be homologous if

$$\int_{\gamma_1} \omega = \int_{\gamma_2} \omega \quad \text{for all closed } \omega \in \mathcal{E}^1(X).$$

This defines an equivalence relation on $Z_1(X)$. The set of equivalence classes, called *homology classes*, is an additive group $H_1(C,\mathbb{Z})$, the first homology group of X.

Since closed homotopic curves are also homologous, there is a group homomorphism $\pi_1(X) \to H_1(X)$. This map is surjective, but not in general injective, since $\pi_1(X)$ is not always abelian.

The image of the map

$$H_1(C,\mathbb{Z}) \longrightarrow \mathbb{C}^g, \quad \gamma \longmapsto \left(\int_\gamma \omega_1, \ldots, \int_\gamma \omega_g\right)$$

is a lattice in $\mathbb{C}^g$, i.e., a discrete subgroup of rank $2g$. Thus, the quotient

$$J(C) = \mathbb{C}^g / H_1(C,\mathbb{Z})$$

is a complex torus, called the *Jacobian variety* of C. It can be shown that $J(C)$ is isomorphic to a projective variety. A complex torus with this property is called an *abelian variety.*

By construction, the integration map $U \to \mathbb{C}^g$ discussed above extends to a holomorphic map

$$\alpha : C \longrightarrow J(C), \quad p \longmapsto \left(\int_{p_0}^p \omega_1, \ldots, \int_{p_0}^p \omega_g\right) \bmod H_1(C,\mathbb{Z}),$$

called the *Abel–Jacobi map.* In these terms, the integration of $(\omega_1, \ldots, \omega_g)$ is essentially equivalent to the following two steps:

(i) Determine the Jacobian variety $J(C)$.

(ii) Describe the Abel–Jacobi map $\alpha : C \to J(C)$.

Unfortunately, this can be done explicitly only in very few cases. However, one can try to study the geometry of the pair $(J(C), \alpha)$, and this may be considered as a substitute for step (i). This method is not restricted to hyperelliptic integrals but applies to holomorphic integrals on any compact Riemann surface.

6.14.2 Meromorphic Functions with Prescribed Divisors

Let X be a compact Riemann surface and $D \in \mathrm{Div}(X)$. We say that $f \in \mathcal{M}(X)$ is a solution of the divisor D if $(f) = D$. A necessary condition for this is that $\deg D = 0$.

Let $X_D := \{x \in X : D(x) \geq 0\}$. A weak solution of D is a function $f \in \mathcal{E}(X_D)$ such that for each $a \in X$ there is a coordinate neighborhood (U, z) with $z(a) = 0$ and a function $\psi \in \mathcal{E}(U)$ with $\psi(a) \neq 0$ such that

$$f = z^k \psi \quad \text{on } U \cap X_D, \quad \text{where } k = D(a). \tag{6.8}$$

Then a weak solution f of D is a solution of D if and only if $f \in \mathcal{O}(X_D)$. If f, g are two weak solutions of D, then there exists a non-vanishing function $\varphi \in \mathcal{E}(X)$ such that $f = \varphi g$.

If f_i is a weak solution of D_i, where $i = 1, 2$, then $f_1 f_2$ is a weak solution of $D_1 + D_2$ and f_1/f_2 is a weak solution of $D_1 - D_2$. It may happen that, for instance, $D_1(a) + D_2(a) \geq 0$ but $D_1(a) < 0$ or $D_2(a) < 0$. Then formally $f_1 f_2$ is not defined at a, but it extends by continuity.

Suppose that f is a weak solution of D. The logarithmic derivative df/f is a smooth 1-form on the complement of the support supp $D := \{x \in X : D(x) \neq 0\}$ of the divisor D. For $a \in \operatorname{supp} D$ and $k = D(a)$, (6.8) implies

$$\frac{df}{f} = k\frac{dz}{z} + \frac{d\psi}{\psi} \tag{6.9}$$

where $d\psi/\psi$ is smooth in a neighborhood of a. For each 1-form $\sigma \in \mathcal{E}^1(X)$ with compact support the integral

$$\int_X \frac{df}{f} \wedge \sigma$$

exists; this can be easily checked in polar coordinates. Note that the 1-form $\bar{\partial} f/f$ is smooth on all of X, since (6.8) implies $\bar{\partial} f/f = \bar{\partial}\psi/\psi$.

Lemma 6.14.1. *Let X be a Riemann surface and let D be a divisor on X with* $\operatorname{supp} D = \{a_1, \ldots, a_n\}$. *Let f be a weak solution of D. Then for every $g \in \mathcal{E}(X)$ with compact support, we have*

$$\frac{1}{2\pi i}\int_X \frac{df}{f} \wedge dg = \sum_{j=1}^{n} D(a_j)\, g(a_j)\,.$$

Proof. There exist disjoint coordinate neighborhoods (U_j, z_j) of the points a_j with $z_j(a_j) = 0$ such that on U_j

$$f = z_j^{k_j}\psi_j \quad \text{for non-vanishing } \psi_j \in \mathcal{E}(U_j) \text{ and } k_j := D(a_j)\,.$$

We may assume that $z_j(U_j) = \mathbb{D} \subseteq \mathbb{C}$ for all j. Let $0 < r_1 < r_2 < 1$. Choose functions $\varphi_j \in \mathcal{E}(X)$ with $\operatorname{supp}\varphi_j \subseteq \{|z_j| < r_2\}$ and $\varphi_j|_{\{|z_j| \leq r_1\}} = 1$. Set $g_j := \varphi_j g, j = 1, \ldots, n$, and $g_0 := g - (g_1 + \cdots + g_n)$. Then g_0 has compact support in $Y = X \setminus \{a_1, \ldots, a_n\}$. Thus, by Theorem 5.13.16,

$$\int_X \frac{df}{f} \wedge dg_0 = -\int_Y d\left(g_0\frac{df}{f}\right) = 0$$

It follows that

$$\int_X \frac{df}{f} \wedge dg = \sum_{j=1}^{n}\int_{U_j} \frac{df}{f} \wedge dg_j = \sum_{j=1}^{n} k_j \int_{U_j} \frac{dz_j}{z_j} \wedge dg_j$$

using (6.9) and again Theorem 5.13.16 By Stokes' theorem,

$$\begin{aligned}\int_{U_j} \frac{dz_j}{z_j} \wedge dg_j &= -\lim_{\epsilon \downarrow 0}\int_{\epsilon \leq |z_j| \leq r_2} d\left(g_j\frac{dz_j}{z_j}\right)\\ &= \lim_{\epsilon \downarrow 0}\int_{|z_j| = \epsilon} g_j\frac{dz_j}{z_j} = 2\pi i g_j(a_j) = 2\pi i g(a_j)\end{aligned}$$

The proof is complete. □

6.14.3 Existence of Weak Solutions of $\partial\gamma$

Lemma 6.14.2. *Let X be a Riemann surface, $\gamma : [0,1] \to X$ a curve, and U a relatively compact open neighborhood of $\gamma([0,1])$. Then there exists a weak solution f of the divisor $\partial\gamma$ with $f|_{X\setminus U} = 1$ such that*

$$\int_\gamma \omega = \frac{1}{2\pi i}\int_X \frac{df}{f}\wedge\omega \quad \textit{for all closed } \omega \in \mathcal{E}^1(X)\,.$$

Proof. The integral on the right-hand side exists, since $df/f = 0$ on $X\setminus U$. Let us first assume that (U, z) is a coordinate chart in X with $z(U) = \mathbb{D}$ and $\gamma([0,1]) \subseteq U$. Let $a = \gamma(0)$ and $b = \gamma(1)$. There exists $r < 1$ such that $\gamma([0,1]) \subseteq \{|z| < r\}$. The function $z \mapsto \log\left(\frac{z-b}{z-a}\right)$ has a well-defined branch in $\{r < |z| < 1\}$. Choose $\psi \in \mathcal{E}(U)$ such that $\psi|_{\{|z|\le r\}} = 1$ and $\psi|_{\{|z|\ge R\}} = 0$, where $r < R < 1$. Define $f_0 \in \mathcal{E}(U\setminus\{a\})$ by

$$f_0(z) := \begin{cases} \frac{z-b}{z-a} & \text{if } |z| \le r \\ \exp\left(\psi(z)\log\frac{z-b}{z-a}\right) & \text{if } r < |z| < 1 \end{cases}$$

Then f_0 equals 1 on $\{R < |z| < 1\}$ and hence can be extended to a function $f \in \mathcal{E}(X\setminus\{a\})$ by setting $f|_{X\setminus U} = 1$. By construction, f is a weak solution of $\partial\gamma$. Let $\omega \in \mathcal{E}^1(X)$ be closed. Then ω has a primitive on U. So there exists $g \in \mathcal{E}(X)$ with compact support such that $dg = \omega$ on $\{|z| \le R\}$. By Lemma 6.14.1,

$$\frac{1}{2\pi i}\int_X \frac{df}{f}\wedge\omega = \frac{1}{2\pi i}\int_X \frac{df}{f}\wedge dg = g(b) - g(a) = \int_\gamma \omega$$

In general there exists a partition $0 = t_0 < t_1 < \cdots < t_n = 1$ and coordinate charts $(U_j, z_j)\,, j = 1, \ldots, n$, such that $\gamma_j := \gamma|_{[t_{j-1}, t_j]}$ lies in $U_j \subseteq U$ and $z_j(U_j) = \mathbb{D}$ for all j. By the special case there exist weak solutions f_j of $\partial\gamma_j$ such that $f_j|_{X\setminus U_j} = 1$ and

$$\int_{\gamma_j} \omega = \frac{1}{2\pi i}\int_X \frac{df_j}{f_j}\wedge\omega \quad \text{for all closed } \omega \in \mathcal{E}^1(X)$$

Then $f := f_1 \cdots f_n$ is as desired. □

6.15 Abel's Theorem

Theorem 6.15.1 (Abel's theorem)**.** *Let D be a divisor on a compact Riemann surface X with* $\deg D = 0$. *Then D has a solution if and only if there exists a 1-chain $\gamma \in C_1(X)$ with $\partial\gamma = D$ such that*

$$\int_\gamma \omega = 0 \quad \textit{for all } \omega \in \mathcal{O}^1(X)\,. \tag{6.10}$$

It is clearly sufficient to verify (6.10) on a basis of $\mathcal{O}^1(X)$. If $\tilde{\gamma} \in C_1(X)$ is an arbitrary 1-chain with $\partial\tilde{\gamma} = D$, then (6.10) can be reformulated as follows: there exists a cycle $\alpha \in Z_1(X)$ (namely, $\alpha = \tilde{\gamma} - \gamma$) such that

$$\int_{\tilde{\gamma}} \omega_i = \int_\alpha \omega_i \quad \text{for a basis } \omega_1, \ldots, \omega_g \text{ of } \mathcal{O}^1(X)\,. \tag{6.11}$$

Proof. Let $\gamma \in C_1(X)$ with $\partial\gamma = D$ such that (6.10) holds. By Lemma 6.14.2, there exists a weak solution f of D such that

$$\int_\gamma \omega = \frac{1}{2\pi i}\int_X \frac{df}{f} \wedge \omega \quad \text{for all closed } \omega \in \mathcal{E}^1(X)\,.$$

By (6.10), for every $\omega \in \mathcal{O}^1(X)$,

$$0 = \int_\gamma \omega = \frac{1}{2\pi i}\int_X \frac{df}{f} \wedge \omega = \frac{1}{2\pi i}\int_X \frac{\bar{\partial} f}{f} \wedge \omega\,.$$

Note that $\bar{\partial} f/f \in \mathcal{E}^{0,1}(X)$. By Corollary 5.14.8, there exists $g \in \mathcal{E}(X)$ such that $\bar{\partial} g = \bar{\partial} f/f$. Then $F := e^{-g}f$ is a weak solution of D, and

$$\bar{\partial} F = -e^{-g} f \bar{\partial} g + e^{-g}\bar{\partial} f = 0\,.$$

This implies that $F \in \mathcal{O}(X_D)$, and hence it is a meromorphic solution of D. Now let $f \in \mathcal{M}(X)$ be a solution of D. We may assume that $D \neq 0$. The function f defines an n-sheeted branched covering $f : X \to \widehat{\mathbb{C}}$ for some positive integer n. Let $a_1, \ldots, a_r \in X$ be the branch points, and set

$$Y := \widehat{\mathbb{C}} \backslash \{f(a_1), \ldots, f(a_r)\}\,.$$

Each $y \in Y$ has an open neighborhood V such that $f^{-1}(V)$ is a disjoint union of open sets $U_1, \ldots, U_n$, and all the maps $f|_{U_j} : U_j \to V$ are biholomorphic, with inverses $\varphi_j := f|_{U_j}^{-1}$. Given a holomorphic 1-form $\omega \in \mathcal{O}^1(X)$, consider the 1-form

$$\varphi_1^*\omega + \cdots + \varphi_n^*\omega$$

on V. If we repeat the same construction on an open neighborhood of another point in Y, then on the intersection we obtain the same 1-form. We thus obtain a holomorphic 1-form $\mathrm{tr}(\omega)$ on all of $\widehat{\mathbb{C}}$, similarly as in Theorem 1.3.11. Since $\mathcal{O}^1(\widehat{\mathbb{C}}) = 0$, we have $\mathrm{tr}(\omega) = 0$.

Let σ be a curve in $\widehat{\mathbb{C}}$ from ∞ to 0 that lies in Y except possibly at the endpoints. The preimage under f of σ consists of n curves $\gamma_1, \ldots, \gamma_n$ which join the poles of f with its zeros. Thus $\gamma = \gamma_1 + \cdots + \gamma_n$ satisfies $\partial\gamma = D$, and

$$\int_\gamma \omega = \int_\sigma \mathrm{tr}(\omega) = 0\,.$$

□

We are now ready to state the second main theorem of this chapter.

Theorem 6.15.2 (Abel). *Let X be a compact Riemann surface of genus g. The kernel of $A : \mathrm{Div}_0(X) \to \mathrm{Jac}(X)$ is precisely $\mathrm{Div}_P(X)$.*

This theorem provides a necessary and sufficient condition for the existence of a meromorphic function with prescribed zeros and poles. Specifically, if we encode the desired zeros and poles in a divisor D, then $\deg D$ must be 0, and D will admit a meromorphic solution if and only if $D \in \operatorname{Ker} A$. Most of the work in proving the theorem involves constructing a meromorphic solution, for which we need the notion of a weak solution of a divisor.

6.16 The Jacobi Inversion Problem

In Abel's theorem, we found necessary and sufficient conditions for a divisor to be principal. In this section, we study the quotient group of divisors of degree zero modulo the subgroup of principal divisors. We shall see that this group is isomorphic to a g-dimensional torus, where g is the genus of the underlying Riemann surface.

6.16.1 n-Dimensional Lattices

Let V be an n-dimensional vector space over $\mathbb{R}$. An additive subgroup $\Lambda \subseteq V$ is called a *lattice* if there exist n linearly independent vectors $\lambda_1, \dots, \lambda_n \in V$ such that

$$\Lambda = \mathbb{Z}\lambda_1 + \cdots + \mathbb{Z}\lambda_n \,.$$

Proposition 6.16.1. *An additive subgroup $\Lambda \subseteq V$ is a lattice if and only if:*

(1) Λ is discrete.

(2) Λ is contained in no proper subspace of V.

Proof. The necessity is clear. Assume that $\Lambda \subseteq V$ satisfies (1) and (2). We will show that there exist $n = \dim V$ linearly independent vectors $\lambda_1, \dots, \lambda_n$ such that

$$\Lambda = \mathbb{Z}\lambda_1 + \cdots + \mathbb{Z}\lambda_n \,.$$

We proceed by induction on n. The statement is trivial for $n = 0$. Let $n > 0$. By (2), there exist n linearly independent vectors $x_1, \dots, x_n \in \Lambda$. Consider $V_1 := \mathrm{span}(x_1, \dots, x_{n-1})$ and $\Lambda_1 := \Lambda \cap V_1$. By the induction hypothesis, there exist $n-1$ linearly independent vectors $y_1, \dots, y_{n-1} \in \Lambda_1$ such that

$$\Lambda_1 = \mathbb{Z}y_1 + \cdots + \mathbb{Z}y_{n-1} \,.$$

Every $x \in \Lambda$ can be written uniquely as

$$x = c_1(x)y_1 + \cdots + c_{n-1}(x)y_{n-1} + c(x)x_n, \qquad c_j(x), c(x) \in \mathbb{R}.$$

Consider the compact parallelotope

$$P = \{t_1y_1 + \cdots + t_{n-1}y_{n-1} + tx_n : t_j, t \in [0,1]\}.$$

By (1), $\Lambda \cap P$ is finite. Then there exists a vector $y_n \in (\Lambda \cap P) \setminus V_1$ such that

$$c(y_n) = \min\{c(x) : x \in (\Lambda \cap P) \setminus V_1\}.$$

Clearly, $0 < c(y_n) \leq 1$.

We now show that $\Lambda = \Lambda_1 + \mathbb{Z}y_n$. Let $x \in \Lambda$. Then there exist integers k_j such that

$$x' := x - \sum_{j=1}^{n} k_j y_j = \sum_{j=1}^{n-1} t_j y_j + t x_n,$$

where $0 \leq t_j < 1$ for $j = 1, \ldots, n-1$ and $0 \leq t < c(y_n)$. But $x' \in \Lambda \cap P$, hence $t = 0$. Consequently, $x' \in \Lambda_1$, so all t_j are integers and therefore zero. Thus $x' = 0$, and the assertion follows. □

6.16.2 Period Lattices

Let X be a compact Riemann surface of genus $g \geq 1$, and let $\omega_1, \ldots, \omega_g$ be a basis of $\mathcal{O}^1(X)$. The *period lattice* of X relative to the basis $\omega_1, \ldots, \omega_g$ is the subgroup of $\mathbb{C}^g$ defined by

$$\begin{aligned} \operatorname{Per}(\omega_1, \ldots, \omega_g) &:= \left\{ \left(\int_\alpha \omega_1, \ldots, \int_\alpha \omega_g \right) \in \mathbb{C}^g : \alpha \in \pi_1(X) \right\} \\ &= \left\{ \left(\int_\alpha \omega_1, \ldots, \int_\alpha \omega_g \right) \in \mathbb{C}^g : \alpha \in H_1(X) \right\}. \end{aligned}$$

To see that $\operatorname{Per}(\omega_1, \ldots, \omega_g)$ is a lattice, we need the following lemma.

Lemma 6.16.2. *Let X be a compact Riemann surface of genus g. There exist g distinct points $a_1, \ldots, a_g \in X$ with the following property: if $\omega \in \mathcal{O}^1(X)$ vanishes at all a_j, then $\omega = 0$.*

Proof. For any $a \in X$, consider the set $H_a := \{\omega \in \mathcal{O}^1(X) : \omega(a) = 0\}$. Each H_a either coincides with $\mathcal{O}^1(X)$ or has codimension one. Since $\bigcap_{a \in X} H_a = \{0\}$ and $\dim \mathcal{O}^1(X) = g$, there exist g points $a_1, \ldots, a_g \in X$ such that $H_{a_1} \cap \cdots \cap H_{a_g} = \{0\}$. The lemma follows. □

Proposition 6.16.3. *We have that* $\operatorname{Per}(\omega_1, \ldots, \omega_g)$ *is a lattice in* $\mathbb{C}^g \cong \mathbb{R}^{2g}$.

Proof. Let $a_1, \dots, a_g \in X$ be the points provided by Lemma 6.16.2. Choose simply connected coordinate neighborhoods (U_j, z_j) of a_j with $z_j(a_j) = 0$ for all $j = 1, \dots, g$. With respect to these coordinates,

$$\omega_i = \varphi_{ij}\, dz_j \quad \text{on } U_j\,.$$

By Lemma 6.16.2, the matrix $A := (\varphi_{ij}(a_j))_{1 \le i,j \le g}$ has rank g. Define a map $F : U_1 \times \cdots \times U_g \to \mathbb{C}^g$ by

$$F(x) = (F_1(x), \dots, F_g(x)), \quad \text{where } F_i(x) := \sum_{j=1}^{g} \int_{a_j}^{x_j} \omega_i\,. \tag{6.12}$$

Here the integral $\int_{a_j}^{x_j} \omega_i$ is taken along any curve in U_j from a_j to x_j (since U_j is simply connected). Then F is complex differentiable with respect to $x_1, \dots, x_g$ and has Jacobian matrix $J_F(x) = (dF_i(x)/dx_j) = (\varphi_{ij}(x_j))$. Thus $J_F(a_1, \dots, a_g) = A$ is invertible. It follows that $W := F(U_1 \times \cdots \times U_g)$ is a neighborhood of $F(a_1, \dots, a_g) = 0 \in \mathbb{C}^g$.

We claim that $\Lambda \cap W = \{0\}$, where $\Lambda := \operatorname{Per}(\omega_1, \dots, \omega_g)$. This implies that Λ is a discrete subgroup of $\mathbb{C}^g$. Suppose there exists a point other than 0 in $\Lambda \cap W$. Then there exists $x \in U_1 \times \cdots \times U_g$, $x \neq a$, such that $F(x) \in \Lambda$. Renumbering if necessary, we may assume that $x_j \neq a_j$ for $1 \le j \le k$ and $x_j = a_j$ for $j > k$, where $1 \le k \le g$. By Abel's Theorem 6.15.1 and Equation (6.11), there exists a meromorphic function f on X with a simple pole at a_j and a simple zero at x_j for $1 \le j \le k$, holomorphic elsewhere (since $F(x) \in \Lambda$). Let c_j/z_j be the principal part of f at a_j; then $c_j \neq 0$ for $1 \le j \le k$. By the residue Theorem 6.7.3,

$$0 = \operatorname{res}(f\omega_i) = \sum_{j=1}^{k} c_j \varphi_{ij}(a_j) \quad \text{for } i = 1, \dots, g\,,$$

which contradicts the fact that $A = (\varphi_{ij}(a_j))$ has rank g. The claim is proved.

To finish the proof we show that Λ is not contained in any proper real linear subspace of $\mathbb{C}^g$. Otherwise, there would be a real non-trivial linear form on $\mathbb{C}^g$ vanishing on Λ. Since every real linear form is the real part of a complex linear form, there is a non-zero vector $(c_1, \dots, c_g) \in \mathbb{C}^g$ such that

$$\operatorname{Re}\left(\sum_{j=1}^{g} c_j \int_\alpha \omega_j \right) = 0 \quad \text{for all } \alpha \in \pi_1(X)$$

This would imply $c_1\omega_1 + \cdots + c_g\omega_g = 0$, by Corollary 5.14.5, a contradiction. □

Corollary 6.16.4. *Let X be a compact Riemann surface of genus $g \ge 1$. Then $H_1(X) \cong \mathbb{Z}^{2g}$.*

Proof. By Proposition 6.16.3, there exist $2g$ closed curves $\alpha_1, \ldots, \alpha_{2g}$ in X such that the vectors

$$\lambda_j = \left(\int_{\alpha_j} \omega_1, \ldots, \int_{\alpha_j} \omega_g \right), \quad j = 1, \ldots, 2g$$

are linearly independent over $\mathbb{R}$ and

$$\mathrm{Per}\,(\omega_1, \ldots, \omega_g) = \mathbb{Z}\lambda_1 + \cdots + \mathbb{Z}\lambda_{2g}$$

It follows that the homology classes of the α_j in $H_1(X)$ are linearly independent over $\mathbb{Z}$ and generate $H_1(X)$. The statement follows. □

6.16.3 The Jacobi Variety and the Picard Group

Let X be a compact Riemann surface of genus g and let $\omega_1, \ldots, \omega_g$ be a basis of $\mathcal{O}^1(X)$. Then

$$\mathrm{Jac}(X) = \mathbb{C}^g / \mathrm{Per}\,(\omega_1, \ldots, \omega_g)$$

is called the *Jacobi variety* of X. It is an abelian group and has the structure of a g-dimensional complex torus. The definition depends on the basis $\omega_1, \ldots, \omega_g$, but a different basis yields an isomorphic $\mathrm{Jac}(X)$.

Let $\mathrm{Div}_0(X) \subsetneq \mathrm{Div}(X)$ be the subgroup of divisors of degree 0 and $\mathrm{Div}_p(X) \subsetneq \mathrm{Div}_0(X)$ the subgroup of principal divisors. The quotient

$$\mathrm{Pic}(X) := \mathrm{Div}(X) / \mathrm{Div}_p(X)$$

is the Picard group of X. We shall be primarily interested in the subgroup

$$\mathrm{Pic}_0(X) := \mathrm{Div}_0(X) / \mathrm{Div}_p(X)\,.$$

Since $\mathrm{Div}(X)/\mathrm{Div}_0(X) = \mathbb{Z}$, we have an exact sequence

$$0 \to \mathrm{Pic}_0(X) \to \mathrm{Pic}(X) \to \mathbb{Z} \to 0\,.$$

Next we define a map $\Phi : \mathrm{Div}_0(X) \to \mathrm{Jac}(X)$ as follows. Let $D \in \mathrm{Div}_0(X)$ and let $\gamma \in C_1(X)$ be a chain with $\partial\gamma = D$. Then the vector

$$\left(\int_\gamma \omega_1, \ldots, \int_\gamma \omega_g \right) \in \mathbb{C}^g$$

is determined uniquely by D up to equivalence modulo $\mathrm{Per}\,(\omega_1, \ldots, \omega_g)$. By definition, $\Phi(D)$ is its equivalence class. Note that Φ is a group homomorphism.

Abel's Theorem 6.15.1 states that $\ker \Phi = \mathrm{Div}_p(X)$, whence we get an injective map

$$j : \mathrm{Pic}_0(X) \to \mathrm{Jac}(X)\,.$$

The Jacobi inversion problem asks if this map is also surjective.

Theorem 6.16.5. *The map* $j : \mathrm{Pic}_0(X) \to \mathrm{Jac}(X)$ *is an isomorphism for every compact Riemann surface* X.

Proof. Let $p \in \mathrm{Jac}(X)$ be represented by $\xi \in \mathbb{C}^g$. If $N \in \mathbb{N}$ is large enough, then $N^{-1}\xi$ lies in the image of the map F from (6.12) (in fact, we saw that the image of F is a neighborhood of 0 in $\mathbb{C}^g$). So there exist points $a_j, x_j \in X$ and curves γ_j from a_j to x_j such that, for $\gamma = \gamma_1 + \cdots + \gamma_g$,

$$\left(\int_\gamma \omega_1, \ldots, \int_\gamma \omega_g\right) = \frac{1}{N}\xi$$

Thus, for the divisor $D = \partial\gamma$,

$$\Phi(D) = \frac{1}{N}\xi \quad \mathrm{mod\,Per}\,(\omega_1, \ldots, \omega_g)$$

If θ is the element of $\mathrm{Pic}_0(X)$ represented by the divisor ND, then $j(\theta) = p$. □

6.16.4 A Sharper Version

Let X be a compact Riemann surface of genus g and let $a_1, \ldots, a_g \in X$ be arbitrarily chosen points. We define a map $\psi : X^g \to \mathrm{Pic}_0(X)$ as follows. Let D_x, for $x \in X$, be the divisor defined by $D_x(x) = 1$ and $D_x(y) = 0$ if $y \neq x$. For $(x_1, \ldots, x_g) \in X^g$ set

$$\psi\,(x_1, \ldots, x_g) := \sum_{j=1}^{g} \left(D_{x_j} - D_{a_j}\right) \quad \mathrm{mod\,Div}_p(X)\,.$$

Let $J := j \circ \psi : X^g \to \mathrm{Jac}(X)$. Then

$$J\,(x_1, \ldots, x_g) = \left(\sum_{j=1}^{g} \int_{a_j}^{x_j} \omega_1, \ldots, \sum_{j=1}^{g} \int_{a_j}^{x_j} \omega_g\right) \quad \mathrm{mod\,Per}\,(\omega_1, \ldots, \omega_g)$$

Theorem 6.16.6. *The map* $J : X^g \to \mathrm{Jac}(X)$ *is surjective.*

Proof. By Theorem 6.16.5, it suffices to prove that $\psi : X^g \to \mathrm{Pic}_0(X)$ is surjective. This means that every divisor $D \in \mathrm{Div}_0(X)$ is equivalent modulo $\mathrm{Div}_p(X)$ to a divisor of the form $\sum_{j=1}^{g}\left(D_{x_j} - D_{a_j}\right)$ for $(x_1, \ldots, x_g) \in X^g$.

Let $D' := D + D_{a_1} + \cdots + D_{a_g}$. Then $\deg D' = g$. By the Riemann–Roch Theorem 6.11.4, $\dim H^0\left(X, \mathcal{L}_{D'}\right) \geq 1$, and hence there exists a non-trivial meromorphic function f on X with $D'' := (f) + D' \geq 0$. Since $\deg D'' = g$, there exist points $x_1, \ldots, x_g \in X$ with $D'' = D_{x_1} + \cdots + D_{x_g}$. Then

$$\sum_{j=1}^{g} \left(D_{x_j} - D_{a_j}\right) = D'' - D' + D = (f) + D$$

as desired. □

6.16.5 Riemann Surfaces of Genus One

Theorem 6.16.7. *The map* $J : X \to \mathrm{Jac}(X)$ *is an isomorphism for every compact Riemann surface of genus one.*

Proof. Let $\omega \in \mathcal{O}^1(X)$ be non-trivial. Let $a \in X$. For all $x \in X$, we have

$$J(x) = \int_a^x \omega \quad \mathrm{mod}\ \mathrm{Per}(\omega)\,.$$

Clearly, J is holomorphic and surjective (by Theorem 6.16.6). Suppose that there exists $y \neq x$ such that $J(y) = J(x)$. Then there exists a 1-cycle $\alpha \in Z_1(X)$ with

$$\int_a^y \omega = \int_a^x \omega + \int_\alpha \omega\,.$$

By Abel's Theorem 6.15.1, this would imply the existence of a meromorphic function f on X having a single pole of order one. In that case X would be isomorphic to $\widehat{\mathbb{C}}$, a contradiction. □

Corollary 6.16.8. *The Riemann surfaces of genus one are precisely the complex tori* $\mathbb{C}/\Lambda$.

Proof. Theorem 6.16.7 and the fact that for any lattice $\Lambda \in \mathbb{C}$ the complex torus $\mathbb{C}/\Lambda$ has genus 1. □

7

Covering Surfaces and Classical Plane Geometries

In the theory of one complex variable the Riemann surface S of $f(z) = \sqrt{z}$ is typically described as a "branched" covering of $\mathbb{C}$ or of the closed Riemann sphere. Only after a deeper investigation of the "branching point" 0 does one realize that in a more abstract sense the complex structure of S near 0 is the same as near every other point on S.

The set $M = \{(z, w) \in \mathbb{C}^2 \mid w^2 - z = 0\}$ is a complex submanifold of $\mathbb{C}^2$ of dimension 1, and it provides a concrete representation of S (without the point at ∞). The projection $\pi_1 : M \to \mathbb{C}$ onto the first coordinate exhibits the familiar branched covering of $M = S$ over $\mathbb{C}$, while the projection $\pi_2 : M \to \mathbb{C}$ onto the second coordinate represents the function $\sqrt{z}$ on M. There is an obvious way, made precise in the following section, in which these functions are holomorphic on M.

Let S_1 and S_2 be two Riemann surfaces and $\pi : S_1 \to S_2$ be a local homeomorphism so that each point on S_2 has an evenly covered neighborhood. We then say that (S_1, S_2, π) is a covering surface or, more commonly, that S_1 is a covering surface of S_2 with cover map or projection π. The number of points in $\pi^{-1}(p)$ is independent of $p \in S_2$ and is called the *number of sheets* of the covering.

Covering surfaces occur classically in the study of algebraic equations with symmetries. For example, the equation $w^2 - z = 0$ admits the symmetry obtained by replacing w by $-w$. In this context the symmetry is called *sheet interchange*. The Riemann surface S of $w^2 - z = 0$ over $\mathbb{C}\backslash\{0\}$ is a covering surface of $\mathbb{C}\backslash\{0\}$. Notice that the sheet interchange cannot be defined to be a covering surface at $z = 0$. Points where the map topologically looks like $z \mapsto z^n$ are called *ramification points.*

Covering surfaces with isolated singularities, all of which are ramification points, are called *ramified covers.*

DOI: 10.1201/9781003713005-7

7.1 Normal Families and Automorphisms

Let $D \subset \mathbb{C}$ denote the open unit disc $\{\zeta \in \mathbb{C} \mid |\zeta| < 1\}$. Also $D(p,r) \subset \mathbb{C}$ denotes the open disc with radius r centered at p. For $r > 0$ we let

$$D^n(0,r) \equiv \underbrace{D(0,r) \times \cdots \times D(0,r)}_{n \text{ times}} .$$

Further, if $p = (p_1, \ldots, p_n) \in \mathbb{C}^n$ and $r > 0$, then

$$D^n(p,r) \equiv D(p_1,r) \times \cdots \times D(p_n,r) .$$

If $f : D \to D \subset \mathbb{C}$ is a holomorphic function with $f(0) = 0$ and $|f'(0)| = 1$, then f has the form $f(z) = f'(0)z$. In particular, if $f \in \mathrm{Aut}(D)$ and if such an f has $f'(0) = 1$, then $f(z) = z$. This is part of the classical Schwarz lemma. The following result is a direct generalization to several variables, and to arbitrary bounded domains. There are many possible generalizations of the Schwarz lemma, some of which will be discussed later on in this book, but this is the one that will play the most direct role in our investigations. For example, it will enable us to see that, if Ω is a bounded domain, then $\mathrm{Aut}(\Omega)$ has compact isotropy group at each point.

Theorem 7.1.1 (H. Cartan). *Suppose that Ω is a bounded domain in $\mathbb{C}^n$. Let $\phi : \Omega \to \Omega$ be holomorphic and suppose that, for some $p \in \Omega, \phi(p) = p$ and $d\phi(p) = id$. [Here $d\phi$ is the n-dimensional complex differential.] Then ϕ is the identity mapping from Ω to itself.*

Remark: Boundedness of Ω is an essential hypothesis: consider the automorphism of $\mathbb{C}^2$ given by $(z_1, z_2) \mapsto (z_1 + z_2^2, z_2)$.

Proof of Theorem 7.1.1. We may assume that $p = \mathbf{0}$ (the origin). For proof by contradiction, assume that ϕ does not coincide with the identity mapping. Expanding ϕ in a power series about $p = \mathbf{0}$ (and remembering that ϕ is vector-valued, hence so is the expansion) yields

$$\phi(z) = z + P_k(z) + O\left(|z|^{k+1}\right)$$

where P_k is the first nonvanishing homogeneous polynomial (of degree k) of order exceeding 1 in the Taylor expansion. Defining $\phi^j(z) = \phi \circ \cdots \circ \phi$ (j times); direct computation then gives that

$$\begin{aligned}
\phi^2(z) &= z + 2P_k(z) + O\left(|z|^{k+1}\right) \\
\phi^3(z) &= z + 3P_k(z) + O\left(|z|^{k+1}\right) \\
&\vdots \\
\phi^j(z) &= z + jP_k(z) + O\left(|z|^{k+1}\right) .
\end{aligned}$$

Choose polydiscs $D^n(0,a) \subseteq \Omega \subseteq D^n(0,b)$. The Cauchy estimates imply then that, for any multi-index $\alpha = (\alpha_1, \ldots, \alpha_n)$ with $|\alpha| := \alpha_1 + \cdots + \alpha_n = k$,

$$j \cdot \left| \left(\frac{\partial}{\partial z} \right)^{\alpha} \phi \bigg|_{\mathbf{0}} \right| = \left| \left(\frac{\partial}{\partial z} \right)^{\alpha} \phi^j \bigg|_{\mathbf{0}} \right| \leq n \cdot \frac{b \cdot \alpha!}{a^k}$$

where

$$\left(\frac{\partial}{\partial z} \right)^{\alpha} = \frac{\partial^{\alpha_1}}{\partial z_1^{\alpha_1}} \cdots \frac{\partial^{\alpha_n}}{\partial z_n^{\alpha_n}} .$$

Note that the rightmost item in this estimate is independent of j. Hence, for each such multi-index α with $|\alpha| = k$, $(\partial/\partial z)^{\alpha} \phi|_{\mathbf{0}} = \mathbf{0}$. Thus $P_k = 0$, a contradiction. □

This argument in particular applies when the dimension $n = 1$ and the domain Ω is the unit disc. There it gives a conceptually direct proof of the corresponding part of the classical Schwarz lemma.

Cartan's result has some further immediate but surprising consequences.

Corollary 7.1.2. *Suppose that Ω is a bounded, circular domain in $\mathbb{C}^n$, that is $(e^{i\theta} z_1, e^{i\theta} z_2, \ldots, e^{i\theta} z_n) \in \Omega$ whenever $(z_1, z_2, \ldots, z_n) \in \Omega$ for every $\theta \in \mathbb{R}$. If $\mathbf{0} \in \Omega$ and $f \in \mathrm{Aut}(\Omega)$ with $f(\mathbf{0}) = \mathbf{0}$, then f is a linear mapping.*

Proof. For $\theta \in \mathbb{R}$ and $z \in \Omega$, let $F(z) = e^{-i\theta} f(e^{i\theta} z)$. Then $F \in \mathrm{Aut}(\Omega)$, since Ω is circular. By the chain rule it follows that

$$d\left(f^{-1} \circ F\right)\big|_0 = id .$$

Hence

$$f^{-1} \circ F = id$$

on Ω, or equivalently $f = F$. If we write $f = (f_1, \ldots, f_n)$, $F = (F_1, \ldots, F_n)$, and

$$f_j(z) = \sum_{|N|=1}^{+\infty} a_N z^N$$

is the Taylor expansion of f_j, then the Taylor expansion of F_j is, by definition of F and by substitution,

$$F_j = \sum_{|N|=1}^{+\infty} e^{-i\theta} a_N e^{i|N|\theta} z^N .$$

But $F_j = f_j$. Therefore $e^{i(|N|-1)\theta} a_N = a_N$ for all multi-indices N and all $\theta \in \mathbb{R}$. This implies[1] that $a_N = 0$ for $|N| \geq 2$. Thus each f_j is linear. □

[1]Here $N = (n_1, \ldots, n_n)$ and $|N| = n_1 + \cdots + n_n$.

It is easy to modify this argument to show that, if Ω_1, Ω_2 are two bounded, circular domains containing the origin $\mathbf{0}$ and if $F : \Omega_1 \to \Omega_2$ is biholomorphic with $F(\mathbf{0}) = \mathbf{0}$, then F is linear. This immediately implies that, when $n \geq 2$, the unit ball $\left\{(z_1, \ldots, z_n) \,\middle|\, |z_1|^2 + \cdots + |z_n|^2 < 1\right\}$ and the unit polydisc $\{(z_1, \ldots, z_n) \mid |z_j| < 1, j = 1, \ldots, n\}$ are not biholomorphic: If there were a biholomorphic map between them, then applying suitable biholomorphic maps to each variable in the unit polydisc separately would produce a biholomorphic map that took $\mathbf{0}$ to $\mathbf{0}$. This would then have to be linear, which is not possible, since, e.g., the ball has smooth boundary and the polydisc does not (when $n \geq 2$). Thus the direct analogue of the Riemann mapping theorem fails in $\mathbb{C}^n, n \geq 2$: (bounded) domains can be homeomorphic to the ball without being biholomorphic to it. This failure, even for small perturbations of the ball, will be explained in much more detail in later chapters.

The second corollary will play an important role in what follows.

Corollary 7.1.3. *If Ω is a bounded domain in $\mathbb{C}^n$ and $p \in \Omega$, then the mapping*

$$f \longmapsto df|_p$$

is an injective homomorphism of the group

$$I_p \equiv \{f \in \mathrm{Aut}(\Omega) : f(p) = p\}$$

into $GL(n, \mathbb{C})$.

Proof. If $df|_p = dg|_p$ for $f, g \in I_p$, then the chain rule gives that $d\left(f^{-1} \circ g\right)|_p = id$, where the identity map id is given by the $n \times n$ identity matrix $I_n \in GL(n, \mathbb{C})$. By Theorem 7.1.1, $f^{-1} \circ g : \Omega \to \Omega$ is the identity mapping. Hence $f \equiv g$. We conclude that $f \mapsto df|_p$ is injective on I_p. The homomorphism property is a special case of the chain rule. □

7.2 The Basic Examples

In this section, we shall collect a number of examples for which the automorphism groups are obtained explicitly. Some of these are well known and elementary, and the derivations of their automorphism groups need be outlined only briefly. But it will be convenient to have them all in one place; and looking at them all at once will suggest various paths of exploration that we follow later.

Example 7.2.1. $\mathrm{Aut}(\mathbb{C}) = \{z \mapsto az + b : a, b \in \mathbb{C}, a \neq 0\}$.

If $f : \mathbb{C} \to \mathbb{C}$ is injective, then the only possible singularity of f at ∞ is a simple pole. If instead ∞ were a removable singularity, then f would be constant by Liouville's theorem. If ∞ were an essential singularity, then f

would not be injective in any neighborhood of ∞. Similarly, a pole at ∞ of higher order than 1 would preclude injectivity in a neighborhood of ∞. Thus the nonconstant injective function f is a polynomial of degree one. That any polynomial of degree one is an automorphism is clear.

Example 7.2.2.

$$\operatorname{Aut}(\mathbb{D}) = \{z \mapsto \omega \cdot (z-a)/(1-\overline{a}z) \mid a, \omega \in \mathbb{C}, |\omega| = 1, |a| < 1\} .$$

That

$$T_a : z \longmapsto \frac{z-a}{1-\overline{a}z}$$

is defined and injective from $\mathbb{D}$ to $\mathbb{D}$ is easy algebra. Also $T_a\left(T_{-a}(z)\right) = z$; hence T_a is surjective.

Conversely, suppose that $f \in \operatorname{Aut}(\mathbb{D})$. Let $a = f^{-1}(0)$. Then $g := f/T_a$ is holomorphic and zero-free on $\mathbb{D}$ and

$$\lim_{|\zeta|\to 1} |g(\zeta)| = \lim_{|\zeta|\to 1} \left| \frac{f(\zeta)}{T_a(\zeta)} \right| = 1 .$$

By the maximum principle applied to both g and $1/g$, we see that $|T_a/f| \equiv 1$ on $\mathbb{D}$, hence $f = \omega T_a$ for some constant[2] ω with $|\omega| = 1$.

Example 7.2.3.

$$\operatorname{Aut}(\mathbb{C}\setminus\{0\}) = \{z \mapsto az^{\epsilon} \mid \epsilon = \pm 1, a \in \mathbb{C}, a \neq 0\} .$$

If $f \in \operatorname{Aut}(\mathbb{C}\setminus\{0\})$, then a connectivity argument shows that $\lim_{z\to 0} f(z) = 0$ or $\lim_{z\to 0} |f(z)| = +\infty$. Composing with an inversion, we may assume that the first alternative holds. But then f, considered as a holomorphic function, has a removable singularity at the origin. Thus the extension $f(0) = 0$ makes f an entire function that is an automorphism of the entire plane. Clearly $f(z) = az$, for some $a \neq 0$. In case $\lim_{z\to 0} f(z) = \infty$, the same reasoning applied to $1/f$ gives $1/f(z) = az$.

Example 7.2.4. Denote the annulus $\{z \in \mathbb{C} \mid 0 < r_1 < |z| < r_2\}$ by A. Then

$$\operatorname{Aut}(A) = \{z \mapsto \omega z \mid \omega \in \mathbb{C}, |\omega| = 1\} \cup \{z \mapsto \omega r_1 r_2/z \mid \omega \in \mathbb{C}, |\omega| = 1\} .$$

By a connectivity argument, for each $f \in \operatorname{Aut}(A)$, either

(a) $\lim_{|z|\to r_2} |f(z)| = r_2$ and $\lim_{|z|\to r_1} |f(z)| = r_1$;

or

---

[2] An alternative argument is to note that $T_a \circ f$ maps the disc to the disc and fixes 0. Then Schwarz's lemma implies that $|(T_a \circ f)(z)| \leq |z|$. Applying the same reasoning to the inverse of this mapping gives $|(T_a \circ f)(z)| \geq |z|$. Hence $|T_a \circ f(z)| \equiv |z|$ on $\mathbb{D}$, and $T_a \circ f$ equals $w \cdot \mathrm{id}$ on $\mathbb{D}$ for some ω with $|\omega| = 1$.

(b) $\lim_{|z|\to r_2} |f(z)| = r_1$ and $\lim_{|z|\to r_1} |f(z)| = r_2$.

In either case, repeated application of Schwarz reflection to the boundary circles extends f to an automorphism $\widehat{f} : \mathbb{C}\setminus\{0\} \to \mathbb{C}\setminus\{0\}$ of $\mathbb{C}\setminus\{0\}$. Thus, by Example (7.2.3), $f(z) = az$ or $f(z) = a/z$ for some nonzero $a \in \mathbb{C}$. The condition $f(A) = A$ tells us then that $a = \omega$ in the first instance and that $a = \omega r_1 r_2$ in the second instance.

Example 7.2.5. Consider

$$\mathrm{Aut}\left(\left\{(z_1, z_2) \in \mathbb{C}^2 \,\middle|\, |z_1|^2 + |z_2|^2 < 1\right\}\right).$$

The set

$$B^2 = \left\{(z_1, z_2) \in \mathbb{C}^2 \,\middle|\, |z_1|^2 + |z_2|^2 < 1\right\}$$

is of course the unit ball in $\mathbb{C}^2$. First notice that $I_{(0,0)} = U(2) \subset GL(2, \mathbb{C})$. Obviously $U(2) \subset I_{(0,0)}$. If $f \in I_{(0,0)}$, then f is $\mathbb{C}$-linear according to Corollary 7.1.2. Since f has to preserve the unit sphere (the boundary of B^2), it is immediate that $f \in U(2)$.

Now a direct calculation, analogous to that for the disc, shows that the mapping

$$T_{(a,0)}(z_1, z_2) \equiv \left(\frac{z_1 - a}{1 - \bar{a}z_1}, \frac{\sqrt{1 - |a|^2}\, z_2}{1 - \bar{a}z_1}\right)$$

sends the ball B^2 into itself. Furthermore, the inverse mapping to $T_{(a,0)}$ is $T_{(-a,0)}$. Thus $T_{(a,0)}$ is an automorphism.

If (z_1, z_2) is any point of B^2, then there is an element $\lambda \in U(2)$ that takes (z_1, z_2) to a point of the form $(a, 0)$. Also $T_{(a,0)}(a, 0) = (0, 0)$. These two pieces of information combined tell us that $\mathrm{Aut}\left(B^2\right)$ acts transitively on B^2: this means that any point of B^2 may be moved to any other by some element of the automorphism group. First, B^2 is homogeneous.

Let G denote the subgroup of $\mathrm{Aut}\left(B^2\right)$ generated by $U(2)$ together with $\left\{T_{(a,0)} : a \in \mathbb{C}, |a| < 1\right\}$. Then the isotropy subgroup of G at the origin obviously contains $U(2)$. Thus it equals $U(2)$. It follows that G is the full automorphism group, by Theorem 7.1.1.[3] For future reference, note that if $\varphi \in \mathrm{Aut}\left(B^2\right)$, then one can always express φ in the form $\mu_1 \circ T_{(b,0)} \circ \mu_2$, where μ_1, μ_2 are unitary rotations. First, let $\lambda_2 =$ a unitary rotation taking $\varphi^{-1}((0,0))$ to a point of the form $(b, 0)$. Then $T_{(b,0)} \circ \lambda_2$ takes $\varphi^{-1}((0,0))$ to $(0, 0)$. Hence $T_{(b,0)} \circ \lambda_2 \circ \varphi$ takes $(0, 0)$ to $(0, 0)$. Thus, from our earlier observations, $T_{(b,0)} \circ \lambda_2 \circ \varphi$ is a unitary rotation, say λ_1. Hence $\varphi = \lambda_2^{-1} \circ T_{(b,0)} \circ \lambda_1$, which has the desired form.

Example 7.2.6.

$$\mathrm{Aut}\left(\left\{(z_1, z_2) \in \mathbb{C}^2 \,\middle|\, |z_1|^4 + |z_2|^4 < 1\right\}\right).$$

[3]Determining the automorphism group of B^2 as a recognizable Lie group requires additional work. It turns out that it is $\mathrm{PSL}(2, \mathbb{C})$. See [83] for more on this matter.

By Corollary 7.1.2, the elements of I_0 (the isotropy group at $0 = (0,0)$) are $\mathbb{C}$-linear. Such a map must take a point of the boundary of the form $(\alpha, 0)$ or $(0, \alpha)$ to another point with one coordinate 0. This is so because boundary points with one coordinate 0 are exactly those boundary points where $\partial\Omega$ makes higher than first-order contact with its complex tangent plane, a condition preserved by invertible complex linear maps. Thus

$$\begin{aligned} I_0 &= \{(z_1, z_2) \mapsto (\omega_1 z_1, \omega_2 z_2) \mid \omega_1, \omega_2 \in \mathbb{C}, |\omega_1| = |\omega_2| = 1\} \\ &\cup \{(z_1, z_2) \mapsto (\omega_1 z_2, \omega_2 z_1) : \omega_1, \omega_2 \in \mathbb{C}, |\omega_1| = |\omega_2| = 1\} . \end{aligned}$$

Next, we claim that any element of $\mathrm{Aut}(\Omega)$ must in fact fix the origin. Let ϕ be an automorphism. By standard results in several complex variables, ϕ and ϕ^{-1} are X^∞ up to the boundary of Ω (see [13]). Weakly pseudoconvex boundary points must consequently be mapped only to weakly pseudoconvex boundary points. So ϕ must take the union of the two circles to itself. Thus ϕ must (after composition with the map permuting the coordinates if necessary) preserve the circle $\{(\alpha, 0) \in \partial\Omega\}$, and it must also preserve the circle $\{(0, \alpha) \in \partial\Omega\}$. By the Cauchy integral formula and continuity of ϕ at the boundary, it follows that ϕ preserves the entire discs $\{(\alpha, 0) : |\alpha| \leq 1\}$ and $\{(0, \alpha) : |\alpha| \leq 1\}$. We conclude that $\phi(0) = 0$. Hence ϕ is linear and in fact $\phi \in I_0$. So we have completely identified all elements of $\mathrm{Aut}(\Omega)$, and this verifies that $\mathrm{Aut}(\Omega) = I_0$.

Example 7.2.7. Now consider

$$\mathrm{Aut}(\Omega) \text{ for } \Omega = \left\{(z_1, z_2) \in \mathbb{C}^2 \,\middle|\, 0 < \alpha < |z_1|^2 + |z_2|^2 < 1\right\} .$$

By the Hartogs extension phenomenon, each element $f \in \mathrm{Aut}(\Omega)$ extends uniquely to a holomorphic mapping $\widehat{f} : B^2 \to B^2$, where B^2 is the unit ball in $\mathbb{C}^2$ as usual. These extensions must all be invertible since clearly $\widehat{f \circ g} = \widehat{f} \circ \widehat{g}$ for all $f, g \in \mathrm{Aut}(\Omega)$ (and of course the extension of the identity map is the identity map). Each such $\widehat{f}, f \in \mathrm{Aut}(\Omega)$, is a unitary rotation. To see this, note that, by the remark at the end of Example (7.2.5), $\widehat{f} = \mu_1 \circ T_{(a,0)} \circ \mu_2$ for some unitary rotations μ_1, μ_2 with $T_{(a,0)}$, as in the discussion there. Both μ_1 and μ_2 preserve Ω, but $T_{(a,0)}$ definitely does not preserve Ω if $a \neq 0$. This point is simple to check algebraically by looking at points of the form $ta/|a|$ with $-1 < t < 1$. Thus $\widehat{f}$ can preserve Ω only if $a = 0$ and, hence, $\widehat{f}$ is a unitary rotation. Consequently, $\mathrm{Aut}(\Omega)$ consists of the restrictions to Ω of the set of unitary rotations around the origin $(0, 0)$.

Example 7.2.8. Next we consider $\mathrm{Aut}(\Omega)$ for

$$\begin{aligned} \Omega = &\left\{(z_1, z_2) \in \mathbb{C}^2 \,\middle|\, 1/100 < |z_1|^2 + |z_2|^2 < 1\right\} \\ &\setminus \left[\left\{(z_1, z_2) \in \mathbb{C}^2 \,\middle|\, |z_1 - 3/4|^2 + |z_2|^2 \leq r_1\right\}\right. \\ &\left.\cup \left\{(z_1, z_2) \in \mathbb{C}^2 \,\middle|\, |z_1|^2 + |z_2 - 7/8|^2 \leq r_2\right\}\right] , \end{aligned}$$

with some small positive numbers r_1 and r_2.

Notice first that each element of $\text{Aut}(\Omega)$ again extends uniquely to an element of $\text{Aut}\left(B^2\right)$, by the Hartogs extension theorem. Then each automorphism of Ω must either preserve the sphere

$$\Sigma = \left\{(z_1, z_2) \;\middle|\; |z_1|^2 + |z_2|^2 = 1/100\right\}$$

or map this sphere to one of the other deleted spheres, by topological considerations. Algebraic considerations show that the image of a Euclidean sphere around the origin under an automorphism of B^2 is a Euclidean sphere only if the automorphism fixes the origin and hence is a rotation.

The algebraic determination that the image of a sphere with a center at the origin is again a sphere only if the origin is fixed can be done conveniently as follows. Consider $T_{(a,0)}$, for $-1 < a < 1$, acting on $S(r)$ = the sphere of radius $0 < r < 1$ around the origin $(0,0)$. Then $T_{(a,0)}(r,0)$ and $T_{(a,0)}(-r,0)$ are diametrically opposite on the image sphere. Again, if the image is a sphere, it then follows that the vector from $T_{(a,0)}(0,r)$ to $T_{(a,0)}(-r,0)$ is perpendicular to the vector from $T_{(a,0)}(0,r)$ to $T_{(a,0)}(r,0)$. But direct calculation shows that the inner product of these two vectors is 0 if and only if $a = 0$.

|As in the arguments for Example 7.2.7 above, f is now an automorphism of B^2 preserving the origin, that is the center of Σ. Consequently, any automorphisms of this Ω must be elements of $U(2)$. Since the elements of $U(2)$ are Euclidean isometries, and since the removed balls around $(3/4, 0)$ and $(0, 7/8)$ have centers that are at different distances from the origin, each of these balls must be mapped to itself. It follows that the automorphism which is an element of $U(2)$ must in fact be the identity mapping. Thus $\text{Aut}(\Omega) = \{id\}$: the automorphism group has just the single element, which is the identity. In this circumstance, we say that the domain Ω is *rigid*.

Example 7.2.9. Now we consider $\text{Aut}(\Omega)$ for

$$\Omega = \left\{(z_1, z_2) \in \mathbb{C}^2 \;\middle|\; |z_1|^2 + |z_2|^{2k} < 1\right\}, \quad k > 1$$

First we note that I_0 is linear from Corollary 7.1.3. Also this isotropy group clearly contains all linear maps of the form

$$(z_1, z_2) \mapsto \left(e^{i\theta_1} z_1, e^{i\theta_2} z_2\right), \quad \theta_1, \theta_2 \in \mathbb{R}$$

By the same logic as in Example (7.2.6), the set $\{(\alpha, 0) \in \Omega\}$ must be mapped to itself by any element of this isotropy group. This and the compactness of I_0 imply that

$$I_0 = \left\{(z_1, z_2) \mapsto \left(e^{i\theta_1} z_1, e^{i\theta_2} z_2\right) \;\middle|\; \theta_1, \theta_2 \in \mathbb{R}\right\}$$

as follows. The invariance of the disc $\{(\alpha, 0) \in \Omega\}$ implies that the matrices in I_0 have the form

$$\begin{pmatrix} \alpha_{11} & \alpha_{12} \\ 0 & \alpha_{22} \end{pmatrix}$$

with $\alpha_{11} \neq 0$ and $\alpha_{22} \neq 0$. If also α_{12} were not zero, then the powers of this matrix (which arise under multiple compositions of the mapping) would not be contained in a compact set in $GL(n, \mathbb{C})$. Thus in fact $\alpha_{12} = 0$.

For $a \in \mathbb{C}, |a| < 1$, consider the mapping

$$S_a : (z_1, z_2) \longmapsto \left(\frac{z_1 - a}{1 - \overline{a} z_1}, \frac{\left(1 - |a|^2\right)^{1/2k}}{\left(1 - \overline{a} z_1\right)^{1/2k}} z_2 \right) .$$

We see that S_a belongs to $\mathrm{Aut}(\Omega)$. This assertion can be easily checked by direct calculation. Also S_{-a} is the inverse mapping of S_a. The orbit of $\mathbf{0}$ under $\mathrm{Aut}(\Omega)$ consequently contains $\{(\alpha, 0) \in \Omega\}$. Again, by the logic of Example 7.2.6, using [13] etc., it follows that the set $\{(\alpha, 0) \in \Omega\}$ is preserved by elements of $\mathrm{Aut}(\Omega)$. Hence the $\mathrm{Aut}(\Omega)$ orbit of $\mathbf{0}$ is equal to $\{(\alpha, 0) \in \Omega\}$. This information then completely determines the automorphism group.

Example 7.2.10. Next look at

$$\mathrm{Aut}\left(\mathbb{D}^2\right), \text{ where } \mathbb{D}^2 = \left\{(z_1, z_2) \in \mathbb{C}^2 \mid |z_1| < 1, |z_2| < 1\right\} .$$

We write $\tau_a(z) = (z - a)/(1 - \overline{a} z)$ for $z \in \mathbb{D} \subseteq \mathbb{C}$. The maps of the form $(z_1, z_2) \mapsto (\tau_{a_1}(z_1), \tau_{a_2}(z_2))$ act transitively on $\mathbb{D}^2$. Also the isotropy subgroup I_0 at the origin $(0, 0)$ consists of linear maps only by Corollary 7.1.2. These linear maps must have the form $(z_1, z_2) \mapsto (\omega_1 z_1, \omega_2 z_2)$ or

$$(z_1, z_2) \mapsto (\omega_2 z_2, \omega_1 z_1) \text{ with } |\omega_1| = |\omega_2| = 1 ,$$

since they must preserve the distinguished boundary $\{(z_1, z_2) \,||z_1| = 1, |z_2| = 1\}$: this set is exactly the points where $\partial \mathbb{D}^2$ is not smooth, and the property of being not smooth is preserved by linear maps. It follows that $\mathrm{Aut}\left(\mathbb{D}^2\right)$ is exactly the group generated by the maps $(z_1, z_2) \mapsto (\tau_{a_1}(z_1), \tau_{a_2}(z_2))$, $(z_1, z_2) \mapsto (\omega_1 z_1, \omega_2 z_2)$ with $|\omega_1| = |\omega_2| = 1$, and $(z_1, z_2) \mapsto (z_2, z_1)$.

Examples (7.2.5) and (7.2.10) yield the following historical theorem of Poincaré, which, as already discussed, shows that the Riemann mapping theorem does not hold in complex dimension higher than 1. The proof of this theorem by Poincaré (see below) demonstrated that automorphism groups could play an important role especially in complex dimensions greater than 1. Of course, we have already shown in the remarks after Corollary 7.1.2 that the ball and the polydisc are not biholomorphic, but Poincaré's proof is of historical interest.

Theorem 7.2.11 (Poincaré). *In complex dimension 2, the ball and the bidisc are not biholomorphic to each other.*

Proof. Suppose that there exists a biholomorphic map $f : B^2 \to \mathbb{D}^2 = \mathbb{D} \times \mathbb{D}$. Composing with an automorphism of $\mathbb{D}^2$, we may assume without loss of generality that f maps the origin to itself. Then the map $f_* : \mathrm{Aut}\left(B^2\right) \to \mathrm{Aut}\left(\mathbb{D}^2\right)$

defined by $f_*(\gamma) \equiv f^{-1} \circ \gamma \circ f$ is a continuous group isomorphism. So, this map generates a group isomorphism between the identity components of the isotropy subgroups at the origin. Note that the identity component of the isotropy subgroup of Aut (B^2) at the origin contains $U(2)$, the group of 2×2 unitary matrices (and indeed $= U(2)$). On the other hand, the identity component of the isotropy subgroup of Aut $(\mathbb{D}^2)$ at the origin is the torus group consisting of rotations in each variable separately together with the map that switches variables. But the latter group is commutative, while $U(2)$ is noncommutative. This is a contradiction. Therefore the desired conclusion follows immediately. □

7.3 Riemann Surfaces and Covering Spaces

In this section, we shall discuss the automorphisms of Riemann surfaces as a preview of the higher-dimensional results to come later. In no sense are we going to try to survey completely the enormous collection of results on the subject obtained in the nineteenth century (cf. [46], and historically [52]) nor the continuing investigation of the subject up to our own time. Even less shall we explore the interaction of the theory of Riemann surface automorphisms with number theory, dynamical systems, and so on. Rather, we are going to focus concretely on the circle of ideas involving invariant metrics, since that subject will be one of our major themes in higher dimensions.

The Riemann surface situation has the attractive property that explicit determination of automorphism-invariant metrics is possible, when they exist, as they do in almost all instances. In particular, when a Riemann surface is a quotient of the unit disc, then the "push-down" of the Poincaré metric will be invariant, as we shall see.

It is worth noting, however, that this particular method of constructing automorphism-invariant metrics on (almost all) Riemann surfaces does not extend to higher dimensions. In particular, the celebrated uniformization theorem (Theorem 7.6.1) of Poincaré and Koebe does not really have a higher-dimensional analogue.

Thus the ideas in the chapter are suggestive of the power and attractiveness of invariant metrics, but the extension to higher dimensions requires an alternative construction. It is interesting in this regard that when the higher-dimensional construction is applied to the case of bounded domains in $\mathbb{C}$, it gives different results, except in the case of the unit disc and its biholomorphic images: first, the Bergman metric of a bounded domain in $\mathbb{C}$ is complete and has constant Gauss curvature if and only if the domain is biholomorphic to the disc.

This is of course possible because in general the automorphism group of a bounded domain is not so large that the invariance of a metric under automorphisms implies anything like the essential uniqueness of the metric (up to a constant factor). Invariance does imply the uniqueness of the metric up to a constant multiple for the unit disc (and its biholomorphic images): every automorphism-invariant metric is a constant multiple of the Poincaré metric. But, in other cases, the automorphism group is too small to force such uniqueness.

7.3.1 Coverings of Riemann Surfaces

Let M be a Riemann surface[4]. As a topological surface, M has a "universal covering space." That is, there is a simply connected surface $\widetilde{M}$ and a "projection" $\pi : \widetilde{M} \to M$ that is a covering space in the topological sense of the word. This means by definition that every point $p \in M$ has a neighborhood U_p with $\pi^{-1}(U_p)$ equaling a disjoint union of open sets $\{V_j\}$ such that $\pi|_{V_j} : V_j \to M$ is a homeomorphism onto U_p. One says here that U_p is *evenly covered.*

The topological surface $\widetilde{M}$ can be given a Riemann surface structure by "pullback," that is, by declaring $\pi|_{V_j}$ to be holomorphic. Thus $\widetilde{M}$ becomes a simply connected Riemann surface and $\pi : \widetilde{M} \to M$ is a holomorphic mapping. Koebe's celebrated uniformization theorem (Theorem 7.6.1) asserts that $\widetilde{M}$ can only be one of three things (up to biholomorphic equivalence): the plane, the unit disc, or the Riemann sphere. And, as we shall see, the plane and the sphere arise only in a few special cases. So the uniformization theorem says that all but a finite number (topologically) of Riemann surfaces have a simply connected covering space biholomorphic to the disc.

It is a standard and easily checked fact from topological covering space theory (for which see, e.g., [55, 61],) that the projection $\pi : \widetilde{M} \to M$ can be interpreted as a quotient by a group action. There is a group Γ, the group of covering transformations, acting as homeomorphisms on $\widetilde{M}$, such that M is the orbit space of the action of Γ. In the Riemann surface case, Γ consists of homeomorphisms which are holomorphic, i.e., elements of $\mathrm{Aut}(\widetilde{M})$.

The possibilities $\widetilde{M} = \mathbb{C}$ and $\widetilde{M} = \mathbb{C} \cup \{\infty\}$ yield few Riemann surfaces. It is again part of general covering space theory that the (nonidentity) elements of Γ act without fixed points, that is, if $\sigma \in \Gamma$ and $\sigma(p) = p$ for some $p \in \widetilde{M}$, then σ is the identity map of $\widetilde{M}$ to $\widetilde{M}$. Now direct calculation shows that the linear fractional transformations that are automorphisms of $\mathbb{C} \cup \{\infty\}$ all have fixed points—this is just elementary algebra. Thus $\mathbb{C} \cup \{\infty\}$ has no covering-space quotients except itself. In other words, in the language of the uniformization theorem, the sphere can cover only the sphere.

When $\widetilde{M} = \mathbb{C}$, quotients are possible. The automorphisms of $\mathbb{C}$,

$$\{z \mapsto az + b \mid a \neq 0, a, b \in \mathbb{C}\} \,,$$

do include maps without fixed points. Specifically, any map of the form $z \mapsto z + b$ has no fixed point; these are in fact the only ones, since, if $a \neq 1$,

[4]A Riemann surface, by definition, is a one-dimensional complex manifold.

then $z = b/(1-a)$ is a fixed point. A familiar analysis shows that the possible Riemann surface quotients that can thus arise are $\mathbb{C}\backslash\{0\}$, topologically a cylinder, and surfaces of genus 1, topologically tori. And these possibilities do indeed occur. But these are the only topological possibilities. All other Riemann surfaces must be quotients of the disc D. Clearly the possibilities for groups acting on D to yield Riemann surface covering quotients must be many and varied. But, as we shall see, much can be said about this situation in spite of its generality.

7.4 Covering Spaces and Invariant Metrics, I: Quotients of $\mathbb{C}$

The examples in Section 7.2 suggest at least two themes that will be prominent in what follows. First, topological complexity tends to make the automorphism group small. The "holes" in Examples (7.2.4), (7.2.7), and (7.2.8) made the group compact; in Example (7.2.8), in fact, the group was the identity alone. Second, Example (7.2.6) illustrates the fact that, in complex dimension 2 or greater, the structure of the boundary can restrict the group even when the topology is as simple as possible. In that case, the isotropy subgroup I_0 at 0 had to be a relatively small subgroup of $U(n)$ because the boundary of the domain had some special boundary points that had to be taken only to other boundary points with the same geometry. Later, we shall see that domains in $\mathbb{C}^n, n \geq 2$, can be homeomorphic to the ball but have no automorphisms other than the identity.

The examples of Section 7.2, however, did not really show anything like the full potential of the idea of invariant metrics. Of course, in accord with general principles described in Section 7.1, invariant metrics exist for bounded domains. These will be constructed explicitly in some generality later, without the use of [141]. But, in the examples considered, the invariant metrics, and the automorphism groups themselves, appeared in rather *ad hoc* and case-specific ways. In the present section we shall describe a much more general construction for invariant metrics in the Riemann surface cases for which such metrics exist. This will be achieved by the systematic use of the uniformization theorem (Theorem 7.6.1) and covering-space techniques.

Since uniformization as such has no analogue in higher dimensions, we shall need other constructions there, as noted in the introduction to this chapter. But the Riemann surface cases are still of interest. As already noted earlier, almost all Riemann surfaces are quotients of the unit disc D rather than of $\mathbb{C}$ or $\mathbb{C} \cup \{\infty\}$. But, as it happens, it will be best to dispose of the latter two possibilities first; this illustrates the general idea in a couple of specific situations in which the calculations are as simple as possible.

First it should be noted that the cases of $\mathbb{C}$ and $\mathbb{C} \cup \{\infty\}$ do not fit into the invariant metric picture[5]. This is clear without any generalities since the maps $\{z \mapsto az : a \in \mathbb{C}\}$ belong to $\mathrm{Aut}(\mathbb{C})$ and, interpreted as taking ∞ to ∞, they also lie in $\mathrm{Aut}(\mathbb{C} \cup \{\infty\})$. These maps fix 0, but obviously preserve no metric at 0 (since they dilate without bound in the infinitesimal sense). Now let us return to Example 7.2.3 of Section 7.2—the punctured plane $\mathbb{C}\setminus\{0\}$. This time around we shall analyze it from the viewpoint of uniformization and covering spaces.

The mapping $\exp : z \mapsto e^z$ takes $\mathbb{C}$ to $\mathbb{C}\setminus\{0\}$. It is easy to see that this is a covering map in the topological sense: For if $z_0 \in \mathbb{C}\setminus\{0\}$, then $\exp^{-1}(\{z : |z - z_0| < |z_0|\})$ consists of disjoint connected open sets, on each of which exp is a homeomorphism onto the disc $\{z : |z - z_0| < |z_0|\}$. This is simply a geometric expression of the familiar fact that the logarithm function has well-defined holomorphic "branches" on discs in $\mathbb{C}\setminus\{0\}$.

Now suppose that $f : \mathbb{C}\setminus\{0\} \to \mathbb{C}\setminus\{0\}$ is an automorphism of $\mathbb{C}\setminus\{0\}$. We shall prove that there exists a holomorphic function $f^* : \mathbb{C} \to \mathbb{C}$ such that

$$f \circ \exp = \exp \circ f^* . \tag{7.1}$$

The meaning of (7.1) is clearest if we think of it in terms of a commutative diagram:

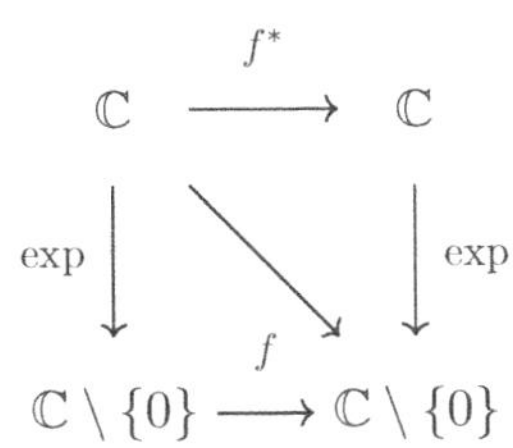

In topological language, f^* is a "lift" of f.

Proof of (7.1). Set $f^*(0) = L(f(1))$, where L is a "holomorphic branch of the logarithm" defined on a neighborhood of $f(1)$, i.e., $\exp(L(z)) \equiv z$ on that neighborhood. The function $L \circ f \circ \exp$ is defined and holomorphic on a neighborhood of 0. And it is easily seen to admit unrestricted analytic continuation (along every curve with initial point 0) in $\mathbb{C}$. By the monodromy theorem, there is a global holomorphic function $f^* : \mathbb{C} \to \mathbb{C}$ such that, in a neighborhood of $0, f^* \equiv L \circ f \circ \exp$. In that neighborhood,

$$\exp \circ f^* = (\exp \circ L) \circ f \circ \exp = f \circ \exp .$$

Hence $\exp \circ f^* = f \circ \exp$ on all of $\mathbb{C}$. □

[5]By an invariant metric, we mean a metric that is invariant under the action of the automorphism group

Remark 7.4.1. We gave the proof of (7.1) in terms of function theory and the monodromy theorem out of respect for tradition and familiarity. But actually the best way to think of it is in terms of pure topology. The mapping

$$f \circ \exp : \mathbb{C} \to \mathbb{C} \backslash \{0\}$$

lifts because $\mathbb{C}$ is simply connected, and every map of a simply connected space lifts to any covering.

Next we want to show that the lift f^* is an automorphism of $\mathbb{C}$. To prove this statement, we first obtain from (7.1) that

$$\begin{array}{ccc} \mathbb{C} & \xrightarrow{f^*} & \mathbb{C} \\ {\scriptstyle \exp}\downarrow & \searrow & \downarrow{\scriptstyle \exp} \\ \mathbb{C}\setminus\{0\} & \xrightarrow{f} & \mathbb{C}\setminus\{0\} \end{array} \quad \text{and} \quad \begin{array}{ccc} \mathbb{C} & \xrightarrow{f^{-1*}} & \mathbb{C} \\ {\scriptstyle \exp}\downarrow & \searrow & \downarrow{\scriptstyle \exp} \\ \mathbb{C}\setminus\{0\} & \xrightarrow{f^{-1}} & \mathbb{C}\setminus\{0\} \end{array}$$

Clearly $f^* \circ \left(f^{-1}\right)^*$ is a lift of the identity map, i.e.,

$$\begin{array}{ccc} \mathbb{C} & \xrightarrow{f^* \circ (f^{-1})^*} & \mathbb{C} \\ {\scriptstyle \exp}\downarrow & \searrow & \downarrow{\scriptstyle \exp} \\ \mathbb{C}\setminus\{0\} & \xrightarrow{\text{id}} & \mathbb{C}\setminus\{0\} \end{array}$$

This means that, for each z, we have

$$\exp\left(f^* \circ \left(f^{-1}\right)^* (z)\right) = \exp(z)\,.$$

It follows that, for each $z, f^* \circ \left(f^{-1}\right)^* (z) - z \in 2\pi i\mathbb{Z}$. By continuity, $f^* \circ \left(f^{-1}\right)^* (z) = z + 2\pi i k$ for some fixed integer k and all $z \in \mathbb{C}$. Similarly, $\left(f^{-1}\right)^* \circ f^*$ is also a translation. We then have that f^* is one-to-one and onto. Hence $f^* \in \text{Aut}(\mathbb{C})$.

Note that the lift $f^* : \mathbb{C} \to \mathbb{C}$ of a given $f \in \text{Aut}(\mathbb{C}\backslash\{0\})$ is far from unique: we were free to choose the "branch" L of the logarithm in the proof of (7.1) arbitrarily. The set of all lifts of the identity map of $\mathbb{C}\backslash\{0\}$ to $\mathbb{C}\backslash\{0\}$ forms a subgroup of $\text{Aut}(\mathbb{C})$. This subgroup is called the *group of covering transformations* of the covering space $\mathbb{C} \xrightarrow{\exp} \mathbb{C}\backslash\{0\}$. Note that a covering transformation is uniquely determined by the image of 0 (or of any pre-chosen point). Moreover, every element of $\exp^{-1}$ is obtainable as an image under a covering transformation of the point 0. These general statements from covering space theory are easy to verify: The covering transformations are exactly the translation mappings $z \mapsto z + 2\pi i k, k \in \mathbb{Z}$, and obviously there is one and only one of these taking 0 to a given element of $\exp^{-1} = 2\pi i\mathbb{Z}$.

Next we want to see which elements of $\text{Aut}(\mathbb{C})$ actually arise as lifts of elements of $\text{Aut}(\mathbb{C}\setminus\{0\})$. One condition for this to happen is clear: an $f^* \in \text{Aut}(\mathbb{C})$ must have the property that

$$\exp\left(f^*(z+2\pi ik)\right) = \exp\left(f^*(z)\right)$$

for all $z \in \mathbb{C}$; this is necessary in order for f^* to arise as the "lift" of some holomorphic map of $\mathbb{C}\setminus\{0\}$ to itself. This necessary condition is also almost sufficient: an element $F \in \text{Aut}(\mathbb{C})$ with $\exp(F(z+2\pi ik)) = \exp(F(z))$ does give rise to some mapping $f : \mathbb{C}\setminus\{0\} \to \mathbb{C}\setminus\{0\}$ by setting $f(w) = \exp(F(z))$ for some $z \in \exp^{-1}(w)$. The condition on F makes this f well defined, i.e., independent of the choice of z in $\exp^{-1}(w)$. And $f^* = F$. But the resulting f may not be an automorphism of $\mathbb{C}\setminus\{0\}$. For instance, if F is the function $z \mapsto 2z$, then the associated f is $z \mapsto z^2$. The answer to this small conundrum is as follows.

Lemma 7.4.2. *An element* $F \in \text{Aut}(\mathbb{C})$ *is of the form* f^* *for some* $f \in \text{Aut}(\mathbb{C}\setminus\{0\})$ *if and only if* F *maps each set of the form* $\{z+2\pi ik \mid k \in \mathbb{Z}\}$ *in a one-to-one, onto fashion onto another set of the same form.*

Proof. Under the condition given in the lemma, $F = f^*$ for some (not necessarily biholomorphic) $f : \mathbb{C}\setminus\{0\} \to \mathbb{C}\setminus\{0\}$. And $F^{-1} = g^*$ for some g. Then $\left(F \circ F^{-1}\right) = (f \circ g)^*$ and $\left(F^{-1} \circ F\right) = (g \circ f)^*$. From this, $f \circ g$ and $g \circ f$ are both the identity on $\mathbb{C}\setminus\{0\}$. □

Now the automorphisms $z \mapsto az+b, a, b \in \mathbb{C}, a \neq 0$, that preserve sets of the form $\{z+2\pi ik : k \in \mathbb{Z}\}$, as required by Lemma 7.4.2, are easily checked to be exactly those of the forms

$$z \mapsto z+b, \quad b \in \mathbb{C}$$

or

$$z \mapsto -z+b, \quad b \in \mathbb{C}\,.$$

The corresponding automorphisms of $\mathbb{C}\setminus\{0\}$ are $z \mapsto e^b z$ and $z \mapsto e^b/z$. These are all the automorphisms of $\mathbb{C}\setminus\{0\}$. Note that this agrees with the result already obtained in Section 7.2, Example (7.2.3).

Since the mappings $z \mapsto z+b$ and $z \mapsto -z+b$ preserve a metric on $\mathbb{C}$ (first, the standard Euclidean metric), it follows that we can obtain an invariant metric on $\mathbb{C}\setminus\{0\}$ by declaring exp to be a local isometry. [This gives a well-defined metric on $\mathbb{C}\setminus\{0\}$ since, in particular, the covering transformations act as isometries.] Since it is Hermitian, this metric on $\mathbb{C}\setminus\{0\}$ is a multiple of the Euclidean metric. This happens first at $z \in \mathbb{C}\setminus\{0\}$,

$$\|\mathbf{v}\| = \left(\frac{1}{|z|}\right) \cdot \|\mathbf{v}\|_{\text{euclid}}\,.$$

(This can be checked by direct calculation.) If we express the assertion in terms of $(\text{length})^2$, then the new metric on $\mathbb{C}\setminus\{0\}$ becomes, in polar coordinates

(r, θ),

$$\frac{1}{r^2}\left(dr^2 + r^2 d\theta^2\right) = \frac{1}{r^2}dr^2 + d\theta^2 .$$

If we reparameterize $\mathbb{C}\backslash\{0\}$ by $R = \ln r$ and θ, then $dR = (1/r)dr$, so the metric becomes $dR^2 + d\theta^2$ for $R \in (-\infty, +\infty)$, $\theta \in [0, 2\pi)$. Thus $\mathbb{C}\backslash\{0\}$ in the new metric is isometric to a right circular cylinder of radius 1.

The automorphism $z \mapsto \left(r_0 e^{i\theta_0}\right) z$ corresponds to translation of R by $\ln r_0$ and rotation in θ by θ_0, i.e., the map$(R, \theta) \mapsto (R + \ln r_0, \theta + \theta_0)$. The automorphism $z \mapsto \left(r_0 e^{i\theta_0}\right)/z$ corresponds to $(R, \theta) \mapsto (-R + \ln r_0, -\theta + \theta_0)$ (the two minus signs make it an orientation-preserving map as required). Thus the automorphisms of $\mathbb{C}\backslash\{0\}$ are visualized in an explicit geometric form as isometries of the cylinder. Moreover, the covering map $\mathbb{C} \to \mathbb{C}\backslash\{0\}$ is visualized as just wrapping the plane around the cylinder.

The other Riemann surfaces covered by $\mathbb{C}$ are the torus family, the Riemann surfaces obtained as quotients by translations in two different directions. Specifically, one defines $z \sim w, z, w \in \mathbb{C}$, if

$$z - w \in \{m\omega_1 + n\omega_2 \mid m, n \in \mathbb{Z}\} ,$$

where ω_1, ω_2 are (fixed) complex numbers linearly independent over $\mathbb{R}$. Then the torus T is obtained as $\mathbb{C}$ modulo this equivalence relation $\sim$. The situation here is much as before: the covering transformations are the maps $z \mapsto z + m\omega_1 + n\omega_2, m, n \in \mathbb{Z}$ fixed. Automorphisms of T lift to automorphisms of $\mathbb{C}$ which preserve the covering-transformation orbits, i.e., preserve the equivalence classes. These automorphisms of $\mathbb{C}$ are exactly the translations $z \mapsto z + b$ and the negation-translations $z \mapsto -z + b$, together with, possibly, some additional elements that arise when there is a $\mathbb{C}$-linear mapping other than $z \mapsto -z$ which takes the lattice

$$\{n\omega_1 + m\omega_2 \mid m, n \in \mathbb{Z}\}$$

in a one-to-one fashion onto itself. Such maps (if any) must be rotations since, if $|c| \neq 1$, then the powers of $z \mapsto cz$ or its inverse would move lattice points $\neq 0 + 0i$ arbitrarily close to $0 + 0i$, a contradiction. The set of lifts of T-automorphisms to $\mathbb{C}$ thus preserves the standard, Euclidean metric on $\mathbb{C}$ so that this metric, pushed down to T, is $\mathrm{Aut}(T)$-invariant. As before, the push-down is well defined because the covering transformations act as isometries.

7.5 Covering Spaces and Invariant Metrics, II: Quotients of D

The construction in the previous section of automorphism-invariant metrics on $\mathbb{C}\backslash\{0\}$ and on 2-tori involved a certain delicate point: while $\mathbb{C}$ has no automorphism-invariant metrics itself, it does possess a metric which is invariant under all the fixed-point-free automorphisms that can potentially be covering transformations for nontrivial covering-space quotients. For the unit disc D, this subtlety does not arise: There is a Riemannian (actually Hermitian) metric on D which is invariant under every element of $\mathrm{Aut}(D)$, fixed-point-free or not. This is of course the well-known Poincaré metric[6] (cf. [105]).

$$\rho = \frac{4}{(1-|z|^2)^2}\left(dx^2+dy^2\right) .$$

In the context of the ideas developed in Section 7.1, the existence of an automorphism-invariant metric on D is anything but a surprise. Indeed, in that context the Poincaré metric arises all but automatically, as follows. Since $I_0 = \{z \mapsto \omega \mid |\omega| = 1, \omega \in \mathbb{C}\}$, there is, up to a constant factor, only one I_0-invariant Riemannian metric at 0—the Euclidean metric dx^2+dy^2. This metric is of course Hermitian. Now, if there is to be an $\mathrm{Aut}(D)$-invariant metric on all of D, then it must be that

$$\begin{aligned} T_a : D &\to D \\ z &\mapsto \frac{z-a}{1-\overline{a}z} \end{aligned}$$

is an isometry[7]. In particular, the length of the real vector $(1,0)$ at a must be equal to the length of $\left.(dT_a)\right|_a (1,0)$ at 0. Now the vector $(dT_a)(1,0)$ has length equal to $|T_a'(a)|$ since the real differential of a holomorphic function f on an open subset of $\mathbb{C}$ acts as a rotation composed with a dilation by the factor $|f'|$. Note that

$$T_a'(a) = \frac{1}{1-|a|^2} .$$

Hence the invariant-metric length of $(1,0)$ at $a \in D$ must be

$$\left(\frac{1}{1-|a|^2}\right) \times (\text{the length of a Euclidean unit vector at } 0) .$$

In other words, the invariant metric in $(\text{length})^2$ form must be, at $a \in D$, given by

$$\frac{c}{(1-|a|^2)^2}\left(dx^2+dy^2\right) .$$

With $c = 4$ we obtain the usual Poincaré metric.

The argument just given shows at first only what the invariant metric must be if it exists. But a similar and more detailed analysis shows that in fact the

---

[6]The constant factor 4 is chosen for geometric convenience: it gives the metric's Gaussian curvature the constant value of -1. For our present purposes, the factor 4 can be regarded as simply historically motivated.

[7]These mappings are often, in the context of function theory, called *Möbius transformations* for all $a \in D$.

metric obtained *must* be invariant. Think of it this way: if $F \in \operatorname{Aut}(D)$ and $F(z) = w$, then F can be written as $T_w^{-1} \circ r \circ T_z$, where r is some element of I_0. This is just group-theoretic formalism since $T_w^{-1} \circ F \circ T_z(0) = 0$. Since T_z, r, and T_w are by construction isometries at $z, 0$, and w, then so is F.

Meanwhile, the following result follows easily from the existence of the $\operatorname{Aut}(D)$-invariant metric and the covering-space ideas introduced in the previous section.

Proposition 7.5.1 (invariant metrics on quotients of D). *Let M be a Riemann surface. If $\pi : D \to M$ is a covering map with π holomorphic, then there is a Hermitian metric H on M such that π is a local isometry (here D is equipped with the Poincaré metric). Moreover, the metric H, which is uniquely determined by the condition that π be a local isometry, has the property that* $\operatorname{Aut}(M)$ *acts on M as orientation-preserving isometries of H. And every orientation-preserving isometry of M with metric H is an element of* $\operatorname{Aut}(M)$.

Outline of the Proof. Since M is the quotient of D by a subgroup Γ of $\operatorname{Aut}(D)$ and since $\operatorname{Aut}(D)$ acts as isometries of the Poincaré metric, the Poincaré metric pushes down to M, as in the earlier discussion. Here we use also the fact that π is a local holomorphic diffeomorphism, and in particular $d\pi$ is nonsingular. An automorphism f of M can be lifted to an automorphism f^* of D, also as before. And since f^* is isometric for the Poincaré metric on D, the original element f must be isometric for H. The final statement follows from the fact that H is Hermitian. □

We turn now to some specific examples.

Example 7.5.2. $M = D\setminus\{0\}$. An explicit universal covering space of M is given by

$$\pi : \{z \in \mathbb{C} : \operatorname{Re} z < 0\} \to M$$
$$z \mapsto \exp z\,.$$

Calculation via the standard Cayley biholomorphic map from the half-plane $\{z \in \mathbb{C} \mid \operatorname{Re} z < 0\}$ to D shows that the Poincaré metric on D transferred to $\{z \in \mathbb{C} \mid \operatorname{Re} z < 0\}$ is $(1/|\operatorname{Re} z|) \cdot$ (the Euclidean metric). First, the usual biholomorphic map of $\{z \in \mathbb{C} \mid \operatorname{Re} z < 0\}$ to D is the composition of $z \mapsto -iz$ (left half-plane to upper half-plane) and $z \mapsto (z-i)/(z+i)$ (upper half-plane to D). So $F : z \mapsto (z+1)/(z-1)$. The derivative of this map is $-2/(z-1)^2$. For F to be an isometry, the Poincaré-induced length of $(1, 0)$ at z must be

$$\begin{aligned}\frac{2\,|F'(z)|}{1-|F(z)|^2} &= \frac{2}{-z-\overline{z}}\\ &= \frac{1}{-\operatorname{Re} z}\\ &= \frac{1}{|\operatorname{Re} z|}\,.\end{aligned}$$

For a similar direct calculation of the induced metric on $D\backslash\{0\}$, note that the length at $\exp z$ of the $d\pi$-image of the real Euclidean vector $(1,0)$ is $(\exp z)(1,0)$. Since the Poincaré length of $(1,0)$ at $z \in \{\zeta \in \mathbb{C} \mid \operatorname{Re}\zeta < 0\}$ is $1/|\operatorname{Re} z|$, the induced-metric H-length of $(1,0)$ at $\exp z \in M$ must be $(1/|\exp z|) \cdot (1/|\operatorname{Re} z|)$. Expressed in terms of z itself,

$$
\begin{aligned}
H\text{-length} &= \frac{1}{|z| \cdot |\ln |z||} \cdot (\text{ Euclidean length }) \\
&= \frac{-1}{|z| \cdot \ln |z|} \cdot (\text{ Euclidean length }).
\end{aligned}
$$

In polar coordinates $z = re^{i\theta}$:

$$
H\text{-length at } z = \left[-\frac{1}{r \ln r}\right] \cdot (\text{ Euclidean length }).
$$

Readers may check directly for themselves that the H-metric has Gauss curvature $\equiv -1$ using the standard formula[8]. In length^2 notation,

$$
\text{Gauss curvature of } e^{2\sigma}\left(dx^2 + dy^2\right) = -\frac{\Delta\sigma}{e^{2\sigma}}.
$$

Here $e^{2\sigma} = \left(1/r^2 \ln^2 r\right)$ so that $\sigma = -\ln r - \ln(|\ln r|)$. Of course, since Gaussian curvature is preserved by local isometry, the fact that the H-metric has Gauss curvature $\equiv -1$ follows from the fact that the Poincaré metric on Δ has Gauss curvature -1. This latter fact can be verified with the same formula but with $\sigma = \ln\left(1 - |z|^2\right) + \ln 2$ on D. This corresponds to the Poincaré metric $\left[4/\left(1 - |z|^2\right)^2\right]\left(dx^2 + dy^2\right)$.

Example 7.5.3. Let $M = \{z \in \mathbb{C} \mid e^{-\alpha} < |z| < e^{\alpha}\}$, some $\alpha > 0$. Here the covering map requires a bit more effort to construct. The exponential $z \mapsto e^z$ maps $U_1 = \{z \in \mathbb{C} \mid -\alpha < \operatorname{Re} z < \alpha\}$ onto M, and this is a covering. On the other hand, $z \mapsto (\pi i/2\alpha)z$ maps U_1 biholomorphically onto $U_2 = \{z \in \mathbb{C} \mid -\pi i/2 < \operatorname{Im} z < \pi i/2\}$, and $z \mapsto \exp z$ takes U_2 biholomorphically to $\{z \in \mathbb{C} : \operatorname{Re} z > 0\}$. The Poincaré-induced metric for this last region is

$$
\text{Poincaré length} = \left(\frac{1}{\operatorname{Re} z}\right) \cdot (\text{ Euclidean length }).
$$

Tracing through the derivatives of these maps gives the H-metric on M:

$$
\begin{aligned}
H\text{-length} &= \left(\frac{\pi/[2\alpha]}{r\cos([\pi \ln r]/[2\alpha])}\right) \cdot (\text{ Euclidean length }) \\
H &= \left(\frac{(\pi/[2\alpha])^2}{r^2 \cos^2([\pi \ln r]/[2\alpha])}\right) \cdot \left(dx^2 + dy^2\right).
\end{aligned}
$$

[8]See [107, p. 184].

It is instructive to write this metric in polar coordinates, where

$$dx^2 + dy^2 = dr^2 + r^2 d\theta^2 \,.$$

Thus

$$H = \left(\frac{(\pi/[2\alpha])^2}{r^2\cos^2((\pi\ln r)/(2\alpha))}\right) dr^2 + \left(\frac{(\pi/[2\alpha])^2}{\cos^2((\pi\ln r)/(2\alpha))}\right) d\theta^2 \,.$$

The homology generator of M which goes once around the point 0 counter-clockwise has as representatives closed curves along which the total change of θ is $+2\pi$. For such a curve that is, say, piecewise X^1 and is written in the form $(r(t), \theta(t)), t \in [0, 1]$, one has

$$\int_0^1 \frac{d\theta}{dt} dt = 2\pi \,.$$

Now the H-length of the curve is, with $\beta = \pi/(2\alpha)$,

$$\begin{aligned}
&\int_0^1 \sqrt{\left(\frac{\beta^2}{r^2\cos^2(\beta\ln r)}\right)\left(\frac{dr}{dt}\right)^2 + \left(\frac{\beta^2}{\cos^2(\beta\ln r)}\right)\left(\frac{d\theta}{dt}\right)^2}\, dt \\
&\geq \int_0^1 \frac{\beta}{\cos(\beta\ln r)} \cdot \left|\frac{d\theta}{dt}\right| dt \\
&\geq \int_0^1 \beta \cdot \frac{d\theta}{dt} dt \\
&= 2\pi\beta \,.
\end{aligned}$$

When $(r(t), \theta(t)) \equiv (1, 2\pi t)$, the inequalities become equalities and that is, up to parameterization, the only case when they do. Thus the "central circle" $r = 1$ is the unique minimum-length curve in its homology class. As such, by standard geometry, it must be a geodesic for the H-metric. Refer to Fig. 7.1.

In particular, this central circle must be taken to itself by an automorphism, except of course its direction may be reversed, since the circle traversed clockwise is the other homology generator.

Now we are able to interpret the automorphisms of

$$\left\{z \in \mathbb{C} \;\middle|\; e^{-\alpha} < |z| < e^{+\alpha}, \alpha > 0\right\} ,$$

as determined in Example 7.2.4 of Section 7.2. The automorphisms, being isometries, come in two classes. The first is that in which the central circle and its orientation are preserved. These are rotations $z \mapsto \omega z$, $|\omega| = 1$, and these act as isometries since the expression for the H-metric has no θ-dependence of its coefficients.

The second class of isometries that preserve orientation and are hence automorphisms is that in which the central circle is mapped to itself with opposite orientation.

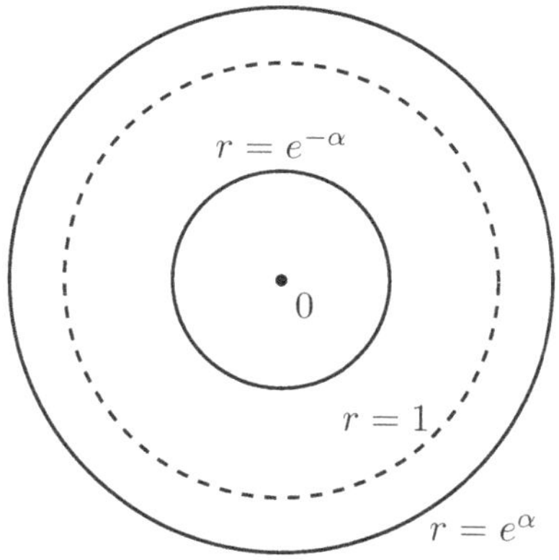

FIGURE 7.1: The central geodesic.

For orientation to be preserved, the outward directions perpendicular to the circle must be mapped to inward directions. The maps have the form $(r, \theta) \mapsto (1/r, -\theta + \theta_0)$ or, equivalently, $z \mapsto \omega/z, |\omega| = 1$. That these are isometries arises from the θ-independence of the H-metric coefficients together with the observations that

$$\frac{\beta}{(1/r)\cos^2\left(\frac{\pi}{2\alpha} \cdot \ln \frac{1}{r}\right)} = r^2 \cdot \frac{\beta}{r\cos^2\left(\frac{\pi}{2\alpha} \cdot \ln r\right)}$$

while $|(1/z)'| = 1/|z|^2$. The reader is invited to trace through the logic to see that consequently $z \mapsto 1/z$ is length-preserving for the H-metric.

One first observes (with $\beta = \pi/[2\alpha]$ as before) that

$$\int_1^{e^{\alpha}} \frac{\beta}{r\cos(\beta \ln r)} dr = +\infty \, .$$

This follows from elementary estimates. Similarly,

$$\int_{e^{-\alpha}}^{1} \frac{\beta}{r\cos(\beta \ln r)} dr = +\infty \, .$$

From this information, and from the H-metric in polar form, it follows that any curve $\gamma : [0, 1) \to \Omega = \{z \in \mathbb{C} : e^{-\alpha} < |z| < e^{+\alpha}\}$ with $|\gamma(0)| = 1$ and $\lim_{t\uparrow 1} |\gamma(t)| = e^{\alpha}$ has infinite length. Similarly, if $\lim_{t\uparrow 1} |\gamma(t)| = e^{-\alpha}$, then γ has infinite length. In an obvious sense, it is "infinite H-distance" to the boundary of Ω from one (and hence any) interior point of Ω. This property is equivalent to the domain being complete in the H-metric in the sense of Riemann-geometric completeness. This completeness property holds for all the metrics that arise in the situation of Proposition 7.5.1. This matter will be discussed in more detail and in greater generality later in the book. Meanwhile, one can check that the H-metric on $\{z \in \mathbb{C} \mid 0 < |z| < 1\}$ also gives infinite distance to the boundary from each interior point.

It is reassuring to be able to visualize the covering maps of Examples 7.5.2 and 7.5.3 geometrically, analogous to visualizing the covering $\mathbb{C} \to \mathbb{C}\backslash\{0\}$ as mapping a plane onto a cylinder. This visualization is in fact almost exactly analogous to the cylinder case.

In detail, the curve in polar coordinates with $\theta = 0$ and r varying from $e^{-\alpha}$ to $e^{+\alpha}$ is, up to parameterization, a geodesic. This is clear since it is a shortest curve joining any two of its points (cf. the argument for why it is infinite distance to the boundary). Alternatively, one can see that this curve is a component of the fixed-point set of the (orientation-reversing, nonholomorphic) isometry $(r,\theta) \mapsto (r,-\theta)$ and hence must be a geodesic. As in basic Riemann surface theory for function elements, we can use this geodesic as a "cut" and obtain the simply connected covering space of $\{z \in \mathbb{C} \mid e^{-\alpha} < |z| < e^{+\alpha}\}$ by connected "sheets" along the cut. Translated into the unit disc D, this process "develops" $\{z \in \mathbb{C} \mid e^{-\alpha} < |z| < e^{+\alpha}\}$ on D isometrically. A fundamental domain, analogous to the strip with parallel sides glued together to form a cylinder, is given by the region between two Poincaré-metric geodesics in D, each perpendicular to a third geodesic, with the points of intersection separated by Poincaré distance $2\pi\beta = \pi/\alpha$. This is of course the length of the "central circle" geodesic $\{z \in \mathbb{C} \mid |z| = 1\}$ of $\{z \in \mathbb{C} \mid e^{-\alpha} < |z| < e^{+\alpha}\}$ in the H-metric.

Without loss of generality, we can take the third geodesic in D to be the x-axis and the points of intersection to be $-a + i0$ and $a + i0$, where $a > 0$. Here a must satisfy

$$\int_{-a}^{a} \frac{2}{1-t^2}dt = \frac{\pi^2}{\alpha}.$$

Therefore $a = \left(e^{\pi^2/[2\alpha]} - 1\right) / \left(e^{\pi^2/[2\alpha]} + 1\right)$. In this situation, the group Γ of covering transformations is generated by $T_a \circ T_{-a}$, which takes $-a + 0i$ to $a + 0i$. [Note also that the map $T_a \circ T_{-a}$ takes z to $(z+b)/(1+bz)$, where $b = 2a/\left(1+a^2\right)$.] Then D/Γ becomes the fundamental domain (the region between the two geodesics perpendicular to the x-axis) with its edges glued together via the identification map $T_a \circ T_{-a}$. As noted, this is truly the analogue of the flat cylinder situation $\mathbb{C} \to \mathbb{C}\backslash\{0\}$. In this setup, the generators of the cylinder, the parallel lines that form the cylinder, are replaced by H-geodesics from the inner edge of the annulus to the outer edge—first those of constant θ value (lying in rays from the origin). And these H-geodesics are images of the Poincaré metric geodesics perpendicular to the x-axis. The boundary of our fundamental domain, consisting of two Poincaré metric geodesics, has each of these Poincaré metric geodesics mapping onto the H-metric geodesic $re^{-i\pi}$ in $\{z \in \mathbb{C} \mid e^{-\alpha} < |z| < e^{+\alpha}\}$, where r varies from $e^{-\alpha}$ to $e^{+\alpha}$. Refer to Fig. 7.2.

The situation of $D \to D\backslash\{0\}$ in Example 7.5.2 is conceptually similar but different in that there is no geodesic "central circle." As before, we can use the, $\theta = 0, r \in (0,1)$ curve (which is a geodesic for the same reasons as before) as a "branch cut." We can begin our development map unwrapping the universal cover of $D\backslash\{0\}$ onto D by sending the $\theta = \pi$ geodesic to the y-axis, with

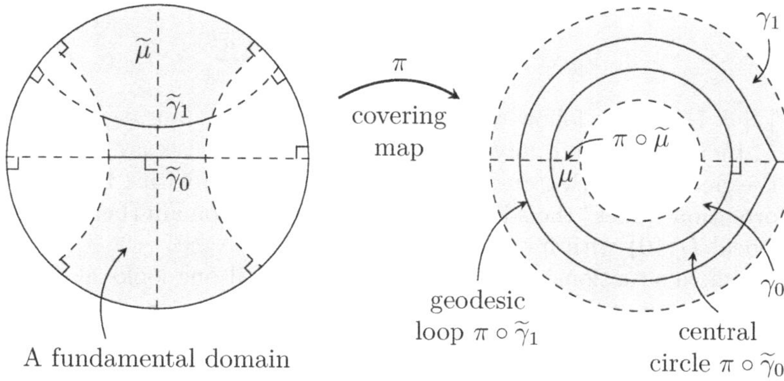

FIGURE 7.2: The mapping $D \to A \equiv \{z \in \mathbb{C} \mid e^{-\alpha} < |z| < e^{\alpha}\}$.

$r \to 1$ corresponding to $y \to 1$: to normalize the precise correspondence, we could, for instance, send $r = 1/2, \theta = -\pi$, to $0 + i0 \in D$. The fundamental domain becomes the region between the two geodesics σ_1, σ_2 as shown, where the distance between the two goes to 0 at the boundary point. See Fig. 7.3.

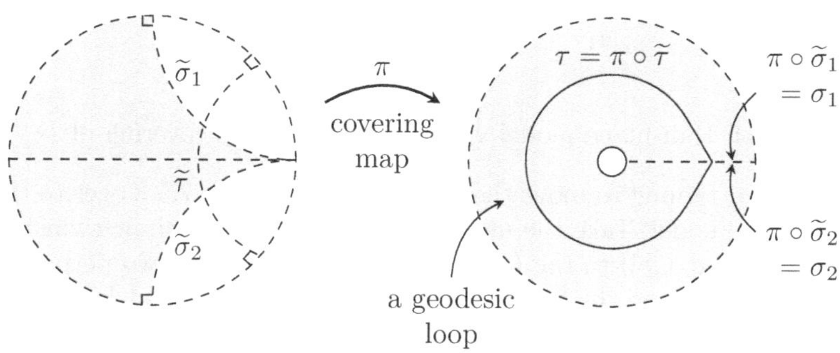

FIGURE 7.3: Fundamental domain for $D \to D\backslash\{0\}$.

The two geodesics σ_1, σ_2 in D are each a lift of the $\theta = 0$ geodesic in $D\backslash\{0\}$ with r varying. The fact that they approach each other in the limit corresponding to $r \to 0^+$ is a consequence of the fact that the H-length of a (Euclidean) circle of radius r around $0 + i0$ goes to 0 as $r \to 0^+$. The H-length of the circle is in fact $2\pi/|\ln r|$.

The group of covering transformations Γ on D needs to have as generator the mapping that moves σ_1 to σ_2 while fixing the boundary point $1 \in \partial D$ at which σ_1 and σ_2 "meet." Thus Γ should be a cyclic subgroup of the classical horocycle transformations at 1.

The form of the transformations, which are supposed to be automorphisms of D, is determined by requiring that $z \mapsto \omega \cdot (z - a)/(1 - \overline{a}z)$ should fix 1, i.e.,

$$1 = \omega \cdot \frac{-i - a}{1 + \overline{a}i} \quad \text{or} \quad \omega = \frac{1 + \overline{a}i}{-i - a}.$$

Since $|\omega| = 1$, it must follow that a is real. Furthermore, for each real a with $|a| < 1$, there is one and only one possible ω value. It is easy to check that there is exactly one such transformation taking σ_1 to σ_2 with 1 fixed, and this transformation "glues" the edges of the fundamental domain (between σ_1 and σ_2) to yield $D \setminus \{0\}$ with its H-metric.

This whole situation is even easier to visualize if one looks at the upper half-plane version of the Poincaré metric: the geodesics σ_1, σ_2 become vertical half-lines emanating from points on the x-axis and meeting "at infinity," and the covering transformation group is generated by a translation $z \mapsto z + c, c$ real. This picture really does look cylindrical. The reader is invited to explore this matter in more detail. Refer to Fig. 7.4.

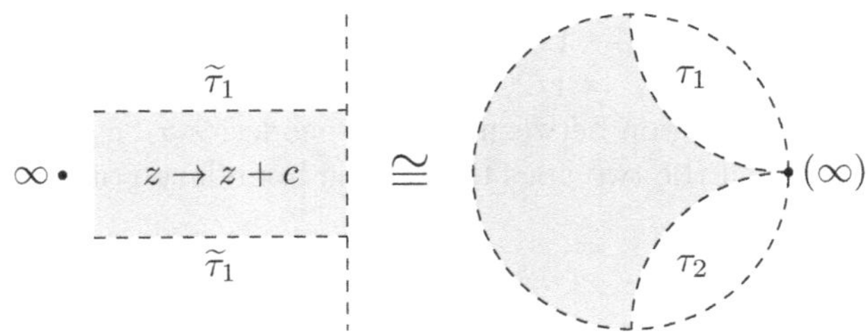

FIGURE 7.4: Half-plane model vs. disc model for the covering of $D \setminus \{0\}$.

It is also intriguing to move the situation of Example 7.5.3 over to the upper half-plane model. Take the unwrapping of the central circle (which is the x axis in the D-model) to the (geodesic) x-axis. Then the two perpendicular geodesics which bound the fundamental domain in our D-model become half-circles around the origin. And the group of covering transformations, again of course a cyclic group, is now generated by $z \mapsto cz, c > 0$, where c is the ratio of the radii of the two half-circles. Again the reader is invited to explore the details.

Covering Spaces and Automorphisms in General

While much of the material of the previous two sections was very specific, underlying the considerations there and illustrated by them were some general principles. In this section, we shall make these principles explicit. This is well worthwhile in spite of the tedium of the notation involved because these principles have wide applications.

Proposition 7.5.4. *Suppose that $\widetilde{M}$ and M are complex manifolds with $\widetilde{M}$ simply connected and further suppose that $\pi : \widetilde{M} \to M$ is a holomorphic mapping which is also a covering in the sense of topology so that, in particular, π is a local holomorphic diffeomorphism. Let Γ be the subgroup of* $\mathrm{Aut}(\widetilde{M})$

consisting of the covering transformations of $\widetilde{M}$ *over* M. *In this case these transformations are necessarily holomorphic. Then:*

1. *For each* $f \in \mathrm{Aut}(M)$, *there is an element* $\widehat{f}$ *of* $\mathrm{Aut}(\widetilde{M})$ *such that* $f \circ \pi \equiv \pi \circ \widehat{f}$. *Any such* $\widehat{f}$ *maps each orbit of* Γ *bijectively onto another orbit of* Γ. *Moreover, given* $p \in M, \widehat{p} \in \widetilde{M}$ *with* $\pi(\widehat{p}) = p$, *and* $\widehat{q} \in \widetilde{M}$ *with* $\pi(\widehat{q}) = f(p)$, *there is exactly one such* $\widehat{f}$ *with* $\widehat{f}(\widehat{p}) = \widehat{q}$.

2. *Conversely, if* $F \in \mathrm{Aut}(\widetilde{M})$ *and if* F *maps each orbit of* Γ *bijectively onto another orbit of* Γ, *then there is exactly one* $f \in \mathrm{Aut}(M)$ *such that* $\pi \circ F \equiv f \circ \pi$.

Proof. If $f : M \to M$ is a holomorphic mapping, whether an automorphism or not, then there is exactly one continuous mapping $\widehat{f} : \widetilde{M} \to \widetilde{M}$ with $f \circ \pi = \pi \circ \widehat{f}$ and $\widehat{f}(\widehat{p}) = \widehat{q}$. This is just the standard lifting argument for covering spaces $\widetilde{M} \to M$ with $\widetilde{M}$ simply connected. Since π is a holomorphic local diffeomorphism, it follows that $\widehat{f}$ must be holomorphic. To see that such an $\widehat{f}$ is an automorphism of $\widetilde{M}$, let $\widehat{g}$ be the lift to $\widetilde{M}$ of $f^{-1} : M \to M$ with $\widehat{g}(\widehat{q}) = \widehat{p}$, i.e., $\pi \circ \widehat{g} = f^{-1} \circ \pi$. Now, since π is a local diffeomorphism and since $g \circ f$ is the identity, it follows that $\widehat{g} \circ \widehat{f}$ is the identity in a neighborhood of $\widehat{p}$. By uniqueness of analytic continuation, $\widehat{g} \circ \widehat{f}$ is the identity on all of $\widetilde{M}$. Similarly, $(\widehat{f} \circ \widehat{g})(\widehat{q}) = \widehat{f}(\widehat{p}) = \widehat{q}$ and $\widehat{f} \circ \widehat{g}$ is also the identity on $\widetilde{M}$. So $\widehat{f} \in \mathrm{Aut}(\widetilde{M})$.

Now, for $x \in \widetilde{M}$ and $\gamma \in \Gamma$,

$$\pi\widehat{f}(\gamma(x)) = f\pi(\gamma(x)) = f(\pi(x)) .$$

So $\widehat{f}(\gamma(x))$ is in the Γ-orbit of $\widehat{f}(x)$ since the orbits of Γ are exactly the sets of the form $\pi^{-1}(\{\text{points of } M\})$. Now, with $\widehat{g}$ as given, the same holds for $\widehat{g}$ since $\widehat{g}$ is a lift of f^{-1}: the map $\widehat{g}$ sends Γ-orbits to Γ-orbits. Since $\widehat{f}$ and $\widehat{g}$ are inverses, the maps by $\widehat{f}$ of Γ-orbits to Γ-orbits must each be bijective.

For the second statement, note that under the hypothesis that $F : \widetilde{M} \to \widetilde{M}$ maps Γ-orbits (that is, sets of the form $\pi^{-1}(\{p\})$) to other sets of the same form, set-theoretic considerations show that F induces a map on M. First, set $f(x) = \pi \circ F(\widehat{x})$ for any $\widehat{x} \in \pi^{-1}(\{x\})$; this f is well defined by hypothesis. The full hypothesis on F, including the bijectivity on inverse image sets, means that F^{-1} also pushes down to a well-defined map $g : M \to M$. It is easy to check that $f \circ g$ and $g \circ f$ are the identity on M, so that $f \in \mathrm{Aut}(M)$. □

This "general nonsense" argument becomes clearer if one thinks of it in terms of the orbits of the covering transformation group Γ acting on $\widetilde{M}$. The hypothesis of the second statement says then that F maps each Γ-orbit bijectively onto another Γ-orbit, and so does F^{-1}. The latter condition is needed, since the orbits are in general infinite sets. As an instance, the map $z \mapsto 2z$ maps orbits of Γ for $\exp : \mathbb{C} \to \mathbb{C}\setminus\{0\}$ into themselves: if $z_1 - z_2 = 2\pi ik$ then

$2z_1 - 2z_2 \in 2\pi i\mathbb{Z}$. But the induced map on $\mathbb{C}\setminus\{0\}$ is $z \mapsto z^2$, which has no inverse (see Section 7.4).

Now the set of all $F \in \mathrm{Aut}(\widetilde{M})$ such that both F and F^{-1} map orbits to orbits is a subgroup of $\mathrm{Aut}(\widetilde{M})$. Moreover, this subgroup is closed. One can see this from the definition of covering space and the group Γ. Suppose that the sequence $\{F_j\}$ converges uniformly on compact sets to some F_0. For fixed $\gamma_0 \in \Gamma$ and $p_0 \in \widetilde{M}$, the images $F_j(\gamma_0(p_0))$ by hypothesis belong to the same orbit of $F_j(p_0)$ so that $F_j(\gamma_0(p_0)) = \gamma_j(F_j(p_0))$ for some unique $\gamma_j \in \Gamma$. Now choose a neighborhood of $F_0(\gamma_0(p_0))$, say U, (sufficiently small) so that the conditions $\sigma_1(U) \cap \sigma_2(U) \neq \emptyset$ and $\sigma_1, \sigma_2 \in \Gamma$ imply $\sigma_1 = \sigma_2$. This is possible by the definition of covering space. Then, for j_0 so large that $j \geq j_0$ implies that $F_j(\gamma_0(p_0)) \in U$, it must be that γ_j is independent of j. For clearly $F_0(p_0) \in \gamma_{j_1}^{-1}(U) \cap \gamma_{j_2}^{-1}(U)$ for any $j_1, j_2 > j_0$. Thus $F_0(\gamma_0(p_0)) = \gamma_\infty(F_0(p_0))$, where γ_∞ is the eventually constant value of γ_j. In particular, F_0 maps orbits to orbits. (A similar argument applies to F^{-1} to show that F^{-1} also maps orbits to orbits.)

Proposition 7.5.5. *With $\pi : \widetilde{M} \to M$ as in Proposition 7.5.4, let G be the subset of* $\mathrm{Aut}(\widetilde{M})$ *consisting of those $F \in \mathrm{Aut}(\widetilde{M})$ such that F maps each orbit of Γ bijectively to another orbit of Γ. The covering transformation group Γ of $\widetilde{M}$ over M is a closed, normal subgroup of G; and* $\mathrm{Aut}(M)$ *is homeomorphically isomorphic to G/Γ.*

Proof. Almost everything follows from arguments already presented. We check only the normality of Γ in G. Suppose that $F \in G$ and $\gamma \in \Gamma$. We need to see that $F \circ \gamma \circ F^{-1}$ is a covering transformation. Obviously $F \circ \gamma \circ F^{-1} \in \mathrm{Aut}(\widetilde{M})$, so we need only check that $\pi \circ (F \circ \gamma \circ F^{-1}) = \pi$. Now, for $p \in \widetilde{M}$, we see that $\gamma \circ F^{-1}(p)$ is an element of the set of Γ-orbits of $F^{-1}(p)$, and hence $F(\gamma(F^{-1}(p)))$ is an element of the Γ-orbit of $F(F^{-1}(p))$. This is so because F applied to a point of an orbit of $q \in \widetilde{M}$ is equal to the orbit of the point $F(q)$ by hypothesis. Note that

$$\pi\left(\Gamma\text{-orbit of } F\left(F^{-1}(p)\right)\right) = \pi(\Gamma\text{-orbit of } p) = \{\pi(p)\}.$$

The remainder of the proof of the proposition, which follows the patterns already established, is left to the reader. □

Corollary 7.5.6. *If* $\mathrm{Aut}(\widetilde{M})$ *is a Lie group, then* $\mathrm{Aut}(M)$ *is a Lie group.*

Proof. Let G be the same as Proposition 7.5.5 Since a closed subgroup of a Lie group is a Lie group, G is a Lie group. A quotient of a Lie group by a closed, normal subgroup is again a Lie group. Hence G/Γ is a Lie group. □

Proposition 7.5.7. *If* $\mathrm{Aut}(\widetilde{M})$ *has compact isotropy subgroups, then so does* $\mathrm{Aut}(M)$.

Proof. Given $p \in M$, choose $\widehat{p} \in \widetilde{M}$ such that $\pi(\widehat{p}) = p$. By Proposition 7.5.4 (a), we see that, for each $f \in I_p$, there is exactly one $\widehat{f} \in I_{\widehat{p}}$ with $\pi \circ \widehat{f} = f \circ \pi$. And $\widehat{f}$ maps each orbit of Γ bijectively onto another orbit. Also, by Proposition 7.5.4 (b), the set of all $F \in I_{\widehat{p}}$ that map each Γ-orbit bijectively to a Γ-orbit is exactly the set of such $\widehat{f}$. Moreover, this set of F is a closed subgroup, say $G_{\widehat{p}}$ of $I_{\widehat{p}}$, by the proof of Proposition 7.5.5. Also $G_{\widehat{p}}$ is compact since $I_{\widehat{p}}$ is, by hypothesis. And with G as in Proposition 7.5.5, $G_{\widehat{p}} \subset G$ maps, under the continuous map $G \to G/\Gamma$, onto a subgroup corresponding to I_p under the isomorphism $\mathrm{Aut}(M) \cong G/\Gamma$. Indeed $G_{\widehat{p}}$ maps isomorphically onto the subgroup of G/Γ corresponding to I_p. In particular, since $G \to G/\Gamma$ is continuous and $G_{\widehat{p}}$ is compact, it follows that I_p is compact. □

In the case when M has an $\mathrm{Aut}(M)$-invariant metric, all the previous considerations become considerably simplified.

Theorem 7.5.8. *Let M be a compact, complex manifold. Suppose that d: $M \times M \to \mathbb{R}$ is a metric space structure on M, with the property that the metric space topology for d is the same as the manifold topology. If d is* $\mathrm{Aut}(M)$*-invariant in the sense that $d(f(p), f(q)) = d(p, q)$ for all $f \in \mathrm{Aut}(M)$ and all $p, q \in M$, then* $\mathrm{Aut}(M)$ *is a compact Lie group.*

Proof. If $\{f_j\}$ is a sequence in $\mathrm{Aut}(M)$ then, by the invariance of d, the sequence is d-equicontinuous and indeed d-Lipschitz continuous with constant 1. By the Arzela–Ascoli theorem and the compactness of M, there is a subsequence $\{f_{j_k}\}$ that converges uniformly on M to a continuous d-isometric map $f_0 : M \to M$. Passing again to a subsequence if necessary, we can suppose that the sequence $\{f_{j_k}^{-1}\}$, which is also d-equicontinuous, converges uniformly on M to a limit $g_0 : M \to M$, also d-isometric. It is now elementary to see that $f_0 \circ g_0$ and $g_0 \circ f_0$ are both the identity map of M to M. Also, d-uniform convergence implies uniform-on-compact-subsets convergence in local coordinates, since the d-topology is equivalent to the manifold topology. So f_0 is holomorphic and invertible, and $f_0 \in \mathrm{Aut}(M)$.

Since $\mathrm{Aut}(M)$ is compact, it is certainly locally compact. By the Bochner–Montgomery theorem, $\mathrm{Aut}(M)$ is a Lie group. □

We turn now to some general circumstances under which a complex manifold M has an $\mathrm{Aut}(M)$-invariant metric.

Proposition 7.5.9. *If $\pi : \widetilde{M} \to M$ is a holomorphic covering space with $\widetilde{M}$ simply connected and if $\widetilde{H}$ is an Hermitian metric on $\widetilde{M}$ that is invariant under the subgroup Γ of* $\mathrm{Aut}(\widetilde{M})$ *consisting of the covering transformations of $\pi : \widetilde{M} \to M$, then there is a Hermitian metric H on M such that π is a local isometry, and H is uniquely determined by this condition. If, in addition, $\widetilde{H}$ is* $\mathrm{Aut}(\widetilde{M})$*-invariant, then H is* $\mathrm{Aut}(M)$*-invariant.*

Proof. The proof of the existence and uniqueness of H follows a by-now-familiar pattern. To determine H at $p \in M$, choose $\widehat{p}$ with $\pi(\widehat{p}) = p$. Then define H at p by requiring $d\pi|_{\widehat{p}}$ to be isometric from $\widetilde{H}$ at $\widehat{p}$ to H at p. Since $d\pi|_{\widehat{p}}$ is a vector space isomorphism, this gives one and only one H at p. The metric H is well defined since any other element of $\pi^{-1}(p)$ has the form $\gamma(\widehat{p})$ for some $\gamma \in \Gamma$ and γ is an isometry of $\widetilde{H}$.

For the second statement, it suffices to apply Proposition 7.5.4. To wit, if $f \in \text{Aut}(M)$, then there is an $\widehat{f} \in \text{Aut}(\widetilde{M})$ with $f \circ \pi = \widehat{f} \circ \pi$. Since $\widehat{f}$ is $\widetilde{H}$-isometric and since π is locally $\widetilde{H}$-to-H isometric, we see that f must be H-isometric. □

Proposition 7.5.9 was more or less obvious. What is somewhat less obvious is that the completeness of $\widetilde{H}$ and of H happen together. An even more general result holds:

Proposition 7.5.10. *Let $\widetilde{M}$ and M be Riemannian manifolds equipped with metrics $\widetilde{H}$ and H, respectively. Suppose that $\pi : \widetilde{M} \to M$ is a covering, and that π is locally isometric. Then the $\widetilde{M}$ metric $\widetilde{H}$ is complete if and only if the M metric is complete.*

Remark 7.5.11. This is a standard result in Riemannian geometry.

For the fact that completeness in the Cauchy sense is equivalent for a Riemannian manifold to infinite extendibility of geodesics see [107]. Geodesics in M can be lifted to geodesics in $\widetilde{M}$ because continuous curves can always be lifted. It is also the case that the image under π of a geodesic in $\widetilde{M}$ is a geodesic in M. Thus infinite extendibility in $\widetilde{M}$ happens precisely when it does so in M.

The most obvious application of Proposition 7.5.10 is to quotients of D with its Poincaré metric.

Corollary 7.5.12. *If $\pi : D \to M$ is a holomorphic covering space of a Riemann surface M, then the H-metric (in our previous terminology) on M is complete.*

The Poincaré metric on the disc D is of course complete. To check the corollary by hand, from the geodesic extendibility viewpoint, it is enough to deal with geodesics through 0. These are, up to parameter, straight lines: the form of the metric in polar coordinates shows immediately that these are minimal-length connections, hence geodesics. With the Poincaré metric arc length parameter, they have the form in polar coordinates given[9] by

$$\sigma_{\theta_0}(t) = \left(\frac{e^t - 1}{e^t + 1}, \theta_0\right), \quad \theta_0 \in [0, 2\pi) .$$

[9] To check for arc length parameterization, calculate

$$\frac{dr}{dt} = \frac{d}{dt}\left(\frac{e^t - 1}{e^t + 1}\right) = \frac{2e^t}{(1 + e^t)^2}$$

In particular, the geodesics σ_{θ_0} are defined on $(-\infty, +\infty)$. So D is complete. This completeness is also clear from the compactness of closed metric balls (the equivalence of all these different notions of completeness is the famous Hopf–Rinow theorem, treated in all Riemannian geometry texts, e.g., [83, 106, 142]). Explicitly

$$\left\{z \in \mathbb{C} \;\middle|\; \mathrm{dis}_{\widetilde{H}}(0, z) \leq r\right\} = \{z \in \mathbb{C} \mid |z| \leq (e^r - 1) / (e^r + 1)\}$$

which is compact in D. So the (quotient) H-metrics on Riemann surfaces covered by D are necessarily complete.

7.6 Compact Quotients of $\mathbb{D}$ and Their Automorphisms

In this section, we shall investigate the automorphism groups of compact Riemann surfaces. We have already discussed the situation for the Riemann sphere and for the torus quotients of $\mathbb{C}$. In actuality, all other compact Riemann surfaces are quotients of $\mathbb{D}$ or, in other words, their universal covering space is biholomorphic to $\mathbb{D}$. We almost have this result already, but not quite. The "not quite" arises from the fact that we need to show that a torus can only arise as a quotient of $\mathbb{C}$; it cannot have $\mathbb{D}$ as a (holomorphic) universal cover. Before proving this statement, we quickly remind the reader of some basic results about the topology of compact surfaces (see, e.g., [126]).

According to nineteenth century work of Jordan and Möbius, every compact orientable surface M (i.e. topological 2-manifold) is homeomorphic either to the 2-sphere S^2 or to the 2-sphere with g handles attached (here g is a finite, positive integer). The number g is uniquely determined; g is called the *genus* of the surface[10]. The homology groups of the surface M are:

- $H_0(M, \mathbb{Z}) \cong \mathbb{Z}$
- $H_1(M, \mathbb{Z}) \cong \mathbb{Z} \oplus \cdots \oplus \mathbb{Z} \quad (2g \text{ times});$
- $H_2(M, \mathbb{Z}) \cong \mathbb{Z}$
- $H_k(M, \mathbb{Z}) = 0$ if $k > 2$.

hence, at the point $\left(\left(e^t - 1\right) / \left(e^t + 1\right), \theta_0\right)$, the length of the vector $\sigma'(t)$ is

$$\left(\frac{2e^t}{(1 + e^t)^2}\right) \Big/ \left(\frac{2}{1 - \left(\frac{e^t - 1}{e^t + 1}\right)^2}\right) = 1\,.$$

[10]By convention, the sphere S^2, having no handles, is considered to have genus 0.

In particular, the Euler characteristic of M is $2 - 2g$.[11] No two surfaces of different genus are homeomorphic or even of the same homotopy type. This assertion follows immediately from the homology results, e.g., because the first homology group is the abelianization of the first homotopy group (i.e. the fundamental group) and that group is clearly a homotopy type invariant. Or one can use directly the homotopy type invariance of homology. If $g = 0$ then the fundamental group of a surface M of genus g is the one-element group $\{0\}$. If $g \geq 1$, then the fundamental group has $2g$ generators $a_1, b_1, \ldots, a_g, b_g$ subject to the one relation

$$a_1 b_1 a_1^{-1} b_1^{-1} a_2 b_2 a_2^{-1} b_2^{-1} \cdots a_g b_g a_g^{-1} b_g^{-1} = 1\,.$$

The group $\pi_1(M)$ is noncommutative if $g \geq 2$.

We now turn to the result which shows the relationship between the topology of the surface and the complex structure of the universal cover.

Theorem 7.6.1 (uniformization of compact Riemann surfaces: Poincaré and Koebe)**.** *Let M be a compact Riemann surface of genus g and $\pi : \widetilde{M} \to M$ its (holomorphic) universal cover. Then:*

1. *If $g = 0$, then $\widetilde{M} = M$ and M is conformally equivalent to the Riemann sphere $\mathbb{C} \cup \{\infty\}$;*

2. *If $g = 1$, then $\widetilde{M}$ is conformally equivalent to $\mathbb{C}$ and M is obtained from $\mathbb{C}$ by quotienting out by a group of the form $\Gamma = \{m\omega_1 + n\omega_2 : m, n \in \mathbb{Z}\}$, with ω_1 and ω_2 being $\mathbb{R}$-linearly independent complex numbers.*

3. *If $g \geq 2$, then $\widetilde{M}$ is conformally equivalent to $\mathbb{D}$ and M is $\mathbb{D}/\Gamma$. Here Γ is a subgroup of* $\mathrm{Aut}(\mathbb{D})$.

Proof. We have already covered all the issues involved with this argument save one: We need to show that a surface of genus 1 cannot arise as a quotient of $\mathbb{D}$. This can be established by working with the geometry of $\mathbb{D}$ and its holomorphic quotients in terms of the group actions of covering transformations, but it is much quicker to use a result of classical differential geometry.

The torus, the topological surface of genus 1, has Euler characteristic equal to 0. The Gauss–Bonnet theorem thus yields that $\int K dA = 0$ for any Riemannian metric on the torus, where K is the Gauss curvature and dA denotes the Riemannian metric area element. Now the Gauss curvature of the Poincaré metric is $\equiv -1$, and hence so is the Gauss curvature of the H-metric on any covering space quotient of $\mathbb{D}$. Thus no topological torus admits such an H-metric. □

[11]Note that when $g = 0$, or M is the sphere, then $H_1(M, \mathbb{Z}) = 0$, so the formulae still apply.

The sudden incursion of a result from pure Riemannian geometry may at first be startling. So we outline how the impossibility of covering a torus holomorphically by the unit disc $\mathbb{D}$ can be proved using only direct ideas about the quotienting possibilities that could arise. Refer to Fig. 7.5.

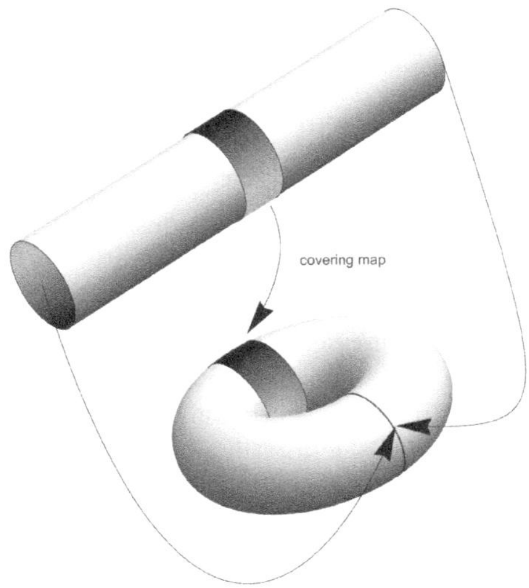

FIGURE 7.5: The covering of the torus.

Specifically, if M is a Riemann surface that is topologically $S^1 \times S^1$, then associated to this situation we could find a surface $\widetilde{M}$ covering $S^1 \times S^1$ where $\widetilde{M}$ is topologically $\mathbb{R} \times S^1$. This is done simply by "unwinding" the first S^1 factor and leaving the second factor alone. Now if M were holomorphically covered by $\mathbb{D}$, then $\widetilde{M}$ could also be taken to be holomorphically covered by $\mathbb{D}$, and the covering $\widetilde{M} \to M$ to be holomorphic as well. Both coverings would be isometric for the "H-metrics" as usual. Now choose a sequence $\{p_j\}$ diverging to infinity in one end of the topological cylinder $\widetilde{M}$ and $\{q_j\}$ diverging to infinity in the other end. Choose a closed curve X in $\widetilde{M}$ going around the cylinder in the obvious sense of those words so that $\widetilde{M} \backslash C$ has two components, both unbounded. Let, for each j, γ_j be an H-metric geodesic from p_j to q_j with minimal length. Then γ_j intersects X at least once. Let ξ_j be one such point of intersection. Now parameterize γ_j by arc length with $\gamma_j(0) = \xi_j$. By compactness of X, there is an arc-length parameter geodesic $\gamma_0 : (-\infty, +\infty) \to \widetilde{M}$ with $\gamma_0(0)$ a point of X which is minimizing between any two of its points. It is in a sense a minimal connection between the two ends of $\widetilde{M}$. [**Note:** To get γ_0, pass to a subsequence γ_{j_k} of the γ_js with ξ_j converging and with $\gamma_j'(0)$ converging. Then let γ_0 be the geodesic with $\gamma_0(0) = \lim \gamma_j(0)$ and $\gamma_0'(0) = \lim \gamma_j'(0)$.] Then $\widetilde{M} \backslash \{\gamma_0(t) : t \in (-\infty, +\infty)\}$ is isometric to an open subset of $\mathbb{D}$ (in the Poincaré metric) bounded by two nonintersecting geodesics, say σ and τ. Then $\widetilde{M}$ is obtained by gluing the two geodesics together in the same way that a circular cylinder is obtained by gluing together two parallel lines that bound a strip in the plane. We can parameterize σ and τ so that $\sigma(t)$ is glued to $\tau(t)$ to give $\gamma_0(t)$. See Fig. 7.6.

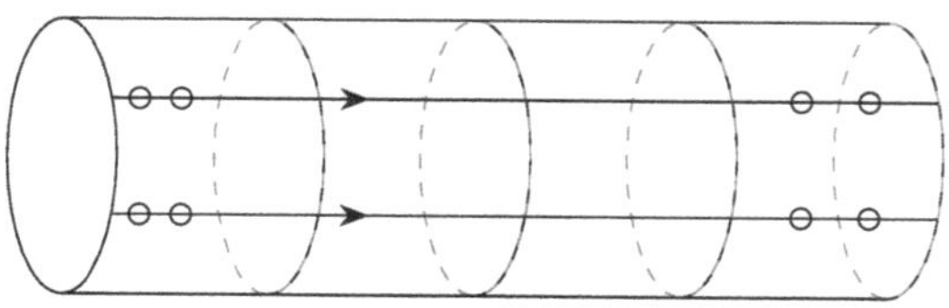

FIGURE 7.6: Behavior of geodesics on $S^1 \times R$.

Two such nonintersecting (Poincaré metric) geodesics in $\mathbb{D}$ can be related in only one of two ways: the geodesics diverge from each other at both ends, or they converge at one end but diverge at the other. These cases correspond to the Euclidean circles perpendicular to the boundary, which are the Poincaré metric geodesics in $\mathbb{D}$ having distinct boundary endpoints or having one boundary endpoint in common. [Two boundary endpoints in common is not possible; the geodesics would then coincide.] In the convergence-at-one-end case, one gets an immediate contradiction of our situation: in this case, one obtains arbitrarily short curves from one point far out on the geodesic (near the common boundary point) to the corresponding point on the "other side" of $\widetilde{M}$ cut open. This corresponds on M to arbitrarily short curves representing the free homotopy class of "once around" the first factor S^1. Such an eventuality is impossible: such a curve cannot be shorter than the injectivity radius of M.

The case of divergence of the pair of geodesics at both ends is also impossible, but the reason is slightly more subtle. Note that, by a simple compactness argument, no covering spaces needed, there is a number L such that, for each $p \in M$, there is a curve starting and ending at p with length $\leq L$ which lies in the free homotopy class of "once around the first factor" of $S^1 \times S^1$. Now choose p very far out on the geodesic γ_0 in the direction of divergence of the two copies σ, τ of γ_0 bounding the region in $\mathbb{D}$ that glues together at the edge to give $\widetilde{M}$. Choose a closed curve of length $\leq L$ at the projection of p into M in the free-homotopy class of "once around" the first S^1 factor. The lift of this curve to $\widetilde{M}$ that starts at a point $\sigma(t)$ projecting to p must end at $\tau(t)$ (or vice versa depending on direction). But this lift has length $\leq L$. This is impossible if p is sufficiently far out on γ since σ and τ diverge from each other. [**Note:** One could use this argument twice to show convergence of σ and τ at both ends and detour around the argument we have used for the impossibility of the convergence-at-one-end case.] See Fig. 7.7.

This is impossible if p is sufficiently far out on γ since σ and τ diverge from each other. See Fig. 7.7.

These arguments are intricate to express in words. But if the reader will consider suitable pictorial representations, then the ideas will become clear.

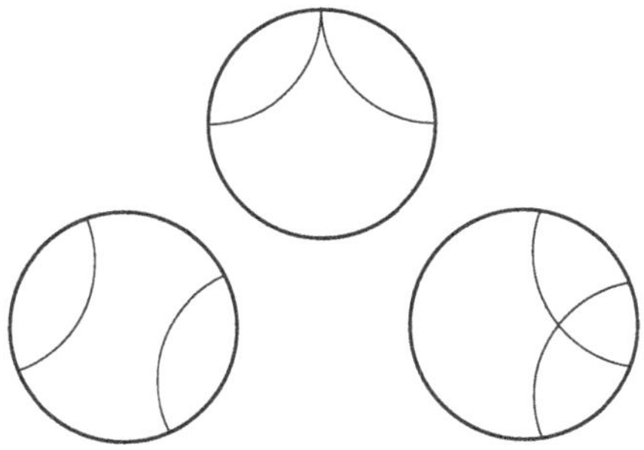

FIGURE 7.7: The behavior of geodesics in $\mathbb{D}$.

The geometric analysis of possible groups Γ of holomorphic covering transformations acting on $\mathbb{D}$ can be pushed much further. It can be shown that, for such a Γ, every abelian subgroup is cyclic. This recovers the part of Theorem 7.6.1 ruling out the covering by $\mathbb{D}$ of a topological torus. For if M were a torus and $\pi : \mathbb{D} \to M$ a holomorphic covering, then the fundamental group $\pi_1(M)$ would be isomorphic to the group Γ of covering transformations. But π_1 of a torus is $\mathbb{Z} \oplus \mathbb{Z}$, abelian but not cyclic. The cyclic nature of abelian subgroups actually holds in much more general situations: it applies to subgroups of the fundamental group of a compact Riemannian manifold of any dimension having negative Riemannian sectional curvature (see Preissmann's theorem [149]), or, for a recent treatment, see [142].

It is not difficult to see by direct construction using hyperbolic geometry that, for each $g = 2, 3, 4, \ldots$, there is a (compact) covering-space quotient of $\mathbb{D}$ with genus g. This amounts just to realizing in hyperbolic geometry the representation of the topological surface of genus g via a polygon of $4g$ sides with suitable identifications. However, this is far from showing that every Riemann surface of genus $g \geq 2$ is holomorphically a covering space quotient of $\mathbb{D}$. There are infinitely many complex structures on a compact surface of genus $g \geq 2$ (or indeed $g \geq 1$). The direct geometric construction of some of them does not a priori guarantee that all are obtained by that means. To see that all are uniformized by $\mathbb{D}$ requires the uniformization theorem (Theorem 7.6.1), or at least some method beyond pure hyperbolic geometry.[12]

We turn now to the subject of the automorphism groups $\mathrm{Aut}(M)$ for compact Riemann surfaces M of genus ≥ 2.

Proposition 7.6.2. *If M is a compact Riemann surface of genus $g \geq 2$, then $\mathrm{Aut}(M)$ is a finite group.*

Remark 7.6.3. A compact Riemann surface can certainly have an infinite automorphism group; even a compact surface with an automorphism-invariant metric can have such a group. For instance, the quotient T of $\mathbb{C}$ by the integer

[12] Proofs that use differential equations without appeal to uniformization are also known—see [16, 99].

lattice $\{m + ni : m, n \in \mathbb{Z}\}$, which is topologically a torus, has a positive-dimensional automorphism group induced by the translations $z \mapsto z + \alpha$, $\alpha \in \mathbb{C}$. Thus the condition that the genus be at least 2 in the proposition is essential.

Proof of the Proposition. By Theorem 7.5.8 and the existence of the invariant H-metric on M (Proposition 7.6.2 with $\widetilde{M} = \mathbb{D}$), $\operatorname{Aut}(M)$ is a compact Lie group. Thus $\operatorname{Aut}(M)$ is a finite group if and only if the identity component of $\operatorname{Aut}(M)$ consists of the identity element alone (i.e. the group must be topologically discrete). By standard Lie group theory, this happens if and only if M contains no nontrivial one-parameter subgroup. For a proof by contradiction, suppose that $\{\varphi_t\}$ is a nontrivial one-parameter subgroup of $\operatorname{Aut}(M)$. Then $X(p) = d\varphi_t(p)/\,dt|_{t=0}$ for all $p \in M$ is a $\mathcal{C}^\infty$ vector field on M. [It is a part of the Bochner–Montgomery theorem [25] that the action of the Lie group $\operatorname{Aut}(M)$ on M is smooth, so that X is indeed defined and $\mathcal{C}^\infty$.] The vector field X on M lifts to a $\mathcal{C}^\infty$ vector field $\widehat{X}$ on $\mathbb{D}$ by setting $\widehat{X}(w)$ equal to the unique vector $\mathbf{v}$ at w such that $d\pi|_w(\mathbf{v}) = X(\pi(w))$, where $\pi : \mathbb{D} \to M$ is the universal covering of M. The vector field $\widehat{X}$ generates a one-parameter group $\{\psi_t\}$ of diffeomorphisms of $\mathbb{D}$; indeed, one obtains (complete) integral curves for $\widehat{X}$ by lifting the integral curves of X. Clearly the elements of the group $\{\psi_t\}$ are isometries of $\mathbb{D}$. In particular, the vector field $\widehat{X}$ can have at most one zero. For if $\widehat{X}(w_0) = 0$, then $\{\psi_t\}$ consists of a one-parameter group of Poincaré rotations around the fixed point w_0, and such a one-parameter group has generator that vanishes only at w_0—as one sees by writing it out explicitly (setting $w_0 = 0$ without loss of generality). On the other hand, if $\widehat{X}(w_0) = 0$, then $\widehat{X}(\gamma(w_0)) = 0$ for all $\gamma \in \Gamma$. Since Γ is infinite by the compactness of M, it must be that $\widehat{X}$, having at most one zero, has in fact no zeros at all. Thus X itself has no zeros. But a compact manifold with a nowhere vanishing vector field has Euler characteristic 0 (by the Poincaré-Hopf theorem, (cf. [134])), while M has Euler characteristic $2 - 2g < 0$. That is a contradiction. □

7.7 The Automorphism Group of a Riemann Surface of Genus at Least 2

The proof in the previous section that the automorphism group $\operatorname{Aut}(M)$ of a compact Riemann surface M of genus $g \geq 2$ is necessarily a finite group involved very general considerations: the group $\operatorname{Aut}(M)$ was compact because the universal cover of M was $\mathbb{D}$ and hence M inherited an $\operatorname{Aut}(M)$-invariant metric from the $\operatorname{Aut}(\mathbb{D})$-invariant Poincaré metric of $\mathbb{D}$. The identity component of $\operatorname{Aut}(M)$ was the identity map alone because otherwise M would have had a zero-free vector field, contradicting the fact that its Euler characteristic

$2 - 2g$ was nonzero. The specific structure of M played no role in these considerations, and the result was correspondingly general: finiteness of $\mathrm{Aut}(M)$ with no explicit estimate of the number of elements in $\mathrm{Aut}(M)$. It is quite surprising in this context that an explicit estimate is possible in the same generality.

Theorem 7.7.1 (Hurwitz). *If M is a compact Riemann surface of genus $g \geq 2$, then the order of the finite group $\mathrm{Aut}(M)$ does not exceed $84(g-1)$.*

This theorem seems mysterious at first sight. And the apparent mystery is deepened by the fact that the estimate is sharp. In fact there are Riemann surfaces with genus $g > 2$ for which the order of $\mathrm{Aut}(M)$ is precisely equal to $84(g-1)$ (see, for instance, [46, p. 325, VII.3.10]). The exact origin of the specific number 84 will become apparent from the proof that follows.

The proof of the Hurwitz theorem that we shall present is closely related to Hurwitz's original argument. The subject has tended meanwhile to become burdened with abstract Riemann surface theory. In our treatment we shall, like Hurwitz himself, stick to the basics. However, we shall use our metric viewpoint rather than Hurwitz's analogous pure function theory. The theorem will then turn out to be completely geometric in nature.

Before beginning the proof itself, we need to develop some general ideas about quotients by group actions. Specifically, we want to consider the orbit space of $\mathrm{Aut}(M)$ on M. An orbit of $\mathrm{Aut}(M)$ is as usual, by definition, for a given point p, the set $\{\gamma(p) : \gamma \in \mathrm{Aut}(M)\}$. The manifold M is a disjoint union of orbits, because $\mathrm{Aut}(M)$ is a group, and "being in the same orbit" is an equivalence relation, with each orbit an equivalence class.

Let $\mathcal{M}$ be the collection of orbits. Then $\mathcal{M}$ has a natural topology: the quotient topology. In other words, a set of orbits is said to be open if and only if the union of the orbits as sets is an open subset of M. For an arbitrary group action on a manifold, the topological orbit space can be complicated. But in our case, where $\mathrm{Aut}(M)$ is finite, $\mathcal{M}$ is a decent topological space; in particular, it is necessarily a Hausdorff space, as is easily checked. This is a special property of orbit spaces for finite groups acting on manifolds (or even just on Hausdorff topological spaces).[13]

Lemma 7.7.2. *If M is a compact, orientable Riemannian 2-manifold and if Γ is a finite group of orientation-preserving isometries of M, then the orbit space M/Γ is an orientable 2-manifold.*

Proof. If two points $p, q \in M$ lie in the same Γ-orbit, then $I_p = \{\gamma \in \Gamma : \gamma(p) = p\}$ and $I_q = \{\gamma \in \Gamma : \gamma(q) = q\}$ are conjugate subgroups of Γ; first, if $\sigma(p) = q, \sigma \in \Gamma$, then $\sigma \circ \gamma \circ \sigma^{-1} \in I_q$ if and only if $\gamma \in I_p$. With this idea in mind, we divide the orbits of Γ (that is to say, the points of M/Γ) into two classes:

[13]However, for a general finite group action on a manifold, the quotient need not be a manifold. For example, the orbit space of the action $\{x \mapsto x, x \mapsto -x\}$ on $\mathbb{R}$ is homeomorphic to $\{x \in \mathbb{R} : x \geq 0\}$. This is not a manifold. This difficulty does not arise in our situation.

(a) *nonsingular orbits* $\mathcal{O}$: For these, $I_p = \{\mathrm{id}_M\}$ if $p \in \mathcal{O}$;

(b) *singular orbits* $\mathcal{O}$: For these, $I_p \neq \{\mathrm{id}_M\}$ if $p \in \mathcal{O}$.

Similarly, we call points of M itself *singular* or *nonsingular* according to whether they belong to singular or nonsingular orbits. These ideas are all well defined by our conjugacy observation made at first.

The next step in the proof of Lemma 7.7.2 is the following subsidiary result.

Lemma 7.7.3. *The singular orbits of M/Γ and the set of singular points of M are both finite sets.*

Proof. By standard results from Riemannian geometry, there is an $\epsilon > 0$ such that, for any two points $p, q \in M$ with $\mathrm{dist}_M(p, q) < \epsilon$, there is a unique geodesic from p to q of length $\mathrm{dist}_M(p, q)$. Now suppose that $p_1, p_2, \ldots$ were an infinite sequence of distinct singular points of M. Since Γ is finite, we can suppose, by passing to a subsequence, that there is a fixed nonidentity element $\gamma_0 \in \Gamma$ such that $\gamma_0 \in I_{p_j}$ for all j. By the compactness of M, there are two points p_{j_1} and p_{j_2} with $\mathrm{dist}_M\,(p_{j_1}, p_{j_2}) < \epsilon$. Clearly γ fixes the (short) geodesic $c(t)$ from p_{j_1} to p_{j_2}. But the only nonidentity isometry that fixes every point of a geodesic segment in dimension 2 is locally given by reflection in that geodesic. Such an isometry is orientation-reversing. Thus by contradiction the set of singular points of Γ in M is finite, and hence so is the set of singular orbits. □

Returning now to the proof of Lemma 7.7.2, we need to exhibit a Euclidean neighborhood of each point of $\mathcal{M}$, that is, a neighborhood that is homeomorphic to an open subset of $\mathbb{R}^2$. For a nonsingular orbit $\mathcal{O}$, this is easy: Choose a point $p \in \mathcal{O} \subset M$ and choose a Euclidean neighborhood U of p in M with every point of U nonsingular (this is possible by Lemma 7.7.3) and with the natural map $U \to \mathcal{M}$ injective. The latter choice is possible by the finiteness of Γ and the fact that p is nonsingular. For if $\{(p_j, q_j)\}$ were Γ-equivalent pairs of distinct points with $\lim_j p_j = \lim_j q_j = p$, then there would be an element $\gamma \neq \mathrm{id}_M$ in I_p.[14] Thus U maps homeomorphically onto a neighborhood of $\mathcal{O}$, giving a Euclidean neighborhood as required.

The situation is more complicated when we want to find a Euclidean neighborhood of a singular orbit $\mathcal{O}$; but it is only a little more complicated. Let $p \in \mathcal{O}$. The group I_p, being a group of orientation-preserving isometries that all fix the point p, has a simple structure—both as a group and as a space acting on a neighborhood of p. Choose a small ball B around p, and let (r, θ) be a geodesic polar coordinate system centered at p on this ball. The group I_p acts in this coordinate system as a finite group of Euclidean rotations. In particular, I_p is cyclic, say of order m_p.

---

[14]More precisely, if $q_j = \gamma_j\,(p_j)\,, \gamma_j \neq id_M$, then by the finiteness of Γ, and passing to a subsequence if necessary, γ_j is independent of j and (by continuity) $\gamma_j \in I_p$.

Recall from Lemma 7.7.3 that, if the ball B around p is small enough, then every point of the ball except p is nonsingular. Moreover, $B \to M/\Gamma$ is exactly m_p-to-1, except on the orbit $\mathcal{O}$ corresponding to p (which has only the pre-image p). The same limit argument as in the nonsingular case shows that, for B small enough, no quotienting occurs on B except via I_p. But it is then clear by Euclidean geometry that $\mathcal{O}$ has a Euclidean neighborhood, since a disc modulo a finite group of rotations around its center is locally Euclidean—it is a one-nappe circular cone with p as vertex.

The orientability is clear since Γ contains only orientation-preserving transformations, and the singular orbits do not separate the collection of orbits $\mathcal{M}$, even locally. □

Geometrically, $\mathcal{M}$ is a $\mathcal{C}^\infty$ manifold with $\mathcal{C}^\infty$ metric pushed down from M except for a finite number of cone-point singularities of the structure already described. Topologically, $\mathcal{M}$ remains a compact, orientable manifold as indicated, although it is likely to be a quite different manifold from M itself. Let g^* denote the genus of $\mathcal{M}$.

Now we want to describe a relationship between the genus g of M, the genus g^* of $\mathcal{M}$, and the action of Γ. To express this idea, we need some notation. Write $|\Gamma|$ for the order of Γ, let $\mathcal{O}_1, \ldots, \mathcal{O}_m$ be the singular orbits (finite in number by Lemma 7.7.3) of Γ, and let $k_1, \ldots, k_m$ be the orders of the corresponding isotropy subgroups $I_{p_1}, \ldots, I_{p_m}$ for $p_1 \in \mathcal{O}_1, p_2 \in \mathcal{O}_2, \ldots, p_m \in \mathcal{O}_m$. [The groups I_{p_j} are determined only up to conjugacy, but their orders are of course uniquely determined.]

Lemma 7.7.4. *In the notation just given,*

$$2 - 2g = |\Gamma| \cdot (2 - 2g^*) - |\Gamma| \cdot \sum_{j=1}^{m} \left(1 - \frac{1}{k_j}\right).$$

Lemma 7.7.4, which will turn out to be the crucial ingredient for the proof of Theorem 7.7.1, is a special case of the well-known Riemann–Hurwitz formula for branched coverings. But an exact description of the Riemann–Hurwitz general setup would be almost as long as the proof in our case. So we give the direct proof instead. In fact, we shall give two proofs. The first one is a bit disingenuous, since we shall assume without proof the existence of a triangulation of $\mathcal{M}$ of a certain type; but it is still quite instructive. The second proof uses the Gauss–Bonnet theorem and is completely self-contained: we push the triangulation question back to the corresponding difficulty in the proof of the Gauss–Bonnet theorem.

Proof I of Lemma 7.7.4. Choose a triangulation of $\mathcal{M}$ such that the singular orbits $\mathcal{O}_1, \ldots, \mathcal{O}_m$ (thought of as points of $\mathcal{M}$) occur as vertices of the triangulation. This triangulation pulls back to a triangulation of M. In this triangulation of M, every edge of the $\mathcal{M}$-triangulation gives rise to $|\Gamma|$ edges and every face from $\mathcal{M}$ to $|\Gamma|$ faces. Let $F_{\mathcal{M}}$ denote the number of faces, $E_{\mathcal{M}}$

the number of edges, and V_M the number of vertices for the M-triangulation. We have similar notation for the $\mathcal{M}$-triangulation. Then

$$E_M = |\Gamma| \cdot E_{\mathcal{M}}$$

and

$$F_M = |\Gamma| \cdot F_{\mathcal{M}}\,.$$

Howeverx the pre-image $\pi^{-1}(\mathcal{O}_j)$ contains not $|\Gamma|$ points but rather $|\Gamma|/k_j$ points. Of course, pre-images of vertices of the $\mathcal{M}$-triangulation that are non-singular still contain $|\Gamma|$ points. Clearly the errors arising from the singular orbits are accounted for correctly by

$$V_M = |\Gamma| \cdot V_{\mathcal{M}} - \sum_{j=1}^{m} \left(|\Gamma| - \frac{|\Gamma|}{k_j} \right).$$

Thus the Euler characteristic $\chi(M) = V_M - E_M + F_M$ is related to $\chi(\mathcal{M}) = V_{\mathcal{M}} - E_{\mathcal{M}} + F_{\mathcal{M}}$ by

$$\chi(M) = |\Gamma| \cdot \chi(\mathcal{M}) - \sum_{j=1}^{m} \left(|\Gamma| - \frac{|\Gamma|}{k_j} \right)$$

or

$$2 - 2g = |\Gamma| \cdot (2 - 2g^*) - |\Gamma| \cdot \sum_{j=1}^{m} \left(1 - \frac{1}{k_j} \right).$$

That ends Proof I.

Proof II of Lemma 7.7.4. We cannot apply the Gauss–Bonnet theorem directly to $\mathcal{M}$ because of the "cone point" singularities of the metric at $\mathcal{O}_1, \ldots, \mathcal{O}_m$. We instead set $\mathcal{M}^\epsilon$ equal to the image under the mapping $M \to M/\Gamma$ of $M \setminus \bigcup_{p \text{ singular}} \mathbb{B}(p, \epsilon)$. Then $\mathcal{M}^\epsilon$ is, for $\epsilon > 0$ small enough, an orientable, compact, $\mathcal{C}^\infty$ manifold with boundary having a $\mathcal{C}^\infty$ metric inherited from M (we use here the fact that the elements of Γ are isometries). The manifold with boundary $\mathcal{M}^\epsilon$ is in effect $\mathcal{M}$ with the cone points and ϵ-discs around them removed. The Gauss–Bonnet theorem does apply to $\mathcal{M}^\epsilon$ as a manifold with boundary. Thus

$$2\pi\chi(\mathcal{M}^\epsilon) = \int_{\mathcal{M}^\epsilon} K dA - \int_{\partial\mathcal{M}^\epsilon} K_g ds$$

where K_g is the geodesic curvature of the boundary (with $K_g > 0$ in our situation). Now $\chi(\mathcal{M}^\epsilon) = \chi(\mathcal{M}) - m$ by standard topology. Also write X_ϵ^j for the component of $\partial\mathcal{M}^\epsilon$ at $\mathcal{O}_j$ and $\widehat{C}_\epsilon^j$ for its pre-image in M at a particular point $p_j \in \mathcal{O}_j$; then

$$\lim_{\epsilon \to 0^+} \int_C K_g ds = \frac{1}{k_j} \lim_{\epsilon \to 0^+} \int_{\widehat{C}} K_g ds = 2\pi \cdot \frac{1}{k_j}.$$

Here, for typographical convenience, $X = C^j_\epsilon$ and $\widehat{C} = \widehat{C}^j_\epsilon$. Combining these results gives

$$\chi(\mathcal{M}) = m - \sum_{j=1}^{m} \frac{1}{k_j} + \lim_{\epsilon \to 0^+} \frac{1}{2\pi} \int_{\mathcal{M}^\epsilon} K dA$$

or

$$\chi(\mathcal{M}) = \sum_{j=1}^{m} \left(1 - \frac{1}{k_j}\right) + \lim_{\epsilon \to 0^+} \frac{1}{2\pi} \int_{\mathcal{M}^\epsilon} K dA .$$

But

$$\lim_{\epsilon \to 0^+} \frac{1}{2\pi} \int_{\mathcal{M}^\epsilon} K dA = \frac{1}{2\pi|\Gamma|} \int_M K dA = \frac{1}{|\Gamma|} \cdot \chi(M) .$$

Thus

$$\chi(M) = |\Gamma| \cdot \chi(\mathcal{M}) - |\Gamma| \sum_{j=1}^{m} \left(1 - \frac{1}{k_j}\right) .$$

That ends Proof II.

With Lemma 7.7.4 now in hand, we are ready to prove Hurwitz's theorem.

Proof of Hurwitz's Theorem 7.7.1. We see that

$$\frac{2-2g}{|\Gamma|} = (2 - 2g^*) - \sum_{j=1}^{m} \left(1 - \frac{1}{k_j}\right)$$

or

$$\frac{2g-2}{|\Gamma|} = (2g^* - 2) + \sum_{j=1}^{m} \left(1 - \frac{1}{k_j}\right) .$$

To get an upper bound for $|\Gamma|$, we need a lower bound for the righthand side of this last equation (for g fixed). Now $2g - 2 > 0$ by hypothesis. So the righthand side is required to be positive. If $g^* \geq 2$, then the righthand side is at least 2. If $g^* = 1$, then $2g^* - 2 = 0$ and there must be at least one term of the form $1 - 1/k_j$, with k_j a positive integer greater than 1. Hence in this case, the righthand side is at least $1/2$. Finally, if $g^* = 0$, so that $2g^* - 2 = -2$, then there must be at least three terms of the form $1 - 1/k_j$ with $k_j \geq 2$. If there are five or more of these, then the righthand side would be at least $-2 + 5(1/2) = 1/2$. If there are four, then the righthand side—in order to be positive—must be at least $-2 + 3(1/2) + 2/3 = 1/6$. Finally, if there are three terms of the form $1 - 1/k_j$, then only one of these can have $k_j = 1/2$. A little further experimentation shows that $-2 + 3 - (1/k_1 + 1/k_2 + 1/k_3)$ has the minimum possible value $1/42$, obtained when $k_1 = 2, k_2 = 3$, and $k_3 = 7$.

Thus the minimum possible positive value in all cases for the righthand side is $1/42$, from which

$$|\Gamma| \leq 42(2g - 2) = 84(g - 1) .$$

□

As already noted earlier in this section, the number 84 is sharp: the Klein quartic is a Riemann surfaces of genus 3 with $168 = 84 \cdot 2$ automorphisms. There is a considerable literature on the subject of when the $84(g-1)$ bound is attained (cf. [122] and [193]). Since the automorphism group of a Riemann surface of genus $g > 1$ is precisely the group of orientation-preserving automorphisms of the "push-down" metric from its uniformization by the unit disc (see [64, Proposition 2.3.1]), the examples where the order of the automorphism group has order $84(g-1)$ also serve as examples for the sharpness of the bound for the order of the group of orientation-preserving isometries.

Another application of the idea that geometric quotients give rise to holomorphic ones is to the question of which compact Riemann surfaces cover others in the sense of covering spaces. This corresponds essentially to Lemma 7.7.4 without branch points. Specifically, one can see either by the triangular argument of Proof I of Lemma 7.7.4 or via the Gauss–Bonnet argument used in Proof II that the following holds:

If $\pi : M \to M'$ is a covering map from one compact Riemann surface to another, then, with g denoting the genus of M and g' denoting the genus of M',

$$2 - 2g = k \cdot (2 - 2g') \ .$$

Here k is a positive integer, and indeed k is the "sheeting number" of π, i.e. the number of points in $\pi^{-1}(\{p\})$ for each $p \in M$.

The necessary condition that, when $g' \geq 2$, then $g' - 1$ is a divisor of $g - 1$, is in fact also sufficient. This is clear geometrically: consider M as a torus of revolution in $\mathbb{R}^3$ with $g - 1$ tubes smoothly attached in such a way that rotation by $2\pi/(g-1)$ takes each one to the next. Let Γ be the group of rotations by multiples of $2\pi/[(g-1)/k]$. Then $M \to M/\Gamma$ is a covering onto a surface with $(g-1)/k = g^* - 1$ tubes attached. That is a Riemann surface of genus $1 + (g^* - 1) = g^*$. This geometric construction is then, according to the principle of associating complex structures to metrics, holomorphic in suitable complex structures. First, there is a unique complex structure for which the metric of M as a surface in $\mathbb{R}^3$ is Hermitian. Since Γ acts as orientation-preserving isometries on M, it acts holomorphically for this complex structure. And thus $M \to M/\Gamma$ is a holomorphic covering space.

We have downplayed the topological questions which arise in the study; the reader could consult any number of good texts for the basic material on manifolds. In the text are mentioned [10, 125, 126, 133, 135], and [166]; the analysis on manifolds is very well done in [133], and [30] has a solid discussion of the topological questions arising specifically for Riemann surfaces.

As to preliminary material, namely the basics on functions of one complex variable, the authors have taught or taken courses using [4, 40, 69, 117], and [23] as texts, and all have their strengths.

Complex manifolds of higher dimension is the subject of [132].

7.8 Automorphisms of Multiply Connected Domains

A simply connected, bounded domain in the plane $\mathbb{C}$ is biholomorphic to the unit disc D—this is the Riemann mapping theorem. A bounded domain U in $\mathbb{C}$ such that $\mathbb{C}\backslash U$ has exactly two connected components is biholomorphic to $D\backslash\{z \mid |z| \le r\}$ for some $r \in [0, 1)$. This assertion follows by a Dirichlet problem argument using boundary values of the absolute value of the desired biholomorphic mapping f. One notes that $\log |f|$ is harmonic with boundary value 0 on one component of the boundary of U and constant value α to be determined on the other component of the boundary of U. Let h be the harmonic function with these boundary values. Choosing this value α correctly, one can arrange that $*dh$ has integral 2π around an oriented homology generator of U, from which it follows that the associated map $z \mapsto \exp\left(h + i \oint *dh\right)$ is well defined and one-to-one onto $D\backslash\{z \mid |z| \le r\}$, where $r = e^{\alpha}$ (α being the boundary value noted above). Strictly speaking, one needs some regularity of the boundary of U to solve the Dirichlet problem here, but this technical difficulty is easily disposed of by approximating U from the inside by domains with smooth boundary and the same connectivity as U (cf. [69] or [4] for further details).

More generally, one can show that any bounded domain U in $\mathbb{C}$ such that $\mathbb{C}\backslash U$ has finitely many connected components is biholomorphic to D with a finite number of closed discs removed (the closed discs can, in principle, be arcs). This result holds even in the case of countably many components of the complement—see [81]. As is standard, we say that the domain has connectivity k if its complement has $k + 1$ connected components.

Understanding the automorphisms of these discs with holes that serve as models is certainly a reasonable goal in itself, and it is treated in detail in [151]. But these interesting one-dimensional results are not a very good indication of what to expect in higher dimensions. The reason for this is very simple: as discussed in Section 7.5, the unit ball in $\mathbb{C}^n$, for $n \ge 2$, with a finite number of closed balls removed (call this domain Ω), has the property (by the Hartogs extension phenomenon) that every automorphism of Ω arises from an automorphism of the entire ball (restricted to Ω). This property holds by the same argument for any connected open set obtained by removing a finite number of compact subsets from an open ball. So the theory of the automorphisms of such open sets—"balls with holes"—is subsumed by the knowledge of the automorphisms of the ball. What we shall discover instead, as the book develops, is that the boundary geometry is the deciding feature in studying automorphism groups of domains; basic topology tells us little.

In spite of all this, there are aspects of the planar situation that offer clues to the higher-dimensional study in general. We now discuss some of these.

Theorem 7.8.1. *If Ω is a bounded planar domain of finite connectivity that is not simply connected, then* $\mathrm{Aut}(\Omega)$ *is compact.*

Actually, we have already proved this result for connectivity 1 by explicit determination of the automorphism group in that case, so the interesting consideration will be for higher connectivity.

Proof of Theorem 7.8.1. Suppose that Ω has connectivity $k \geq 1$ and that $\text{Aut}(\Omega)$ is noncompact. Choose $q \in \Omega$. Then the orbit $\{f(q) : f \in \text{Aut}(\Omega)\}$ is noncompact. So there is a sequence $\{f_j\} \subseteq \text{Aut}(\Omega)$ such that $f_j(q)$ converges to a point $p \in \partial\Omega$. Passing to a subsequence if necessary, we can assume that $\{f_j\}$ converges uniformly on compact subsets of Ω to some limit function $f_0 : \Omega \to$ (closure of Ω). The functions $z \mapsto f_j(z) - p$ have no zeros in Ω but their limit $f_0 - p$ is 0 at p. By Hurwitz's theorem, $f_0 - p$ is identically 0. Thus $\{f_j\}$ converges uniformly to the constant function with value 0 on each compact subset of Ω.

Now choose a collection of k closed curves, $k =$ connectivity of $\Omega, k \geq 1$, which form a generating set for the homology of Ω. For example, we could choose $C_1, C_2, \ldots, C_k$ where each $X_\ell, \ell = 1, 2, \ldots, k$, winds once around points of the ℓth bounded component of $\mathbb{C}\backslash\Omega$ and does not wind around points of the other bounded components. To choose these conveniently, we can and do assume that Ω has the "standard form" $D\backslash$ (k closed discs or points). The image, for each $j, f_j\left(C_\ell\right), \ell = 1, \ldots, k$, is again a homology basis for Ω. But when j is large enough, these images all lie in an arbitrarily small neighborhood of p. In particular, if p is not in the ℓth bounded component of $\mathbb{C}\backslash\Omega$, then none of these images wind around any point of the ℓth bounded component when j is large.

Clearly this situation is impossible if $k > 1$. When $k = 1$, it is possible only if the one bounded component of Ω (with Ω in "standard form") is a point, since to wind around points of a closed disc, a curve cannot lie in an arbitrarily small neighborhood of any given point. But we have already determined that $\text{Aut}(\Omega)$, where Ω is the disc D with one point removed, is compact. First, $\text{Aut}(\Omega) = \{T_{-a} \circ R_\omega \circ T_a \mid |\omega| = 1\}$ if a is the removed point, where R_ω is the map from D to D defined by $R_\omega(z) = \omega z$ and T_a is the map $z \mapsto (z - a)/(1 - \overline{a}z)$ already discussed. □

We shall have occasion frequently in the sequel to use this type of argument: an automorphism sequence which pushes one interior point to a given boundary point will tend to push all interior points to that same boundary point, and this will yield topological restrictions. Note, however, that in higher dimensions this idea does not apply automatically as it does in one dimension. Consider, for example, a sequence of automorphisms of $D \times D$ in $\mathbb{C}^2$ acting on the first variable only and converging in that first variable to the constant 1. The accumulation set in the boundary is the whole set $\{1\} \times D$.

Just as genus 1 Riemann surfaces and those of genus $g > 1$ differ as to possible automorphism behavior, so do domains of connectivity 1 (which have an infinite group of automorphisms) differ from those of connectivity > 1. But the reason is now different. For compact surfaces, the difference arises from change of the universal cover-from $\mathbb{C}$ to D. But connectivity 1 bounded domains are covered by D just as those of connectivity 2 are. So the difference lies elsewhere.

Theorem 7.8.2. *If* Ω *is a bounded, planar domain of connectivity* > 1, *then* $\text{Aut}(\Omega)$ *is finite.*

This result was proved, for instance, by Heins [82]. The proof here is different, being more geometric and topological, as preparation for later use of the ideas involved.

Proof of the Theorem. Suppose to the contrary that $\text{Aut}(\Omega)$ is in fact infinite. Since $\text{Aut}(\Omega)$ is compact (and a Lie group), as already shown, the identity component must be positive dimensional. In particular, the identity component of $\text{Aut}(\Omega)$ must contain a subgroup G that is isomorphic to the circle, that is, the group $\{\omega \in \mathbb{C} \mid |\omega| = 1\}$ with multiplication as the binary group operation. This is standard Lie group theory: the closure of any nontrivial one-parameter subgroup is a compact abelian subgroup of positive dimension and hence contains a circle subgroup.

The remainder of the proof consists of formalizing the intuition that Ω cannot be a union of the kinds of orbits—circles or points—that would arise in this situation, except in case of connectivity 0 or 1.

We begin by noting that, since the action of G is via orientation-preserving isometries of Ω relative to what we are calling the H-metric of Ω, the fixed points of the action are isolated. That is, the associated vector field that generates the circle action has isolated zeros. [This can also be seen function-theoretically without reference to metrics.]

For each $p \in \Omega$, we define a continuous map $\gamma_p : G = S^1 \to \Omega$ by $\gamma_p(g) = g(p)$. This collection of closed curves in Ω is parameterized continuously by p. In particular, they all belong to the same free homotopy class. Note that any two such curves, say γ_{p_1} and γ_{p_2}, have either disjoint images in Ω or the same image in Ω, and in the latter case, they are the same curve up to a rotation of the circle in itself.

Suppose that, for some $p_0 \in \Omega$, the image $\gamma_{p_0}(G)$ consists of the point p_0 alone. Since G acts by isometries (of the H-metric), and acts nontrivially, it follows that, for p near p_0 but unequal to p_0, the curve γ_p represents a nontrivial element of the free homotopy classes of curves in $\mathbb{C} \setminus \{p_0\}$. Indeed, γ_p is the (possibly multiple) transversal of an H-metric circle around p_0 in a fixed direction, the associated G-action near p_0 being rotation when expressed in H-metric geodesic polar coordinates.

The previous paragraphs imply immediately that there is at most one point p_0 such that the image of γ_{p_0} is p_0 alone. For if there were two such points, say p_0 and q_0, then, for q near q_0 and p near p_0, the curves γ_p and γ_q would be freely homotopic in the G-invariant set $\Omega \setminus \{p_0, q_0\}$ (since this set is connected and γ_x depends continuously on x in this set). Note that the image of γ_x lies in $\Omega \setminus \{p_0, q_0\}$ for each $x \in \Omega \setminus \{p_0, q_0\}$. But clearly this free homotopy is not the case: γ_p winds around p_0 but not around q_0 and γ_q winds around q_0 but not around p_0.

We now consider the two (exhaustive) cases:

(1) γ_p does not have a onepoint image for any $p \in \Omega$;

(2) there is exactly one $p_0 \in \Omega$ such that the image of γ_{p_0} is p_0 alone.

Looking at case (2) first, consider a neighborhood of an orbit $\mathcal{O}_p$, that is in fact equal to image γ_p, for some fixed p. Nearby orbits are labeled uniquely by their (unique) point of intersection with a (short) H-metric geodesic γ perpendicular to $\mathcal{O}_p$ at p: here we use the fact that $\mathcal{O}_p$ is a smooth curve. That G acts by isometries shows that the G-action on a neighborhood of $\mathcal{O}_p$ is a tubular-neighborhood product action; diffeomorphically it is exactly the rotation action of S^1 on an annulus around the origin (possibly multiply covered): once around G may rotate the annulus multiple times but the multiplicity is constant—the same for each orbit. This gives the space of orbits the structure of a 1-manifold, hence an open interval (since it is connected and open). It follows that the domain Ω is homeomorphic to $(0, 1) \times S^1$ and hence that the fundamental group of Ω is the same as that of S^1; it is just $\mathbb{Z}$. So Ω has connectivity 1. This is a contradiction.

For the first case, one applies the same reasoning to $\Omega \setminus \{p_0\}$, which is G invariant, to conclude that $\Omega \setminus \{p_0\}$ has connectivity 1. But this is impossible since Ω already has connectivity at least 1 and removing p_0 adds 1 to the connectivity. [Note here that rotation is an S^1 action on D with a single "degenerate" orbit. So case (1) occurs—but only when Ω is simply connected.]

□

The reader approaching this subject from the strictly analytical viewpoint may find the great use of geometry and topology in these proofs a bit overwhelming. But, for our purposes, it illustrates well the utility of the ideas of metric geometry in this subject. That G operates as isometries gives useful information in determining the possibilities for its action. This way of understanding key ideas will continue to be important in our later development.

Exercises

1. For each $g \geq 0$ give an example of a Riemann surface of genus g.

2. If R_1 is a Riemann surface of genus g_1 and having automorphism group of order $84(g_1 - 1)$ and if R_2 is a Riemann surface of genus g_2 and having automorphism group of order $84(g_2 - 1)$, then what can you say about the automorphism group of $R_1 \times R_2$?

3. It is a remarkable fact that the only simply connected Riemann surfaces are the disc, the plane, and the sphere. What are all the simply connected, smooth, closed 2-surfaces?

4. Of course, the standard, and essentially the only, invariant metric on the disc is the Poincaré metric. Is there an invariant metric on the plane? Is there an invariant metric on the sphere?

5. It is a decisive result of Maurice Heins that the automorphism group of a planar domain of connectivity greater than two is finite. Sketch a proof of this result using the idea of invariant metrics.

6. Give an example of a Riemann surface, other than the disc or the plane, that has non-compact automorphism group.

7. Describe the Riemann surface for the function $f(z) = \log z$.

8

The Uniformization Theorem

This chapter assembles the potential-theoretic tools for the Uniformization theorem. The preliminary section develops the analytic framework via harmonic and subharmonic functions and the Perron family. We then introduce Green's functions and record the properties needed later. Finally, we present bipolar Green's functions as a refined construction. The presentation moves from foundational notions to the specific objects used throughout the chapter.

The Riemann mapping theorem asserts that any non-empty, simply connected, proper open subset of the complex plane is conformally equivalent to the open unit disc. The uniformization theorem extends this result, stating that every simply connected Riemann surface is conformally equivalent to the open unit disc, the complex plane, or the Riemann sphere. Felix Klein [103] and Henri Poincaré [148] conjectured the uniformization theorem for (the Riemann surfaces of) algebraic curves. Henri Poincaré [148] extended this to arbitrary multivalued analytic functions and gave informal arguments in its favor. The first rigorous proofs of the general uniformization theorem were given by Poincaré [147] and Paul Koebe [108–110]. Paul Koebe later gave several more proofs and generalizations. The history is described in Gray [60]; a complete account of uniformization up to the 1907 papers of Koebe and Poincaré is given with detailed proofs in de Saint-Gervais [162] (the Bourbaki-type pseudonym of the group of fifteen mathematicians who jointly produced this publication).

In this chapter we present a proof of the uniformization theorem. The exposition follows the approach of the final chapter of T. Gamelin's *Complex Analysis* [53]. The proof of the uniformization theorem given here is based on solving Laplace's equation via Perron's method and Green's functions. The Riemann mapping theorem is taken as known.

8.1 Preliminary Results

Proposition 8.1.1. *Let u be a real-valued harmonic function on a simply connected domain $\Omega \subseteq \mathbb{C}$. Then there exists a holomorphic function $F : \Omega \to \mathbb{C}$ such that* $\mathrm{Re}(F) = u$. *Furthermore, the imaginary part of this function is uniquely determined up to an additive real constant.*

DOI: 10.1201/9781003713005-8

Certain sequences of harmonic functions behave nicely and allow us to extract a subsequence that converges to a harmonic function.

Proposition 8.1.2. *Let $\{u_n\}$ be a sequence of harmonic functions on Ω that is uniformly bounded on each compact subset of Ω. Then $\{u_n\}$ has a convergent subsequence that converges uniformly to a harmonic function on each compact subset of Ω.*

Proposition 8.1.3 (Harnack's inequality). *Let K be a compact subset of a domain Ω, and u a non-negative harmonic function. There is a constant $C > 1$ such that*

$$\frac{1}{C} \leq \frac{u(x)}{u(y)} \leq C$$

for all $x, y \in K$.

Proposition 8.1.4 (Harnack's Principle). *Let $\{u_n\}$ be a pointwise increasing sequence of harmonic function on a domain Ω. Then either $u_n(x) \to \infty$ for all $x \in \Omega$, or $\{u_n\}$ converges uniformly on compact subsets of Ω to a harmonic function.*

Proposition 8.1.5. *Every Riemann surface X has a countable basis of precompact coordinate discs.*

Proof. We only sketch the proof. For a more detailed proof, we refer the reader to [119]. Each point in X is contained in a chart $(U_\alpha, \varphi_\alpha)$, and we can use these charts to form an open cover $\{U_\alpha\}$ of X. Since X is second countable, we can obtain a countable subcover $\{U_i\}$. For each (U_i, φ_i), obtain a countable basis of precompact discs in the image $\varphi_i(U_i)$ by considering balls of rational center and radius. Pulling this basis back to X for each chart (U_i, φ_i) yields a countable basis of precompact coordinate discs. □

Definition 8.1.6. Let X be a Riemann surface. A real-valued function $u : X \to \mathbb{R}$ is harmonic at $p \in X$ if there exists a coordinate disc (Δ, φ) containing p such that $u \circ \varphi^{-1} : \mathbb{D} \to \mathbb{R}$ is a *harmonic function*. If u is harmonic at each $p \in X$, we say u is a harmonic function on X.

We note that harmonicity at p is invariant with respect to the choice of coordinate disc. Indeed, if (Δ_1, φ_1) and (Δ_2, φ_2) are two coordinate discs containing p and $u \circ \varphi_1^{-1}$ is harmonic, then $u \circ \varphi_2^{-1} = (u \circ \varphi_1^{-1}) \circ (\varphi_1 \circ \varphi_2^{-1})$ is the composition of a harmonic function with a holomorphic function, hence is harmonic.

Definition 8.1.7. A continuous function $u : X \to [-\infty, \infty)$ is *subharmonic* if for every coordinate disc (Δ, φ), if $v : \overline{\Delta} \to \mathbb{R}$ is a harmonic function such that $u(p) \leq v(p)$ for all $p \in \partial\Delta$, then $u(p) \leq v(p)$ for all $p \in \Delta$.

Equivalently, we can define a continuous function $u : X \to [-\infty, \infty)$ to be subharmonic if for each $p \in X$, there exists a coordinate disc (Δ, φ) containing p such that $u \circ \varphi^{-1} : \mathbb{D} \to \mathbb{R}$ is a subharmonic function. As expected, every harmonic function on X is subharmonic on X.

Proposition 8.1.8. *If $f : \Omega \to \mathbb{C}$ is a holomorphic function defined on a domain Ω, then $\log|f(z)|$ is a subharmonic function.*

Proposition 8.1.9 (maximum principle)**.** *Let u be a subharmonic function on a Riemann surface X. If u attains a maximum at $p \in X$, then $u \equiv$ const.*

Proof. Let M be the supremum of u. Consider the set $\Gamma = \{p \in X : u(p) = M\}$. Then $\Gamma = u^{-1}(\{M\})$ is closed by continuity of u. Let $q \in \Gamma$. There is a coordinate disc (Δ, φ) containing q such that $u \circ \varphi^{-1} : \mathbb{D} \to \mathbb{R}$ is a subharmonic function. If we let $z_0 = \varphi(q)$, we see that $u \circ \varphi^{-1}$ is a subharmonic function on $\mathbb{D}$ that achieves its maximum at $z_o \in \mathbb{D}$. By the maximum principle for subharmonic functions on domains in the complex plane, $u \circ \varphi^{-1} \equiv$ const. Since φ is bijective, $u \equiv M$ in Δ. This shows that Γ is open. Since Γ is non-empty, open, and closed, by connectedness of X we have $R = \Gamma$.

Let (Δ, φ) be a coordinate disc, and u be a subharmonic function on X. By using the Poisson integral, we can solve the Dirichlet problem

$$\begin{cases} \frac{\partial^2 w}{\partial x^2} + \frac{\partial^2 w}{\partial y^2} = 0 & \text{in } \mathbb{D} \\ w = u \circ \varphi^{-1} & \text{on } \partial\mathbb{D} \end{cases}$$

We then define

$$u_\Delta(p) = \begin{cases} u(p) & p \notin \Delta \\ w \circ \varphi(p) & p \in \Delta \end{cases}$$

By construction, u_Δ is harmonic in Δ. Notice that $u_\Delta \geq u$ and that the function u_Δ is subharmonic. □

It is immediate that if u, v are subharmonic functions, then $\max\{u, v\}$ is a subharmonic function. Therefore the following definition makes sense:

Definition 8.1.10. A *Perron family* on X is a collection $\mathcal{F}$ of subharmonic functions such that

(1) If $u_1, u_2 \in \mathcal{F}$, then $\max\{u_1, u_2\} \in \mathcal{F}$.

(2) If $u \in \mathcal{F}$, then $u_\Delta \in \mathcal{F}$.

Taking the supremum of a Perron family is a useful method of obtaining a harmonic function, and will be used in our definition of the Green's function.

Proposition 8.1.11. *Let $\mathcal{F}$ be a Perron family on X. Then*

$$u(p) = \sup\{v(p) : v \in \mathcal{F}\}$$

is either harmonic, or $u(p) = +\infty$ for all $p \in X$.

Proof. First, we show that the set $\{z \in X : u(z) < \infty\}$ is open and u is harmonic in this set.

Let $p \in X$ be such that $u(p) < \infty$ and let (Δ, φ) be a coordinate disc containing p. There exists a sequence $\{f_n'\} \in \mathcal{F}$ such that $f_n'(p)$ converges to $u(p)$.

If we define $f_n = \max\{f'_1, \ldots, f'_n\}$, we obtain a sequence of increasing subharmonic functions. Therefore, $\{(f_n)_\Delta\}$ is a sequence of increasing functions that are harmonic on Δ.

By Harnack's theorem, $f = \sup\{(f_n)_\Delta\}$ is either $+\infty$ or harmonic in Δ. But since $(f_n)_\Delta \in \mathcal{F}$, we have $(f_n)_\Delta \leq u$, hence $f \leq u$. Therefore $f(p) < \infty$, and so f is harmonic in Δ. Note that $f(p) = u(p)$. Indeed, $f_n(p)$ is an increasing sequence converging to $u(p)$, so

$$f(p) = \sup_n \{(f_n)_\Delta (p)\} \geq \sup_n \{f_n(p)\} = u(p)\,.$$

We now show that $f = u$ in Δ, and hence $u < \infty$ in Δ and u is harmonic in Δ. Suppose $f(q) < u(q)$ for some $q \in \Delta$. There exists a sequence $\{g'_n\} \in \mathcal{F}$ such that $g'_n(q)$ converges to $u(q)$. If we define $g_n = \max\{g'_1, \ldots, g'_n, f'_1, \ldots, f'_n\}$, we obtain a sequence of increasing subharmonic functions such that $g_n \geq f_n$. Proceeding as above, we obtain $g = \sup\{(g_n)_\Delta\}$, a harmonic function in Δ such that $g(p) = u(p), g(q) = u(q)$, and $g \geq f$. But since $g(p) = f(p)$ and $g(q) > f(q)$, then $(f - g)$ is a non-constant harmonic function on Δ that achieves its maximum at p. Hence we have a contradiction to the maximum principle.

We have shown that the set $\{z \in X \mid u(z) < \infty\}$ is open and u is harmonic in this set. Its complement, $\{z \in X \mid u(z) = \infty\}$ is also open. If $u(p) = \infty$, then using the above notation, $f = \infty$ in Δ by Harnack's theorem. But $f \leq u$ in Δ, so $u = \infty$ in Δ.

Therefore, $\{z \in X \mid u(z) < \infty\}$ is both open and closed, hence is either equal to X or $\emptyset$. If it is equal to X, then u is harmonic, else $u = \infty$. □

8.2 Green's Functions

The goal of this section is to develop the machinery of Green's functions on a Riemann surface.

Fix a point $q \in X$, and let (Δ, φ) be a coordinate disc containing q such that $\varphi(q) = 0$. Let $\mathcal{P}_q$ be the family of subharmonic functions on $R \setminus \{q\}$ satisfying:

(1) every $u \in \mathcal{P}_q$ has compact support, and

(2) for every $u \in \mathcal{P}_q$, the function $v(p) = u(p) + \log|\varphi(p)|$ is subharmonic on Δ.

This definition makes $\mathcal{P}_q$ a Perron family on $R \setminus \{q\}$. The set is non-empty since $u = 0 \in \mathcal{P}_q$; indeed, $\log|\varphi(p)|$ is subharmonic by Proposition 8.1.8.

Definition 8.2.1. Suppose $\sup\{u(p) : u \in \mathcal{P}_q\} < \infty$ for some $p \in X$. A *Green's function* for X with a pole at q is defined as

$$g(p, q) = \sup\{u(p) : u \in \mathcal{P}_q\}, \qquad p \in X \setminus \{q\}.$$

If the supremum is identically infinite, we say that a Green's function for X with a pole at q does not exist. The Green's function will be a central tool in the proof of the uniformization theorem. We now collect some important facts about Green's functions.

Proposition 8.2.2. *Let $g(p, q)$ be a Green's function for X with a pole at q. Then $g(p, q) > 0$ and $g(p, q)$ is harmonic on $X \setminus \{q\}$. Furthermore, if (Δ, φ) is a coordinate disc such that $\varphi(q) = 0$, then*

$$h(p) = g(p, q) + \log|\varphi(p)|$$

is harmonic on Δ.

Proof. By Proposition 8.1.11, $g(p, q)$ is harmonic on $R \setminus \{q\}$. Since $u = 0 \in \mathcal{P}_q$, we have $g(p, q) \geq 0$. By the maximum principle, $-g$ cannot attain zero on $R \setminus \{q\}$, so $g(p, q) > 0$ for all $p \in X \setminus \{q\}$.

Next, set $h(p) = g(p, q) + \log|\varphi(p)|$. On $\Delta \setminus \{q\}$ we compute

$$h \circ \varphi^{-1}(z) = g \circ \varphi^{-1}(z) + \log|z|,$$

which is harmonic on the punctured unit disc. If h is bounded, Riemann's removable singularity theorem allows us to conclude that h is harmonic on Δ.

Let M denote the maximum of $g(p, q)$ on the compact boundary set

$$\partial\Delta = \{p \in X : |\varphi(p)| = 1\}.$$

For any $u \in \mathcal{P}_q$, we have $u \leq M$ on $\partial\Delta$. Since $u + \log|\varphi(p)|$ is subharmonic on Δ and $\log|\varphi(p)| = 0$ on $\partial\Delta$, the maximum principle yields

$$u(p) + \log|\varphi(p)| \leq M \qquad \text{for all } p \in \Delta.$$

Taking the supremum, we obtain $h(p) \leq M$.

On the other hand, consider the function $v : X \to \mathbb{R}$ defined by

$$v(p) = \begin{cases} -\log|\varphi(p)| & p \in \Delta, \\ 0 & p \notin \Delta. \end{cases}$$

This belongs to $\mathcal{P}_q$, hence $g(p, q) \geq -\log|\varphi(p)|$ for all $p \in \Delta$. Therefore,

$$0 \leq h(p) \leq M, \qquad p \in \Delta,$$

completing the proof. ◻

Proposition 8.2.3. *Let $g(p,q)$ be a Green's function on X with a pole at q. Then*

$$\inf_{p\in X} g(p,q) = 0\,.$$

Proof. Let (Δ,φ) be a coordinate disc containing q, with $\varphi(q)=0$. Suppose instead that $\inf_{p\in X} g(p,q) = a > 0$. Then $g(p,q) - a > 0$ for all $p \in X\setminus\{q\}$, and $g(p,q) - a + \log|\varphi(p)|$ is harmonic on Δ.

For $u \in \mathcal{P}_q$, the function $u(p)+\log|\varphi(p)|$ is subharmonic on Δ. Subtracting $g(p,q) - a + \log|\varphi(p)|$ yields $u(p) - (g(p,q) - a)$, which is subharmonic on Δ. Since u has compact support, this function is negative outside a compact set. The maximum principle then implies $u(p) < g(p,q) - a$ for all $p \in X$. Taking the supremum gives $g(p,q) \le g(p,q) - a$, a contradiction. □

Proposition 8.2.4. *Suppose a Green's function $g(p,q_0)$ with a pole at q_0 exists for some $q_0 \in X$. Then a Green's function with a pole at q exists for every $q \in X$.*

Proof. Let $\Gamma \subseteq X$ be the set of points q for which a Green's function $g(p,q)$ with a pole at q exists.

First, we show that Γ is open. Fix $q_0 \in \Gamma$, and let (U,φ) be a chart centered at q_0 with $\varphi(q_0)=0$. Choose $r>0$ small enough so that $\Delta_r \subset U$. Define

$$\mathcal{F} = \{v : X\setminus\Delta_r \to \mathbb{R} \mid v \text{ subharmonic},\ v \le 1,\ v \text{ compactly supported}\}\,,$$

and let

$$\tilde{v}(p) = \sup\{v(p) \mid v \in \mathcal{F}\}\,.$$

By Proposition 8.1.11, $\tilde{v}$ is harmonic on $X\setminus\Delta_r$.

If $g(p,q_0)$ exists, then $\tilde{v} \in (0,1)$. Indeed, choose $a>0$ such that $g(p,q_0) > a$ on $\partial\Delta_r$. Then $\tilde{v}(p) \le 1 < g(p,q_0)/a$ on $\partial\Delta_r$, and by the maximum principle the inequality extends to $X\setminus\Delta_r$. Since $\inf g(p,q_0) = 0$, it follows that $\inf\tilde{v} = 0$, hence $\tilde{v} \not\equiv 1$, so $\tilde{v} \in (0,1)$.

Now assume $\tilde{v} \in (0,1)$. We claim that then $g(p,q)$ exists for every $q \in \Delta_r$. Let $q \in \Delta_r$, and choose Δ_s with $\Delta_r \subset \Delta_s \subset U$. For any $u \in \mathcal{P}_q$, the function $u(p)+\log|\varphi(p)-\varphi(q)|$ is subharmonic on Δ_s. Estimating its boundary values and applying the maximum principle, one shows that u is uniformly bounded above on Δ_s, independently of u. Taking the supremum then yields existence of the Green's function at q. Hence Γ is open.

Finally, suppose $q_0 \in X\setminus\Gamma$. Then, by the same reasoning, $\tilde{v} \equiv 1$ on $X\setminus\Delta_r$, and no Green's function exists in Δ_r. Thus $X\setminus\Gamma$ is open as well.

Since Γ is both open and closed in X, and X is connected, we conclude that $\Gamma = X$ or $\Gamma = \varnothing$. But Γ is nonempty by assumption, hence $\Gamma = X$. □

Proposition 8.2.5. *Let S be a Riemann surface, and let $X \subset S$ be a connected open subset. Suppose X admits an atlas consisting of finitely many precompact coordinate discs in S. Then for every $q \in X$ there exists a Green's function on X with a pole at q.*

Proof. Cover X by finitely many precompact coordinate discs $\{U_j\}_{j=1}^N$. Fix $q \in X$, and choose a disc U_j containing q. On U_j, define

$$v(p) = \max\{0, -\log|\varphi_j(p) - \varphi_j(q)|\}\,.$$

This function is subharmonic, compactly supported in U_j, and has the desired logarithmic singularity at q. Extending v by zero outside U_j, we obtain an element of $\mathcal{P}_q$. Hence $\mathcal{P}_q$ is nonempty, and the Green's function $g(p,q)$ exists by Definition 8.2.1. □

Lemma 8.2.6. *Suppose X is a simply connected Riemann surface such that a Green's function $g(p,q)$ with a pole at q exists. Then there exists an analytic function $F : X \to \mathbb{C}$ such that*

$$|F(p)| = e^{-g(p,q)}\,.$$

Proof. Set $h(p) = g(p,q) + \log|\varphi(p)|$ on a local coordinate disc (Δ, φ) centered at q. By Proposition 8.2.2, h is harmonic on Δ. Hence locally there exists a harmonic conjugate $\tilde{h}$, and we may define a holomorphic function

$$f(p) = h(p) + i\tilde{h}(p)\,.$$

On overlaps of coordinate discs, the real parts of these local functions agree, so their difference is purely imaginary and constant. Therefore, the local functions glue together up to multiplicative factors of modulus one. Since X is simply connected, analytic continuation produces a global holomorphic function F with

$$|F(p)| = e^{-g(p,q)}\,.$$

□

Theorem 8.2.7. *Let X be a simply connected Riemann surface such that a Green's function $g(p,q)$ exists. Then there exists a conformal map from X onto the open unit disc.*

Proof. By Lemma 8.2.6, there exists a holomorphic function $F : X \to \mathbb{C}$ such that $|F(p)| = e^{-g(p,q)}$. Since $g(p,q) > 0$ on $X \setminus \{q\}$, we have $0 < |F(p)| < 1$ for $p \neq q$, hence $F(X) \subseteq \mathbb{D}$.

Near q, $g(p,q) \sim -\log|\varphi(p)|$, so F has a simple zero at q. Thus F is nonconstant. By the open mapping theorem, $F(X)$ is open in $\mathbb{D}$. Since X is simply connected, the map F is a universal covering of its image. But the unit disc $\mathbb{D}$ admits no nontrivial simply connected coverings other than itself. Therefore F maps X conformally onto $\mathbb{D}$. □

Corollary 8.2.8. *Let X be a Riemann surface admitting a Green's function. Then for all distinct $p, q \in X$, the Green's function is symmetric:*

$$g(p,q) = g(q,p)\,.$$

Proof. Symmetry was already observed in the proof of Theorem 8.2.7: the modulus of the associated holomorphic function is independent of the order of the arguments. Hence $g(p,q) = g(q,p)$. □

8.3 Bipolar Green's Functions

In this section, we classify simply connected Riemann surfaces X that do not admit a Green's function. We will show that in this case X is conformally equivalent either to the Riemann sphere (if X is compact) or to the complex plane (otherwise).

Definition 8.3.1. Let q_1, q_2 be distinct points of a Riemann surface X. Let (Δ_1, φ_1) and (Δ_2, φ_2) be coordinate discs such that $\Delta_1 \cap \Delta_2 = \emptyset$ and $\varphi_1(q_1) = \varphi_2(q_2) = 0$. A harmonic function $G(p, q_1, q_2)$ on $X \setminus \{q_1, q_2\}$ is called a *bipolar Green's function* for X with poles at q_1 and q_2 if it satisfies:

- $G(p, q_1, q_2) + \log|\varphi_1(p)|$ is harmonic at q_1,
- $G(p, q_1, q_2) - \log|\varphi_2(p)|$ is harmonic at q_2, and
- $G(p, q_1, q_2)$ is bounded on $X \setminus (\Delta_1 \cup \Delta_2)$.

Proposition 8.3.2. *For all distinct points $q_1, q_2 \in X$, there exists a bipolar Green's function $G(p, q_1, q_2)$.*

Proof. Let S be a Riemann surface with an atlas consisting of finitely many precompact coordinate discs, whose closures lie inside a Riemann surface X. Further assume that X is an open subset of a Riemann surface R', and that X has an atlas of finitely many precompact coordinate discs. We first derive results under these assumptions, and then pass to the general case by taking countable unions of coordinate discs.

Let q_1, q_2 be distinct points in S, and let (U_1, φ_1) and (U_2, φ_2) be charts such that $q_i \in U_i$, $\varphi_i(q_i) = 0$, and $U_1 \cap U_2 = \emptyset$. Choose $r > 0$ such that

$$\Delta_i = \{p \in U_i : |\varphi_i(p)| \leq r\}$$

is well-defined. Select $s > 0$ with $r > s$, and set

$$\sigma_i = \{p \in U_i : |\varphi_i(p)| \leq s\}.$$

By Proposition 8.2.5, Green's functions exist on X.

The first objective is to bound the difference of the Green's functions $g_R(p, q_1)$ and $g_R(p, q_2)$ for $p \in X \setminus (\Delta_1 \cup \Delta_2)$, in a way that is independent of X. Let

$$M_i = \max_{p \in \partial\sigma_i} g_R(p, q_i).$$

Since $\inf_{p \in X} g(p, q_i) = 0$, we have $g(p, q_i) \to 0$ as $p \to \partial R$. The function $g(p, q_i)$ is harmonic on $R \setminus (\sigma_1 \cup \sigma_2)$, so by the maximum principle we obtain $g_R(p, q_i) \leq M_i$ for $p \in X \setminus (\sigma_1 \cup \sigma_2)$. In particular, $M_i - g_R(p, q_i) \geq 0$ for $p \in S \setminus (\sigma_1 \cup \sigma_2)$.

By compactness of $\partial\Delta_i$, there exists $z_i \in \partial\Delta_i$ such that $g(p, q_i) \leq g(z_i, q_i)$ for all $p \in \partial\Delta_i$. Since $g_R(p, q_i) + \log|\varphi_i(p)|$ is harmonic on Δ_i, the maximum principle yields

$$g_R(p, q_i) + \log|\varphi_i(p)| \leq g(z_i, q_i) + \log r$$

for all $p \in \Delta_i$. Taking the maximum over $\partial\sigma_i$ gives

$$M_i - g(z_i, q_i) \leq \log(r/s)\,.$$

Since $M_i - g_R(p, q_i)$ is a nonnegative harmonic function on $S \setminus (\sigma_1 \cup \sigma_2)$ and $\partial\Delta_1 \cup \partial\Delta_2$ is compact, Harnack's inequality provides a constant $C > 0$ such that

$$\frac{M_i - g_R(p, q_i)}{M_i - g_R(z_i, q_i)} \leq C$$

for all $p \in \partial\Delta_1 \cup \partial\Delta_2$. Hence

$$M_i - g_R(p, q_i) \leq C\,(M_i - g_R(z_i, q_i)) \leq C\log(r/s) =: C_0$$

for $p \in \partial\Delta_1 \cup \partial\Delta_2$, where C_0 is independent of X. Thus

$$M_i - C_0 \leq g_R(p, q_i) \leq M_i$$

for all $p \in \partial\Delta_1 \cup \partial\Delta_2$. Therefore,

$$|g_R(p, q_1) - g_R(p, q_2)| \leq (M_1 - M_2) + C_0$$

for $p \in \partial\Delta_1 \cup \partial\Delta_2$.

Now, since $g(p, q_i) \to 0$ as $p \to \partial R$, applying the maximum principle extends the estimate to all $p \in X \setminus (\Delta_1 \cup \Delta_2)$. However, the M_i depend on X, so we refine the argument.

Because $g_R(p, q_1)$ is harmonic on Δ_2, the maximum principle gives

$$M_1 - C_0 \leq g_R(p, q_1) \leq M_1, \qquad p \in \Delta_2\,,$$

and in particular

$$M_1 - C_0 \leq g_R(q_2, q_1) \leq M_1\,.$$

Similarly,

$$M_2 - C_0 \leq g_R(q_1, q_2) \leq M_2\,.$$

Since $g_R(q_1, q_2) = g_R(q_2, q_1)$, we conclude

$$|M_1 - M_2| \leq C_0\,,$$

independent of X. Hence

$$|g_R(p, q_1) - g_R(p, q_2)| \leq 2C_0$$

for all $p \in X \setminus (\Delta_1 \cup \Delta_2)$.

Now consider an arbitrary Riemann surface M. Let $q_1, q_2 \in M$ be distinct points, and let (Δ_1, φ_1), (Δ_2, φ_2) be coordinate discs such that $\Delta_1 \cap \Delta_2 = \emptyset$, $\varphi_1(q_1) = 0$, $\varphi_2(q_2) = 0$. By Proposition 8.1.5,

$$M = \bigcup_{i=1}^{\infty} S_i$$

where each S_i is a Riemann surface with an atlas of finitely many precompact coordinate discs. By taking finite unions and relabeling, we may assume $\Delta_1 \cup \Delta_2 \subset S_1$ and $\overline{S_i} \subseteq S_{i+1}$.

For each S_n, let $g_n(p, q_1)$ and $g_n(p, q_2)$ be Green's functions, and define

$$G_n(p, q_1, q_2) = g_n(p, q_1) - g_n(p, q_2)\,.$$

As shown above, $|G_n(p, q_1, q_2)| \leq 2C_0$ for all n, so $\{G_n\}$ is a sequence of uniformly bounded harmonic functions. By Proposition 8.1.2, there exists a subsequence $\{G_{n_k}\}$ converging to a harmonic function $G(p, q_1, q_2)$ on $M \setminus (\Delta_1 \cup \Delta_2)$.

Moreover,

$$g_{n_k}(p, q_1) + \log|\varphi_1(p)| - g_{n_k}(p, q_2) = G_{n_k}(p, q_1, q_2) + \log|\varphi_1(p)|$$

is harmonic in Δ_1. On $\partial\Delta_1$ we have the uniform bound

$$|G_{n_k}(p, q_1, q_2) + \log|\varphi_1(p)|| = |g_{n_k}(p, q_1) - g_{n_k}(p, q_2)| \leq 2C_0\,,$$

which extends to Δ_1 by the maximum principle. Therefore, $G_{n_k}(p, q_1, q_2) + \log|\varphi_1(p)|$ converges to a harmonic function on Δ_1. The same reasoning applies to Δ_2. Hence the limit $G(p, q_1, q_2)$ is a bipolar Green's function for M. □

Lemma 8.3.3. *Let X be a Riemann surface. Suppose there exists a bounded, nonconstant holomorphic function $h : X \to \mathbb{C}$. Then X admits a Green's function.*

Proof. Fix $q \in X$. Since h is bounded, say $|h| < B_0$, define

$$\varphi(p) = \frac{h(p) - h(q)}{B_0}\,.$$

Then φ is holomorphic, $\varphi(q) = 0$, and $|\varphi| < 1$. For any $u \in \mathcal{P}_q$, the function $u(p) + \log|\varphi(p)|$ is subharmonic and strictly negative outside the compact support of u. By the maximum principle,

$$u(p) < -\log|\varphi(p)|\,.$$

Since this holds for all $u \in \mathcal{P}_q$, the supremum is not identically infinite. Hence a Green's function exists. □

Theorem 8.3.4. *Let X be a simply connected Riemann surface such that a Green's function $g(p,q)$ does not exist. Then X is conformally equivalent either to the complex plane or to the Riemann sphere.*

Proof. Fix distinct points $q_1, q_2 \in X$, and let $G(p, q_1, q_2)$ be a bipolar Green's function. By a modification of the proof of Lemma 8.2.6, there exists a meromorphic function $F : X \to \mathbb{C} \cup \{\infty\}$ such that

$$|F(p)| = e^{-G(p,q_1,q_2)}.$$

Note that F has a simple zero at q_1 and a simple pole at q_2.

We show that F is injective. Choose $q_0 \in X \setminus \{q_1, q_2\}$, and let $H : X \to \mathbb{C} \cup \{\infty\}$ be meromorphic with

$$|H(p)| = e^{-G(p,q_0,q_2)}.$$

From the definition of bipolar Green's functions, F is bounded away from $\{q_1, q_2\}$ and H is bounded away from $\{q_0, q_2\}$. Moreover, $F(q_1) = 0$, $F(q_2) = \infty$, and $H(q_0) = 0$, $H(q_2) = \infty$.

Since all poles and zeros are simple, the function

$$\varphi(p) = \frac{F(p) - F(q_0)}{H(p)}$$

is well-defined, analytic, and bounded. By Lemma 8.3.3, φ must be constant. Since $\varphi(q_1) = -F(q_0)/H(q_1) \neq 0$, it follows that φ is a nonzero constant. Hence $F(p) = F(q_0)$ if and only if $p = q_0$. As q_0 was arbitrary, F is injective.

Let $\Omega = F(X) \subseteq \hat{\mathbb{C}}$, where $\hat{\mathbb{C}} = \mathbb{C} \cup \{\infty\}$. The set Ω is open and simply connected, and $F : X \to \Omega$ is conformal. If $\hat{\mathbb{C}} \setminus \Omega$ has more than one point, a Möbius transformation moves one of them to ∞, so Ω becomes a proper subset of $\mathbb{C}$. By the Riemann mapping theorem, Ω is then conformally equivalent to the unit disc. Hence X admits a nonconstant bounded holomorphic function, contradicting Lemma 8.3.3. Therefore $\hat{\mathbb{C}} \setminus \Omega$ has at most one point.

If $\hat{\mathbb{C}} \setminus \Omega$ is a single point, we can move it to ∞ by a MÖbius transformation, and thus $\Omega \cong \mathbb{C}$. If $\hat{\mathbb{C}} \setminus \Omega = \varnothing$, then $\Omega = \hat{\mathbb{C}}$, so F maps X conformally onto the Riemann sphere. $\square$

9

Analytic Capacity

The idea of analytic capacity is an old one. It was formally introduced by L. Ahlfors in the 1940s in his study of removable sets for bounded, analytic functions. But these ideas were anticipated by work of Painlevé in the nineteenth century.

Analytic capacity is analogous to measure theory, in that it assigns a numerical "size" to a compact set in the plane. But capacity is much trickier to study. For example, it is only recently that it was proved that analytic capacity is subadditive. It is generally challenging to calculate the capacity of any given set.

It is natural to ask how analytic capacity relates to length, to Hausdorff dimension, and to Lebesgue measure. Considerable effort has been devoted to studying all these questions.

Analytic capacity is also a natural language for studying questions in analytic approximation theory.

This chapter introduces the key ideas in the theory of analytic capacity.

9.1 Preliminaries

It is well known to every student that, if U is a domain (even the entire complex plane) and $P \in U$, then a bounded, holomorphic function on $U \setminus \{P\}$ continues analytically to all of U. This is a classical result of Riemann.[1] It is of interest to know which compact sets can replace P and give the same result. In studying this question in 1947, Lars Ahlfors [2] formulated the idea of analytic capacity.

Definition 9.1.1. A set $K \subseteq \mathbb{C}$ is said to be *removable* if, whenever U is a domain containing K, then every holomorphic function on $U \setminus K$ has a holomorphic continuation to all of U.

These ideas suggest the following question:

Given a compact set $E \subseteq \mathbb{C}$, when does there exist a non-constant, bounded, holomorphic function f on $\mathbb{C} \setminus E$?

[1] Less well known is the fact that it is sufficient for f to be square integrable.

DOI: 10.1201/9781003713005-9

A partial answer to the question was given by Painlevé in the nineteenth century.

Prelude: We provide now one of the key technical ideas in the theory of removable sets. This idea dates back to the nineteenth century.

Theorem 9.1.2 (Painlevé). *Let $E \subseteq \mathbb{C}$ be compact. Assume that, for every $\epsilon > 0$, the set E can be covered by a collection of open discs whose radii do not exceed ϵ. Then the set of bounded, analytic functions on $\mathbb{C} \setminus E$ consists only of the constant functions.*

Proof. For each $\epsilon > 0$, cover E by a collection of discs D_j of radius r_j such that $\sum_j r_j < \epsilon$. Set $D_\epsilon = \cup_j D_j$. Let $\Gamma_\epsilon = \partial D_\epsilon$. Let f be a bounded, holomorphic function on $\mathbb{C} \setminus E$. Now, if $z \notin \overline{D_\epsilon}$, then

$$f'(z) = -\frac{1}{2\pi i} \oint_{\Gamma_\epsilon} \frac{f'(\zeta)}{(z-\zeta)^2} d\zeta \,.$$

Then

$$|f'(z)| \leq \frac{2\pi\epsilon \cdot \sup_{\Gamma_\epsilon} |f'(\zeta)|}{2\pi \left[d\left(z, \Gamma_\epsilon\right)\right]^2} \,.$$

As $\epsilon \to 0$, we see that $|f(z)| = 0$. So f is constant. □

If $U \subseteq \mathbb{C}$ then we let $H^\infty(U)$ be the space of bounded, holomorphic functions on U. Now we have Ahlfors's definition:

Definition 9.1.3. Let $E \subseteq \mathbb{C}$. Let $U = \mathbb{C} \setminus E$. We define the *analytic capacity* $\gamma(E)$ of E by

$$\gamma(E) = \sup \{|f'(\infty)| : f \in H^\infty(U) \text{ and } \|f\|_{\text{sup}} \leq 1\}$$

Here $f'(\infty)$ is calculated relative to the local coordinate $z = 1/\zeta$ on the Riemann sphere as

$$f'(\infty) = \lim_{\zeta \to 0} f'(1/\zeta) = \lim_{\zeta \to 0} \frac{f(1/\zeta) - f(\infty)}{\zeta - 1/\zeta} = \lim_{z \to \infty} z \cdot (f(z) - f(\infty))$$

Remark 9.1.4. Use the notation from the definition. Consider the Möbius transformation

$$\varphi(z) = \frac{f(z) - f(\infty)}{1 - \overline{f(\infty)} f(z)}$$

This function is certainly a bounded, analytic function on $\mathbb{C} \setminus E$. Thus we need only consider functions f with $f(\infty) = 0$ because $\varphi(\infty) = 0$.

Analytic capacity is interesting because it characterizes which domains support a non-constant bounded analytic function. Any proper open set of the Riemann sphere has a non-constant holomorphic function defined on it, but if we consider bounded analytic functions then this is not always true. Riemann's theorem says that an isolated singularity of a bounded holomorphic function is removable. Therefore if f is bounded and holomorphic on the complement of a finite number of points, then it can be extended to be bounded and holomorphic on the whole plane. Hence it is constant by Liouville's theorem.

It follows easily from the definition of analytic capacity that $H^\infty(\Omega)$ has nonconstant functions iff $\gamma(\Omega^c) \neq 0$.

Now we have

Proposition 9.1.5. *These properties hold for the analytic capacity:*

(a) If $f(\infty) = 0$, then γ has this invariance property:

$$\gamma(aE + b) = |a|\gamma(E)$$

(b) If $E \subseteq F$, then $\gamma(E) \leq \gamma(F)$.

Proof. The first statement follows from the equation

$$\lim_{z\to\infty} zf(az + b) = af'(\infty)$$

The second statement is obvious. □

The next result is important for calculations.

Proposition 9.1.6. *Assume that E is connected, simply connected, but not a singleton. Let g be the conformal map of $\widehat{\mathbb{C}}\setminus E$ to the unit disc D satisfying $g(\infty) = 0, g'(\infty) > 0$. Then $\gamma(E) = |g'(\infty)|$.*

Proof. Clearly $|g'(\infty)| \leq \gamma(E)$ by the definition of γ. If f is any other candidate mapping for the analytic capacity of E, then apply Schwarz's lemma to $F \equiv f \circ g^{-1}$. We thus obtain $|f(\infty)| \leq |g'(\infty)|$ for any such f. Thus $\gamma(E) \leq |g'(\infty)|$, completing the proof. □

9.2 Calculating Analytic Capacity

Capsule: Analytic capacity is most meaningful in the context of concrete examples. And there is always some interesting function theory involved in working the examples.

Example 9.2.1. Let E be the real interval $[-2, 2] = \{x + i0 \mid -2 \leq x \leq 2\}$. Define $g(z) = z + 1/z$. Then g is a conformal mapping onto the complement of E that takes the unit circle to E. Thus $\gamma(E) = \left(g^{-1}\right)'(\infty)$, so that we have

$$\gamma(E) = \frac{1}{g'(g(\infty))} = \frac{1}{1 - 1/\infty} = 1$$

More generally, let E be the real interval $[a, b] = \{x + i0 \mid a \le x \le b\}$. Then, using results of the last section, we see that

$$\gamma(E) = \gamma\left(\frac{b-a}{4}[-2,2] + \frac{a+b}{2}\right) = \frac{b-a}{4}\gamma([-2,2]) = \frac{b-a}{4}$$

It is actually possible to show (we shall omit the details) that, if $E \subseteq \mathbb{R} \subseteq \mathbb{C}$ is any Boreal measurable set, then $\gamma(E)$ is $1/4$ the measure of E.

Example 9.2.2. Let E be the disc $\{z \in \mathbb{C} \mid |z - z_0| \le r\}$. Then

$$\widehat{\mathbb{C}} \backslash E = \{z \in \mathbb{C} \mid |z - z_0| > r\} ,$$

and the mapping

$$z \longmapsto \frac{r}{z - z_0}$$

is a conformal map of $\widehat{\mathbb{C}} \backslash E$ to the disc D.

If we set $g(z) = r/(z - z_0)$, then, using the local coordinate $z = 1/\xi$, we have

$$g(1/\xi) = \frac{r}{1/\xi - z_0} = \frac{r\xi}{1 - z_0\xi}$$

Now differentiation shows us that

$$g'\left(\frac{1}{\xi}\right) = \frac{r}{1 - z_0\xi} - \frac{r\xi}{(1 - z_0\xi)^2}$$

Thus, if we send ξ to 0, we obtain

$$g'(\infty) = \lim_{\xi \to 0} g'(1/\xi) = r$$

We conclude that the analytic capacity of a disc of radius r is r.

9.3 Analytic Capacity and Removability

Capsule: The whole idea of analytic capacity is that it gives a measure of "how removable" a set is. If a set has analytic capacity zero, then it is removable for bounded, holomorphic functions. If it has positive analytic capacity, then it is nonremovable—and the capacity gives a measure of how nonremovable it is.

Ahlfors's original motivation for conceiving of and studying analytic capacity was the characterization of removable sets. In that vein we have the following essential result.

Prelude: This result shows explicitly how to understand analytic capacity in terms of the concept of removable set. It is a fundamental theorem.

Theorem 9.3.1. *Let $E \subseteq \mathbb{C}$ be a compact set. Then the following assertions are equivalent:*

(a) $\gamma(E) = 0$.

(b) Every bounded, holomorphic function $f : \mathbb{C} \backslash E \to \mathbb{C}$ is constant.

(c) E is removable for bounded, analytic functions.

Proof. It is clear that (b) implies (a). Also (c) implies (b) by Liouville's theorem.

For (a) $\Rightarrow$ (b), Suppose that there exists a nonconstant, bounded, holomorphic function $f : \mathbb{C} \backslash E \to \mathbb{C}$ with $f(\infty) = 0$ and $f(z_0) = 0$ for some $z_0 \in \mathbb{C} \backslash E$. Let us define

$$g(z) = \begin{cases} \dfrac{f(z) - f(z_0)}{z - z_0} & \text{if } z \in \mathbb{C} \backslash E \text{ and } z \neq z_0 \\ f'(z_0) & \text{if } z = z_0 \end{cases}$$

We see that g is a bounded, holomorphic function on $\mathbb{C} \backslash E$ and that $g'(\infty) = f(z_0) \neq 0$. Thus $\gamma(E) > 0$, and that is a contradiction.

Finally, for (b) $\Rightarrow$ (c), suppose that E satisfies (b). We claim that E must be totally disconnected. If not, then the Riemann Mapping theorem tells us that there is a nonconstant, bounded, analytic function $f : \mathbb{C} \backslash E_0 \to D$ for some subset $E_0 \subseteq E$. That would be a contradiction.

Now let U be an open set that contains E. Let f be a bounded, holomorphic function on $U \backslash E$, and fix a point $z_0 \in U \backslash E$. Since E is totally disconnected, there are two simple closed curves γ_1, γ_2 in $U \backslash E$ so that z_0 is in the domain bounded by γ_1 but not in the domain bounded by γ_2. Then the Cauchy integral formula tells us that

$$\oint_{\gamma_1} \frac{f(z)}{z - z_0} dz + \oint_{\gamma_2} \frac{f(z)}{z - z_0} dz$$

defines an analytic extension of f to all of U. So (c) holds for E. ☐

Prelude: This is a very basic, and intuitively appealing, result about analytic capacity. It relates capacity to area.

Theorem 9.3.2. *Let $E \subseteq \mathbb{C}$ have positive area, and let U be the complement of E in the Riemann sphere. Then there are nonconstant, holomorphic functions on U that extend continuously to the sphere.*

Proof. Define

$$F(z) = \iint_E \frac{d\xi d\eta}{\zeta - z}$$

with $\zeta = \xi + i\eta$. Then it is clear that F is holomorphic on the complement U of E. Since F is the convolution of the locally integrable function $1/\zeta$ and the characteristic function of E, we see that F is continuous on the entire complex plane. Since $\lim_{z\to\infty} F(z) = 0$, we see that F extends continuously to the entire Riemann sphere. Also F is not constant since $\lim_{z\to\infty} zF(z) = -\operatorname{Area}(E)$. That completes the proof. □

Prelude: Again, we relate capacity to area. This is one of the key theorems in the subject.

Theorem 9.3.3. *The analytic capacity of E is bounded below by*

$$\frac{1}{2}\sqrt{\frac{\operatorname{Area}(E)}{\pi}}$$

Proof. Let F be as in the previous result. At infinity, F has the expansion

$$F(z) = -\frac{\operatorname{Area}(E)}{z} + \frac{a_2}{z^2} + \cdots$$

Define the positive number R by

$$\pi R^2 = \operatorname{Area}(E)$$

Fix z and let Δ be the disc $\{\zeta \in \mathbb{C} \mid |\zeta - z| \leq R\}$. Then

$$|F(z)| \leq \iint_E \frac{d\xi d\eta}{|\zeta - z|} = \iint_{E\cap\Delta} \frac{d\xi d\eta}{|\zeta - z|} + \iint_{E\setminus\Delta} \frac{d\xi d\eta}{|\zeta - z|}$$

Since E and Δ have the same area, so do $E\setminus\Delta$ and $\Delta\setminus E$. But the integrand on the righthand side of the last displayed equation is larger on $\Delta\setminus E$ than it is on $E\setminus\Delta$. So

$$\iint_{E\setminus\Delta} \frac{d\xi d\eta}{|\zeta - z|} \leq \iint_{\Delta\setminus E} \frac{d\xi d\eta}{|\zeta = z|}$$

Hence

$$F(z) \leq \iint_{\Delta} \frac{d\xi d\eta}{|\zeta - z|} = 2\pi R .$$

This gives us a bounded function

$$g(z) = \frac{F(z)}{2\pi R}$$

on U such that $\|g\| \leq 1$ and

$$g(z) = \frac{b_1}{z} + \frac{b_2}{z^2} + \cdots ,$$

where

$$|b_1| \geq \frac{R}{2} = \frac{1}{2}\sqrt{\frac{\operatorname{Area}(E)}{\pi}} .$$

In other words, we have estimated the analytic capacity of E from below. If U is a domain in $\mathbb{C}$, then we let $H^\infty(U)$ denote the space of bounded, analytic functions on U. And we let $A(U)$ denote the elements of $H^\infty(U)$ that possess continuous extensions to the Riemann sphere. If $E \subseteq \mathbb{C}$, and U is the complement of E, then define

$$\alpha(E) = \sup \{|f'(\infty)| \mid f \in A(U), \|f\| \leq 1\} .$$

It is useful to compare and contrast $\alpha(E)$ with $\gamma(E)$. □

Prelude: This is our opportunity to compare the original analytic capacity γ with the new capacity α. The result is elegant and useful.

Theorem 9.3.4. *Let $E \subseteq \mathbb{C}$, and let U be the complement of E. The space $H^\infty(U)$ consists only of the constants if and only if $\gamma(E) = 0$. The space $A(U)$ consists only of the constants if and only if $\alpha(E) = 0$.*

Proof. If $\gamma(E) > 0$ then $H^\infty(U)$ contains nonconstant functions. On the other hand, if $H^\infty(U)$ is nontrivial, then there is an $f \in H^\infty(U)$ with $f(\infty) = 0$ and $f(z_0) \neq 0$ for some $z_0 \neq \infty$. Then the function

$$g(z) = \frac{f(z) - f(z_0)}{z_0 - z}$$

is in $H^\infty(U)$ and has derivative $f(z_0)$ at ∘; hence $\gamma(E) > 0$. □

The very same argument shows that $\alpha(U)$ is nontrivial if and only if $\alpha(E) > 0$.

Theorem 9.3.5. *For any set E,*

$$\alpha(E) \leq \gamma(E) \leq \operatorname{diam}(E) .$$

If E is connected, then

$$\gamma(E) \geq \frac{\operatorname{diam}(E)}{4}$$

Proof. The first set of inequalities follows from monotonicity of the analytic capacity, just because E lies in a disc of radius equal to $diam(E)$. For the second inequality, we may assume that E is not a singleton. Let

$$g(z) = \frac{\gamma(E)}{z} + \frac{a_2}{z^2} + \cdots$$

be the Riemann mapping of the unbounded component of $U = \mathbb{C} \setminus E$ onto the unit disc. Fix $z_0 \in E$, and write

$$f(w) = \frac{\gamma(E)}{g^{-1}(w) - z_0}$$

Then f is univalent on $\{|w| < 1\}, f(0) = 0$, and $f(0) = 1$. Now the Koebe–Bieberbach theorem tells us that the range of f contains the disc $\{|z| < 1/4\}$. Thus, if $z_1 \in E$, then

$$\frac{\gamma(E)}{|z_1 - z_0|} \geq \frac{1}{4}$$

One conclusion that we may draw from the last proposition is that, if $\gamma(E) = 0$, then E is totally disconnected. □

Prelude: Again we relate capacity to removability for sets. The ideas are fundamental and important.

Theorem 9.3.6. *Let E be a relatively closed subset of an open set $U \subseteq \mathbb{C}$ and assume that $\gamma(E) = 0$. If $f \in H^\infty(U \backslash E)$, then f has a unique extension in $H^\infty(U)$.*

Proof. The uniqueness is trivial because E is nowhere dense.

Let $z_0 \in E$. By the last proposition, E is totally disconnected. So there is an analytic, simple, closed curve Γ in $U \backslash E$ which encloses z_0. Let U be the domain bounded by Γ.

Using the Cauchy integral formula, we may write

$$f = f_1 + f_2$$

in a neighborhood of Γ, where $f_1 \in H^\infty(U)$ and $f_2 \in H^\infty(\widehat{\mathbb{C}} \backslash (E \cap U))$. Since $\gamma(E \cap U) = 0$, we see that f_2 is constant. Hence f extends analytically to U. If $E \subseteq \mathbb{C}$ and $M > 0$, then define

$$A(E, M) = \{f \in H^\infty(U) \mid \|f\| \leq M, f(\infty) = 0\} .$$

□

Prelude: Now we are estimating the size of a function in terms of analytic capacity. The theory is becoming richer.

Theorem 9.3.7. *Let $E \subseteq \mathbb{C}$. Let $f \in A(E, 1)$. Then, for $z_0 \in U = \mathbb{C} \backslash E$,*

$$|f(z_0)| \leq \frac{\gamma(E)}{\operatorname{dist}(z_0, E)} .$$

If f has a zero of multiplicity m at ∞, and E has diameter d, then

$$|f(z_0)| \leq \frac{d^{m-1}\gamma(E)}{(\operatorname{dist}(z_0, E))^m} .$$

Proof. Consider the function

$$g(z) = \frac{1}{z - z_0} \cdot \frac{f(z) - f(z_0)}{1 - \overline{f(z_0)} f(z)} .$$

Then $g \in H^{\infty}(U), g(\infty) = 0$, and

$$|g(z)| \leq \frac{1}{\operatorname{dist}(z_0, E)}.$$

Then

$$|f(z_0)| = |g'(\infty)| \leq \frac{\gamma(E)}{\operatorname{dist}(z_0, E)}.$$

If f has order m at ∞ and $z_1 \in E$ with $|z_1 - z_0| = \operatorname{dist}(z_0, E)$, then

$$h(z) \equiv \left(\frac{z - z_0}{d}\right)^{m-1} f(z)$$

is in $A(E, 1)$. Hence

$$|f(z_0)| \leq \left(\frac{d}{|z - z_0|}\right)^{m-1} \cdot |h(z_0)| \leq \frac{d^{m-1}\gamma(E)}{(\operatorname{dist}(z_0, E))^m}.$$

□

Prelude: Here we learn more about the structure of an extremal function. We are developing in new directions.

Theorem 9.3.8. *Let U be a domain such that $\infty \in U$. Assume that ∂U consists of pairwise disjoint curves $\Gamma_2, \ldots, \Gamma_k$. Let $E = \widehat{\mathbb{C}} \backslash U$. Then*

(a) $\alpha(E) = \gamma(E)$.

If $f \in A(E, 1)$ is an extremal function, then

(b) f is analytic across ∂U.

(c) $|f| = 1$ on ∂U.

(d) f has k zeros on U.

Proof. Let σ be a measure on ∂U of measure norm $\|\sigma\| = \alpha(E)$ and so that

$$\int f d\sigma = f'(\infty)$$

for all $f \in A(U)$. The measure

$$\mu = \frac{dz}{2\pi i} - \sigma$$

is orthogonal to $A(U)$. Hence, by the F. and M. Riesz theorem,

$$\sigma = \psi(z) dz$$

for

$$\psi(z) = \frac{1}{2\pi i} + \frac{b_2}{z^2} + \cdots \in H^1(U)$$

The function ψ solves the dual extremal problem

$$\int_{\partial U} |\psi(\zeta)| ds = \inf \left\{ \int_{\partial U} |h(\zeta)| ds \;\middle|\; h \in H^1(U), h(\infty) = \frac{1}{2\pi} \right\}$$

We call ψ the *Garabedian function.*

If $g \in A(E, 1)$, then $g \cdot \psi \in H^1(U)$ and $|g\psi| \leq |\psi|$ on ∂U. Thus

$$|g'(\infty)| = \left| \int_{\partial U} g(\zeta)\psi(\zeta) d\zeta \right| \leq \int_{\partial U} |\psi| ds = \alpha(E)$$

Taking $g'(\infty) = \gamma(E)$, we now obtain $\gamma(E) \leq \alpha(E)$, so that (a) follows.

Now let $f \in A(E, 1)$ satisfy $f(\infty) = \gamma(E)$. Then

$$\int_{\partial U} f(\zeta)\psi(\zeta) d\zeta = \gamma(E) = \int_{\partial U} |\psi(\zeta)| ds$$

This equality implies that

(i) $f(\zeta)\psi(\zeta)d\zeta \geq 0$ on ∂U;

and, because ψ cannot vanish on a set of positive measure,

(ii) $|f(\zeta)| = 1$ a.e. on ∂U.

Condition (i) and the analyticity of each Γ_j imply that $f(z)\psi(z)$ is analytic across ∂U. For, if $\zeta(w)$ is analytic on $\{r < |w| < 1/r\}$, with $\zeta(\{|w| = 1\}) = \Gamma_j$, then (for r near 1) $g(w) = f(\zeta)\psi(\zeta)$ is in H^1 of the annulus $A = \{1 < |w| < 1/r\}$. Also $g(w)\zeta'(w) \geq 0$ on $\{|w| = 1\}$. Schwarz reflection and Morera's theorem can now be used to extend $f(w)\zeta'(w)$ to the larger annulus $\{r < |w| < 1/r\}$ so that $f(\zeta)\varphi(\zeta)$ is analytic on ∂U. Now the argument principle and condition (i) above imply that

(iii) $f(z)\psi(z)$ has k zeros on U.

For each $\zeta \in \Gamma$, take a neighborhood V of ζ such that $V \cap U$ is simply connected, $V \cap \Gamma$ is an arc, and $f(z)\psi(z)$ has no zeros on $V \cap U$. After applying a conformal map $z = z(w)$, we have nonvanishing functions $\widetilde{f}(w)$ and $\widetilde{\psi}(w)$ on the unit disc D and an arc J in ∂D (corresponding to $V \cap \Gamma$) on which $\widetilde{f}(w)\widetilde{\psi}(w)$ is continuous. Write $\widetilde{f} = S_1 F_1$ and $\widetilde{\psi} = S_2 F_2$, where each S_j is a singular inner function and each F_j is outer (see [85] for these ideas). Then, because $\widetilde{f}\widetilde{\psi}$ is analytic across J, the same holds for $S_1 \cdot S_2$ and hence for S_1. Since

$$F_1(w) = \exp\left(\frac{1}{2\pi} \int_{\partial U} \frac{e^{i\theta} + w}{e^{i\theta} - w} \log \left| F_1\left(e^{i\theta}\right) \right| d\theta \right)$$

and $|F_1| = 1$ on J by (ii), we find that F_1 is analytic across J also. Therefore $\widetilde{f}$ is analytic across J and $|\widetilde{f}| = 1$ on J. Returning to V, we see that we have proved (b) and (c).

Finally, (b) and (c) tell us that f has at least k zeros on U while (iii) ensures that there can be no more than k zeros. That ends the proof. □

Corollary 9.3.9. *When ∂U consists of finitely many pairwise disjoint, analytic Jordan curves then the Garabedian function ψ is analytic across ∂U and* $\log \psi$ *is single-valued on U.*

Prelude: This result tells us a way to find extremal functions for analytic capacity.

Theorem 9.3.10. *For any compact set $E \subseteq \mathbb{C}$ there is a unique function $f \in A(E, 1)$ such that $f'(\infty) = \gamma(E)$.*

Proof. Let E_j be compact sets decreasing to E so that ∂E_j consists of finitely many pairwise disjoint, analytic curves. Let

$$f_j(z) = \frac{\gamma(E_j)}{z} + \cdots$$

be any Ahlfors function for E_j and let

$$\psi_j(z) = \frac{1}{2\pi i} + \cdots$$

be a Garabedian function for E_j. Then, for $g \in A(E_j, 1)$ and for $z \in U_j \equiv \mathbb{C} \setminus E_j$, we have (by Cauchy's theorem)

$$\int_{\partial U_j} \frac{g(z) - g(\zeta)}{z - \zeta} \psi_j(\zeta) d\zeta = g(z)$$

Therefore

$$g(z) = \left(\oint \frac{\psi_j(\zeta)}{\zeta - z} d\zeta - 1 \right)^{-1} \cdot \oint \frac{g(\zeta)\psi_j(\zeta)}{\zeta - z} d\zeta .$$

Choose k_j so that the sequence $\{f_{kj}\}$ converges pointwise to a function $f \in A(E, 1)$. The sequence $\{\psi_{kj}(\zeta)d\zeta\}$ then converges in the weak-* topology to a measure σ on E. Also the sequence

$$\tau_{k_j} = \{f_{k_j}(\zeta)\psi_{k_j}(\zeta)d\zeta\} = \{|\psi_k(\zeta)|\,|d\zeta|\}$$

converges in the weak-* topology to a positive measure τ on E. Thus

$$\begin{aligned} f(z) &= \lim_{j\to\infty} \left(\oint \frac{\psi_{k_j}(\zeta)d\zeta}{\zeta - z} - 1 \right)^{-1} \cdot \oint \frac{f_{k_j}(\zeta)\psi_{k_j}(\zeta)}{\zeta - z} d\zeta \\ &= \left(\int_E \frac{d\sigma(\zeta)}{\zeta - z} - 1 \right)^{-1} \cdot \int_E \frac{d\tau(\zeta)}{\zeta - z} \end{aligned}$$

Let g be any extremal function for E with $g \in A(E,1), g'(\infty) = \gamma(E)$. Then

$$g(z) = \left(\int_E \frac{d\sigma(\zeta)}{\zeta - z} - 1\right)^{-1} \cdot \lim_{j\to\infty} \oint \frac{g(\zeta)\psi_{k_j}(\zeta)}{\zeta - z} d\zeta$$

We assert that the measures $\lambda_j = g(\zeta)\psi_{kj}(\zeta)d\zeta$ converge weak-* to ψ. That clearly implies that $g = f$. Since

$$\left|g(\zeta)\psi_{k_j}(\zeta)d\zeta\right| \leq \left|\psi_{k_j}(\zeta)\right| |d\zeta|$$

we have $|\lambda_j| \leq \tau_{kj}$. On the other hand

$$\int d\lambda_j = \int g(\zeta)\psi_{k_j}(\zeta)d\zeta = \gamma(E) = \lim_{j\to\infty} \int d\tau_{k_j} = \int d\tau$$

Thus if λ is a weak-* limit of $\{\lambda_j\}$, then $|\lambda| \leq \tau$ and $\int d\lambda = \int d\tau$. Therefore $\lambda = \tau$. That means that λj converges to τ. □

Prelude: As previously noted, additivity properties for analytic capacity are hard to come by. So this is a fundamental result.

Theorem 9.3.11 (Pommerenke). *Let $E_1, E_2, \ldots, E_k$ be pairwise disjoint, compact sets in $\mathbb{C}$ and set $K = E_1 + E_2 + \cdots + E_k$ be their Minkowski sum.*[2] *Then*

$$\gamma\left(\bigcup_{j=1}^{k} E_j\right) \leq \gamma(K)$$

Proof. Let U be the complement of $E \equiv \cup_{j=1}^{k} E_j$. Let $f \in A(E,1)$ satisfy $f(\infty) = \gamma(E)$. Conformally mapping U to a domain with real analytic boundary (for instance, a large disc with smaller discs removed), we see by results of Theorem 9.3.8 above that

(i) f is a k-to-1 covering map of D.

(ii) f is proper (in the sense of topology).

That is to say, if $U \ni z_j \to z \in E$, then $|f(z_j)| \to 1$.

Choose a number $\rho, 0 < \rho < 1$, such that $L \equiv f^{-1}(\{|w| = \rho\})$ consists of k analytic curves and such that there are inverses $\varphi_1, \varphi_2, \ldots, \varphi_k$ of f mapping $\{\rho < |w| < 1\}$ onto the k annular components of $\{z : \rho < |f(z)| < 1\}$. Now Cauchy's theorem tells us that

$$\frac{1}{2\pi i} \oint_L z(f(z))^n f'(z)dz = -\gamma(E)\delta_{n,0}$$

[2]The plus signs denote the Minkowski sum of sets. Here the Minkowski sum $A + B$ of two sets A and B is defined to be $A + B = \{a + b : \alpha \in A, b \in B\}$

for $n \geq 0$. (Here $\delta_{n,0}$ is the Kronecker delta.) Thus

$$\sum_{j=1}^{k} \frac{1}{2\pi i} \oint_{|w|=\rho} \varphi_j(w) w^n dw = \gamma(E) \cdot \delta_{n,0}$$

Also

$$\Phi(w) \equiv \sum_{j=1}^{k} \varphi_j(w)$$

extends to be meromorphic on $\{|w| < 1\}$ with a simple pole at 0:

$$\Phi(w) = \frac{\gamma(E)}{w} + a_0 + a_1 w + \cdots$$

Let $T = \{\lim \Phi(w_k) : |w_k| \to 1\}$. Then T is a compact subset of K. Let U be the component of $D \backslash \Phi^{-1}(T)$ containing the origin. Then Φ is a proper mapping of U onto $\Phi(U)$ and Φ has a simple pole at 0. Thus, by the argument principle, Φ is univalent. Also $\partial\Phi(U) = T \subset K$ and Φ^{-1} is in $A(T, 1)$ with derivative $\gamma(E)$ at ∞. Thus $\gamma(E) \leq \gamma(T) \leq \gamma(K)$. □

Prelude: Now we have a very concrete result about analytic capacity for subsets of the real line.

Theorem 9.3.12. *Let E be a subset of the real axis $\mathbb{R}$. Then*

$$\gamma(E) = \frac{\ell(E)}{4}$$

where $\ell(E)$ is the linear measure of E.

Proof. We already know this result for E an interval. Now assume that E is the pairwise disjoint union of closed intervals $\{E_j\}_{j=1}^{k}$. Then

$$K \equiv E_1 + E_2 + \cdots + E_k$$

is a closed interval of length $\ell(E)$. By Pommerenke's theorem,

$$\gamma(E) \leq \gamma(K) = \frac{\ell(E)}{4}$$

Approximating from above yields the same inequality for all compact sets E; approximating from below yields the result for all E. Conversely, let E be compact, and write

$$f(z) = \frac{1}{2} \int_E \frac{dt}{t - z}.$$

Then $f(\infty) = 0, f(\infty) = -\ell(E)/2$, and

$$|\operatorname{Im} f(z)| = \frac{|y|}{2} \int_E \frac{dt}{(t - x)^2 + y^2}, \quad z = x + iy.$$

As a result,

$$|\operatorname{Im} f(z)| = \frac{|y|}{2} \int_E \frac{dt}{(t-x)^2 + y^2}, \qquad z = x + iy\,.$$

Thus $\operatorname{Re} e^{f(z)} > 0$ and

$$|\operatorname{Im} f(z)| \le \frac{1}{2} \int_{-\infty}^{\infty} \frac{du}{1+u^2} = \frac{\pi}{2}$$

satisfies $|F(z)| \le 1$. Therefore $\gamma(E) \ge \ell(E)/4$, and that is our result. □

Exercises

1. Let $E = [a, b]$ be an interval in the real line. Calculate the analytic capacity of E.

2. Let $E \subseteq \mathbf{R}$ be any Borel measurable set. Show that the analytic capacity of E is $1/4$ of the measure of E.

3. Calculate the analytic capacity of a disc of radius r.

4. Let $E_1, E_2, \ldots, E_k$ be pairwise disjoint compact sets in the complex plane. Relate the analytic capacity of the union of the E_j to the analytic capacity of the Minkowski sum of the E_j.

5. Relate the idea of analytic capacity of a set to the idea of a removable singularity for a holomorphic function on the complement.

10

The Bergman Kernel

We have already noted that it is difficult to create an explicit integral formula, with holomorphic reproducing kernel, for holomorphic functions on a domain in $\mathbb{C}^n$.

We now examine one of several non-constructive approaches to this problem. This circle of ideas, due to S. Bergman [19] and to G. Szegö [175], (some of the ideas presented here were anticipated by the thesis of S. Bochner [24]) will later be seen to have profound applications to the boundary regularity of holomorphic mappings.

In this chapter we will see some of the invariance properties of the Bergman kernel. This will lead to the definition of the Bergman metric (in which all biholomorphic mappings become isometries) and to such other canonical constructions as representative domains. The Bergman kernel has certain extremal properties that make it a powerful tool in the theory of partial differential equations (see S. Bergman and M. Schiffer [18]).

10.1 Bergman Space

Let $\Omega \subset \mathbb{C}^n$ be a domain. Define the Bergman space

$$A^2(\Omega) = \left\{ f \text{ holomorphic on } \Omega : \left(\int_\Omega |f(z)|^2 dV(z) \right)^{1/2} \equiv \|f\|_{A^2(\Omega)} < \infty \right\}$$

Lemma 10.1.1. *Let $K \subset \Omega$ be compact. There is a constant $C_K > 0$, depending on K and on n, such that*

$$\sup_{z \in K} |f(z)| \le C_K \|f\|_{A^2(\Omega)}, \qquad \textit{all } f \in A^2(\Omega)$$

Proof. Since K is compact, there is an $r(K) = r > 0$ so that for any $z \in K, B(z,r) \subseteq \Omega$. Therefore, for each $z \in K$ and $f \in A^2(\Omega)$, we have

DOI: 10.1201/9781003713005-10

that

$$\begin{aligned}|f(z)| &= \frac{1}{V(B(z,r))}\left|\int_{B(z,r)} f(t)dV(t)\right| \\ &\le (V(B(z,r)))^{-1/2}\|f\|_{L^2(B(z,r))} \\ &\le C(n)r^{-n}\|f\|_{A^2(\Omega)} \\ &\equiv C_K\|f\|_{A^2(\Omega)}\end{aligned}$$

□

Lemma 10.1.2. *The space $A^2(\Omega)$ is a Hilbert space with the inner product $\langle f, g\rangle \equiv \int_\Omega f(z)\overline{g(z)}dV(z)$.*

Proof. Everything is clear except for completeness. Let $\{f_j\} \subseteq A^2$ be a sequence that is Cauchy in norm. Since L^2 is complete there is an L^2 limit function f. We need to see that f is holomorphic. But norm convergence implies normal convergence. Moreover holomorphic functions are closed under normal limits. Therefore f is holomorphic and $A^2(\Omega)$ is complete. □

Lemma 10.1.3. *For each fixed $z \in \Omega$, the functional*

$$\Phi_z : f \mapsto f(z), \quad f \in A^2(\Omega)$$

is a continuous linear functional on $A^2(\Omega)$.

Proof. This is immediate from Lemma 10.1.1 if we take K to be the singleton $\{z\}$. □

We may now apply the Riesz representation theorem to see that there is an element $k_z \in A^2(\Omega)$ such that the linear functional Φ_z is represented by inner product with k_z: if $f \in A^2(\Omega)$ then for all $z \in \Omega$ then we have

$$f(z) = \langle f, k_z\rangle$$

Definition 10.1.4. The Bergman kernel is the function

$$K(z,\zeta) = \overline{k_z(\zeta)}, z, \zeta \in \Omega\,.$$

It has the reproducing property

$$f(z) = \int K(z,\zeta)f(\zeta)dV(\zeta), \quad \forall f \in A^2(\Omega)$$

Proposition 10.1.5. *The Bergman kernel $K(z,\zeta)$ is conjugate symmetric: $K(z,\zeta) = \overline{K(\zeta,z)}$*

Proof. By its very definition, $\overline{K(\zeta,\cdot)} \in A^2(\Omega)$ for each fixed ζ. Therefore the reproducing property of the Bergman kernel gives

$$\int_\Omega K(z,t)\overline{K(\zeta,t)}dV(t) = \overline{K(\zeta,z)}$$

On the other hand,

$$\begin{aligned}\int_\Omega K(z,t)\overline{K(\zeta,t)}dV(t) &= \overline{\int K(\zeta,t)\overline{K(z,t)}dV(t)} \\ &= \overline{\overline{K(z,\zeta)}} = K(z,\zeta)\end{aligned}$$

□

Proposition 10.1.6. *The Bergman kernel is uniquely determined by the properties that it is an element of $A^2(\Omega)$ in z, is conjugate symmetric, and reproduces $A^2(\Omega)$.*

Proof. Let $K'(z,\zeta)$ be another such kernel. Then

$$\begin{aligned}K(z,\zeta) = \overline{K(\zeta,z)} &= \int K'(z,t)\overline{K(\zeta,t)}dV(t) \\ &= \overline{\int K(\zeta,t)\overline{K'(z,t)}dV(t)} \\ &= \overline{\overline{K'(z,\zeta)}} = K'(z,\zeta)\end{aligned}$$

Since $L^2(\Omega)$ is a separable Hilbert space then so is its subspace $A^2(\Omega)$. Thus there is a complete orthonormal basis $\{\phi_j\}_{j=1}^\infty$ for $A^2(\Omega)$. □

Proposition 10.1.7. *Let K be a compact subset of Ω. Then the series*

$$\sum_{j=1}^\infty \phi_j(z)\overline{\phi_j(\zeta)}$$

sums uniformly on $K \times K$ to the Bergman kernel $K(z,\zeta)$.

Proof. By the Riesz–Fischer and Riesz representation theorems, we obtain

$$\begin{aligned}\sup_{z\in K}\left(\sum_{j=1}^\infty |\phi_j(z)|^2\right)^{1/2} &= \sup_{z\in K}\left\|\{\phi_j(z)\}_{j=1}^\infty\right\|_{\ell^2} \qquad (10.1)\\ &= \sup_{\substack{\|\{a_j\}\|_{\ell^2}=1\\ z\in K}}\left|\sum_{j=1}^\infty a_j\phi_j(z)\right| \\ &= \sup_{\substack{\|f_n\|_{A^2=1}\\ z\in K}} |f(z)| \le C_K\end{aligned}$$

In the last inequality we have used Lemma 10.1.1. Therefore

$$\sum_{j=1}^\infty \left|\phi_j(z)\overline{\phi_j(\zeta)}\right| \le \left(\sum_{j=1}^\infty |\phi_j(z)|^2\right)^{1/2}\left(\sum_{j=1}^\infty |\phi_j(\zeta)|^2\right)^{1/2}$$

and the convergence is uniform over $z, \zeta \in K$. For fixed $z \in \Omega$, (10.1) shows that $\{\phi_j(z)\}_{j=1}^{\infty} \in \ell^2$. Hence we have that $\sum \phi_j(z)\overline{\phi_j(\zeta)} \in \overline{A^2(\Omega)}$ as a function of ζ. Let the sum of the series be denoted by $K'(z,\zeta)$. Notice that K' is conjugate symmetric by its very definition. Also, for $f \in A^2(\Omega)$, we have

$$\int K'(\cdot,\zeta)f(\zeta)dV(\zeta) = \sum \hat{f}(j)\phi_j(\cdot) = f(\cdot)$$

where convergence is in the Hilbert space topology. [Here $\hat{f}(j)$ is the jth Fourier coefficient of f with respect to the basis $\{\phi_j\}$.] But Hilbert space convergence dominates pointwise convergence (Lemma 10.1.1) so

$$f(z) = \int K'(z,\zeta)f(\zeta)dV(\zeta), \qquad \text{all } f \in A^2(\Omega)$$

Therefore K' is the Bergman kernel. □

Remark 10.1.8. It is worth noting explicitly that the proof of Proposition 10.1.7 shows that

$$\sum \phi_j(z)\overline{\phi_j(\zeta)}$$

equals the Bergman kernel $K(z,\zeta)$ no matter what the choice of complete orthonormal basis $\{\phi_j\}$ for $A^2(\Omega)$.

Proposition 10.1.9. *If Ω is a bounded domain in $\mathbb{C}^n$ then the mapping*

$$P : f \mapsto \int_{\Omega} K(\cdot,\zeta)f(\zeta)dV(\zeta)$$

is the Hilbert space orthogonal projection of $L^2(\Omega, dV)$ onto $A^2(\Omega)$.

Proof. Notice that P is idempotent and self-adjoint and that $A^2(\Omega)$ is precisely the set of elements of L^2 that are fixed by P. □

Definition 10.1.10. Let $\Omega \subset \mathbb{C}^n$ be a domain and let $f : \Omega \to \mathbb{C}^n$ be a holomorphic mapping, that is $f(z) = (f_1(z), \ldots, f_n(z))$ with $f_1, \ldots, f_n$ holomorphic on Ω. Let $w_j = f_j(z), j = 1, \ldots, n$. Then the holomorphic Jacobian matrix of f is the matrix

$$J_{\mathbb{C}}f = \frac{\partial(w_1, \ldots, w_n)}{\partial(z_1, \ldots, z_n)}$$

Write $z_j = x_j + iy_j, w_k = \xi_k + i\eta_k, j, k = 1, \ldots, n$. Then the real Jacobian matrix of f is the matrix

$$J_{\mathbb{R}}f = \frac{\partial(\xi_1, \eta_1, \ldots, \xi_n, \eta_n)}{\partial(x_1, y_1, \ldots, x_n, y_n)}$$

Proposition 10.1.11. *With notation as in the definition, we have*

$$\det J_{\mathbb{R}} f = |\det J_{\mathbb{C}} f|^2$$

whenever f is a holomorphic mapping.

Proof. We exploit the functoriality of the Jacobian. Let

$$w = (w_1, \ldots, w_n) = f(z) = (f_1(z), \ldots, f_n(z)) \,.$$

Write $z_j = x_j + iy_j, w_j = \xi_j + i\eta_j, j = 1, \ldots, n$. Then, using the fact that f is holomorphic,

$$d\xi_1 \wedge d\eta_1 \wedge \cdots \wedge d\xi_n \wedge d\eta_n = (\det J_{\mathbb{R}} f(x,y))\, dx_1 \wedge dy_1 \wedge \cdots \wedge dx_n \wedge dy_n \quad (10.2)$$

On the other hand,

$$\begin{aligned} & d\xi_1 \wedge d\eta_1 \wedge \cdots \wedge d\xi_n \wedge d\eta_n \\ &= \frac{1}{(2i)^n} d\bar{w}_1 \wedge dw_1 \wedge \cdots \wedge d\bar{w}_n \wedge dw_n \\ &= \frac{1}{(2i)^n} \overline{(\det J_{\mathbb{C}} f(z))}\, (\det J_{\mathbb{C}} f(z))\, d\bar{z}_1 \wedge dz_1 \wedge \cdots \wedge d\bar{z}_n \wedge dz_n \\ &= |\det J_{\mathbb{C}} f(z)|^2\, dx_1 \wedge dy_1 \wedge \cdots \wedge dx_n \wedge dy_n \qquad (10.3) \end{aligned}$$

Equating (10.2) and (10.3) gives the result. □

Exercise for the Reader: Prove Proposition 10.1.11 using only matrix theory (no differential forms). This will give rise to a great appreciation for the theory of differential forms (see L. Bers [21, Chapter 7] for help).

Now we can prove the holomorphic implicit function theorem:

Theorem 10.1.12. *Let $f_j(w,z), j = 1, \ldots, m$, be holomorphic functions of $(w,z) = ((w_1, \ldots, w_m), (z_1, \ldots, z_n))$ near a point $(w^0, z^0) \in \mathbb{C}^m \times \mathbb{C}^n$. Assume that*

$$f_j\left(w^0, z^0\right) = 0, \quad j = 1, \ldots, m$$

and that

$$\det\left(\frac{\partial f_j}{\partial w_k}\right)_{j,k=1}^{m} \neq 0 \text{ at } \left(w^0, z^0\right)$$

Then the system of equations

$$f_j(w,z) = 0 \quad , \quad j = 1, \ldots, m$$

has a unique holomorphic solution $w(z)$ in a neighborhood of z^0 that satisfies $w\left(z^0\right) = w^0$.

Proof. We rewrite the system of equations as

$$\operatorname{Re} f_j(w,z) = 0, \quad \operatorname{Im} f_j(w,z) = 0$$

for the $2m$ real variables $\operatorname{Re} w_k, \operatorname{Im} w_k, k = 1, \ldots, m$. The determinant of the Jacobian over $\mathbb{R}$ of this new system is the modulus squared of the determinant of the Jacobian over $\mathbb{C}$ of the old system. By our hypothesis, this number is non-vanishing at the point (w^0, z^0). Therefore the classical implicit function theorem (see W. Rudin [160]) implies that there exist C^1 functions $w_k(z), k = 1, \ldots, m$, with $w(z^0) = w^0$ and that solve the system. Our job is to show that these functions are in fact holomorphic. When properly viewed, this is purely a problem of geometric algebra:

Applying exterior differentiation to the equations

$$0 = f_j(w(z), z), \quad j = 1, \ldots, m$$

yields that

$$0 = df_j = \sum_{k=1}^{m} \frac{\partial f_j}{\partial w_k} dw_k + \sum_{k=1}^{m} \frac{\partial f_j}{\partial z_k} dz_k$$

There are no $d\bar{z}_j$'s and no $d\bar{w}_k$'s because the f_j's are holomorphic.

The result now follows from linear algebra only: the hypothesis on the determinant of the matrix $(\partial f_j / \partial w_k)$ implies that we can solve for dw_k in terms of dz_j. Therefore w is a holomorphic function on z. □

A holomorphic mapping $f : \Omega_1 \to \Omega_2$ of domains $\Omega_1 \subset \mathbb{C}^n, \Omega_2 \subset \mathbb{C}^m$ is said to be *biholomorphic* if it is one-to-one, onto, and $\det J_{\mathbb{C}} f(z) \neq 0$ for every $z \in \Omega_1$.

Remark 10.1.13. It is true, but not at all obvious, that the non-vanishing of the Jacobian determinant is a superfluous condition in the definition of "biholomorphic mapping;" that is, the non-vanishing of the Jacobian follows from the univalence of the mapping.

In what follows we denote the Bergman kernel for a given domain Ω by K_Ω.

Proposition 10.1.14. *Let Ω_1, Ω_2 be domains in $\mathbb{C}^n$. Let $f : \Omega_1 \to \Omega_2$ be biholomorphic. Then*

$$\det J_{\mathbb{C}} f(z) K_{\Omega_2}(f(z), f(\zeta)) \det \overline{J_{\mathbb{C}} f(\zeta)} = K_{\Omega_1}(z, \zeta)$$

Proof. Let $\phi \in A^2(\Omega_1)$. Then, by change of variable,

$$\begin{aligned}
&\int_{\Omega_1} \det J_{\mathbb{C}} f(z) K_{\Omega_2}(f(z), f(\zeta)) \det \overline{J_{\mathbb{C}} f(\zeta)} \phi(\zeta) dV(\zeta) \\
&= \int_{\Omega_2} \det J_{\mathbb{C}} f(z) K_{\Omega_2}(f(z), \tilde{\zeta}) \det \overline{J_{\mathbb{C}} f\left(f^{-1}(\tilde{\zeta})\right)} \phi\left(f^{-1}(\tilde{\zeta})\right) \\
&\quad \times \det J_{\mathbb{R}} f^{-1}(\tilde{\zeta}) dV(\tilde{\zeta})
\end{aligned}$$

Clear that this simplifies to

$$\det J_{\mathbb{C}} f(z) \int_{\Omega_2} K_{\Omega_2}(f(z), \tilde{\zeta}) \left\{ \left(\det J_{\mathbb{C}} f\left(f^{-1}(\tilde{\zeta})\right)\right)^{-1} \phi\left(f^{-1}(\tilde{\zeta})\right) \right\} dV(\tilde{\zeta})$$

By change of variables, the expression in braces is an element of $A^2(\Omega_2)$. So the reproducing property of K_{Ω_2} applies and the last line equals

$$= \det J_{\mathbb{C}} f(z) \left(\det J_{\mathbb{C}} f(z)\right)^{-1} \phi\left(f^{-1}(f(z))\right) = \phi(z).$$

By the uniqueness of the Bergman kernel, the proposition follows. □

Proposition 10.1.15. *For $z \in \Omega \subset\subset \mathbb{C}^n$ it holds that $K_\Omega(z, z) > 0$.*

Proof. Now

$$K_\Omega(z, z) = \sum_{j=1}^{\infty} |\phi_j(z)|^2 \geq 0$$

If in fact $K(z, z) = 0$ for some z then $\phi_j(z) = 0$ for all j hence $f(z) = 0$ for every $f \in A^2(\Omega)$. This is absurd. □

Definition 10.1.16. For any $\Omega \subset \mathbb{C}^n$ we define a Hermitian metric on Ω by

$$g_{ij}(z) = \frac{\partial^2}{\partial z_i \partial \bar{z}_j} \log K(z, z), \quad z \in \Omega$$

This means that the square of the length of a tangent vector $\xi = (\xi_1, \ldots, \xi_n)$ at a point $z \in \Omega$ is given by

$$|\xi|_{B,z} = \sum_{i,j} g_{ij}(z) \xi_i \bar{\xi}_j$$

The metric that we have defined is called the *Bergman metric.*

In a Hermitian metric $\{g_{ij}\}$, the length of a C^1 curve $\gamma : [0, 1] \to \Omega$ is given by

$$\ell(\gamma) = \int_0^1 \left(\sum_{i,j} g_{i,j}(\gamma(t)) \gamma_i'(t) \overline{\gamma_j'}(t) \right)^{1/2} dt$$

If P, Q are points of Ω then their distance $d_\Omega(P, Q)$ in the metric is defined to be the infimum of the lengths of all piecewise C^1 curves connecting the two points.

It is not a priori obvious that the Bergman metric for a bounded domain Ω is given by a positive definite matrix at each point.

Proposition 10.1.17. *Let $\Omega_1, \Omega_2 \subset \mathbb{C}^n$ be domains and let $f : \Omega_1 \to \Omega_2$ be a biholomorphic mapping. Then f induces an isometry of Bergman metrics:*

$$|\xi|_{B,z} = |(J_{\mathbb{C}} f)\, \xi|_{B,f(z)}$$

for all $z \in \Omega_1, \xi \in \mathbb{C}^n$. Equivalently, f induces an isometry of Bergman distances in the sense that

$$d_{\Omega_2}(f(P), f(Q)) = d_{\Omega_1}(P, Q)$$

Proof. This is a formal exercise but we include it for completeness: From the definitions, it suffices to check that

$$\sum g_{i,j}^{\Omega_2}(f(z))\,(J_{\mathbb{C}} f(z)w)_i \left(\overline{J_{\mathbb{C}} f(z)w}\right)_j = \sum_{i,j} g_{ij}^{\Omega_1}(z) w_i \bar{w}_j$$

for all $z \in \Omega, w = (w_1, \ldots, w_n) \in \mathbb{C}^n$. But by Proposition 10.1.14,

$$\begin{aligned}
g_{ij}^{\Omega_1}(z) &= \frac{\partial^2}{\partial z_i \bar{z}_j} \log K_{\Omega_1}(z, z) \\
&= \frac{\partial^2}{\partial z_i \bar{z}_j} \log \left\{ |\det J_{\mathbb{C}} f(z)|^2 K_{\Omega_2}(f(z), f(z)) \right\} \\
&= \frac{\partial^2}{\partial z_i \bar{z}_j} \log K_{\Omega_2}(f(z), f(z))
\end{aligned}$$

since $\log |\det J_{\mathbb{C}} f(z)|^2$ is locally

$$\log\left(\det J_{\mathbb{C}} f\right) + \log\left(\overline{\det J_{\mathbb{C}} f}\right) + C$$

hence is annihilated by the mixed second derivative. But this is nothing other than

$$\sum_{\ell,m} g_{\ell,m}^{\Omega_2}(f(z)) \frac{\partial f_\ell(z)}{\partial z_i} \frac{\partial \overline{f_m(z)}}{\partial \bar{z}_j}$$

and the result follows. □

Proposition 10.1.18. *Let $\Omega \subset\subset \mathbb{C}^n$ be a domain. Let $z \in \Omega$. Then*

$$K(z, z) = \sup_{f \in A^2(\Omega)} \frac{|f(z)|^2}{\|f\|_{A^2}^2} = \sup_{\|f\|_{A^2(\Omega)} = 1} |f(z)|^2$$

Proof. Now

$$\begin{aligned}
K(z, z) &= \sum |\phi_j(z)|^2 \\
&= \left(\sup_{\|\{a_j\}\|_{\ell^2} = 1} \left| \sum \phi_j(z) a_j \right| \right)^2 \\
&= \sup_{\|f\|_{A^2} = 1} |f(z)|^2
\end{aligned}$$

by the Riesz–Fischer theorem,

$$= \sup_{f \in A^2} \frac{|f(z)|^2}{\|f\|_{A^2}^2}$$

□

We shall use this proposition in a moment. Meanwhile, we should like to briefly mention some open problems connected with the Bergman kernel:

The Lu Qi-Keng Conjecture: We have already noticed that $K_\Omega(z,z) > 0$, all $z \in \Omega$, any bounded Ω. It is reasonable to ask whether $K_\Omega(z,\zeta)$ is ever equal to zero. In fact various geometric constructions connected with the Bergman metric and associated biholomorphic invariants (which involve division by K) make it particularly desirable that K be non-vanishing.

If $\Omega = D$, the unit disc, then explicit calculation (which we perform below) shows that

$$K(z,\zeta) = \frac{1}{\pi} \frac{1}{(1 - z\bar{\zeta})^2}$$

hence $K(z,\zeta)$ is non-vanishing on $D \times D$. Proposition 10.1.14 and the Riemann mapping theorem then show that the Bergman kernel for any proper, simply connected subdomain of $\mathbb{C}$ is non-vanishing.

The Bergman kernel for the annulus was studied in M. Skwarczynski [167] and was seen to vanish at some points. It is shown in N. Suita and A. Yamada [174] that that if $\Omega \subset \mathbb{C}$ is a multiply connected domain with smooth boundary then K_Ω must vanish—this is proved by an analysis of differentials on the Riemann surface consisting of the double of Ω. By using the easy fact that the Bergman kernel for a product domain is the product of the Bergman kernels (exercise), we may conclude that any domain in $\mathbb{C}^2$ of the form $A \times \Omega$, where A is multiply connected, has a Bergman kernel with zeroes. The Lu Qi-Keng conjecture can be formulated as

Conjecture: A topologically trivial domain in $\mathbb{C}^n$ has non-vanishing Bergman kernel.

It is known (R.E. Greene and S.G. Krantz [63, 68]) that a domain that is C^∞ sufficiently close to the ball in $\mathbb{C}^n$ has non-vanishing Bergman kernel. Also, if a domain Ω has Bergman kernel that is bounded from zero (and satisfies a modest geometric condition) then all nearby domains have Bergman kernel that is bounded from zero. Thus it came as a bit of a surprise when in H. Boas [22] it was shown that there exist topologically trivial domains—even ones with real analytic boundary and satisfying all reasonable additional geometric conditions—for which the Bergman kernel has zeroes. See also J. Wiegerinck [192], where interesting ideas contributing to the solution of this problem first arose.

Exercise for the Reader: The set of smoothly bounded domains for which the Lu Qi-Keng conjecture is true is closed in the Hausdorff topology on domains.

10.2 Smoothness to the Boundary of K_Ω

It is of interest to know whether K_Ω is smooth on $\overline{\Omega} \times \overline{\Omega}$. We can see from the above formula for the Bergman kernel of the disc that $K_{D(z,z)}$ blows up as $z \to 1^-$. In fact this property of blowing up prevails at any boundary point of a domain at which there is a peaking function (apply Proposition 10.1.18 to a high power of the peaking function). The reference T. Gamelin [54] contains background information on peaking functions.

However there is strong evidence that—as long as Ω is smoothly bounded—on compact subsets of

$$\overline{\Omega} \times \overline{\Omega} \setminus ((\partial\Omega \times \partial\Omega) \cap \{z = \zeta\}),$$

the Bergman kernel will be smooth. For strongly convex domains (all boundary curvatures are positive), this statement is true; its proof (see N. Kerzman [100]) uses deep and powerful methods of partial differential equations. Perhaps the most central open problem in the function theory of several complex variables is to prove that a biholomorphic mapping of two smoothly bounded domains extends to a diffeomorphism of the closures. It is known (see S. Bell and H. Boas [14]) that a sufficient condition for this problem to have an affirmative answer on a smoothly bounded domain $\Omega \subseteq \mathbb{C}^n$ is that for any multi-index α there are constants $C = C_\alpha$ and $m = m_\alpha$ such that the Bergman kernel K_Ω satisfies

$$\sup_{z \in \Omega} \left| \frac{\partial^\alpha}{\partial z^\alpha} K_\Omega(z, \zeta) \right| \leq C \cdot \delta_\Omega(\zeta)^{-m}$$

for all $\zeta \in \Omega$. Here $\delta_\Omega(w)$ denotes the distance of the point $w \in \Omega$ to the boundary of the domain.

10.3 Calculating the Bergman Kernel

The Bergman kernel can almost never be calculated explicitly; unless the domain Ω has a great deal of symmetry—so that a useful orthonormal basis for $A^2(\Omega)$ can be determined—there are few techniques for determining K_Ω. In 1974 C. Fefferman (C. Fefferman [49]) introduced a new technique for obtaining an asymptotic expansion for the Bergman kernel on a large class of domains. (For an alternate approach see L. Boutet de Monvel and J. Sjöstrand [28].) This work enabled rather explicit estimations of the Bergman metric and opened up an entire branch of analysis on domains in $\mathbb{C}^n$ (see C. Fefferman [48], S.S. Chern and J. Moser [35], P. Klembeck [104] and R.E. Greene and S.G. Krantz [62, 63, 65–68, 70–74]for example). The Bergman theory that we have presented here would be a bit hollow if we did not at least calculate the kernel in a few instances. We complete the section by addressing that task. Restrict attention to the ball $B \subset \mathbb{C}^n$. The functions z^α, α a multi-index, are each in $A^2(B)$ and are pairwise orthogonal by the symmetry of the ball. By the uniqueness of the power series expansion for an element of $A^2(B)$, the elements z^α form a complete orthogonal system on B (their closed linear span is $A^2(B)$). Setting

$$\gamma_\alpha = \int_B |z^\alpha|^2 \, dV(z)$$

we see that $\{z^\alpha / \sqrt{\gamma_\alpha}\}$ is a complete orthonornal system in $A^2(B)$. Thus, by Proposition 10.1.7,

$$K(z, \zeta) = \sum_\alpha \frac{z^\alpha \bar{\zeta}^\alpha}{\gamma_\alpha} \tag{10.4}$$

If we want to calculate the Bergman kernel for the ball in closed form, we need to calculate the γ_α's. This requires some lemmas from real analysis. These lemmas will be formulated and proved on $\mathbb{R}^N$ and $B_N = \{x \in \mathbb{R}^N : |x| < 1\}$.

Lemma 10.3.1. *We have that*

$$\int_{\mathbb{R}^N} e^{-\pi|x|^2} dx = 1$$

Proof. The case $N = 1$ is familiar from calculus (or see E.M. Stein and G. Weiss [172]). For the N-dimensional case, write

$$\int_{\mathbb{R}^N} e^{-\pi|x|^2} dx = \int_{\mathbb{R}} e^{-\pi x_1^2} dx_1 \cdots \int_{\mathbb{R}} e^{-\pi x_N^2} dx_N$$

and apply the one dimensional result. □

Let σ be the unique rotationally invariant area measure on $S_{N-1} = \partial B_N$ (see Appendix II) and let $\omega_{N-1} = \sigma(\partial B)$.

Lemma 10.3.2. *We have*

$$\omega_{N-1} = \frac{2\pi^{N/2}}{\Gamma(N/2)}$$

where

$$\Gamma(x) = \int_0^\infty t^{x-1} e^{-t} dt$$

is Euler's function.

Proof. Introducing polar coordinates we have

$$1 = \int_{\mathbb{R}^N} e^{-\pi|x|^2} dx = \int_{S^{N-1}} d\sigma \int_0^\infty e^{-\pi r^2} r^{N-1} dr$$

or

$$\frac{1}{\omega_{N-1}} = \int_0^\infty e^{-\pi r^2} r^N \frac{dr}{r}$$

Letting $s = r^2$ in this last integral and doing some obvious manipulations yields the result. Now we return to $B\mathbb{C}^n$. We set

$$\eta(k) = \int_{\partial B} |z_1|^{2k}\, d\sigma, \quad N(k) = \int_B |z_1|^{2k}\, dV(z) \quad , \quad k = 0, 1, \ldots$$

□

Lemma 10.3.3. *We have*

$$\eta(k) = \pi^n \frac{2(k!)}{(k+n-1)!} \quad , \quad N(k) = \pi^n \frac{k!}{(k+n)!} .$$

Proof. Polar coordinates show easily that $\eta(k) = 2(k+n)N(k)$. So it is enough to calculate $N(k)$. Let $z = (z_1, z_2, \ldots, z_n) = (z_1, z')$. We write

$$\begin{aligned}
N(k) &= \int_{|z|<1} |z_1|^{2k}\, dV(z) \\
&= \int_{|z'|<1} \left(\int_{|z_1| \le \sqrt{1-|z'|^2}} |z_1|^{2k}\, dV(z_1) \right) dV(z') \\
&= 2\pi \int_{|z'|<1} \int_0^{\sqrt{1-|z'|^2}} r^{2k} r dr dV(z') \\
&= 2\pi \int_{|z'|<1} \frac{\left(1-|z'|^2\right)^{k+1}}{2k+2} dV(z') \\
&= \frac{\pi}{k+1}\omega_{2n-3} \int_0^1 \left(1-r^2\right)^{k+1} r^{2n-3} dr \\
&= \frac{\pi}{k+1}\omega_{2n-3} \int_0^1 (1-s)^{k+1} s^{n-1} \frac{ds}{2s} \\
&= \frac{\pi}{2(k+1)}\omega_{2n-3}\beta(n-1, k+2)
\end{aligned}$$

where β is the classical beta function of special function theory (see G. Carrier, M. Crook, and C. Pearson [33] or E. Whittaker and G. Watson [191]). By a standard identity for the beta function we then have

$$\begin{aligned}
N(k) &= \frac{\pi}{2(k+1)}\omega_{2n-3}\frac{\Gamma(n-1)\Gamma(k+2)}{\Gamma(n+k+1)} \\
&= \frac{\pi}{2(k+1)} \frac{2\pi^{n-1}}{\Gamma(n-1)} \frac{\Gamma(n-1)\Gamma(k+2)}{\Gamma(n+k+1)} \\
&= \frac{\pi^n k!}{(k+n)!}
\end{aligned}$$

This is the desired result. □

Lemma 10.3.4. *Let $z \in B \subset \mathbb{C}^n$ and $0 < r < 1$. The symbol $\mathbb{1}$ denotes the point $(1, 0, \ldots, 0)$. Then*

$$K_B(z, r\mathbb{1}) = \frac{n!}{\pi^n} \frac{1}{(1 - rz_1)^{n+1}}$$

Proof. Refer to the formula (10.4) preceding Lemma 10.3.1. Then

$$\begin{aligned} K_B(z, r\mathbb{1}) &= \sum_\alpha \frac{z^\alpha (r\mathbb{1})^\alpha}{\gamma_\alpha} = \sum_{k=0}^\infty \frac{z_1^k r^k}{N(k)} \\ &= \frac{1}{\pi^n} \sum_{k=0}^\infty (rz_1)^k \cdot \frac{(k+n)!}{k!} \\ &= \frac{n!}{\pi^n} \sum_{k=0}^\infty (rz_1)^k \binom{k+n}{n} \\ &= \frac{n!}{\pi^n} \cdot \frac{1}{(1 - rz_1)^{n+1}} \end{aligned}$$

This is the desired result. □

Theorem 10.3.5. *If $z, \zeta \in B$ then*

$$K_B(z, \zeta) = \frac{n!}{\pi^n} \frac{1}{(1 - z \cdot \bar{\zeta})^{n+1}}$$

where $z \cdot \bar{\zeta} = z_1\bar{\zeta}_1 + z_2\bar{\zeta}_2 + \cdots + z_n\bar{\zeta}_n$.

Proof. Let $z = r\tilde{z} \in B$, where $r = |z|$ and $|\tilde{z}| = 1$. Also, fix $\zeta \in B$. Choose a unitary rotation ρ such that $\rho\tilde{z} = \mathbb{1}$. Then, by Proposition 10.1.14 and Lemma 10.3.4 we have

$$\begin{aligned} K_B(z, \zeta) &= K_B(r\tilde{z}, \zeta) = K\left(r\rho^{-1}\mathbb{1}, \zeta\right) \\ &= K(r\mathbb{1}, \rho\zeta) = \overline{K(\rho\zeta, r\mathbb{1})} \\ &= \frac{n!}{\pi^n} \cdot \frac{1}{\left(1 - r(\overline{\rho\zeta})_1\right)^{n+1}} \\ &= \frac{n!}{\pi^n} \cdot \frac{1}{(1 - (r\mathbb{1}) \cdot (\overline{\rho\zeta}))^{n+1}} \\ &= \frac{n!}{\pi^n} \cdot \frac{1}{\left(1 - (r\rho^{-1}\mathbb{1}) \cdot \bar{\zeta}\right)^{n+1}} \\ &= \frac{n!}{\pi^n} \cdot \frac{1}{(1 - z \cdot \bar{\zeta})^{n+1}} \end{aligned}$$

□

Proposition 10.3.6. *The Bergman metric for the ball* $B = B(0,1)\mathbb{C}^n$ *is given by*

$$g_{ij}(z) = \frac{n+1}{\left(1-|z|^2\right)^2}\left[\left(1-|z|^2\right)\delta_{ij} + \bar{z}_i z_j\right]$$

Proof. Since $K(z,z) = n!/\left(\pi^n\left(1-|z|^2\right)^{n+1}\right)$, this is a routine computation that we leave to the reader. □

Corollary 10.3.7. *The Bergman metric for the disc (i.e. the ball in dimension one) is*

$$g_{ij}(z) = \frac{2}{\left(1-|z|^2\right)^2}, i = j = 1$$

This is the well-known Poincaré, or Poincaré-Bergman, metric.

Proposition 10.3.8. *The Bergman kernel for the polydisc* $D^n(0,1)\mathbb{C}^n$ *is the product*

$$K(z,\zeta) = \frac{1}{\pi^n}\prod_{j=1}^{n}\frac{1}{\left(1-z_j\bar{\zeta}_j\right)^2}$$

Proof. Exercise. Use the uniqueness property of the Bergman kernel. □

Exercise for the Reader: Calculate the Bergman metric for the polydisc.

10.4 The Poincaré-Bergman Metric on the Disc

If $D \subset \mathbb{C}$ is the unit disc, $z \in D$, then Corollary 10.3.7 shows that

$$|w|_{B,z} = \left\{\frac{2|w|^2}{\left(1-|z|^2\right)^2}\right\}^{1/2} = \frac{\sqrt{2}|w|}{1-|z|^2}$$

where the subscript B indicates that we are working in the Bergman metric. We now use this formula to derive an explicit expression for the *Poincaré distance* from $0 \in D$ to $r + i0 \in D, 0 < r < 1$. Call this distance $d(0,r)$. Then

$$d(0,r) = \inf\left\{\int_0^1 |\gamma'(t)|_{B,\gamma(t)}\,dt : \gamma \text{ is a curve in } D, \gamma(0) = 0, \gamma(1) = r + i0\right\}.$$

Elementary comparisons show that, among curves of the form

$$\psi(t) = t + iw(t), 0 \le t \le 1,$$

the curve $\gamma(t) = tr + i0$ is the shortest in the Poincaré metric. Further elementary arguments show that a general curve of the form $\psi(t) = v(t) + iw(t)$ is always longer than some corresponding curve of the form $t + i\tilde{w}(t)$. We leave the details of these assertions to the reader.

Thus

$$\begin{aligned} d(0,r) &= \int_0^1 \frac{\sqrt{2}r}{(1-(rt)^2)}dt \\ &= \sqrt{2}\int_0^r \frac{1}{1-t^2}dt \\ &= \frac{1}{\sqrt{2}}\log\left(\frac{1+r}{1-r}\right) \end{aligned}$$

Since rotations are conformal maps of the disc we may next conclude that

$$d\left(0, re^{i\theta}\right) = \frac{1}{\sqrt{2}}\log\left(\frac{1+r}{1-r}\right)$$

Finally, if w_1, w_2 are arbitrary, then the Möbius transformation

$$\phi : z \mapsto \frac{z - w_1}{1 - \bar{w}_1 z}$$

satisfies $\phi(w_1) = 0, \phi(w_2) = (w_2 - w_1)/(1 - \bar{w}_1 w_2)$. Then Proposition 10.1.17 yields that

$$\begin{aligned} d(w_1, w_2) &= d\left(0, \frac{w_2 - w_1}{1 - \bar{w}_1 w_2}\right) \\ &= \frac{1}{\sqrt{2}}\left(\frac{1 + \left|\frac{w_2 - w_1}{1 - \bar{w}_1 w_2}\right|}{1 - \left|\frac{w_2 - w_1}{1 - \bar{w}_1 w_2}\right|}\right) \end{aligned}$$

We note in passing that the expression $\rho(w_1, w_2) \equiv |(w_1 - w_2)/(1 - \bar{w}_1 w_2)|$ is called the *pseudohyperbolic distance.* It is also conformally invariant, but it does not arise from integrating an infinitesimal metric (i.e. lengths of tangent vectors at a point). A fuller discussion of both the Poincaré metric and the pseudohyperbolic metric on the disc may be found in J. Garnett [56].

10.5 Appendix to Section 10.3: The Biholomorphic Inequivalence of the Ball and the Polydisc

Theorem 10.5.1. *There is no biholomorphic map of the bidisc $D^2(0,1)$ to the ball $B(0,1) \in \mathbb{C}^2$.*

Proof. Suppose, seeking a contradiction, that there is such a map. Since Möbius transformations act transitively on the disc, pairs of them act transitively on the bidisc. Therefore we may compose ϕ with a self-map of the bidisc and assume that ϕ maps 0 to 0. If $Y \in \partial B$ then the disc $d_Y = \{z \in B : z = \zeta Y, \zeta \in \mathbb{C}, |\zeta| < 1\}$ is a totally geodesic submanifold of B (informally, this means that if P, Q are points of d_Y then the geodesic connecting them in the Riemannian manifold d_Y is the same as the geodesic connecting them in the Riemannian manifold B—see S. Kobayashi and K. Nomizu [106, 107]). By our discussion in the calculation of the Poincaré metric, we may conclude that the geodesics, or paths of least length, emanating from the origin in the ball are the rays $\tau_Y : t \mapsto tY$. (This assertion may also be derived from symmetry considerations.) Likewise, if $\alpha, \beta \in \mathbb{C}, |\alpha| = 1, |\beta| = 1$, then the disc $e_0 = \{(\zeta\alpha, \zeta\beta) : \zeta \in D\} \subseteq D^2(0,1)$ is a totally geodesic submanifold of $D^2(0,1)$. Again we may apply our discussion of the Poincaré metric on the disc to conclude that the geodesic curve emanating from the origin in the bidisc in the direction $X = (\alpha, \beta)$ is $\psi_{\alpha\beta} : t \mapsto tX$. A similar argument shows that the curve $t \mapsto (t, 0)$ is a geodesic in the bidisc. Now if $t \mapsto tX$ is one of the above mentioned geodesics on the bidisc then it will be mapped under ϕ to a geodesic $t \mapsto tY$ in the ball. If $0 < t_1 < t_2 < 1$ then the points $t_1X, t_2X \in D^2$ will be mapped to points $t_1'Y, t_2'Y \in B$ and it must be that $0 < t_1' < t_2' < 1$ since ϕ is an isometry hence must map the point t_2X to a point further from the origin than it maps t_1X (because t_2X is further from the origin than t_1X). It follows that the limit

$$\lim_{t \to 1^-} \phi(tX)$$

exists for every choice of X and the limit lies in ∂B. After composing ϕ with a rotation we may suppose that $\{\phi(t(1,0))\}$ terminates at $(1,0)$. Now consider the function $f(z_1, z_2) = (z_1 + 1)/2$ on B. This function has the property that $f(1,0) = 1, f$ is holomorphic on a neighborhood of $\bar{B}$, and $|f(z)| < 1$ for $z \in \bar{B} \setminus \{(1,0)\}$. For $0 < r < 1$ we invoke the mean value property for a harmonic function to write

$$\frac{1}{2\pi} \int_0^{2\pi} f \circ \phi \left(r, re^{i\theta}\right) d\theta = f \circ \phi(r, 0) \tag{10.5}$$

As $r \to 1^-$ the right hand side tends to $\lim_{t \to 1^-} f(t, 0) = 1$. However each of the paths $r \to \left(r, re^{i\theta}\right)$ is a geodesic in the bidisc, as discussed above, and for different $\theta \in [0, 2\pi)$ they are distinct. Thus the curves $r \to \phi\left(r, re^{i\theta}\right)$ have distinct limits in ∂B, and these limits will be different from the point $(1,0) \in \partial B$. In particular, $\lim_{r \to 1^-} f \circ \phi\left(r, re^{i\theta}\right)$ exists for each $\theta \in [0, 2\pi)$ and assumes a value of modulus strictly less than 1. By the Lebesgue dominated convergence theorem we may pass to the limit as $r \to 1^-$ in the left side of (10.5) to obtain a limit that must be strictly less than one in absolute value. That is the required contradiction. □

11

Appendix

We collect some remarks that do not fit neatly into the above discussion, but that might nonetheless be helpful in dispelling confusion.

11.1 Covering Spaces

We will assume that all topological spaces are locally path-connected.

Definition 11.1.1.

(a) A continuous map $\pi : X \to Y$ is said to be a *covering map* if, for any $y \in Y$, there exists an open neighborhood U of y in Y such that $\pi^{-1}(U)$ is a disjoint union of open sets $V_i \subset X$, each of which is mapped homeomorphically onto U by π. Such a neighborhood U is said to be an *evenly covered neighborhood.* The sets V_i are called the *sheets* of π over U.

(b) Let Y be a topological space. A *covering space* of Y is a topological space X together with a covering map $\pi : X \to Y$.

(c) If $\pi : X \to Y$ is a covering map, then for any $y \in Y$, the set $\pi^{-1}(y)$ is called the *fiber* over y.

Definition 11.1.2. Let $X_1 \xrightarrow{\pi_1} Y$ and $X_2 \xrightarrow{\pi_2} Y$ be two covering spaces of Y. A *morphism of covering spaces* is a continuous map $F : X_1 \to X_2$ such that $\pi_2 \circ F = \pi_1$, i.e., the diagram below is commutative.

$$\begin{array}{ccc} X_1 & \xrightarrow{\quad F \quad} & X_2 \\ & {\scriptstyle \pi_1}\searrow \quad \swarrow{\scriptstyle \pi_2} & \\ & Y & \end{array}$$

Definition 11.1.3. If $X \xrightarrow{\pi} Y$ is a covering space then its automorphisms are called *deck transformations.* The deck transformations form a group denoted by $\mathrm{Deck}(X, \pi)$.

DOI: 10.1201/9781003713005-11

Example 11.1.4. Since $\mathbb{C}$ is simply connected, $\exp : \mathbb{C} \to \mathbb{C}^*$ is the universal covering of $\mathbb{C}^*$. Let $\tau_n : \mathbb{C} \to \mathbb{C}$ denote the translation $\tau_n(z) = z + 2\pi i n$. Then τ_n is a deck transformation for every $n \in \mathbb{Z}$. Suppose that σ is any deck transformation. Then $\exp(\sigma(0)) = \exp(0) = 1$ and hence $\sigma(0) = 2\pi i n$ for some $n \in \mathbb{Z}$. It follows that $\sigma = \tau_n$. That shows that $\mathrm{Deck}(\exp : \mathbb{C} \to \mathbb{C}^*) = \{\tau_n \mid n \in \mathbb{Z}\}$, and we may conclude by Theorem 11.1.18 that

$$\pi_1(\mathbb{C}^*) \cong \mathbb{Z}\,.$$

Example 11.1.5. (complex tori, III). Let $\Lambda = \mathbb{Z}w_1 + \mathbb{Z}w_2$ be a lattice in $\mathbb{C}$. Then the quotient projection $\mathbb{C} \to \mathbb{C}/\Lambda$ is the universal covering of the torus $\mathbb{C}/\Lambda$. Analogous to the previous example $\mathrm{Deck}(\mathbb{C} \to \mathbb{C}/\Lambda) = \{\tau_w : w \in \Lambda\}$, where τ_w is the translation by w. It follows that

$$\pi_1(\mathbb{C}/\Lambda) \cong \Lambda \cong \mathbb{Z} \times \mathbb{Z}\,.$$

Definition 11.1.6. Let $X \xrightarrow{\pi} Y$ be a covering space and $f : Z \to Y$ a continuous map. A *lift* of f is a continuous map $F : Z \to X$ such that $\pi \circ F = f$, i.e., the diagram below is commutative.

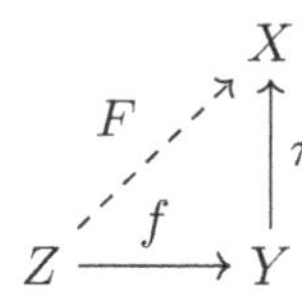

Proposition 11.1.7 (unique path lifting)**.** *Let $X \xrightarrow{\pi} Y$ be a covering map, $\gamma : [0,1] \to Y$ a path in Y, and x_0 a point in the fiber over $y_0 = \gamma(0)$, i.e., $x_0 \in \pi^{-1}(y_0)$. Then there exists at most one lift $\Gamma : [0,1] \to X$ of γ such that $\Gamma(0) = x_0$.*

Proof. We argue by contradiction. Assume there exist two such lifts, $\Gamma_1, \Gamma_2 : [0,1] \to X$. Set

$$S := \{t \in [0,1] \mid \Gamma_1(t) = \Gamma_2(t)\}\ .$$

The set S is nonempty since $0 \in S$. Clearly, S is closed, so it suffices to prove that it is also open. We will show that for every $s \in S$, there exists $\varepsilon > 0$ such that $[s-\varepsilon, s+\varepsilon] \cap [0,1] \subset S$. For simplicity, we consider the case $s = 0$. The general situation is entirely similar.

We will show that there exists $r_0 > 0$ such that $[0, r_0] \subset S$. Pick a small open neighborhood U of x_0 such that π restricts to a homeomorphism onto $\pi(U)$. There exists $r_0 > 0$ such that $\Gamma_i([0,r_0]) \subset U$, for $i = 1, 2$. Since $\pi \circ \Gamma_1 = \pi \circ \Gamma_2$, we deduce that $\Gamma_1|_{[0,r_0]} = \Gamma_2|_{[0,r_0]}$. This completes the proof. □

Theorem 11.1.8. *Let $X \xrightarrow{\pi} Y$ be a covering space, and $f : Z \to Y$ a continuous map, where Z is a connected space. Fix $z_0 \in Z$ and $x_0 \in \pi^{-1}(y_0)$, where $y_0 = f(z_0)$. Then there exists at most one lift $F : Z \to X$ of f such that $F(z_0) = x_0$.*

Proof. For each $z \in Z$, let α_z be a continuous path connecting z_0 to z. If F_1 and F_2 are two lifts of f such that $F_1(z_0) = F_2(z_0) = x_0$, then for any $z \in Z$, the paths $\Gamma_1 = F_1(\alpha_z)$ and $\Gamma_2 = F_2(\alpha_z)$ are two lifts of $\gamma = f(\alpha_z)$ starting at the same point. From Proposition 11.1.7, we deduce that $\Gamma_1 \equiv \Gamma_2$, i.e., $F_1(z) = F_2(z)$ for any $z \in Z$. □

11.1.1 Homotopy Lifting Property

Theorem 11.1.9 (homotopy lifting property). *Let $X \xrightarrow{\pi} Y$ be a covering space, $f : Z \to Y$ be a continuous map, and $F : Z \to X$ be a lift of f. If*

$$h : [0,1] \times Z \to Y \quad (t, z) \mapsto h_t(z)$$

is a homotopy of f ($h_0(z) \equiv f(z)$), then there exists a unique lift of h

$$H : [0,1] \times Z \to X \quad (t, z) \mapsto H_t(z)$$

such that $H_0(z) \equiv F(z)$.

Proof. For each $z \in Z$ we can find an open neighborhood U_z of $z \in Z$, and a partition $0 = t_0 < t_1 < \cdots < t_n = 1$, depending on z, such that h maps $[t_{i-1}, t_i] \times U_z$ into an evenly covered neighborhood of $h_{t_{i-1}}(z)$. Following this partition, we can now successively lift $h|_{I \times U_z}$ to a continuous map $H = H^z : I \times U_z \to X$ such that $H_0(\zeta) = F(\zeta)$, for all $\zeta \in U_z$. By unique lifting property, the liftings on $I \times U_{z_1}$ and $I \times U_{z_2}$ agree on $I \times (U_{z_1} \cap U_{z_2})$, for any $z_1, z_2 \in Z$, and hence we can glue all these local lifts together to obtain the desired lift H on $I \times Z$. □

Corollary 11.1.10 (path lifting property). *Suppose that $X \xrightarrow{\pi} Y$ is a covering map, $y_0 \in Y$, and $\gamma : [0,1] \to Y$ is a continuous path starting at y_0. Then, for every $x_0 \in \pi^{-1}(y_0)$ there exists a unique lift $\Gamma : [0,1] \to X$ of γ starting at x_0.*

Proof. Use the previous theorem with $f : \{pt\} \to Y, f(pt) = \gamma(0)$, and $h_t(pt) = \gamma(t)$. □

Corollary 11.1.11. *Let $X \xrightarrow{\pi} Y$ be a covering space, and $y_0 \in Y$. If $\gamma_0, \gamma_1 \in \Omega(Y, y_0)$ are homotopic relative to y_0, then any lifts Γ_0, Γ_1 which start at the same point also end at the same point, i.e., $\Gamma_0(1) = \Gamma_1(1)$.*

Proof. Lift the homotopy connecting γ_0 to γ_1 to a homotopy in X. By unique lifting property this lift connects Γ_0 to Γ_1. We thus get a continuous path $\Gamma_t(1)$ inside the fiber $\pi^{-1}(y_0)$ which connects $\Gamma_0(1)$ to $\Gamma_1(1)$. Since the fibers are discrete, this path must be constant.

Let $X \xrightarrow{\pi} Y$ be a covering space, and $y_0 \in Y$. Then, for every $x \in \pi^{-1}(y_0)$, and any $\gamma \in \Omega(Y, y_0)$, denote by Γ_x the unique lift of γ starting at x. Set

$$x \cdot \gamma := \Gamma_x(1)$$

By Corollary 11.1.11, if $\omega \in \Omega(Y, y_0)$ is homotopic to γ rel y_0, then

$$x \cdot \gamma = x \cdot \omega$$

Hence, $x \cdot \gamma$ depends only upon the equivalence class $[\gamma] \in \pi_1(Y, y_0)$. Clearly

$$x \cdot ([\gamma] \cdot [\omega]) = (x \cdot [\gamma]) \cdot [\omega]$$

and

$$x \cdot \mathfrak{e}_{y_0} = x$$

so that the correspondence

$$\pi^{-1}(y_0) \times \pi_1(Y, y_0) \ni (x, \gamma) \mapsto x \cdot \gamma \in x \cdot \gamma \in \pi^{-1}(y_0)$$

defines a right action of $\pi_1(Y, y_0)$ on the fiber $\pi^{-1}(y_0)$. This action is called the *monodromy of the covering.* The map $x \mapsto x \cdot \gamma$ is called the *monodromy along* γ. Note that when Y is simply connected, the monodromy is trivial. The map π induces a group morphism

$$\pi_* : \pi_1(X, x_0) \to \pi_1(Y, y_0) \quad x_0 \in \pi^{-1}(y_0)$$

□

Proposition 11.1.12. *π_* is injective.*

Proof. Indeed, let $\gamma \in \Omega(X, x_0)$ such that $\pi(\gamma)$ is trivial in $\pi_1(Y, y_0)$. The homotopy connecting $\pi(\gamma)$ to $\mathfrak{e}_{y_0}$ lifts to a homotopy connecting γ to the unique lift of $\mathfrak{e}_{y_0}$ at x_0, which is $\mathfrak{e}_{x_0}$. □

11.1.2 On the Existence of Lifts

Theorem 11.1.13. *Let $X \xrightarrow{\pi} Y$ be a covering space, $x_0 \in X$, $y_0 = \pi(x_0) \in Y$, $f : Z \to Y$ a continuous map and $z_0 \in Z$ such that $f(z_0) = y_0$. Assume the spaces Y and Z are connected (and thus path connected). The function f admits a lift $F : Z \to X$ such that $F(z_0) = x_0$ if and only if*

$$f_*(\pi_1(Z, z_0)) \subset \pi_*(\pi_1(X, x_0)) \,. \tag{11.1}$$

Proof. Necessity. If F is such a lift then, using the functoriality of the fundamental group construction, we deduce $f_* = \pi_* \circ F_*$. This implies the inclusion (11.1).

Sufficiency. For any $z \in Z$, choose a path γ_z from z_0 to z. Then $\alpha_z = f(\gamma_z)$ is a path from y_0 to $y = f(z)$. Denote by A_z the unique lift of α_z starting at x_0, and set $F(z) = A_z(1)$. We claim that F is a well-defined map.

Indeed, let ω_z be another path in Z connecting z_0 to z. Set $\lambda_z := f(\omega_z)$, and denote by Λ_z its unique lift in X starting at x_0. We have to show that $\Lambda_z(1) = A_z(1)$. Construct the loop based at z_0

$$\beta_z = \omega_z * \gamma_z^-$$

Then $f(\beta_z)$ is a loop in Y based at y_0. From (11.1) we deduce that the lift B_z of $f(\beta_z)$ at $x_0 \in X$ is a closed path, i.e., the monodromy along $f(\beta_z)$ is trivial. We now have

$$\Lambda_z(1) = B_z(1/2) = A_z(0) = A_z(1)$$

This proves that F is a well-defined map. We still have to show that this map is also continuous.

Pick $z \in Z$. Since f is continuous, for every arbitrarily small, evenly covered neighborhood U of $f(z) \in Y$ there exists a path connected neighborhood V of $z \in Z$ such that $f(V) \subset U$. For any $\zeta \in V$ pick a path $\sigma = \sigma_\zeta$ in V connecting z to ζ. Let ω denote the path $\omega = \gamma_z * \sigma_\zeta$ (go from z_0 to z along γ_z, and then from z to ζ along σ_ζ). Then $F(\zeta) = \Omega(1)$, where Ω is the unique lift of $f(\omega)$ starting at x_0. Since $f(\zeta) \in U$, we deduce that $\Omega(1)$ belongs to the local sheet Σ, containing $F(z)$, which homeomorphically covers U. We have thus proved $z \in V \subset F^{-1}(\Sigma)$. The proof is complete since the local sheets Σ form a basis of neighborhoods of $F(z)$. □

Definition 11.1.14. Let Y be a connected space. A covering space $X \xrightarrow{\pi} Y$ is said to be *universal* if X is simply connected.

Corollary 11.1.15. *Let* $X_i \xrightarrow{p_i} Y (i = 0, 1)$ *be two covering spaces of* Y. *Fix* $x_i \in X_i$ *such that* $p_0(x_0) = p(x_1) = y_0 \in Y$. *If* X_0 *is universal, then there exists a unique covering morphism* $F : X_0 \to X_1$ *such that* $F(x_0) = x_1$.

Proof. A bundle morphism $F : X_0 \to X_1$ can be viewed as a lift of the map $p_0 : X_0 \to Y$ to the total space of the covering space defined by p_1. The corollary follows immediately from Theorem 11.1.13 and the unique lifting property. □

Corollary 11.1.16. *Every space admits at most one universal covering space (up to isomorphism).*

Theorem 11.1.17. *Let* Y *be a connected, locally path connected space such that each of its points admits a simply connected neighborhood. Then* Y *admits an (essentially unique) universal covering space.*

Sketch of proof. Assume for simplicity that Y is a metric space. Fix $y_0 \in Y$. Let $\mathcal{P}_{y_0}$ denote the collection of continuous paths in Y starting at y_0. It can be topologized using the metric of uniform convergence in Y. Two paths in $\mathcal{P}_{y_0}$ are said to be *homotopic rel* endpoints if they we can deform one to the other while keeping the endpoints fixed. This defines an equivalence relation on $\mathcal{P}_{y_0}$. We denote the space of equivalence classes by $\tilde{Y}$, and we endow it with the quotient topology. Define $p : \tilde{Y} \to Y$ by

$$p([\gamma]) = \gamma(1) \quad \forall \gamma \in \mathcal{P}_{y_0}.$$

Then $(\tilde{Y}, p)$ is a universal covering space of Y. □

11.1.3 The Universal Cover and the Fundamental Group

Theorem 11.1.18. *Let $\tilde{X} \xrightarrow{p} X$ be the universal cover of a space X. Then*

$$\pi_1(X, \cdot) \cong \mathrm{Deck}(\widetilde{X} \to X) .$$

Proof. Fix $\xi_0 \in \widetilde{X}$ and set $x_0 = p(\xi_0)$. There exists a bijection

$$\mathrm{Ev} : \mathrm{Deck}(\widetilde{X}) \to p^{-1}(x_0) ,$$

given by the evaluation

$$\mathrm{Ev}(F) := F(\xi_0) .$$

For any $\xi \in \pi^{-1}(x_0)$, let γ_ξ be a path connecting ξ_0 to ξ. Any two such paths are homotopic rel endpoints since $\widetilde{X}$ is simply connected (check this). Their projections on the base X determine identical elements in $\pi_1(X, x_0)$. We thus have a natural map

$$\Psi : \mathrm{Deck}(\widetilde{X}) \to \pi_1(X, x_0) \quad F \mapsto p\left(\gamma_{F(\xi_0)}\right) .$$

The map Ψ is clearly a group morphism. (Think monodromy!) The injectivity and the surjectivity of Ψ are consequences of the lifting properties of the universal cover. □

Corollary 11.1.19. *If the space X has a compact universal cover, then $\pi_1(X, \cdot)$ is finite.*

Proof. Indeed, the fibers of the universal cover have to be both discrete and compact. Hence they must be finite. The map Ev in the above proof is a bijection onto Deck $(\widetilde{X})$. □

Example 11.1.20. Let (M, g) be a complete Riemann manifold such that Ricci curvature satisfies the inequality

$$\mathrm{Ric}(X, X) \geq \text{ const } \cdot |X|_g^2 , \tag{11.2}$$

where const denotes a strictly positive constant. By Myers theorem M is compact see [137]. Using the previous exercise we deduce that the universal cover $\widetilde{M}$ is a complete Riemann manifold locally isometric with (M, g). Hence the inequality (11.2) continues to hold on the covering $\widetilde{M}$. Myers theorem implies again that the universal cover $\widetilde{M}$ is compact t! In particular, the universal cover of a semisimple, compact Lie group is compact!!!

Corollary 11.1.21 (H. Weyl). *The fundamental group of a compact semisimple group is finite.*

Proof. Indeed, we deduce from the above example that the universal cover of such a group is compact. □

11.2 The Sheaf of Germs of Holomorphic Functions

Let X be a Riemann surface, and let $a \in X$. We consider pairs (U, f), where U is an open neighborhood of a and f is a holomorphic function on U. Two such pairs (U, f) and (V, g) are said to be equivalent, and define the same germ of a holomorphic function at a, if there exists an open neighborhood W of a, with $W \subset U \cap V$, such that $f|_W = g|_W$. An equivalence class is called a germ of a holomorphic function at a; the class of a pair (U, f) is called the *germ* of f at a and denoted by f_a. The value of f_a at a is defined by $f_a(a) = f(a)$ for any pair (U, f) defining f_a.

If we choose a chart (U, φ) with $a \in U$ and $\varphi(a) = 0$, and identify functions on subsets of U with functions on subsets of $\varphi(U)$, then we can also speak of the values of the derivatives of a germ f_a at a:

$$f_a^{(k)}(a) = \left(\frac{d}{dz}\right)^k f \circ \varphi^{-1}\Big|_{z=0} \qquad \text{if } (V, f) \text{ is a pair}$$

defining f_a.

Let $\mathcal{O}_a$ be the set of all germs at a. The set $\mathcal{O}_a$ is clearly a ring, even a $\mathbb{C}$-algebra. The set m_a of germs f with $f(a) = 0$ is an ideal; the complement $\mathcal{O}_a \setminus m_a$ consists of the units of $\mathcal{O}_a$ (f_a has an inverse in $\mathcal{O}_a \iff f_a(a) \neq 0$), so that m_a is the unique maximal ideal in $\mathcal{O}_a$.

If we choose a chart as above, then the map $f_a \longmapsto \sum_{n=0}^{\infty} \frac{1}{n!} f_a^{(n)}(a) z^n$ is an isomorphism (as $\mathbb{C}$-algebras) of $\mathcal{O}_a$ with the ring $\mathbb{C}\{z\}$ of power series with a non-zero radius of convergence.

Let $\mathcal{O}_X = \bigcup_{a \in X} \mathcal{O}_a$ (disjoint union). We will sometimes write simply $\mathcal{O}$ for $\mathcal{O}_X$. We define a map $p : \mathcal{O}_X \to X$ by $p(f) = a$ if $f \in \mathcal{O}_X$.

Let now $f_a \in \mathcal{O}_X$, and (U, f) a pair defining the germ f_a. We set $N(U, f) = \{f_x \mid x \in U\}$, the set of all germs defined by f at the different points of U. We define a topology on $\mathcal{O}_X$ by the condition that the sets $\{N(U, f)\}$ form a fundamental system of neighborhoods of f_a when (U, f) runs over all pairs defining f_a.

Lemma 11.2.1. *With this topology, $\mathcal{O}_X$ is Hausdorff and the map $p : \mathcal{O}_X \to X$ is a local homeomorphism.*

Proof. Let $f_a, \underline{g}_b \in \mathcal{O}_X$, and suppose that $f_a \neq \underline{g}_b$. We must show that they can be separated by open neighborhoods, and consider two cases.

1. $a \neq b$. Let (U, f), (V, g) be pairs defining $f_a, \underline{g}_b$ respectively ($a \in U, b \in V$). We can find such pairs with $U \cap V = \emptyset$, in which case $N(U, f) \cap N(V, g) = \emptyset$.

2. $a = b$. Let U be a connected open set, $a \in U$, and f, g holomorphic functions on U so that the pairs (U, f), (U, g) define $f_a, \underline{g}_a$ respectively.

We claim that $N(U,f) \cap N(U,g) = \emptyset$. In fact, if $\underline{h}_x$ ($x \in U$) is a germ in the intersection, then both f and g induce the germ $\underline{h}_x$ at x, hence coincide in some neighborhood of x. Since U is connected, the principle of analytic continuation implies that $f \equiv g$, so that $f_a = \underline{g}_a$, a contradiction.

Thus, $\mathcal{O}_X$ is Hausdorff.

Moreover, if U is open in X and f is holomorphic on U, then $p(N(U,f)) = U$ and $p|_{N(U,f)}$ is injective, with inverse $x \mapsto f_a$ (the germ of f at x). The properties of p stated in the lemma follow from this.

□

Remark 11.2.2. If X is given with a countable base for its open sets, it can be proved directly that any connected component of $\mathcal{O}_X$ has a countable base. This is a consequence of the so-called Poincaré–Volterra theorem (for a statement and proof see [139]).

11.3 Sheaves

A *sheaf* (plural: *sheaves*) is a tool for systematically tracking data (such as sets, abelian groups, or rings) attached to the open sets of a topological space and defined locally with respect to them. For example, for each open set, the data could be the ring of continuous functions defined on that open set. Such data are well-behaved in that they can be restricted to smaller open sets, and the data assigned to an open set are equivalent to all collections of compatible data assigned to collections of smaller open sets covering the original open set (intuitively, every datum is determined by its constituent data).

The field of mathematics that studies sheaves is called sheaf theory.

Sheaves are understood conceptually as general and abstract objects. Their precise definition is rather technical. They are specifically defined as sheaves of sets or sheaves of rings, for example, depending on the type of data assigned to the open sets.

There are also maps (or morphisms) from one sheaf to another. Sheaves (of a specific type, such as sheaves of abelian groups), together with their morphisms on a fixed topological space, form a category. On the other hand, for each continuous map there are associated both a direct image functor, taking sheaves and their morphisms on the domain to sheaves and morphisms on the codomain, and an inverse image functor operating in the opposite direction. These functors, and certain variants of them, are essential parts of sheaf theory.

Due to their general nature and versatility, sheaves have several applications in topology and especially in algebraic and differential geometry. First, geometric structures such as those of a differentiable manifold or a scheme can be expressed in terms of a sheaf of rings on the space. In such contexts, several geometric constructions, such as vector bundles or divisors, are naturally specified in terms of sheaves. Second, sheaves provide the framework for a very general cohomology theory, known as *sheaf cohomology*, which also encompasses the usual topological cohomology theories, such as singular cohomology. Especially in algebraic geometry and the theory of complex manifolds, sheaf cohomology provides a powerful link between the topological and geometric properties of spaces. Sheaves also provide the basis for the theory of *D-modules*, which have applications to the theory of differential equations. In addition, generalizations of sheaves to settings more general than topological spaces—such as the notion of a sheaf on a category with respect to some Grothendieck topology—have found applications in mathematical logic and number theory.

11.3.1 Presheaves and Sheaves

Let X be a space with topology $\mathfrak{T}$ (i.e., $\mathfrak{T}$ is the system of open sets in X). A *presheaf of abelian groups* on X is a pair $(\mathcal{F}, \rho)$ consisting of a family $\mathcal{F} = (\mathcal{F}(U))_{U\in\mathfrak{T}}$ of abelian groups and a family $\rho = (\rho_V^U)_{U,V\in\mathfrak{T},\, V\subseteq U}$ of group homomorphisms $\rho_V^U : \mathcal{F}(U) \to \mathcal{F}(V)$ such that

1. $\rho_U^U = \mathrm{id}_{\mathcal{F}(U)}$ for all $U \in \mathfrak{T}$,
2. $\rho_W^V \circ \rho_V^U = \rho_W^U$ for all $W \subseteq V \subseteq U$.

The homomorphisms ρ_V^U are called *restriction homomorphisms*. Often we simply write $f|_V$ for $\rho_V^U(f)$ when $f \in \mathcal{F}(U)$.

Analogously, one defines presheaves of vector spaces, rings, sets, etc. A presheaf $\mathcal{F}$ on a topological space X is called a *sheaf* if, for every open $U \subseteq X$ and every family of open subsets $U_i \subseteq U$, $i \in I$, with $U = \bigcup_{i\in I} U_i$, the following conditions are satisfied:

(1) If $f, g \in \mathcal{F}(U)$ satisfy $f|_{U_i} = g|_{U_i}$ for all $i \in I$, then $f = g$.

(2) Let $f_i \in \mathcal{F}(U_i)$, $i \in I$, be such that $f_i|_{U_i\cap U_j} = f_j|_{U_i\cap U_j}$ for all $i, j \in I$. Then there exists $f \in \mathcal{F}(U)$ such that $f|_{U_i} = f_i$ for all $i \in I$.

The element f in (2) is unique by (1).

Example 11.3.1.

(1) Let X be a topological space. For each open $U \subseteq X$, let $\mathcal{C}(U)$ denote the vector space of continuous functions $f : U \to \mathbb{C}$. Then $\mathcal{C}$, with the usual restriction mapping, is a sheaf on X.

(2) Let X be a Riemann surface. Let $\mathcal{O}(U)$ be the ring of holomorphic functions on the open subset $U \subseteq X$. Using the usual restriction mapping, we obtain the sheaf $\mathcal{O}$ of holomorphic functions on X.

(3) Similarly, we obtain the sheaf $\mathcal{M}$ of meromorphic functions on a Riemann surface X.

(4) Let X be a Riemann surface. Denote by $\mathcal{O}^*(U)$ the multiplicative group of all holomorphic functions $f : U \to \mathbb{C}^*$. With the usual restriction map, we obtain a sheaf $\mathcal{O}^*$ on X.

(5) Similarly, we obtain the sheaf $\mathcal{M}^*$. Here $\mathcal{M}^*(U)$ consists of all $f \in \mathcal{M}(U)$ such that f has no zeros.

(6) Let X be a topological space. For open $U \subseteq X$, let $\mathbb{C}(U)$ denote the vector space of locally constant functions $f : U \to \mathbb{C}$. With the usual restriction map, this defines the sheaf $\mathbb{C}$ on X, whereas the constant functions form only a presheaf.

11.3.2 Stalks

Let $\mathcal{F}$ be a presheaf of sets on a topological space X and let $a \in X$. The stalk $\mathcal{F}_a$ of $\mathcal{F}$ at a is defined as the inductive limit

$$\mathcal{F}_a := \varinjlim_{U \ni a} \mathcal{F}(U)$$

over all open neighborhoods U of a. This means the following. Consider the disjoint union

$$\bigsqcup_{U \ni a} \mathcal{F}(U)$$

with the following equivalence relation: $f \in \mathcal{F}(U)$ and $g \in \mathcal{F}(V)$ are equivalent, $f \sim g$, if there is an open set W with $a \in W \subseteq U \cap V$ such that $f|_W = g|_W$. Then $\mathcal{F}_a$ is the set of equivalence classes,

$$\mathcal{F}_a = \varinjlim_{U \ni a} \mathcal{F}(U) := \left(\bigsqcup_{U \ni a} \mathcal{F}(U) \right) / \sim$$

If $\mathcal{F}$ is a presheaf of abelian groups, then the stalk $\mathcal{F}_a$ is also an abelian group in a natural way (similarly, for presheaves of vector spaces, rings, etc.).

Let U be an open neighborhood of a. Let $\rho_a : \mathcal{F}(U) \to \mathcal{F}_a$ denote the map which assigns to $f \in \mathcal{F}(U)$ its equivalence class modulo $\sim$. Then $\rho_a(f) =: f_a$ is called the *germ* of f at a.

Example 11.3.2. Consider the sheaf $\mathcal{O}$ of holomorphic functions on a region $X \subseteq \mathbb{C}$ (a region is a connected domain). The stalk $\mathcal{O}_a$ at $a \in X$ is isomorphic to the ring $\mathbb{C}\{z - a\}$ of convergent power series in $z - a$. Analogously, the stalk $\mathcal{M}_a$ is isomorphic to the ring of all convergent Laurent series with finite principal part

$$\sum_{k=m}^{\infty} c_k(z-a)^k, \quad m \in \mathbb{Z}, \quad c_k \in \mathbb{C}.$$

11.3.3 The Topological Space Associated with a Presheaf

Let $\mathcal{F}$ be a presheaf on a topological space X. Define

$$|\mathcal{F}| := \bigsqcup_{x \in X} \mathcal{F}_x$$

as the disjoint union of all stalks. Let $p : |\mathcal{F}| \to X$ be defined by $\mathcal{F}_x \ni \varphi \mapsto x$. For any open $U \subseteq X$ and $f \in \mathcal{F}(U)$, let

$$(U, f) := \{\rho_x(f) : x \in U\}.$$

Theorem 11.3.3. *The system $\mathfrak{B}$ of all sets (U, f), where U is open in X and $f \in \mathcal{F}(U)$, is a basis for a topology on $|\mathcal{F}|$. With respect to this topology, $p : |\mathcal{F}| \to X$ is a local homeomorphism.*

Proof. We first check that $\mathfrak{B}$ is a basis for a topology on $|\mathcal{F}|$. Clearly, every $\varphi \in |\mathcal{F}|$ is contained in at least one (U, f). We need to verify that if $\varphi \in (U, f) \cap (V, g)$, then there exists $(W, h) \in \mathfrak{B}$ such that $\varphi \in (W, h) \subseteq (U, f) \cap (V, g)$. Let $x = p(\varphi)$. Then $x \in U \cap V$ and $\varphi = \rho_x(f) = \rho_x(g)$. Hence there exists an open neighborhood W of x in $U \cap V$ such that $f|_W = g|_W =: h$. This proves the claim.

To see that $p : |\mathcal{F}| \to X$ is a local homeomorphism, let $\varphi \in |\mathcal{F}|$ and $x = p(\varphi)$. There exists $(U, f) \in \mathfrak{B}$ with $\varphi \in (U, f)$. Then (U, f) is an open neighborhood of φ and U is an open neighborhood of x. The restriction $p|_{(U,f)} : (U, f) \to U$ is a homeomorphism. □

We say that a presheaf $\mathcal{F}$ on a topological space X *satisfies the identity theorem* if the following holds: Let $U \subseteq X$ be a connected open set, and let $f, g \in \mathcal{F}(U)$ be such that $\rho_a(f) = \rho_a(g)$ for some $a \in U$. Then $f = g$.

Theorem 11.3.4. *Let X be a locally connected Hausdorff space. Let $\mathcal{F}$ be a presheaf on X satisfying the identity theorem. Then $|\mathcal{F}|$ is Hausdorff.*

Proof. Let $\varphi_1 \neq \varphi_2 \in |\mathcal{F}|$. Set $x_i = p(\varphi_i)$ for $i = 1, 2$. If $x_1 \neq x_2$, then there exist disjoint neighborhoods U_1, U_2 of x_1 and x_2, respectively, since X is Hausdorff, and $p^{-1}(U_1)$, $p^{-1}(U_2)$ are disjoint neighborhoods of φ_1 and φ_2.

Suppose $x_1 = x_2 =: x$. Let $f_i \in \mathcal{F}(U_i)$ be representatives of the germs φ_i. Let $U \subseteq U_1 \cap U_2$ be a connected open neighborhood of x. Then $(U, f_i|_U)$ is a neighborhood of φ_i. Suppose $\psi \in (U, f_1|_U) \cap (U, f_2|_U)$, and let $y = p(\psi)$. Then $\psi = \rho_y(f_1) = \rho_y(f_2)$. By assumption, $f_1|_U = f_2|_U$, and thus $\varphi_1 = \varphi_2$, a contradiction. □

Corollary 11.3.5. *Let X be a Riemann surface, and let $\mathcal{O}$ (resp. $\mathcal{M}$) be the sheaf of holomorphic (resp. meromorphic) functions on X. Then $|\mathcal{O}|$ (resp. $|\mathcal{M}|$) is Hausdorff.*

11.4 Polynomials

The ring of complex polynomials in one variable is denoted by $\mathbb{C}[X]$. If P and Q are two such polynomials,

$$P(X) = a_n X^n + a_{n-1} X^{n-1} + \cdots + a_0 \,,$$
$$Q(X) = b_m X^m + b_{m-1} X^{n-1} + \cdots + b_0, \quad a_n b_m \neq 0, n \geq m \,,$$

then there are polynomials C and R such that $\deg C = n - m$, and

$$P(X) = C(X)Q(X) + R(X), \quad \deg R < \deg Q \,. \tag{11.3}$$

In fact (11.3) is a system of $n - m + 1$ linear equations for the $n - m + 1$ coefficients of c that can be solved sequentially. At the first step we choose the leading coefficient c_{n-m} equal to the quotient a_n / b_m of the leading coefficients of p and q. At the next step we want

$$a_{n-1} = c_{n-m-1} b_m + c_{n-m} b_{m-1} = c_{n-m-1} b_m + \frac{a_n}{b_m} b_{m-1}$$

which determines c_{n-m-1}, and so on. The coefficients of C and of R in (11.3) are rational functions of the coefficients of p and q.

Lemma 11.4.1. *Suppose that P and Q are complex polynomials with no common factor. Then there are polynomials R and S such that*

$$R(X)P(X) + S(X)Q(X) \equiv 1 \tag{11.4}$$

Proof. We may assume that $\deg P \geq \deg Q$. Set $P_1 = P, P_2 = Q$, and carry out the division algorithm (11.3) to obtain terms P_k of strictly decreasing degree:

$$P_1 = c_2 P_2 + P_3$$
$$P_2 = c_3 P_3 + P_4$$
$$\cdots$$
$$P_{m-3} = c_{m-2} P_{m-2} + P_{m-1}$$
$$P_{m-2} = c_{m-1} P_{m-1} + P_m$$

where P_m is constant. If $P_m = 0$, then P_{m-1}, which has positive degree, divides P_{m-2}. The preceding equation shows, then, that P_{m-1} divides P_{m-3}.

Continuing up the chain of equations, P_{m-1} divides both $P_2 = q$ and $P_1 = p$, a contradiction. Thus P_m is a constant $a \neq 0$. Starting with

$$\begin{aligned} a &= P_{m-2} - c_{m-1}P_{m-1} = P_{m-2} - c_{m-1}\left[P_{m-3} - c_{m-2}P_{m-2}\right] \\ &= \left[1 - c_{m-2}\right]P_{m-2} - c_{m-1}P_{m-3} \end{aligned}$$

we proceed up the previous chain of equations to reach $P_2 = q$ and $P_1 = p$. Dividing the resulting equation by a gives us (11.4). □

We have already spoken of irreducibility for polynomials in two variables. The same concept applies in one variable: $P \in \mathbb{C}[X]$ is irreducible if it is not the product of two non-constant elements of $\mathbb{C}[X]$.

Lemma 11.4.2. *If an irreducible complex polynomial that belongs to $\mathbb{C}[X]$ divides the product of two polynomials that belong to $\mathbb{C}[z]$, then it divides one (or both) of the factors.*

Proof. Suppose that P is irreducible and does not divide either of Q_1 or Q_2. Then it has no common factors with either, so there are polynomials R_j, S_j such that $R_jP + S_jQ_j = 1$. Then

$$S_1S_2Q_1Q_2 = (1 - R_1P)(1 - R_2P) = R_1R_2P^2 - (R_1 + R_2)P + 1$$

Therefore P does not divide the left side. □

Lemma 11.4.3. *Suppose polynomials $P(X,Y)$ and $Q(X,Y)$ have no common factors over $\mathbb{C}[X]$. Then there are polynomials $R(X,Y), S(X,Y)$, and $r(X)$ such that r is not identically zero and*

$$R(X,Y)P(X,Y) + S(X,Y)Q(X,Y) = r(X)$$

Proof. Consider $P(X,Y), Q(X,Y) \in \mathbb{C}[X,Y]$. The condition that P and Q have no common factor over $\mathbb{C}[X]$ means that, when viewed as elements of $\mathbb{C}(X)[Y]$, they are relatively prime.

Since $\mathbb{C}(X)[Y]$ is a Euclidean domain, there exist polynomials $R_1(Y)$, $S_1(Y)$ in $\mathbb{C}(X)[Y]$ such that

$$R_1(Y)P(X,Y) + S_1(Y)Q(X,Y) = 1\,.$$

Write R_1, S_1 with coefficients in $\mathbb{C}(X)$. Let $r(X) \in \mathbb{C}[X]\setminus\{0\}$ be a common denominator for these coefficients. Multiplying through by $r(X)$ gives

$$R(X,Y)P(X,Y) + S(X,Y)Q(X,Y) = r(X)\,,$$

where $R, S \in \mathbb{C}[X,Y]$.

Hence such polynomials R, S, r exist, with $r(X) \neq 0$. □

11.4.1 Fundamental Theorem of Algebra

It is well known that the complex numbers have the property that every nonconstant polynomial in $\mathbb{C}[x]$ has a zero in $\mathbb{C}$. This is known as the fundamental theorem of algebra. The reader is undoubtedly already familiar with the fundamental theorem of algebra. The proof can be found in most complex analysis books, for instance, in [114, Theorem 3.2]. Interestingly, the simplest and shortest proofs of the Fundamental Theorem of Algebra are not algebraic proofs. There are much shorter and easier-to-follow proofs using either analysis or topology. However, behind both the analytical and the topological proofs lies a significant amount of machinery, so perhaps comparing the proofs is not completely fair. At any rate, we include a short proof for students who have studied functions of a complex variable and are familiar with Liouville's theorem.

Theorem 11.4.4 (fundamental theorem of algebra). *The field $\mathbb{C}$ of complex numbers is an algebraically closed field.*

Proof. The first proof. Let the polynomial $f(z) \in \mathbb{C}[z]$ have no zero in $\mathbb{C}$. Then $1/f(z)$ gives an entire function; that is, $1/f(z)$ is analytic everywhere. Also if $f(z) \notin \mathbb{C}, \lim_{|z|\to\infty} |f(z)| = \infty$, so $\lim_{|z|\to\infty} |1/f(z)| = 0$. Thus $1/f(z)$ must be bounded in the plane. Hence by Liouville's theorem of complex function theory, $1/f(z)$ is constant, and thus $f(z)$ is constant. Therefore, a nonconstant polynomial in $\mathbb{C}[z]$ must have a zero in $\mathbb{C}$, so $\mathbb{C}$ is algebraically closed.

The second proof. For the proof it is first necessary to pass from the plane of complex numbers to a compact manifold. Consider the unit sphere $S^2 \subset \mathbb{R}^3$ and the stereographic projection

$$\Phi_1 : S^2 - \{(0,0,1)\} \to \mathbb{R}^2 \times 0 \subset R^3$$

from the "north pole" $(0,0,1)$ of S^2. (See Fig. 5.2) We will identify $\mathbb{R}^2 \times 0$ with the plane of complex numbers. The polynomial map P from $\mathbb{R}^2 \times 0$ itself corresponds to a map f from S^2 to itself; where

$$f(x) = \Phi_1^{-1} P \Phi_1(x) \quad \text{for} \quad x \neq (0,0,1)$$
$$f(0,0,1) = (0,0,1)\,.$$

It is well known that this resulting map f is smooth, even in a neighborhood of the north pole. To see this we introduce the stereographic projection Φ_0 from the south pole $(0,0,-1)$ and set

$$Q(z) = \Phi_0 f \Phi_0^{-1}(z)\,.$$

Note, by elementary geometry, that

$$\Phi_1 \Phi_0^{-1}(z) = z/|z|^2 = 1/\bar{z}\,.$$

Now if $P(z) = a_0 z^n + a_1 z^{n-1} + \cdots + a_n$, with $a_0 \neq 0$, then a short computation shows that

$$Q(z) = z^n / \left(\bar{a}_0 + \bar{a}_1 z + \cdots + \bar{a}_n z^n\right)\,.$$

Thus Q is smooth in a neighborhood of 0, and it follows that $f = \Phi_0^{-1} Q \Phi_0$ is smooth in a neighborhood of $(0, 0, 1)$.

Next observe that f has only a finite number of critical points; for P fails to be a local diffeomorphism only at the zeros of the derivative polynomial $P'(z) = \sum a_{n-j} j z^{i-1}$, and there are only finitely many zeros since P' is not identically zero. The set of regular values of f, being a sphere with finitely many points removed, is therefore connected. Hence the locally constant function $\# f^{-1}(y)$ must actually be constant on this set. Since $\# f^{-1}(y)$ can't be zero everywhere, we conclude that it is zero nowhere. Thus f is an onto mapping, and the polynomial P must have a zero. □

The existence of at least one zero for every nonconstant complex polynomial already implies that complex polynomials decompose into linear factors and that real polynomials decompose into linear and quadratic factors. These consequences of the fundamental theorem are completely elementary, and are a result of the simple fact that a polynomial with a zero at c always has the factor $z - c$.

Lemma 11.4.5 (Factorization Lemma). *If $c \in \mathbb{C}$ is a zero of the polynomial $f \in \mathbb{C}[X]$ of degree n, then there is just one polynomial $g \in \mathbb{C}[X]$ of degree $n-1$, such that $f(Z) = (Z - c)g(Z)$.*

Proof. Let $f = a_0 + a_1 X + \cdots + a_n X^n, a_n \neq 0$. Since $Z^\nu - c^\nu = (X - c) q_\nu(X)$ with $q_\nu(X) := X^{\nu-1} + X^{\nu-2} c + \cdots + c^{\nu-1}$ it follows that

$$f(X) = f(X) - f(c) = \sum_1^n a_\nu \left(X^\nu - c^\nu\right) = (X - c) g(X)$$

where

$$g(X) := \sum_1^n a_\nu q_\nu(X)$$

It is clear that f is of degree $n-1$: since $g(X) = (X - c)^{-1} f(X), z \neq c, g$ is uniquely determined by f and c. □

The factorization lemma holds for all commutative rings, provided that one gives up the uniqueness of g. By induction on n we at once obtain the

Corollary 11.4.6. *A polynomial $f \in \mathbb{C}[X]$ of degree n has at most n zeros.*

Theorem 11.4.7 (factorization of complex polynomials). *Every complex polynomial $f \in \mathbb{C}[X]$ of degree $n \geq 1$ is, disregarding the order of the factors, uniquely representable in the form*

$$f(X) = a\left(X - c_1\right)^{n_1}\left(X - c_2\right)^{n_2} \cdot \cdots \cdot \left(X - c_r\right)^{n_r},$$

where $a \in \mathbb{C}^x$; $r \in \mathbb{N}, c_1, \ldots, c_r \in \mathbb{C}$ are distinct from one another, and $n_1, \ldots, n_r \in \mathbb{N} \backslash \{0\}$ with $n_1 + n_2 + \cdots + n_r = n$.

Proof. We use induction on n, the case $n = 1$ being clearly true. Suppose $n > 1$. By the Fundamental theorem of algebra there exists a $c_1 \in \mathbb{C}$ for which f vanishes. By Lemma 11.4.5, $f(X) = (X - c_1)\, g(Z)$, where $g(Z) \in \mathbb{C}[X]$ is of degree $n - 1$. By the inductive hypothesis there is a unique factorization

$$g(X) = a\,(X - c_1)^{n_1 - 1}\,(X - c_2)^{n_2} \cdot \cdots \cdot (X - c_r)^{n_r}$$

with $n_1 \geq 1, \ldots, n_r \geq 1, n_1 - 1 + n_2 + \cdots + n_r = n - 1; c_1, \ldots, c_r \in \mathbb{C}$ distinct from one another, and $a \in \mathbb{C}^\star$. Consequently (11.4.7) holds. □

The theorem just proved is often stated in the form:

Every complex polynomial of the nth degree has precisely n zeros where each of the zeros c_j is counted according to its multiplicity n_j.

11.4.2 Resultants and Discriminants

Let

$$f(X) = a_0 X^n + a_1 X^{n-1} + \cdots + a_{n-1} x + a_n\,,$$
$$g(X) = b_0 X^m + b_1 X^{m-1} + \cdots + b_{m-1} x + b_m$$

be two polynomials with coefficients in $\mathbb{C}$. Here $n > 0, m > 0$, and we allow the possibility that $a_0 = 0$ or $b_0 = 0$. Over an algebraically closed field, f and g have a common divisor if and only if they have a common root.

Definition 11.4.8. The *resultant* $\mathrm{Res}(f, g)$ of f and g is the homogeneous polynomial in the coefficients of f and g (having degree m in $a_0, \ldots, a_n$ and degree n in $b_0, \ldots, b_m$) given by the determinant of the Sylvester matrix

$$Res(f,g) := \begin{bmatrix} a_0 & a_1 & \cdots & \cdots & a_m & 0 & \cdots & 0 \\ 0 & a_0 & a_1 & \cdots & \cdots & a_m & \ddots & \vdots \\ \vdots & \ddots & \ddots & & & & \ddots & 0 \\ 0 & \cdots & 0 & a_0 & \cdots & \cdots & a_{m-1} & a_m \\ b_0 & b_1 & \cdots & \cdots & b_n & 0 & \cdots & 0 \\ 0 & b_0 & b_1 & \cdots & \cdots & b_n & \ddots & \vdots \\ \vdots & \ddots & \ddots & & & & \ddots & 0 \\ 0 & \cdots & 0 & b_0 & \cdots & \cdots & b_{n-1} & b_n \end{bmatrix}.$$

Clearly, $R(f, g)$ is a homogeneous polynomial of degree m with respect to indeterminates a_i and of degree n with respect to indeterminates b_j. The polynomials f and g have a common divisor if and only if the determinant of the system considered vanishes, i.e., $R(f, g) = 0$.

We now derive the basic properties of the resultant.

Lemma 11.4.9. *We have that Res $(f, g) = 0$ if and only if either $a_0 = 0 = b_0$ or else f and g have a common factor of degree > 0 in $\mathbb{C}[X]$.*

Proof. We first show that the condition $a_0 = 0 = b_0$ or else f and g have a common factor in $\mathbb{C}[X]$ of degree > 0 holds if and only if there exist polynomials f_1 and g_1 not both zero, such that

$$fg_1 + f_1 g = 0, \quad \deg f_1 < n, \quad \deg g_1 < m \quad . \tag{11.5}$$

To see this, let h be the greatest common divisor of f and g and suppose that $\deg h > 0$. Then $f = hf_1$, $g = -hg_1$, and hence $fg_1 + gf_1 = 0$. In addition, $\deg f_1 < n$ and $\deg g_1 < m$, so that (11.5) holds. If $a_0 = 0 = b_0$, we can set $f_1 = f, g_1 = -g$.

Conversely, suppose that (11.5) holds. If 1 is the greatest common divisor of f and g, then since $\mathbb{C}[X]$ is a unique factorization domain, the identity $fg_1 = -gf_1$ implies $f \mid f_1$, $g \mid g_1$.

We now prove that (11.5) is equivalent to $\mathrm{Res}(f, g) = 0$. If we set

$$\begin{aligned} f_1 &= c_0 x^{n-1} + c_1 x^{n-2} + \cdots + c_{n-1}\,, \\ g_1 &= d_0 s^{m-1} + d_1 x^{m-2} + \cdots + d_{m-1} \end{aligned}$$

and use the rules for operating with polynomials to compute the coefficients of $fg_1 + gf_1$, which has degree $\leq n + m - 1$, we can write the condition (11.5) in the form of a square homogeneous system of linear equations with $n + m$ unknowns $d_0, d_1, \ldots, d_{m-1}$, $c_0, c_1, \ldots, c_{n-1}$:

$$\begin{array}{llr} a_0 d_0 + & \cdots + b_0 c_0 & \cdots = 0\,, \\ a_1 d_0 + a_0 d_1 + & \cdots + b_1 c_0 + b_0 c_1 & \cdots = 0\,, \\ a_2 d_0 + a_1 d_1 + a_0 d_2 & \cdots + b_2 c_0 + b_1 c_1 + b_0 c_2 & \cdots = 0\,, \\ & \cdots\,. & \end{array} \tag{11.6}$$

The determinant of the system (11.6) (more precisely, the determinant of the transpose of the matrix of (11.6)) is exactly $\mathrm{Res}(f, g)$. Thus, (11.6) has a non-trivial solution if and only if $\mathrm{Res}(f, g) = 0$, and any non-zero solution to (11.6) gives us a pair of polynomials f_1, g_1 □

Lemma 11.4.10. *Suppose that the polynomials f and g split completely into linear factors*

$$\begin{aligned} f(X) &= a_0 (X - \alpha_1) \ldots (X - \alpha_n)\,, \\ g(X) &= b_0 (X - \beta_1) \ldots (X - \beta_m)\,. \end{aligned}$$

Then

$$\mathrm{Res}(f, g) = a_0^m \prod_{i=1}^{n} g(\alpha_i) = (-1)^{mn} b_0^n \prod_{j=1}^{m} f(\beta_j) = a_0^m b_0^n \prod_{i,j} (\alpha_i - \beta_j)$$

which is to say the resultant is the product of either polynomial evaluated at each root, including multiplicity, of the other polynomial

Proof. It is clear that these formulas, if valid, must possess a *universal* character—that is, they should not depend on the specific properties of f and g. This simple *philosophical* principle, which we will neither justify nor discuss in detail here, allows us to focus on the *general case*: we assume that all of $g(\alpha_1), \ldots, g(\alpha_n)$ are distinct, and likewise, all of $f(\beta_1), \ldots, f(\beta_m)$ are distinct.

Next, since $\operatorname{Res}(g, f) = (-1)^{mn} \operatorname{Res}(f, g)$ (see the definition), it suffices to verify the equality $\operatorname{Res}(f, g) = a_0^m \prod g(\alpha_i)$. To do this we introduce a new variable Y and consider the polynomials $f(X)$ and $g(X) - Y$ over the field of rational functions $\mathbb{C}(Y)$. Using the definition of the resultant and replacing b_m by $b_m - Y$, we find that

$$\operatorname{Res}(f,\ g - Y) = (-1)^n a_0^m Y^n + \cdots + \operatorname{Res}(f,\ g)$$

is a polynomial of degree n in Y with leading coefficient $(-1)^n a_0^m$ and with constant term $\operatorname{Res}(f, g)$. The polynomials $f(X)$ and $g(X) - g(\alpha_i)$ have the root α_i in common, and so both are divisible by $X - \alpha_i$. Using Lemma 11.4.9, we have $\operatorname{Res}(f, g - g(\alpha_i)) = 0$. By Bézout's theorem, the polynomial $\operatorname{Res}(f, g - Y)$ must be divisible by $g(\alpha_i) - Y$, $1 \leq i \leq n$. Since we have assumed that all of the $g(\alpha_i)$ are distinct, it follows that $\operatorname{Res}(f, g - Y) = a_0^m \prod_{i=1}^{n} (g(\alpha_i) - Y)$. Setting $Y = 0$, we obtain the required equality. □

From Lemma 11.4.10 follows, that

$$R(P, P_w)(z) = (-1)^{\frac{n(n-1)}{2}} \prod_{i=1}^{n} \prod_{j=1}^{n-1} (w_i(z) - \tilde{w}_j(z))$$

where $w_i(z), i = 1, \ldots, n$, are the roots of the polynomials $P(z, w)$ and $\tilde{w}_j(z), j = 1, \ldots, n-1$ are the roots of the polynomials $P_w(z, w)$ where z is considered as a parameter.

The resultant has many different applications. For example, given polynomial relations $P(x, z) = 0$ and $Q(y, z) = 0$ we can, with the help of the resultant, obtain a polynomial relation of the form $R(x, y) = 0$, i.e., eliminate z. Indeed, consider the given polynomials $P(x, z)$ and $Q(y, z)$ as polynomials in z regarding x and y as constants. Then the vanishing of the resultant of these polynomials is exactly the relation desired: $R(x, y) = 0$.

The resultant also allows one to reduce the solution of the system of algebraic equations to the search for roots of polynomials. Indeed, let $P(x_0, y_0) = 0$ and $Q(x_0, y_0) = 0$. Consider $P(x, y)$ and $Q(x, y)$ as polynomials in y. For $x = x_0$, they have a common root y_0. Therefore their resultant $R(x)$ vanishes at $x = x_0$.

Let $x_1, \ldots, x_n$ be the roots of the polynomial $f(x) = a_0 x^n + \cdots + a_n$, where $a_0 \neq 0$. The quantity

$$D(f) = a_0^{2n-2} \prod_{i<j} (a_i - a_j)^2$$

is called the *discriminant* of f.

Since the discriminant is zero only if a polynomial has a multiple root, and, equivalently if the polynomial and its derivative share a common factor, there is evidence to suggest the discriminant and the resultant are in some way related.

Theorem 11.4.11. $R(f, f') = \pm a_0 D(f)$.

Proof. By Lemma 11.4.10 we have $R(f, f') = a_0^{n-1} \prod_i f'(x_i)$. It is easy to verify that $f'(x_i) = a_0 \prod_{j \neq i} (x_j - x_i)$. Therefore

$$R(f, f') = a_0^{2n-1} \prod_{j \neq i} (x_j - x_i) = \pm a_0^{2n-1} \prod_{i<j} (x_j - x_i)^2 .$$

□

Remark 11.4.12. It is not difficult to show that

$$R(f, f') = -R(f', f) = (-1)^{n(n-1)/2} a_0 D(f) .$$

At this point, it should not come as much of a surprise that the determinant is defined in terms of the resultant.

Corollary 11.4.13. *The discriminant of f is a polynomial in $a_0, \ldots, a_n$ with integer coefficients.*

See e.g., [39, Resultants and Discriminants. §3.3.2] or [116, §5 The resultant, pp. 200–205]

11.4.3 Bézout's Theorems

An important feature of the projective plane is that any two lines meet. In fact, it has a much stronger property:

Theorem 11.4.14 (weak Bézout theorem). *Any two algebraic curves in $\mathbb{P}^2$ intersect.*

Proof. Assume that the curves are given by homogeneous polynomials $F(x, y, z)$ and $G(x, y, z)$ respectively. After translating the line at infinity if necessary, we can assume that the polynomials $f(x, y) = F(x, y, 1)$ and $g(x, y) = G(x, y, 1)$ are both nonconstant in x and y. Treating these as polynomials in y with coefficients in $\mathbb{C}[x]$, the resultant $\mathrm{Res}(f, g)(x)$ can be regarded as a nonconstant polynomial in x. Since $\mathbb{C}$ is algebraically closed, $\mathrm{Res}(f, g)(x)$ must have a root, say a. Then $f(a, y) = 0$ and $g(a, y) = 0$ have a common solution. □

The most important result of elementary curve theory is Bézout's theorem, which gives the number of points of intersection of two algebraic curves. We can now state the strong form of Bézout's theorem.

Theorem 11.4.15 (Bézout's theorem). *Let Γ and X be two projective curves defined by the homogenous polynomials $Q(X,Y,Z)$ and $F(X,Y,Z)$ of degree d and m respectively. If Γ and X do not have a common component, then they intersect in dm points counting multiplicity.*

Proof. The proof follows directly from the definition of the resultant and involves a tedious but straightforward calculation or see [180, Theorem 10.3, p. 80]. □

Lemma 11.4.16. *Let $Q(X,Y,Z)$ and $F(X,Y,Z)$ be two homogeneous polynomials of degree d and m respectively. Suppose that $Q(0,0,Z) \neq 0$ and $F(0,0,Z) \neq 0$. Then the resultant*

$$\mathrm{R}\,(Q_Z, F_Z)\,(X,Y)$$

is a homogeneous polynomial in X and Y of degree dm.

Proof of Bézout's Theorem. By Lemma 11.4.16, Γ and X have a common component if and only if their resultant is identically zero. Next we consider the case in which Γ and X do not have a common component. We assume that $[0:0:1]$ does not belong to both curves. With this assumption $Q(X,Y,Z) = q_0(X,Y)Z^d + q_1(x,Y)Z^{d-1} + \cdots + q_d(X,Y)$ where $q_j(X,Y)$ are homogeneous polynomials of degree j in X and $Y, j = 0,\ldots,d$ and $q_0 \neq 0$. In the same way $F(X,Y,Z) = f_0(X,Y)Z^m + f_1(X,Y)Z^{m-1} + \cdots + f_m(X,Y)$ where $f_j(X,Y)$ are homogeneous polynomials of degree $j, j = 0,\ldots,m$ and $f_0 \neq 0$. Therefore the resultant is a homogeneous polynomial of degree md by Lemma 11.4.16 and it has md zeros counting their multiplicity. □

11.4.4 Analytic Subsets

Definition 11.4.17. Let Ω be an open subset of a complex manifold M. A set X is called an *analytic set* in Ω when, for every $\xi \in \Omega$, there is an open neighborhood U of ξ in Ω and a finite collection of holomorphic functions $h_1,\ldots,h_s \in \mathcal{O}(U)$ such that

$$X \cap U = \{x \in U : h_1(x) = \cdots = h_s(x) = 0\} = h_1^{-1}(0) \cap \cdots \cap h_s^{-1}(0)\,.$$

X is called *locally analytic* in Ω when there are a neighborhood U and functions $h_1,\ldots,h_s$ as above for every $\xi \in X$ (but not necessarily for $\xi \in \Omega\backslash X$).

Remark 11.4.18.

1. Since an open subset of a manifold is a manifold itself, we can simply speak of [locally] analytic subsets of manifolds.

2. The set X is analytic in Ω iff X is locally analytic and closed in Ω.

3. Every locally analytic set X in Ω is analytic in some open $U \subset \Omega$ (for instance, in $\Omega\backslash(\bar{X}\backslash X)$). The set $\bar{X}\backslash X$ is called the *frontier* of X.

Example 11.4.19. The subset $X = \{(x, y) \in \mathbb{C}^2 : |x|^2 + |y|^2 < 1\} \subset \mathbb{C}^2$ is locally analytic and not analytic in $\mathbb{C}^2$: for every $\xi \in X$, $X \cap U = h^{-1}(0)$, for U an open disc centered at ξ and contained in X, and $h \equiv 0$.

Remark 11.4.20. One also defines globally analytic sets: X is globally analytic in M when there is a finite collection of holomorphic functions $h_1, \dots, h_s \in \mathcal{O}(M)$ such that $X = \{x \in M : h_1(x) = \cdots = h_s(x) = 0\}$.

Example 11.4.21.

1. The union of a collection of some connected components of a manifold M is analytic in M.

2. A closed submanifold N of M is analytic in M.

3. An algebraic subset of $\mathbb{C}^n$ (i.e., the locus of common zeros of a collection of polynomials in n complex variables) is globally analytic in $\mathbb{C}^n$.

4. Nonempty proper analytic subsets of $\mathbb{C}$ (and, in general, of one-dimensional manifolds) are precisely the sets of isolated points without an accumulation point. Locally analytic subsets are the sets of isolated points (e.g., $\{1/n : n \in \mathbb{N}_+\}$).

Definition 11.4.22. An analytic subset X of a manifold M is called *reducible* when there exist nonempty analytic proper subsets X_1, X_2 of X such that $X = X_1 \cup X_2$. Otherwise, X is called *irreducible.*

Proposition 11.4.23. *Let $f : M \to N$ be a holomorphic mapping of manifolds, let X be analytic in M and such that $Y = \varphi(X)$ is analytic in N. If X is irreducible, then so is Y.*

Proof. Suppose $Y = Y_1 \cup Y_2$ for some nonempty analytic $Y_j \neq Y \quad (j = 1, 2)$. Then $f^{-1}(Y_j)$ are analytic in M,

$$X = X \cap f^{-1}(Y) = \left(X \cap f^{-1}(Y_1)\right) \cup \left(X \cap f^{-1}(Y_2)\right),$$

and $X = X \cap f^{-1}(Y) \neq X \cap f^{-1}(Y_j) \quad (j = 1, 2)$; a contradiction. □

Example 11.4.24.

1. The affine algebraic curve $X = \{xy = 0\} \subset \mathbb{C}^2$ is reducible.

2. The affine algebraic curve $X = \{x^2 = y^3\} \subset \mathbb{C}^2$ is irreducible, by Proposition 11.4.23, as the image of (irreducible) $\mathbb{C}$ under the parametrization $\mathbb{C} \ni t \mapsto (t^3, t^2) \in \mathbb{C}^2$.

The algebraic submanifolds of projective space might seem like a restrictive special class, but in fact they are quite general, thanks to the following remarkable theorem proved in 1949 by Wei-Liang Chow [37]: Every closed complex submanifold of $\mathbb{CP}^n$ is algebraic.

A projective algebraic variety is a subset $V \subseteq \mathbb{CP}^n$ of the form

$$V = \{[w] \in \mathbb{CP}^n : p_1(w) = \cdots = p_k(w) = 0\}$$

where $p_1, \ldots, p_k : \mathbb{C}^{n+1} \to \mathbb{C}$ are homogeneous holomorphic polynomials of various degrees. A projective variety is said to be nonsingular or smooth if the affine variety $\widetilde{V} \subseteq \mathbb{C}^{n+1}$ defined by $p_1, \ldots, p_k$ has no singular points other than perhaps the origin. It follows from the next lemma that every nonsingular projective algebraic variety is actually a complex submanifold of $\mathbb{CP}^n$. Projective algebraic varieties are the main objects of study in complex algebraic geometry.

Lemma 11.4.25. *Suppose that* $M \subseteq \mathbb{CP}^n$ *is a subset and* $\widetilde{M} = \pi^{-1}(M) \subseteq \mathbb{C}^{n+1}\backslash\{0\}$, *where* $\pi : \mathbb{C}^{n+1}\backslash\{0\} \to \mathbb{CP}^n$ *is the canonical projection. Then* M *is a complex* k *submanifold of* $\mathbb{CP}^n$ *if and only if* $\tilde{M}$ *is a complex* $(k+1)$*-dimensional submanifold of* $\mathbb{C}^{n+1} \setminus \{0\}$.

Closing Remarks

Complex analysis is one of the grand old ladies of modern mathematics. Stretching back almost four hundred years, it has developed into a diverse and multi-faceted collection of ideas and techniques.

The purpose of this book has been to introduce the reader to a handful of these insights—mostly those with a geometric flavor. We hope that we have inspired both students and professional mathematicians to further study and reflection.

Bibliography

[1] Niels Henrik Abel. *Oeuvres complètes de Niels Henrik Abel: Nouvelle édition.* Ed. by L. Sylow and S. Lie. Cambridge University Press, Aug. 2012. ISBN: 9781108050579. DOI: 10.1017/cbo9781139245807.

[2] Lars V. Ahlfors. "Bounded analytic functions". In: *Duke Math. J.* 14 (1947), pp. 1–11. ISSN: 0012-7094,1547-7398. URL: http://projecteuclid.org/euclid.dmj/1077473984.

[3] Lars V. Ahlfors. *Complex analysis.* Third. International Series in Pure and Applied Mathematics. McGraw-Hill Book Co., New York, 1978, pp. xi+331. ISBN: 0-07-000657-1.

[4] Lars V. Ahlfors. *Complex analysis: An introduction of the theory of analytic functions of one complex variable.* Second. McGraw-Hill Book Co., New York-Toronto-London, 1966, pp. xiii+317.

[5] Lars V. Ahlfors. *Conformal invariants: topics in geometric function theory.* McGraw-Hill Series in Higher Mathematics. McGraw-Hill Book Co., New York-Düsseldorf-Johannesburg, 1973, pp. ix+157.

[6] Lars V. Ahlfors and Leo Sario. *Riemann surfaces.* Vol. No. 26. Princeton Mathematical Series. Princeton University Press, Princeton, NJ, 1960, pp. xi+382.

[7] Selman Akbulut and John D. McCarthy. *Casson's invariant for oriented homology* 3*-spheres.* Vol. 36. Mathematical Notes. Princeton University Press, Princeton, NJ, 1990, pp. xviii+182. ISBN: 0-691-08563-3. DOI: 10.1515/9781400860623.

[8] Paul Appell and Édouard Goursat. *Théorie des fonctions algébriques et de leurs intégrales. I: Étude des fonctions analytiques sur une surface de Riemann. 2. ed. revue et augmentee par P. Fatou.* French. Paris: Gauthier-Villars. XXXV, 526 p (1929). 1929.

[9] E. Arbarello et al. *Geometry of algebraic curves. Vol. I.* Vol. 267. Grundlehren der mathematischen Wissenschaften [Fundamental Principles of Mathematical Sciences]. Springer-Verlag, New York, 1985, pp. xvi+386. ISBN: 0-387-90997-4. DOI: 10.1007/978-1-4757-5323-3.

[10] Mark Anthony Armstrong. *Basic topology.* Undergraduate Texts in Mathematics. Springer-Verlag, New York-Berlin, 1983, pp. xii+251. ISBN: 0-387-90839-0.

[11] Heinrich Behnke and Friedrich Sommer. *Theorie der analytischen Funktionen einer komplexen Veränderlichen.* Vol. Band LXXVII. Die Grundlehren der mathematischen Wissenschaften in Einzeldarstellungen mit besonderer Berücksichtigung der Anwendungsgebiete. Springer-Verlag, Berlin-Göttingen-Heidelberg, 1955, pp. x+582.

[12] Heinrich Behnke and Karl Stein. "Entwicklung analytischer Funktionen auf Riemannschen Flächen". In: *Math. Ann.* 120 (1949), pp. 430–461. ISSN: 0025-5831,1432-1807. DOI: 10.1007/BF01447838.

[13] Steven R. Bell. "Biholomorphic mappings and the $\bar{\partial}$-problem". In: *Ann. of Math. (2)* 114.1 (1981), pp. 103–113. ISSN: 0003-486X. DOI: 10.2307/1971379.

[14] Steven R. Bell and Harold P. Boas. "Regularity of the Bergman projection in weakly pseudoconvex domains". In: *Math. Ann.* 257.1 (1981), pp. 23–30. ISSN: 0025-5831,1432-1807. DOI: 10.1007/BF01450652.

[15] Carlos A. Berenstein and Roger Gay. *Complex variables.* Vol. 125. Graduate Texts in Mathematics. Springer-Verlag, New York, 1991, pp. xii+650. ISBN: 0-387-97349-4. DOI: 10.1007/978-1-4612-3024-3.

[16] Melvyn S. Berger. "Riemannian structures of prescribed Gaussian curvature for compact 2-manifolds". In: *J. Differential Geometry* 5 (1971), pp. 325–332. ISSN: 0022-040X,1945-743X. URL: http://projecteuclid.org/euclid.jdg/1214429996.

[17] Jöran Bergh and Jörgen Löfström. *Interpolation spaces. An introduction.* Vol. No. 223. Grundlehren der Mathematischen Wissenschaften. Springer-Verlag, Berlin-New York, 1976, pp. x+207.

[18] Stefan Bergman and M. Schiffer. *Kernel functions and elliptic differential equations in mathematical physics.* Academic Press, Inc., New York, 1953, pp. xiii+432.

[19] Stefan Bergmann. "Über die Entwicklung der harmonischen Funktionen der Ebene und des Raumes nach Orthogonalfunktionen". In: *Math. Ann.* 86.3-4 (1922), pp. 238–271. ISSN: 0025-5831,1432-1807. DOI: 10.1007/BF01457987.

[20] L. Bers. *Topology, 1956-57. Notes by E. Rodlitz and J. Lewis.* Courant Institute of Mathematical Sciences, New York Univ., 1956.

[21] Lipman Bers. *Introduction to Several Complex Variables.* New York: Courant Institute of Mathematical Sciences, New York University, 1964.

[22] Harold P. Boas. "Counterexample to the Lu Qi-Keng conjecture". In: *Proc. Amer. Math. Soc.* 97.2 (1986), pp. 374–375. ISSN: 0002-9939,1088-6826. DOI: 10.2307/2046535.

[23] Ralph P. Boas. *Invitation to complex analysis.* Second. MAA Textbooks. Mathematical Association of America, Washington, DC, 2010, pp. xiv+327. ISBN: 978-0-88385-764-9.

[24] S. Bochner. "Über orthogonale Systeme analytischer Funktionen". In: *Math. Z.* 14.1 (1922), pp. 180–207. ISSN: 0025-5874,1432-1823. DOI: 10.1007/BF01215898.

[25] Salomon Bochner and Deane Montgomery. "Locally compact groups of differentiable transformations". In: *Ann. of Math. (2)* 47 (1946), pp. 639–653. ISSN: 0003-486X. DOI: 10.2307/1969226.

[26] J. A. Bondy and U. S. R. Murty. *Graph theory with applications.* American Elsevier Publishing Co., Inc., New York, 1976, pp. x + 264.

[27] Raoul Bott and Loring W. Tu. *Differential forms in algebraic topology.* Vol. 82. Graduate Texts in Mathematics. Springer-Verlag, New York-Berlin, 1982, pp. xiv+331. ISBN: 0-387-90613-4.

[28] L. Boutet de Monvel and J. Sjöstrand. "Sur la singularité des noyaux de Bergman et de Szegö". In: *Journées: Équations aux Dérivées Partielles de Rennes (1975).* Vol. No. 34–35. Astérisque. Soc. Math. France, Paris, 1976, pp. 123–164.

[29] Ronald Brown. *Topology.* Second. Ellis Horwood Series: Mathematics and its Applications. Ellis Horwood Ltd., Chichester; Halsted Press [John Wiley & Sons, Inc.], New York, 1988, pp. xviii + 460. ISBN: 0-7458-0231-1.

[30] Peter Buser. *Geometry and spectra of compact Riemann surfaces.* Vol. 106. Progress in Mathematics. Birkhäuser Boston, Inc., Boston, MA, 1992, pp. xiv+454. ISBN: 0-8176-3406-1.

[31] A.-P. Calderón. "Intermediate spaces and interpolation, the complex method". In: *Studia Math.* 24 (1964), pp. 113–190. ISSN: 0039-3223,1730-6337. DOI: 10.4064/sm-24-2-113-190.

[32] Robert Campbell. *Les intégrales eulériennes et leurs applications. Étude approfondie de la fonction gamma.* Vol. XX. Collection Universitaire de Mathématiques. Dunod, Paris, 1966, pp. xxv+268.

[33] George F. Carrier, Max Krook, and Carl E. Pearson. *Functions of a complex variable: Theory and technique.* McGraw-Hill Book Co., New York-Toronto-London, 1966, pp. ix+438.

[34] Renzo Cavalieri and Eric Miles. *Riemann surfaces and algebraic curves.* Vol. 87. London Mathematical Society Student Texts. Cambridge University Press, Cambridge, 2016, pp. xii+183. ISBN: 978-1-316-60352-9; 978-1-107-14924-3. DOI: 10.1017/CBO9781316569252.

[35] S. S. Chern and J. K. Moser. "Real hypersurfaces in complex manifolds". In: *Acta Math.* 133 (1974), pp. 219–271. ISSN: 0001-5962,1871-2509. DOI: 10.1007/BF02392146.

[36] Shiing-shen Chern. "An elementary proof of the existence of isothermal parameters on a surface". In: *Proc. Amer. Math. Soc.* 6 (1955), pp. 771–782. ISSN: 0002-9939,1088-6826. DOI: 10.2307/2032933.

[37] Wei-Liang Chow. "On compact complex analytic varieties". In: *Amer. J. Math.* 71 (1949), pp. 893–914. ISSN: 0002-9327,1080-6377. DOI: 10.2307/2372375.

[38] C. Herbert Clemens. *A scrapbook of complex curve theory.* University Series in Mathematics. Plenum Press, New York-London, 1980, pp. ix+186. ISBN: 0-306-40536-9.

[39] Henri Cohen. *A course in computational algebraic number theory.* Vol. 138. Graduate Texts in Mathematics. Springer-Verlag, Berlin, 1993, pp. xii+534. ISBN: 3-540-55640-0. DOI: 10.1007/978-3-662-02945-9.

[40] John B. Conway. *Functions of one complex variable.* Second. Vol. 11. Graduate Texts in Mathematics. Springer-Verlag, New York-Berlin, 1978, pp. xiii+317. ISBN: 0-387-90328-3.

[41] R. Dedekind and H. Weber. "Theorie der algebraischen Functionen einer Veränderlichen". In: *J. Reine Angew. Math.* 92 (1882), pp. 181–290. ISSN: 0075-4102,1435-5345. DOI: 10.1515/crll.1882.92.181.

[42] Jean Dieudonné. *History of algebraic geometry.* Wadsworth Mathematics Series. Wadsworth International Group, Belmont, CA, 1985, pp. xii+186. ISBN: 0-534-03723-2.

[43] P. H. Doyle and D. A. Moran. "A short proof that compact 2-manifolds can be triangulated". In: *Invent. Math.* 5 (1968), pp. 160–162. ISSN: 0020-9910,1432-1297. DOI: 10.1007/BF01425546.

[44] B. A. Dubrovin, A. T. Fomenko, and S. P. Novikov. *Modern geometry—methods and applications. Part II.* Vol. 104. Graduate Texts in Mathematics. Springer-Verlag, New York, 1985, pp. xv+430. ISBN: 0-387-96162-3. DOI: 10.1007/978-1-4612-1100-6.

[45] J. Duncan. *The elements of complex analysis.* John Wiley & Sons, London-New York-Sydney, 1968, pp. ix+313.

[46] H. M. Farkas and I. Kra. *Riemann surfaces.* Second. Vol. 71. Graduate Texts in Mathematics. Springer-Verlag, New York, 1992, pp. xvi+363. ISBN: 0-387-97703-1. DOI: 10.1007/978-1-4612-2034-3.

[47] Herbert Federer. *Geometric measure theory.* Vol. Band 153. Die Grundlehren der mathematischen Wissenschaften. Springer-Verlag New York, Inc., New York, 1969, pp. xiv+676.

[48] Charles Fefferman. "Parabolic invariant theory in complex analysis". In: *Adv. in Math.* 31.2 (1979), pp. 131–262. ISSN: 0001-8708. DOI: 10.1016/0001-8708(79)90025-2.

[49] Charles Fefferman. "The Bergman kernel and biholomorphic mappings of pseudoconvex domains". In: *Invent. Math.* 26 (1974), pp. 1–65. ISSN: 0020-9910,1432-1297. DOI: 10.1007/BF01406845.

[50] Otto Forster. *Lectures on Riemann surfaces*. Vol. 81. Graduate Texts in Mathematics. Springer-Verlag, New York-Berlin, 1981, pp. viii+254. ISBN: 0-387-90617-7.

[51] George K. Francis and Jeffrey R. Weeks. "Conway's ZIP proof". In: *Amer. Math. Monthly* 106.5 (1999), pp. 393–399. ISSN: 0002-9890,1930-0972. DOI: 10.2307/2589143.

[52] R. Fricke and F. Klein. *Vorlesungen über die Theorie der automorphen Functionen. Erster Band; Die gruppentheoretischen Grundlagen.* German. Leipzig: B. G. Teubner. XIV + 634 S. (1897). 1897. URL: resolver.library.cornell.edu/math/2237592.

[53] Theodore W. Gamelin. *Complex analysis*. Undergraduate Texts in Mathematics. Springer-Verlag, New York, 2001, pp. xviii+478. ISBN: 0-387-95093-1; 0-387-95069-9. DOI: 10.1007/978-0-387-21607-2.

[54] Theodore W. Gamelin. *Uniform algebras*. Prentice-Hall, Inc., Englewood Cliffs, NJ, 1969, pp. xiii+257.

[55] Theodore W. Gamelin and Robert Everist Greene. *Introduction to topology*. Second. Dover Publications, Inc., Mineola, NY, 1999, pp. xiv+234. ISBN: 0-486-40680-6.

[56] John B. Garnett. *Bounded analytic functions*. Vol. 96. Pure and Applied Mathematics. Academic Press, Inc. [Harcourt Brace Jovanovich, Publishers], New York-London, 1981, pp. xvi+467. ISBN: 0-12-276150-2.

[57] John B. Garnett and Donald E. Marshall. *Harmonic measure*. Vol. 2. New Mathematical Monographs. Cambridge University Press, Cambridge, 2005, pp. xvi+571. ISBN: 978-0-521-47018-6; 0-521-47018-8. DOI: 10.1017/CBO9780511546617.

[58] C. G. Gibson. *Elementary geometry of algebraic curves: an undergraduate introduction*. Cambridge University Press, Cambridge, 1998, pp. xvi+250. ISBN: 0-521-64140-3; 0-521-64641-3. DOI: 10 . 1017 / CBO9781139173285.

[59] Jeremy Gray. "On the history of the Riemann mapping theorem". English. In: *Studies in the history of modern mathematics, I*. Palermo: Circolo Matemático di Palermo, 1994, pp. 47–94.

[60] Jeremy Gray. "On the history of the Riemann mapping theorem". In: *Rend. Circ. Mat. Palermo (2) Suppl.* 34 (1994), pp. 47–94. ISSN: 1592-9531.

[61] Marvin J. Greenberg and John R. Harper. *Algebraic topology*. Vol. 58. Mathematics Lecture Note Series. Benjamin/Cummings Publishing Co., Inc., Advanced Book Program, Reading, MA, 1981, xi+311 pp. (loose errata). ISBN: 0-8053-3558-7.

[62] R. E. Greene and S. G. Krantz. "Invariants of Bergman geometry and the automorphism groups of domains in $\mathbb{C}^n$". In: *Geometrical and algebraical aspects in several complex variables (Cetraro, 1989).* Vol. 8. Sem. Conf. EditEl, Rende, 1991, pp. 107–136.

[63] R. E. Greene and Steven G. Krantz. "Stability properties of the Bergman kernel and curvature properties of bounded domains". In: *Recent developments in several complex variables (Proc. Conf., Princeton Univ., Princeton, N. J., 1979).* Vol. No. 100. Ann. of Math. Stud. Princeton Univ. Press, Princeton, NJ, 1981, pp. 179–198. ISBN: 0-691-08285-5.

[64] Robert E. Greene, Kang-Tae Kim, and Steven G. Krantz. *The geometry of complex domains.* Vol. 291. Progress in Mathematics. Birkhäuser Boston, Ltd., Boston, MA, 2011, pp. xiv+303. ISBN: 978-0-8176-4139-9. DOI: 10.1007/978-0-8176-4622-6.

[65] Robert E. Greene and Steven G. Krantz. "Biholomorphic self-maps of domains". In: *Complex analysis, II (College Park, Md., 1985–86).* Vol. 1276. Lecture Notes in Math. Springer, Berlin, 1987, pp. 136–207. ISBN: 3-540-18357-4. DOI: 10.1007/BFb0078959.

[66] Robert E. Greene and Steven G. Krantz. "Characterization of complex manifolds by the isotropy subgroups of their automorphism groups". In: *Indiana Univ. Math. J.* 34.4 (1985), pp. 865–879. ISSN: 0022-2518,1943-5258. DOI: 10.1512/iumj.1985.34.34048.

[67] Robert E. Greene and Steven G. Krantz. "Characterizations of certain weakly pseudoconvex domains with noncompact automorphism groups". In: *Complex analysis (University Park, Pa., 1986).* Vol. 1268. Lecture Notes in Math. Springer, Berlin, 1987, pp. 121–157. ISBN: 3-540-18094-X. DOI: 10.1007/BFb0097301.

[68] Robert E. Greene and Steven G. Krantz. "Deformation of complex structures, estimates for the $\bar{\partial}$ equation, and stability of the Bergman kernel". In: *Adv. in Math.* 43.1 (1982), pp. 1–86. ISSN: 0001-8708. DOI: 10.1016/0001-8708(82)90028-7.

[69] Robert E. Greene and Steven G. Krantz. *Function theory of one complex variable.* Third. Vol. 40. Graduate Studies in Mathematics. American Mathematical Society, Providence, RI, 2006, pp. x+504. ISBN: 0-8218-3962-4. DOI: 10.1090/gsm/040.

[70] Robert E. Greene and Steven G. Krantz. "Normal families and the semicontinuity of isometry and automorphism groups". In: *Math. Z.* 190.4 (1985), pp. 455–467. ISSN: 0025-5874,1432-1823. DOI: 10.1007/BF01214745.

[71] Robert E. Greene and Steven G. Krantz. "Stability of the Carathéodory and Kobayashi metrics and applications to biholomorphic mappings". In: *Complex analysis of several variables (Madison, Wis., 1982)*. Vol. 41. Proc. Sympos. Pure Math. Amer. Math. Soc., Providence, RI, 1984, pp. 77–93. ISBN: 0-8218-1446-X. DOI: 10.1090/pspum/041/740874.

[72] Robert E. Greene and Steven G. Krantz. "Techniques for studying automorphisms of weakly pseudoconvex domains". In: *Several complex variables (Stockholm, 1987/1988)*. Vol. 38. Math. Notes. Princeton Univ. Press, Princeton, NJ, 1993, pp. 389–410. ISBN: 0-691-08579-X.

[73] Robert E. Greene and Steven G. Krantz. "The automorphism groups of strongly pseudoconvex domains". In: *Math. Ann.* 261.4 (1982), pp. 425–446. ISSN: 0025-5831,1432-1807. DOI: 10.1007/BF01457445.

[74] Robert E. Greene and Steven G. Krantz. "The stability of the Bergman kernel and the geometry of the Bergman metric". In: *Bull. Amer. Math. Soc. (N.S.)* 4.1 (1981), pp. 111–115. ISSN: 0273-0979,1088-9485. DOI: 10.1090/S0273-0979-1981-14874-6.

[75] Phillip Griffiths and Joseph Harris. *Principles of algebraic geometry*. Wiley Classics Library. John Wiley & Sons, Inc., New York, 1994, pp. xiv+813. ISBN: 0-471-05059-8. DOI: 10.1002/9781118032527.

[76] Phillip A. Griffiths. *Introduction to algebraic curves*. Vol. 76. Translations of Mathematical Monographs. American Mathematical Society, Providence, RI, 1989, pp. x+221. ISBN: 0-8218-4530-6. DOI: 10.1090/mmono/076.

[77] R. C. Gunning and Raghavan Narasimhan. "Immersion of open Riemann surfaces". In: *Math. Ann.* 174 (1967), pp. 103–108. ISSN: 0025-5831,1432-1807. DOI: 10.1007/BF01360812.

[78] Joe Harris. *Algebraic geometry*. Vol. 133. Graduate Texts in Mathematics. Springer-Verlag, New York, 1992, pp. xx+328. ISBN: 0-387-97716-3. DOI: 10.1007/978-1-4757-2189-8.

[79] Robin Hartshorne. *Algebraic geometry*. Vol. No. 52. Graduate Texts in Mathematics. Springer-Verlag, New York-Heidelberg, 1977, pp. xvi+496. ISBN: 0-387-90244-9.

[80] Allen Hatcher. *Algebraic topology*. Cambridge University Press, Cambridge, 2002, pp. xii+544. ISBN: 0-521-79160-X; 0-521-79540-0.

[81] Zheng-Xu He and Oded Schramm. "Fixed points, Koebe uniformization and circle packings". In: *Ann. of Math. (2)* 137.2 (1993), pp. 369–406. ISSN: 0003-486X,1939-8980. DOI: 10.2307/2946541.

[82] Maurice Heins. "On the number of $1-1$ directly conformal maps which a multiply-connected plane region of finite connectivity $p(>2)$ admits onto itself". In: *Bull. Amer. Math. Soc.* 52 (1946), pp. 454–457. ISSN: 0002-9904. DOI: 10.1090/S0002-9904-1946-08590-0.

[83] Sigurður Helgason. *Differential geometry and symmetric spaces.* Vol. Vol. XII. Pure and Applied Mathematics. Academic Press, New York-London, 1962, pp. xiv+486.

[84] Morris W. Hirsch. *Differential topology.* Vol. No. 33. Graduate Texts in Mathematics. Springer-Verlag, New York-Heidelberg, 1976, pp. x+221.

[85] Kenneth Hoffman. *Banach spaces of analytic functions.* Prentice-Hall Series in Modern Analysis. Prentice-Hall, Inc., Englewood Cliffs, NJ, 1962, pp. xiii+217.

[86] Lars Hörmander. *An introduction to complex analysis in several variables.* Third. Vol. 7. North-Holland Mathematical Library. North-Holland Publishing Co., Amsterdam, 1990, pp. xii+254. ISBN: 0-444-88446-7.

[87] Christian Houzel. *La géométrie algébrique.* Collection Sciences dans l'Histoire. [Science in History Collection]. Librairie Scientifique et Technique Albert Blanchard, Paris, 2002, pp. vi+365. ISBN: 2-85367-222-0.

[88] John Hamal Hubbard. *Teichmüller theory and applications to geometry, topology, and dynamics. Vol. 1.* Matrix Editions, Ithaca, NY, 2006, pp. xx+459. ISBN: 978-0-9715766-2-9; 0-9715766-2-9.

[89] Daniel Huybrechts. *Complex geometry.* Universitext. Springer-Verlag, Berlin, 2005, pp. xii+309. ISBN: 3-540-21290-6.

[90] G. J. O. Jameson. *A first course on complex functions.* Chapman and Hall, Ltd., London; distributed in the U.S.A. by Barnes & Noble Inc., place of publication not identified, 1970, pp. xii+148.

[91] Gareth A. Jones and David Singerman. *Complex functions.* Cambridge University Press, Cambridge, 1987, pp. xiv+342. ISBN: 0-521-30893-3. DOI: 10.1017/CBO9781139171915.

[92] C. Jordan. *Cours d'analyse de l'École Polytechnique.* 2e *édition, entièrement refondue. Tome III. Calcul intégral. (Équations différentielles.)* French. Paris: Gauthier-Villars et Fils. XI + 543 S. 8°. (1896). 1896.

[93] C. Jordan. *Cours d'analyse de l'École Polytechnique.* 3e *édition, revue et corrigée. Tome II: Calcul intégral.* French. Paris: Gauthier-Villars. 712 S. 8°. (1913). 1913.

[94] C. Jordan. *Cours d'analyse de l'École Polytechnique.* 3e *édition, revue et corrigée. Vol. I.* French. Paris: Gauthier-Villars. XVI u. 621 S. 8° (1909). 1909.

[95] Jürgen Jost. *Compact Riemann surfaces.* Third. Universitext. Springer-Verlag, Berlin, 2006, pp. xviii+277. ISBN: 978-3-540-33065-3; 3-540-33065-8. DOI: 10.1007/978-3-540-33067-7.

[96] Egbert R. Van Kampen. "On the Fundamental Group of an Algebraic Curve". In: *Amer. J. Math.* 55.1-4 (1933), pp. 255–260. ISSN: 0002-9327,1080-6377. DOI: 10.2307/2371128.

[97] A. A. Karatsuba and S. M. Voronin. *The Riemann zeta-function.* Vol. 5. De Gruyter Expositions in Mathematics. Walter de Gruyter & Co., Berlin, 1992, pp. xii+396. ISBN: 3-11-013170-6. DOI: 10.1515/9783110886146.

[98] Yitzhak Katznelson. *An introduction to harmonic analysis.* John Wiley & Sons, Inc., New York-London-Sydney, 1968, pp. xiv+264.

[99] Jerry L. Kazdan and F. W. Warner. "Curvature functions for compact 2-manifolds". In: *Ann. of Math. (2)* 99 (1974), pp. 14–47. ISSN: 0003-486X. DOI: 10.2307/1971012.

[100] Norberto Kerzman. "The Bergman kernel function. Differentiability at the boundary". In: *Math. Ann.* 195 (1972), pp. 149–158. ISSN: 0025-5831,1432-1807. DOI: 10.1007/BF01419622.

[101] Robion C. Kirby and Laurence C. Siebenmann. *Foundational essays on topological manifolds, smoothings, and triangulations.* Vol. No. 88. Annals of Mathematics Studies. Princeton University Press, Princeton, NJ; University of Tokyo Press, Tokyo, 1977, pp. vii+355.

[102] F. Klein. *Ueber Riemann's Theorie der algebraischen Functionen und ihrer Integrale.* German. Leipzig. Teubner (1882). 1882. URL: resolver.library.cornell.edu/math/1797666.

[103] Felix Klein. "Neue Beiträge zur Riemann'schen Functionentheorie". In: *Math. Ann.* 21.2 (1883), pp. 141–218. ISSN: 0025-5831,1432-1807. DOI: 10.1007/BF01442920.

[104] Paul F. Klembeck. "Kähler metrics of negative curvature, the Bergmann metric near the boundary, and the Kobayashi metric on smooth bounded strictly pseudoconvex sets". In: *Indiana Univ. Math. J.* 27.2 (1978), pp. 275–282. ISSN: 0022-2518,1943-5258. DOI: 10.1512/iumj.1978.27.27020.

[105] Shoshichi Kobayashi. *Hyperbolic manifolds and holomorphic mappings.* Vol. 2. Pure and Applied Mathematics. Marcel Dekker, Inc., New York, 1970, pp. ix+148.

[106] Shoshichi Kobayashi and Katsumi Nomizu. *Foundations of differential geometry. Vol I.* Interscience Publishers (a division of John Wiley & Sons, Inc.), New York-London, 1963, pp. xi+329.

[107] Shoshichi Kobayashi and Katsumi Nomizu. *Foundations of differential geometry. Vol. II.* Vol. No. 15. Interscience Tracts in Pure and Applied Mathematics. Interscience Publishers John Wiley & Sons, Inc., New York-London-Sydney, 1969, pp. xv+470.

[108] P. Koebe. "Über die Uniformisierung beliebiger analytischer Kurven." German. In: *Nachr. Ges. Wiss. Göttingen, Math.-Phys. Kl.* 1907 (1907), pp. 191–210. URL: https://eudml.org/doc/58678.

[109] P. Koebe. "Über die Uniformisierung beliebiger analytischer Kurven. Zweite Mitteilung." German. In: *Nachr. Ges. Wiss. Göttingen, Math.-Phys. Kl.* 1907 (1907), pp. 633–669. URL: https://eudml.org/doc/58703.

[110] P. Koebe. "Über die Uniformisierung reeller algebraischer Kurven." German. In: *Nachr. Ges. Wiss. Göttingen, Math.-Phys. Kl.* 1907 (1907), pp. 177–190. URL: https://eudml.org/doc/58677.

[111] Paul Koosis. *Introduction to H_p spaces.* Second. Vol. 115. Cambridge Tracts in Mathematics. Cambridge University Press, Cambridge, 1998, pp. xiv+289. ISBN: 0-521-45521-9.

[112] Steven G. Krantz. *A panorama of harmonic analysis.* Vol. 27. Carus Mathematical Monographs. Mathematical Association of America, Washington, DC, 1999, pp. xvi+357. ISBN: 0-88385-031-1.

[113] Steven G. Krantz. *Function theory of several complex variables.* AMS Chelsea Publishing, Providence, RI, 2001, pp. xvi+564. ISBN: 0-8218-2724-3. DOI: 10.1090/chel/340.

[114] Steven G. Krantz and Peter V. Dovbush. "The central role of Montel's theorem in complex function theory". In: *Complex Variables and Elliptic Equations* 70.9 (2025), pp. 1615–1625. DOI: 10.1080/17476933.2024.2417344.

[115] Ernst Kunz. *Introduction to plane algebraic curves.* German. Birkhäuser Boston, Inc., Boston, MA, 2005, pp. xiv+293. ISBN: 978-0-8176-4381-2; 0-8176-4381-8.

[116] Serge Lang. *Algebra.* third. Vol. 211. Graduate Texts in Mathematics. Springer-Verlag, New York, 2002, pp. xvi+914. ISBN: 0-387-95385-X. DOI: 10.1007/978-1-4613-0041-0.

[117] Serge Lang. *Complex analysis.* Second. Vol. 103. Graduate Texts in Mathematics. Springer-Verlag, New York, 1985, pp. xiv+367. ISBN: 0-387-96085-6. DOI: 10.1007/978-1-4757-1871-3.

[118] Herbert Lange. *Abelian varieties over the complex numbers—a graduate course.* Grundlehren Text Editions. Springer, Cham, 2023, pp. xii+384. ISBN: 978-3-031-25569-4; 978-3-031-25570-0. DOI: 10.1007/978-3-031-25570-0.

[119] John M. Lee. *Introduction to smooth manifolds.* Second. Vol. 218. Graduate Texts in Mathematics. Springer, New York, 2013, pp. xvi+708. ISBN: 978-1-4419-9981-8.

[120] John M. Lee. *Introduction to topological manifolds.* Vol. 202. Graduate Texts in Mathematics. Springer-Verlag, New York, 2000, pp. xviii+385. ISBN: 0-387-98759-2; 0-387-95026-5.

[121] J. L. Lions and Jaak Peetre. "On a class of interpolation spaces". French. In: *Publ. Math., Inst. Hautes Étud. Sci.* 19 (1964), pp. 5–68. ISSN: 0073-8301. DOI: 10.1007/BF02684796.

[122] A. Lucchini, M. C. Tamburini, and J. S. Wilson. "Hurwitz groups of large rank". In: *J. London Math. Soc. (2)* 61.1 (2000), pp. 81–92. ISSN: 0024-6107,1469-7750. DOI: `10.1112/S0024610799008467`.

[123] David Madore. *Éléments de Géométrie Algébrique — Table des Matières*. 2005. URL: `https://perso.telecom-paristech.fr/madore/ega-toc.pdf` (visited on 08/23/2025).

[124] Ryuji Maehara. "The Jordan curve theorem via the Brouwer fixed point theorem". In: *Amer. Math. Monthly* 91.10 (1984), pp. 641–643. ISSN: 0002-9890,1930-0972. DOI: `10.2307/2323369`.

[125] William S. Massey. *A basic course in algebraic topology*. Vol. 127. Graduate Texts in Mathematics. Springer-Verlag, New York, 1991, pp. xvi+428. ISBN: 0-387-97430-X.

[126] William S. Massey. *Algebraic topology: an introduction*. Vol. Vol. 56. Graduate Texts in Mathematics. Springer-Verlag, New York-Heidelberg, 1977, pp. xxi+261. ISBN: 0-387-90271-6.

[127] S. N. Mergelyan. "Uniform approximations of functions of a complex variable". In: *Uspehi Matem. Nauk (N.S.)* 7.2(48) (1952), pp. 31–122.

[128] John W. Milnor. *Topology from the differentiable viewpoint*. Princeton Landmarks in Mathematics. Princeton University Press, Princeton, NJ, 1997, pp. xii+64. ISBN: 0-691-04833-9.

[129] Rick Miranda. *Algebraic curves and Riemann surfaces*. Vol. 5. Graduate Studies in Mathematics. American Mathematical Society, Providence, RI, 1995, pp. xxii+390. ISBN: 0-8218-0268-2. DOI: `10.1090/gsm/005`.

[130] Edwin E. Moise. *Geometric topology in dimensions* 2 *and* 3. Vol. Vol. 47. Graduate Texts in Mathematics. Springer-Verlag, New York-Heidelberg, 1977, pp. x+262.

[131] Hugh L. Montgomery and Robert C. Vaughan. *Multiplicative number theory. I. Classical theory*. Vol. 97. Cambridge Studies in Advanced Mathematics. Cambridge University Press, Cambridge, 2007, pp. xviii+552. ISBN: 978-0-521-84903-6; 0-521-84903-9.

[132] James Morrow and Kunihiko Kodaira. *Complex manifolds*. Holt, Rinehart and Winston, Inc., New York-Montreal, Que.-London, 1971, pp. vii+192.

[133] James R. Munkres. *Analysis on manifolds*. Addison-Wesley Publishing Company, Advanced Book Program, Redwood City, CA, 1991, pp. xiv+366. ISBN: 0-201-51035-9.

[134] James R. Munkres. *Elementary differential topology*. Revised. Vol. No. 54. Annals of Mathematics Studies. Princeton University Press, Princeton, NJ, 1966, pp. xi+112.

[135] James R. Munkres. *Elements of algebraic topology*. Addison-Wesley Publishing Company, Menlo Park, CA, 1984, pp. ix+454. ISBN: 0-201-04586-9.

[136] James R. Munkres. *Topology: a first course*. Prentice-Hall, Inc., Englewood Cliffs, NJ, 1975, pp. xvi+413.

[137] S. B. Myers. "Riemannian manifolds with positive mean curvature". In: *Duke Math. J.* 8 (1941), pp. 401–404. ISSN: 0012-7094,1547-7398. URL: http://projecteuclid.org/euclid.dmj/1077492655.

[138] R. Narasimhan. *Analysis on real and complex manifolds*. Vol. 35. North-Holland Mathematical Library. North-Holland Publishing Co., Amsterdam, 1985, pp. xiv+246. ISBN: 0-444-87776-2.

[139] Raghavan Narasimhan. *Complex analysis in one variable*. Birkhäuser Boston, Inc., Boston, MA, 1985, pp. xvi+266. ISBN: 0-8176-3237-9. DOI: 10.1007/978-1-4757-1106-6.

[140] Donald J. Newman. *Analytic number theory*. Vol. 177. Graduate Texts in Mathematics. Springer-Verlag, New York, 1998, pp. viii+76. ISBN: 0-387-98308-2.

[141] Richard S. Palais. "On the existence of slices for actions of non-compact Lie groups". In: *Ann. of Math. (2)* 73 (1961), pp. 295–323. ISSN: 0003-486X. DOI: 10.2307/1970335.

[142] Peter Petersen. *Riemannian geometry*. Second. Vol. 171. Graduate Texts in Mathematics. Springer, New York, 2006, pp. xvi+401. ISBN: 978-0387-29246-5.

[143] E. Picard. *Traité d'analyse. Tome II. Fonctions harmoniques et fonctions analytiques. Introduction à la théorie des équations différentielles. Intégrales abéliennes et surfaces de Riemann. 2e édition, revue et augmentée.* French. Paris: Gauthier-Villars. XV u. 587 S. gr. 8° (1905). 1905.

[144] E. Picard. *Traité d'analyse. Vol. 3. 2e édition.* French. Paris: Gauthier-Villars VII u. 607 S. 8° (1908). 1908.

[145] Émile Picard. *Traité d'Analyse. Tome I. Intégrales simples et multiples. L'équation de Laplace et ses applications. Développements en séries. Applications géométriques du calcul infinitésimal.* French. Paris. Gauthier-Villars et Fils. XII u. 457 S. gr. 8° (1891). 1891.

[146] M. J. Piff and R. J. Webster. "65.12 Winding numbers". In: *The Mathematical Gazette* 65.431 (Mar. 1981), pp. 50–51. ISSN: 2056-6328. DOI: 10.2307/3617941.

[147] H. Poincaré. "Sur l'uniformisation des fonctions analytiques". In: *Acta Math.* 31.1 (1908), pp. 1–63. ISSN: 0001-5962,1871-2509. DOI: 10.1007/BF02415442.

[148] H. Poincaré. "Sur un théorème de la théorie générale des fonctions". In: *Bull. Soc. Math. France* 11 (1883), pp. 112–125. ISSN: 0037-9484. URL: http://www.numdam.org/item?id=BSMF_1883__11__112_1.

[149] Alexandre Preissmann. "Quelques propriétés globales des espaces de Riemann". In: *Comment. Math. Helv.* 15 (1943), pp. 175–216. ISSN: 0010-2571,1420-8946. DOI: 10.1007/BF02565638.

[150] T. Radó. "Über den Begriff der Riemannschen Fläche." German. In: *Acta Litt. Sci. Szeged* 2 (1925), pp. 101–121. URL: www.maths.ed.ac.uk/~v1ranick/papers/rado.pdf.

[151] Reinhold Remmert. *Classical topics in complex function theory.* Vol. 172. Graduate Texts in Mathematics. Springer-Verlag, New York, 1998, pp. xx+349. ISBN: 0-387-98221-3. DOI: 10.1007/978-1-4757-2956-6.

[152] Reinhold Remmert. "From Riemann surfaces to complex spaces". In: *Matériaux pour l'histoire des mathématiques au XXe siècle (Nice, 1996).* Vol. 3. Sémin. Congr. Soc. Math. France, Paris, 1998, pp. 203–241. ISBN: 2-85629-065-5. DOI: 10.1007/978-1-4757-2956-6_9.

[153] Georges de Rham. *Variétés différentiables. Formes, courants, formes harmoniques.* Vol. III. Publications de l'Institut de Mathématique de l'Université de Nancago. Hermann, Paris, 1973, x+198 pp. (loose errata).

[154] B. Riemann. "Theorie der Abel'schen Functionen". In: *J. Reine Angew. Math.* 54 (1857), pp. 115–155. ISSN: 0075-4102,1435-5345. DOI: 10.1515/crll.1857.54.115.

[155] Bernhard Riemann. *Collected papers.* German. Kendrick Press, Heber City, UT, 2004, pp. x+555. ISBN: 0-9740427-2-2; 0-9740427-3-0.

[156] Bernhard Riemann. *Collected papers. Translated from the 1892 German edition by Roger Baker, Charles Christenson and Henry Orde.* English. Heber City, UT: Kendrick Press, 2004. ISBN: 0-9740427-2-2; 0-9740427-3-0.

[157] Bernhard Riemann. *Gesammelte mathematische Werke, wissenschaftlicher Nachlass und Nachträge.* the. Vol. Suppl. 1. Teubner-Archiv zur Mathematik [Teubner Archive on Mathematics]. BSB B. G. Teubner Verlagsgesellschaft, Leipzig; Springer-Verlag, Berlin, 1990, pp. vi+911. ISBN: 3-322-00376-0. DOI: 10.1007/978-3-663-10149-9.

[158] Georg Friedrich Bernhard Riemann. *Œuvres mathématiques de Riemann.* Librairie Scientifique et Technique Albert Blanchard, Paris, 1968, pp. xxxv+453.

[159] G. Roch. "Ueber die Anzahl der willkürlichen Constanten in algebraischen Functionen." German. In: *J. Reine Angew. Math.* 64 (1865), pp. 372–376. ISSN: 0075-4102. DOI: 10.1515/crll.1865.64.372.

[160] Walter Rudin. *Principles of mathematical analysis.* Third. International Series in Pure and Applied Mathematics. McGraw-Hill Book Co., New York-Auckland-Düsseldorf, 1976, pp. x+342.

[161] Walter Rudin. *Real and complex analysis.* McGraw-Hill Book Co., New York-Toronto-London, 1966, pp. xi+412.

[162] Henri Paul de Saint-Gervais. *Uniformization of Riemann surfaces.* Heritage of European Mathematics. European Mathematical Society (EMS), Zürich, 2016, pp. xxx+482. ISBN: 978-3-03719-145-3. DOI: 10.4171/145.

[163] Herbert Seifert and William Threlfall. *Seifert and Threlfall: a textbook of topology.* German. Vol. 89. Pure and Applied Mathematics. Academic Press, Inc. [Harcourt Brace Jovanovich, Publishers], New York-London, 1980, pp. xvi+437. ISBN: 0-12-634850-2.

[164] J. G. Semple and L. Roth. *Introduction to algebraic geometry.* Oxford Science Publications. The Clarendon Press, Oxford University Press, New York, 1985, pp. xviii+454. ISBN: 0-19-853363-2.

[165] C. L. Siegel. *Topics in complex function theory. Vol. I: Elliptic functions and uniformization theory.* Vol. No. 25. Interscience Tracts in Pure and Applied Mathematics. Wiley-Interscience [A division of John Wiley & Sons, Inc.], New York-London-Sydney, 1969, pp. ix+186.

[166] Allan J. Sieradski. *An introduction to topology and homotopy.* The Prindle, Weber & Schmidt Series in Advanced Mathematics. PWS-KENT Publishing Co., Boston, MA, 1992, pp. xiv+479. ISBN: 0-534-92960-5.

[167] M. Skwarczyński. "The invariant distance in the theory of pseudo-conformal transformations and the Lu Qi-keng conjecture". In: *Proc. Amer. Math. Soc.* 22 (1969), pp. 305–310. ISSN: 0002-9939,1088-6826. DOI: 10.2307/2037043.

[168] Michael Spivak. *A comprehensive introduction to differential geometry. Vol. One.* M. Spivak, Brandeis Univ., Waltham, MA, 1970, iii+656 pp. (not consecutively paged) paperbound.

[169] Michael Spivak. *Calculus on manifolds. A modern approach to classical theorems of advanced calculus.* W. A. Benjamin, Inc., New York-Amsterdam, 1965, pp. xii+144.

[170] George Springer. *Introduction to Riemann surfaces.* Addison-Wesley Publishing Co., Inc., Reading, MA, 1957, pp. viii+307.

[171] Elias M. Stein and Rami Shakarchi. *Complex analysis.* Vol. 2. Princeton Lectures in Analysis. Princeton University Press, Princeton, NJ, 2003, pp. xviii+379. ISBN: 0-691-11385-8.

[172] Elias M. Stein and Guido Weiss. *Introduction to Fourier analysis on Euclidean spaces.* Vol. No. 32. Princeton Mathematical Series. Princeton University Press, Princeton, NJ, 1971, pp. x+297.

[173] Ian Stewart and David Tall. *Complex analysis.* Cambridge University Press, Cambridge-New York, 1983, pp. ix+290. ISBN: 0-521-24513-3.

[174] Nobuyuki Suita and Akira Yamada. "On the Lu Qi-keng conjecture". In: *Proc. Amer. Math. Soc.* 59.2 (1976), pp. 222–224. ISSN: 0002-9939,1088-6826. DOI: 10.2307/2041472.

[175] G. Szegö. "Über Orthogonalsysteme von Polynomen". In: *Math. Z.* 4.1-2 (1919), pp. 139–151. ISSN: 0025-5874,1432-1823. DOI: 10.1007/BF01203393.

[176] Nico M. Temme. *Special functions.* A Wiley-Interscience Publication. John Wiley & Sons, Inc., New York, 1996, pp. xiv+374. ISBN: 0-471-11313-1. DOI: 10.1002/9781118032572.

[177] E. C. Titchmarsh. *The theory of the Riemann zeta-function.* Second. The Clarendon Press, Oxford University Press, New York, 1986, pp. x+412. ISBN: 0-19-853369-1.

[178] G. Valiron. *Cours d'analyse mathématique. I. Théorie des fonctions.* French. 522 p. Paris, Masson (1942). 1942.

[179] Claire Voisin. *Hodge theory and complex algebraic geometry. II.* Vol. 77. Cambridge Studies in Advanced Mathematics. Cambridge University Press, Cambridge, 2003, pp. x+351. ISBN: 0-521-80283-0. DOI: 10.1017/CBO9780511615177.

[180] Robert J. Walker. *Algebraic Curves.* Vol. vol. 13. Princeton Mathematical Series. Princeton University Press, Princeton, NJ, 1950, pp. x+201.

[181] C. T. C. Wall. *A geometric introduction to topology.* Addison-Wesley Publishing Co., Reading, Mass.-London-Don Mills, Ont., 1972, pp. vii+168.

[182] Roger Webster. "90.69 Continuous choice of argument". In: *The Mathematical Gazette* 90.519 (Nov. 2006), pp. 463–466. ISSN: 2056-6328. DOI: 10.1017/s0025557200180350.

[183] Elias Wegert. *Visual complex functions.* Birkhäuser/Springer Basel AG, Basel, 2012, pp. xiv+360. ISBN: 978-3-0348-0179-9; 978-3-0348-0180-5. DOI: 10.1007/978-3-0348-0180-5.

[184] Karl Weierstrass. *Mathematische werke von Karl Weierstrass. Herausgegeben unter mitwirkung einer von der Königlich preussischen akademie der wissenschaften eingesetzten commission.* Vol. I. In the digital collection University of Michigan Historical Math Collection. Berlin,: Mayer & Müller, 1894-1927. URL: https://name.umdl.umich.edu/AAN8481.0001.001.

[185] Raymond O. Wells Jr. *Differential analysis on complex manifolds.* Third. Vol. 65. Graduate Texts in Mathematics. Springer, New York, 2008, pp. xiv+299. ISBN: 978-0-387-73891-8. DOI: 10.1007/978-0-387-73892-5.

[186] John Wermer. *Seminar über Funktionen-Algebren*. Vol. 1. Lecture Notes in Mathematics. Springer-Verlag, Berlin-Göttingen-Heidelberg, 1964, pp. iv+30.

[187] Hermann Weyl. *Die Idee der Riemannschen Fläche*. B. G. Teubner Verlagsgesellschaft, Stuttgart, 1955, pp. viii+162.

[188] Hermann Weyl. *The concept of a Riemann surface*. German. ADIWES International Series in Mathematics. Addison-Wesley Publishing Co., Inc., Reading, Mass.-London, 1964, pp. xi+191.

[189] J. H. C. Whitehead. "On C^1-complexes". In: *Ann. of Math. (2)* 41 (1940), pp. 809–824. ISSN: 0003-486X. DOI: 10.2307/1968861.

[190] Hassler Whitney. *Complex analytic varieties*. Addison-Wesley Publishing Co., Reading, Mass.-London-Don Mills, Ont., 1972, pp. xii+399.

[191] E. T. Whittaker and G. N. Watson. *A course of modern analysis*. Cambridge Mathematical Library. Cambridge University Press, Cambridge, 1996, pp. vi+608. ISBN: 0-521-58807-3. DOI: 10.1017/CBO9780511608759.

[192] Jan J. O. O. Wiegerinck. "Domains with finite-dimensional Bergman space". In: *Math. Z.* 187.4 (1984), pp. 559–562. ISSN: 0025-5874,1432-1823. DOI: 10.1007/BF01174190.

[193] Robert A. Wilson. "The Monster is a Hurwitz group". In: *J. Group Theory* 4.4 (2001), pp. 367–374. ISSN: 1433-5883,1435-4446. DOI: 10.1515/jgth.2001.027.

[194] Oscar Zariski and Pierre Samuel. *Commutative algebra, Volume I*. The University Series in Higher Mathematics. D. Van Nostrand Co., Inc., Princeton, NJ, 1958, pp. xi+329.

Index